This monograph describes the theory and practice of electron spectrometry using synchrotron radiation.

The book is in three parts. After a short review of background theory, neon is used to elucidate the principles of the photoelectron and Auger spectra. The second part of the book looks at experimental aspects, including characteristic features of electrostatic analysers, detectors, lenses, disturbances, and optimization, and then illustrates theory and experiment with details of experiments. The third part provides useful reference data, including wavefunctions, special theory, polarization and special aspects of instrumentation. A detailed reference list completes the volume. The study of electron spectrometry using synchrotron radiation is a growing field of research driven by the increasing availability of advanced synchrotron radiation light sources and improved theoretical methods for solving the many-electron problems in atoms. This balanced account will be of value to both theorists and experimentalists working in this area.

Atomic, molecular and chemical physicists, and physical chemists will find this book of interest.

CAMBRIDGE MONOGRAPHS ON ATOMIC, MOLECULAR AND CHEMICAL PHYSICS

ELECTRON SPECTROMETRY OF ATOMS USING SYNCHROTRON RADIATION

CAMBRIDGE MONOGRAPHS ON ATOMIC, MOLECULAR AND CHEMICAL PHYSICS

1. R. Schinke **Photodissociation Dynamics**
2. L. Frommhold **Collision-Induced Absorption in Gases**
3. T. F. Gallagher **Rydberg Atoms**
4. M. Auzinsh and R. Ferber **Optical Polarization of Molecules**
5. I. E. McCarthy and W. Weigold **Electron-Atom Collisions**
6. V. Schmidt **Electron Spectrometry of Atoms using Synchrotron Radiation**

Electron Spectrometry of Atoms using Synchrotron Radiation

VOLKER SCHMIDT
University of Freiburg

CAMBRIDGE UNIVERSITY PRESS
Cambridge, New York, Melbourne, Madrid, Cape Town, Singapore, São Paulo

Cambridge University Press
The Edinburgh Building, Cambridge CB2 2RU, UK

Published in the United States of America by Cambridge University Press, New York

www.cambridge.org
Information on this title: www.cambridge.org/9780521550536

First published 1997
This digitally printed first paperback version 2005

A catalogue record for this publication is available from the British Library

Library of Congress Cataloguing in Publication data

Schmidt, Volker.
Electron spectrometry of atoms using synchrotron radiation / Volker Schmidt.
p. cm. – (Cambridge monographs on atomic, molecular, and chemical physics; 6)
Includes bibliographical references and index.
ISBN 0-521-55053-X
1. Neon – Spectra. 2. Electron spectroscopy. 3. Synchrotron radiation.
4. Photoelectron spectroscopy. 5. Auger effect.
6. Wavefunctions. I. Title. II. Series.
QC462.N5S36 1997
539.7′028′7–dc20 96-19435 CIP

ISBN-13 978-0-521-55053-6 hardback
ISBN-10 0-521-55053-X hardback

ISBN-13 978-0-521-67561-1 paperback
ISBN-10 0-521-67561-8 paperback

Contents

Corrigenda and Addenda

Page 4 eq.(1.3b): replace ∇_i by ∇_i^2

Page 12 after (iv) and page 323 add: A. Dalgarno and H.R. Sadeghpour, Phys.Rev. A46 (1992) R3591; K. Hino et al., Phys. Rev. A48 (1993) 1271

Page 78 eq.(3.1c): replace $^{2S+1}L^o$ by $^{2S+1}L^e$

Page 82 line 7 and p.91 line 29: replace 'substrates' by 'substates'

Page 83 eq.(3.9e): replace + by −

Page 84 eq.(3.14b): multiply with 2

Page 98 after [UDM94] add: J. Ullrich et al., J.Phys.B 30 (1997) 2917 (Topical Review); R. Dörner et al., Physics Reports 330 (2000) 95

Page 151 and page 407 add: J.E. Pollard et al., Rev. Sci. Instrum. 52 (1981) 1837; J. Ullrich et al., J.Phys.B 30 (1997) 2917

Page 238 Figure 5.22 and discussion must be revised; see B. Schmidtke et al., J. Phys. B 33 (2000) 2451 and 5225

Page 292 eqs. (7.39): replace (-) by (-1)

Page 321 eqs. (8.13), (8.14a) and page 322 line 16 from below: replace e^{kr} by e^{ikr}

Page 323 part 8.1.3: note that 'H' means 'H_{atom}'

Page 328 eq. (8.37): replace in { } 'zero' by the angular momentum 'ℓ'

Page 331 eqs.(8.48), (8.49): replace 'np' by '2p'

Page 354 denominator in eq. (8.110c) must be the same as in eq. (8.128)

Page 356 eq. (8.117b): replace summation index 'α' by 'κ'

Page 397 eq.(10.63c): replace E^o_{kin} by $E^0_{kin,i}$

Page 415 replace [Aug95] by [Aug25]

Page 420 [KFe92]: replace 'Feht' by 'Fehr'; [KJG95]: replace Vol. '15' by '75'

Page 422 [PNS82]: replace Hordgren by Nordgren

Dear Reader,

You hold in your hand a book on electron spectrometry using synchrotron radiation which is rather unusual in three respects. First, it is restricted to the study of atoms and indeed only to selected topics in atomic photoionization. However, due to their simplicity atoms provide a natural introduction to the field of electron spectrometry using synchrotron radiation, and an extension to photoprocesses in molecules and/or other applications of electron spectroscopy is straightforward. Second, it is a mixture of experimental and theoretical aspects where the latter are formulated from the viewpoint of an experimentalist. In this context I would like to point out that the close interplay between experiment and theory is one of the most striking and stimulating features of this field of research. Third, specific themes are repeated in Parts A, B, and C, but with increasing specialization. Redundancy ensures that, depending on their backgrounds, readers may start at any place.

The particular organization of the material presented in the three different parts results from my personal experience of working over the years with undergraduate, graduate, and Ph.D. students. This book is based on seminars and lectures given at the University of Freiburg, Germany, and on lecture series presented at the University of Aarhus, Denmark (1976), the University of Lausanne, Switzerland (1983), the University of Lincoln, Nebraska, USA (1987), and the Los Alamos (New Mexico, USA) Summer School (1988). Hence it is my sincere wish to thank all my students and colleagues for their fruitful discussions on these occasions since their contributions have helped me considerably from the didactical point of view to prepare the material. Naturally, I have also been greatly influenced by the many colleagues who have taught me physics, given me advice, and have stimulated me by exchange of ideas. My sincere thanks go to Professor Werner Mehlhorn who introduced me to the broad field of electron spectrometry, encouraged me to do experiments with synchrotron radiation, and has continuously supported my research in this field.

Electron spectrometry of atoms using synchrotron radiation has become an increasing and important field of fundamental physics within the last two decades. The increasing availability of dedicated facilities with tunable synchrotron radiation has allowed detailed exploration of the atom–photon interaction which must take into account the electron–electron interactions, usually termed electron correlations. The parallel experimental and theoretical developments have permitted rather sophisticated investigations of the response of the atomic many-electron system to photon impact over the whole range of energies, from outer-shell excitations to deep inner-shell ionizations. New development of even more advanced light sources (wigglers, undulators, free electron lasers), more dedicated instruments (electron, ion, fluorescence spectrometers adapted to operation with monochromatized synchrotron radiation, and the inclusion of coincidence techniques), as well as new theoretical approaches attacking the many-particle problem, make it evident that the zenith of this rapidly expanding field still lies in the future.

One of the most important experimental tools for the investigation of the structure and dynamics of an atom interacting with a single photon is electron spectrometry using synchrotron radiation, and this special field will be treated in this book. It is not my aim to compete with the many excellent presentations already published on this subject. Instead, as already expressed above, I will present material which I have found to be important when introducing students to this field. The book is split into three parts of increasing difficulty. In Part A a general introduction to the field is given, followed by a detailed analysis of the photoelectron and the K–LL Auger spectrum of neon. Emphasis is placed on a clear and compact presentation without too many formulas. Hence, it should be well suited for students interested in this field and/or looking for an interesting application of quantum mechanics. In Part B the experimental aspects of electron spectrometry are described, including the characteristic features of electrostatic analysers, detectors, lenses, disturbances and optimization. Further, recent examples of electron spectrometry with synchrotron radiation are presented in order to elucidate the power of results obtainable with this method. These topics are addressed to both those physicists who want to use electron spectrometers, whether or not for applications different from those described here, and those who wish to learn more about recent results on the photoionization of free atoms with synchrotron radiation. Part C provides all the information necessary to complete the discussions of Parts A and B. Even though these topics are essential for specialists in this field, they are also general enough to be of importance to the non-specialist, but interested, reader.

I hope that this book fulfils the expectations of its readers. I would finally like to thank many colleagues for fruitful comments and discussions on the manuscript, in particular, Dr Stephen J. Schaphorst and Maureen Storey, and Barbara Müller and Helga Müller for their help in preparing the book and last, but not least, my wife Annemarie and our daughters Verena and Sigune for their patience and understanding when I have spent much of my time in the preparation of this book.

Volker Schmidt

Freiburg
1996

Part A

Background and basic principles

1
Introduction

1.1 Theoretical background and general aims

In the non-relativistic limit, the electronic structure of an atom is determined by the Coulomb interaction between the electrons and the nucleus and the Coulomb interaction between the electrons themselves. In the relativistic case, other interactions have to be added, of which the spin–orbit interaction represents the largest contribution. The complete and exact description of these forces in the atom follows from quantum electrodynamics which is nowadays a well-established theory. Therefore, structure studies in atoms as compared to other systems (nuclei or elementary particles) have the advantage of involving forces which are known exactly. However, even for an ideal case it is extremely difficult accurately to calculate the atomic parameters for a many-electron system. As an example the structure of the helium atom in its ground state wavefunction will be discussed, first within the model of independent particles and then for two types of wavefunction which take into account electron correlations, i.e., the correlated motions of the electrons. The fundamental features demonstrated for this relatively simple case can then also be applied to the more complicated dynamical process of photoionization. Here the observed effects of electron–electron interactions and their theoretical treatment brought a renaissance of atomic physics with exciting new insight into the structure and dynamics of atoms interacting with photons, and this aspect will appear in many places throughout the book.

1.1.1 Atomic structure

In order to understand atomic structure, some results from quantum mechanics have to be recalled. For simplicity, it is sufficient to consider the non-relativistic case; however, the existence of the electron spin must be taken into account. The Hamiltonian H for an atom with Z electrons is given by (using atomic units which are defined in Section 6.1)

$$H = -\tfrac{1}{2}\sum_i \nabla_i^2 - Z\sum_i \frac{1}{r_i} + \sum_{i<j} \frac{1}{r_{ij}}. \tag{1.1}$$

The electronic structure of the ground state follows from the properties of the ground state wavefunction $\tilde{\Psi}$ which is the solution of the stationary Schrödinger equation

$$H\tilde{\Psi} = E\tilde{\Psi}, \tag{1.2}$$

where E is the ground state energy (minimum energy value). The tilde on the wavefunction $\tilde{\Psi}$ indicates that the wavefunction must be antisymmetric with respect to the interchange of any two electrons because electrons are fermions, with spin 1/2. (Fermions are named after E. Fermi who studied their properties in great detail.)

Approximating the Coulomb interaction between the electrons by a mean spherical potential $V(r)$, it follows that

$$H \approx H^0 = \sum_i h_i \tag{1.3a}$$

with

$$h_i = -\tfrac{1}{2}\nabla_i - Z\frac{1}{r_i} + V(r_i), \tag{1.3b}$$

and the Schrödinger equation can easily be solved. First, one solves the single-particle Schrödinger equation with the operator h_i for electron i, which is just the Hamiltonian operator for the hydrogen atom (examples of the solution are given in Section 7.1)

$$h_i \varphi_i = \varepsilon_i \varphi_i. \tag{1.4}$$

Second, one solves the approximate Hamiltonian H^0 for the Z-electron problem

$$H^0 \tilde{\Psi}^0 = E^0 \tilde{\Psi}^0. \tag{1.5}$$

The energy E^0 is simply the sum of the energies ε_i of the individual electrons,

$$E^0 = \varepsilon_1 + \varepsilon_2 + \cdots + \varepsilon_Z. \tag{1.6}$$

The wavefunction $\tilde{\Psi}^0$ then follows as an antisymmetrized product built from the single-particle functions $\varphi_i(\mathbf{r}, m_s)$ for the Z electrons (Slater determinantal wavefunction, see below and Section 7.2), where $\mathbf{r}$ is the spatial vector and m_s the spin magnetic quantum number.

This approach yields the *shell model* of the atom in which, under the restrictions of the Pauli principle† and according to the aufbau principle,‡ the electrons i are placed in the spin-orbitals $\varphi_i(\mathbf{r}, m_s)$. For example, the shell structure of the magnesium atom is sketched schematically in Fig. 1.1.

Due to the properties of determinants, a Slater determinantal wavefunction $\tilde{\Psi}^0$ automatically fulfils the Pauli principle and takes care of the antisymmetric character of fermions. If written explicitly in terms of the single-particle orbitals,

† The *Pauli* exclusion principle [Pau25] states that occupied electron orbitals must differ in at least one of the quantum numbers n, ℓ, m_ℓ, m_s.

‡ In the *aufbau* (German: building-up) principle the electrons of the atoms are placed in the lowest unoccupied orbitals, starting with the 1s orbital and filling the other orbitals one after the other (see, for example [Som19] Vol. I, p. 168). The irregularities which exist for the filling of nd and nf shells are not considered here (see Section 5.3.1).

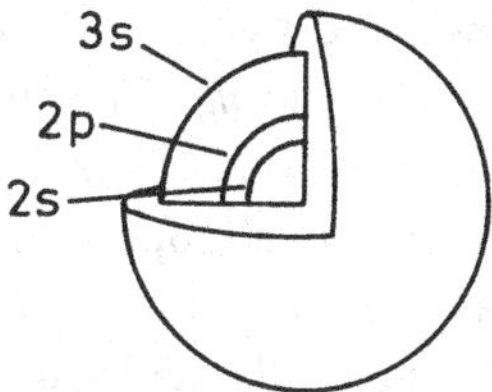

Figure 1.1 Schematic sketch of the shell structure of the magnesium atom. The atom has spherical symmetry, and one quadrant of the sphere has been cut out in order to show the shells which are indicated as part of a circular orbit for 3s, 2p and 2s electrons (the inner 1s shell is not visible).

$\tilde{\Psi}^0$ is defined as

$$\tilde{\Psi}^0(1,\ldots,Z) = \{\varphi_1,\ldots,\varphi_Z\} = \frac{1}{\sqrt{Z!}}\begin{vmatrix}\varphi_1(1) & \cdots & \varphi_1(Z)\\ \vdots & & \vdots \\ \varphi_Z(1) & \cdots & \varphi_Z(Z)\end{vmatrix}, \tag{1.7a}$$

where the subscripts 1–Z stand for the four single-electron quantum numbers:

n = principal quantum number

ℓ = orbital angular momentum quantum number

m_ℓ = magnetic quantum number of the orbital angular momentum (component of ℓ along a preferred direction which is usually called the quantization axis or z-axis)

m_s = spin magnetic quantum number (component of spin along the preferred direction).

(Usually, the values $\ell = 0, 1, 2, 3$ are termed s, p, d, f, respectively, the names coming from observations in the alkali spectra where *sharp*, *principal*, *diffuse*, and *fundamental* series have been distinguished [Ryd89]; higher ℓ-values are then named in alphabetical order.) The numbers within the brackets label the electrons from 1 to Z. Hence, the most compact form which will be used in discussions can be written as

$$\tilde{\Psi}^0(1,\ldots,Z) = \{n_1\ell_1 m_{\ell_1}^{m_{s_1}},\ldots,n_Z\ell_Z m_{l_Z}^{m_{s_Z}}\}. \tag{1.7b}$$

For example, the ground state $\tilde{\Psi}^0$ of the magnesium atom for which 12 electrons must be placed in the spin-orbitals $1s0^{m_s}$, $2s0^{m_s}$, $2pm_\ell^{m_s}$,$3s0^{m_s}$ is represented by the Slater determinantal wavefunction

$$\{1s0^+, 1s0^-, 2s0^+, 2s0^-, 2p1^+, 2p1^-, 2p0^+, 2p0^-, 2p{-1}^+, 2p{-1}^-, 3s0^+, 3s0^-\}.$$

The electronic structure of the atom then follows from the properties of $\tilde{\Psi}^0$. For a short characterization one quotes the *electron configuration* and the observables of the state considered:

$$\tilde{\Psi}^0\text{: electron configuration, state numbers.} \tag{1.8}$$

The electron configuration is described by the contributing spin-orbitals omitting the projection quantum numbers m_ℓ and m_s. The state numbers characterize the resulting state and follow from the solution of the stationary Schrödinger equation, i.e., they represent the eigenvalues of all operators which commute with the Hamiltonian, because these quantities can be measured simultaneously. In general these quantities are the energy E, the total angular momentum J and its projection M along a preferred direction (e.g., z-axis), and the parity π.

Since in many cases the Coulomb interaction between the electrons dominates their spin–orbit interaction (see the Hamiltonian in equ. (1.1)),

$$\sum_{i<j} \frac{1}{r_{ij}} \gg \sum_i \xi(r_i)\boldsymbol{\ell}_i \cdot \mathbf{s}_i, \tag{1.9}$$

and the Coulomb interaction preserves the spin and orbital angular momentum, one can also include in the state numbers the angular momenta L and S which are defined by the vector sum of individual angular momenta

$$\mathbf{L} = \sum \boldsymbol{\ell}_i \quad \text{and} \quad \mathbf{S} = \sum \mathbf{s}_i. \tag{1.10}$$

From these, one gets the total angular momentum J from the coupling

$$\mathbf{J} = \mathbf{L} + \mathbf{S}. \tag{1.11}$$

Summarizing, the short characterization for the electronic structure of a given state follows from

$$\tilde{\Psi}^0\text{: (electron configuration) } {}^{2S+1}L_J^\pi. \tag{1.12}$$

In this expression $2S + 1$ is called the *multiplicity* because if the spin–orbit interaction of the electrons is taken into account in a perturbative approach, the pure LS state splits in energy for the different couplings of L with S leading to J (the LSJ-coupling case), and the number of term splittings is given for $L \geq S$ by $2S + 1$. It is therefore common to say 'singlet', 'doublet', 'triplet' and so on for $2S + 1$ equal to 1, 2, 3, ... For $L < S$ the number of possible terms is smaller than the multiplicity, e.g., ${}^3S_1$ has only one term even though it is called a triplet, ${}^5P_{3,2,1}$ has only three terms even though it is called a quintet.

The parity π of the resulting state reflects the behaviour of the wavefunction with respect to inversion through the origin, i.e.,

$$\Pi\, \tilde{\Psi}^0(\mathbf{r}_1, \ldots, \mathbf{r}_Z) = \tilde{\Psi}^0(-\mathbf{r}_1, \ldots, -\mathbf{r}_Z) = \pi\, \tilde{\Psi}^0(\mathbf{r}_1, \ldots, \mathbf{r}_Z). \tag{1.13a}$$

Since two operations of Π restore the arguments, the parity operator can have only two eigenvalues, $+1$ (called *even* parity) and -1 (called *odd* parity):

$$\pi = \pm 1. \tag{1.13b}$$

For the spherical symmetric systems considered here, the eigenvalue π follows from the property of the spherical harmonics

$$\Pi\, Y_{\ell m}(\vartheta, \varphi) = Y_{\ell m}(\pi - \vartheta, \pi + \varphi) = (-1)^\ell\, Y_{\ell m}(\vartheta, \varphi) \tag{1.14a}$$

attached to the orbital angular momentum quantum numbers ℓ and m of the electron. Hence, one gets for the parity π of many electrons with angular momenta ℓ_i

$$\pi = (-1)^{\Sigma \ell_i}. \tag{1.14b}$$

The eigenvalue $\pi = +1$ is indicated by the upper symbol 'e' (even), the eigenvalue $\pi = -1$ by the symbol 'o' (odd).

With these definitions, the ground state of the magnesium atom is then represented by the electron configuration for the orbitals 1s, 2s, 2p, and 3s (see Fig. 1.1) and the symbols for the angular momenta and parity as

$$\tilde{\Psi}^0\text{: } 1s^2 2s^2 2p^6 3s^2\ {}^1S_0^e.$$

The *LSJ*-coupling scheme introduced above is called the Russell–Saunders coupling scheme [RSa25]. It is based on the validity of equ. (1.9). The other extreme coupling case follows if the spin–orbit interaction dominates the Coulomb interaction between the electrons. This is called the *jjJ*-coupling scheme and requires that

$$\mathbf{j}_i = \boldsymbol{\ell}_i + \mathbf{s}_i \quad \text{and} \quad \mathbf{J} = \sum \mathbf{j}_i. \tag{1.15}$$

In reality, intermediate coupling is more common, and lies somewhere between these two limits. An example is discussed in Section 7.4.3 and applied to K–LL Auger transitions in Section 3.1.3. As a rough rule, one can use *LSJ*-coupling for the outer shells in low-*Z* elements, and *jjJ*-coupling for inner shells in large-*Z* elements.

1.1.2 Ground state wavefunctions (helium)

Apart from the demands of the Pauli principle, the motion of electrons described by the wavefunction $\tilde{\Psi}^0$ attached to the Hamiltonian H^0 is independent. This situation is called the *independent particle* or *single-particle* picture. Examples of single-particle wavefunctions are the hydrogenic functions $\varphi_i(\mathbf{r}, m_s)$ introduced above, and also wavefunctions from a Hartree–Fock (HF) approach (see Section 7.3). HF wavefunctions follow from a *self-consistent* procedure, i.e., they are derived from an *ab initio* calculation without any adjustable parameters. Therefore, they represent the *best* wavefunctions within the independent particle model. As mentioned above, the description of the *Z*-electron system by independent particle functions then leads to the shell model. However, if the Coulomb interaction between the electrons is taken more accurately into account (not by a mean-field approach), this simplified picture changes and the electrons are subject to a correlated motion which is not described by the shell model. This correlated motion will be explained for the simplest correlated system, the ground state of helium.

In the independent particle picture, the ground state of helium is given by $1s^2\ {}^1S_0^e$. For this two-electron system it is always possible to write the Slater determinantal wavefunction as a product of space- and spin-functions with certain symmetries. In the present case of a singlet state, the spin function has to be antisymmetric, χ_a, and one gets

$$\tilde{\Psi}^0(1s^2\ {}^1S_0^e) = \Phi^0(\mathbf{r}_1, \mathbf{r}_2)\chi_a, \tag{1.16a}$$

where

$$\Phi^0(\mathbf{r}_1, \mathbf{r}_2) = \varphi_{1s0}(\mathbf{r}_1)\varphi_{1s0}(\mathbf{r}_2) \tag{1.16b}$$

and

$$\chi_a = \frac{1}{\sqrt{2}}[\chi_{1/2}^+(1)\chi_{1/2}^-(2) - \chi_{1/2}^+(2)\chi_{1/2}^-(1)], \tag{1.16c}$$

where the superscripts on the spinor-functions $\chi_{1/2}^{m_s}$ characterize the $\pm 1/2$ spin projections parallel or antiparallel to the selected quantization axis.

There are many ways to improve this independent-particle model by incorporating electron correlation in the spatial part $\Phi^0(\mathbf{r}_1, \mathbf{r}_2)$ of the wavefunction. Here the Hylleraas function [Hyl29] and the method of configuration interaction (CI) will be used as illustrations.

The fundamental idea of Hylleraas was that the attractive force between the nuclear charge and each of the electrons is well accounted for in the single-particle orbitals $\varphi_{1s0}(\mathbf{r}_1)$ and $\varphi_{1s0}(\mathbf{r}_2)$ by their exponential functions with negative exponents in the coordinates $\mathbf{r}_i$ (see Section 7.1.1):

$$\varphi_{1s0}(\mathbf{r}) = \frac{2}{\sqrt{4\pi}} Z^{3/2}\, e^{-Zr}. \tag{1.17}$$

Hence, one expects that the mutual repulsion of the electrons at the positions $\mathbf{r}_1$ and $\mathbf{r}_2$ can be described by an exponential function with a positive exponent in the relative coordinate r_{12}:

$$e^{r_{12}} \quad \text{with} \quad r_{12} = |\mathbf{r}_1 - \mathbf{r}_2| = \sqrt{(r_1^2 + r_2^2 - 2r_1r_2\cos\vartheta_{12})}. \tag{1.18}$$

As a consequence, elliptical coordinates

$$s = r_1 + r_2, \qquad t = r_2 - r_1 \qquad \text{and} \qquad u = r_{12} \tag{1.19}$$

are introduced, and the correlated wavefunction is expanded in terms of these coordinates. The condition that E_g, the energy of the ground state, has to be a minimum then allows the determination of the unknown expansion coefficients by a variational procedure. In the lowest approximation one gets the three-parameter Hylleraas function,

$$\Phi_{Hyll}(\mathbf{r}_1, \mathbf{r}_2) = N\, e^{-1.815s}(1 + 0.30u + 0.13t^2), \tag{1.20}$$

with the normalization factor $N = 1.32135$ and the energy eigenvalue

$$E_g(\text{Hyll.}) = -2.90244\ \text{au}. \tag{1.21}$$

Before the individual parts of this function are discussed, the energy eigenvalue will be considered. The ground state energy E_g of the helium atom is just the energy value for double-ionization which can be determined accurately by several different kinds of experiments. Before the experimental value can be compared with the calculated one, some small corrections (for the reduced mass effect, mass polarization, relativistic effects, Lamb shift) are necessary which, for simplicity, are not considered here. Having established the corrected $E_g(\text{exp.}) = -2.90372$ au, this can be compared to the theoretical results: in the present context to $E_g(\text{3-param. Hyll.}) = -2.90244$ au and, for comparison, also to the Hartree–Fock value $E_g(\text{HF}) = -2.862$ au (see Section 7.3.1). There is rather close agreement between E_g(3-param. Hyll.) and E_g(exp.), but E_g(HF) is larger. This indicates a considerable improvement of the Hylleraas wavefunction in equ. (1.20) as compared to the HF function.†

The Hylleraas function, with its improved properties as compared to a Hartree–Fock function, is called a *correlated* wavefunction, $\Phi_{\text{corr.}}(\mathbf{r}_1, \mathbf{r}_2)$, because it takes into account the mutual electron–electron interaction much better, and the motion of electrons beyond a mean-field average is termed *correlated motion* or the effect of *electron correlations.* (The definition of electron correlation is used here in the strict terminology. The mean-field average of electron–electron interactions is frequently also called electron correlation.) Comparing equ. (1.20) with equ. (1.16b) one has

$$\Phi_{\text{corr.}}(\mathbf{r}_1, \mathbf{r}_2) = \Phi_{\text{Hyll.}}(\mathbf{r}_1, \mathbf{r}_2) = \Phi^0(\mathbf{r}_1, \mathbf{r}_2) f_{\text{corr.}}(\mathbf{r}_1, \mathbf{r}_2), \tag{1.22a}$$

where the uncorrelated function is now described by screened hydrogenic wavefunctions (effective nuclear charge $Z = 1.815$ instead of the value $Z = 2$ for the bare nucleus),

$$\Phi^0(\mathbf{r}_1, \mathbf{r}_2) = N\,\mathrm{e}^{-1.815(r_1 + r_2)} \tag{1.22b}$$

and a correction or, equivalently, a correlation factor is introduced:

$$f_{\text{corr.}}(\mathbf{r}_1, \mathbf{r}_2) = 1 + 0.30 r_{12} + 0.13(r_2 - r_1)^2. \tag{1.22c}$$

The correlation factor can be analysed with respect to two different correlated motions of the electrons (see [Sla60] and Fig. 1.2):

(i) *Angular* correlation: Here one assumes both electrons to be on the same radius, and one obtains

$$f_{\text{corr.}}(r_1 = r_2 = r; \vartheta_{12}) = 1 + 0.30 r_{12} = 1 + 0.30\sqrt{[2r^2(1 - \cos\vartheta_{12})]}. \tag{1.23a}$$

This factor is responsible for an increase in the correlated wavefunction amplitude and hence in the charge distribution of both electrons when these electrons are on opposite sides of the nucleus ($\vartheta_{12} = 180°$; see Fig. 1.2(*b*)).

(ii) *Radial* correlation: Here one assumes both electrons to be on the same radius vector ($\vartheta_{12} = 0°$) and analyses equ. (1.22c) for $r_1 + r_2 = \text{const} = 1$ au. This corresponds to the case in which the wavefunction has a large amplitude, i.e., in which there is a high probability of finding electrons. The correction factor then gives

$$f_{\text{corr.}}(r_1, r_2; \vartheta_{12} = 0°) = 1 + 0.30 r_{12} + 0.13 r_{12}^2. \tag{1.23b}$$

† An extension of the Hylleraas function, slightly generalized, and with 39 parameters leads to $E_g(\text{theor.}) = -2.9037225$ au [Kin57]. If the necessary corrections are applied to this numerical value, one gets perfect agreement with the experimental result (for a correlated wavefunction of the ground state of helium which leads to E_g with an accuracy of 13 digits, $E_g(\text{theor.}) = -2.903724377033$ au, see [FPe66]).

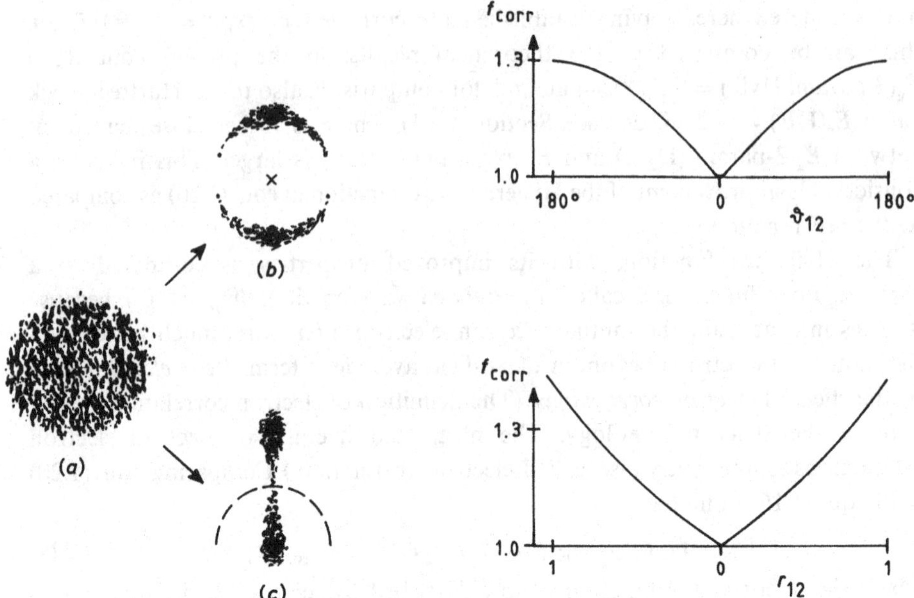

Figure 1.2 Sketch of the charge density distribution for (*a*) uncorrelated and (*b*) and (*c*) correlated motion of the two 1s-electrons in helium. Angular correlation is shown in (*b*) for $r_1 = r_2 = 0.5$ au; radial correlation in (*c*) for $\vartheta_{12} = 0°$ and $r_1 + r_2 = 1$ au. Even though all visual pictures of the electron motion in an atom fail in detail, one can get an impression of what is meant by the correlated motions as follows: by looking at the uncorrelated motion of the electrons with half-closed eyes one obtains the picture in (*a*) with a fluctuating granular structure, but overall homogeneity. Focusing then on one specific region (following the arrows in (*a*)), the homogeneity is destroyed and a certain spatial structure becomes apparent which reflects the correlated motion of the electrons, namely, angular correlation ((*b*): the cross indicates the positon of the nucleus) and radial correlation ((*c*): the centre of the circle indicates the position of the nucleus). The quantitative description of both types of correlation is given by the correlation factors $f_{corr.}$ which follow from equ. (1.23a) and (1.23b), respectively, and this factor is shown at the right-hand side ((*b*) angular correlation; (*c*) radial correlation).

It can be seen that the correlated wavefunction has a minimum for $r_{12} = 0$ which corresponds to $r_1 = r_2 = 0.5$ au, i.e., higher values of the wavefunction are obtained when the electrons are further apart (see Fig. 1.2(*c*)).

The other approach most frequently used to describe a correlated wavefunction beyond the independent-particle model is based on *configuration interaction* (CI). (If the expansion is made on grounds of other basis sets, the approach is often called *superposition of configurations*, SOC, in order to distinguish it from the CI method.) According to the general principles of quantum mechanics, the exact wavefunction which is a solution of the full Hamiltonian H can be obtained as an expansion in any complete set of basis functions which have the same symmetry properties:

$$\tilde{\Psi}_{\text{corr.}} = \sum\!\!\!\!\!\!\int_{\nu} a_\nu \tilde{\Psi}^0_\nu, \tag{1.24}$$

where $\sum\!\!\!\!\!\int$ indicates summation over discrete and integration over continuous basis functions. In atoms, it is the Coulomb interaction between the nuclear charge and the electrons which is responsible for a strong central potential. Hence, it is natural and convenient to use as a complete set of basis functions the solutions of H^0 following from equ. (1.5). The set of basis functions $\tilde{\Psi}^0_\nu$ then has to be constructed to implement different electron configurations. This is the reason why the approach is called configuration interaction.

Following these ideas, the ground state of helium which includes electron correlations can then be represented as

$$\tilde{\Psi}_{\text{corr.}}(\text{`}1s^2\,{}^1S^e_0\text{'}) = A_1\tilde{\Psi}^0(1s^2\,{}^1S^e_0) + a_2\tilde{\Psi}^0(1s2s\,{}^1S^e_0)$$

$$+ a_3\tilde{\Psi}^0(2s^2\,{}^1S^e_0) + a_4\tilde{\Psi}^0(2p^2\,{}^1S^e_0) + \cdots, \qquad (1.25a)$$

where the admixed wavefunctions $\tilde{\Psi}^0_\nu$ belong to different electron configurations whose weight is described by the absolute value squared of the respective mixing coefficients $A_1, a_2, a_3, a_4, \ldots$. Several comments have to be made concerning this expansion:

(i) The expansion theorem given above requires an expansion into a *complete* set of basis functions. This requirement cannot be fulfilled in reality, because for practical reasons all expansions have to be truncated to a finite number. Therefore, the correlated wavefunction $\tilde{\Psi}_{\text{corr.}}$ can only approach the exact wavefunction $\tilde{\Psi}_{\text{exact}}$, and its quality will depend on the kind and number of selected basis functions. Ultimately this approximate treatment can be traced back to the many-body problem of classical quantum mechanics which also can be solved only approximately. In this general context of the many-body problem, atoms can provide a convenient test case for different theoretical models, because all interaction forces are known and the system can be varied from a three-body case (helium) to a full many-body system (heavier atoms).

(ii) The admixed electron configurations should not be confused with a state, even though they are characterized by ${}^{2S+1}L^\pi_J$. This, however, is a consequence of the requirement that the admixed electron configurations must have the same symmetry properties. (How the state function for such a correlated function can be derived from the admixed functions is demonstrated in Section 7.4.) The admixed functions are *basis* functions which are needed to modify the uncorrelated wavefunction in such a way that the correlated motion of the electrons is taken into account. In cases where $|A_1|$ is larger than $|a_i|$, the whole state might be still named after this dominant electron configuration, and this is indicated in the present example by writing '$1s^2\,{}^1S^e_0$'.

(iii) To each admixed electron configuration is attached a different state which might also be related to an observable state. (This depends strongly on the selected kind and number of basis functions (see Section 7.4.) In the present example, some excited states of helium are

$$\tilde{\Psi}_{\text{corr.}}(\text{`1s2s } ^1\text{S}_0^\text{e}\text{'}) = a_{21}\tilde{\Psi}^0(1\text{s}^2\ ^1\text{S}_0^\text{e}) + A_{22}\tilde{\Psi}^0(1\text{s}2\text{s}\ ^1\text{S}_0^\text{e}) + a_{23}\tilde{\Psi}^0(2\text{s}^2\ ^1\text{S}_0^\text{e}) + a_{24}\tilde{\Psi}^0(2\text{p}^2\ ^1\text{S}_0^\text{e}) + \cdots, \quad (1.25\text{b})$$

$$\tilde{\Psi}_{\text{corr.}}(\text{`2s}^2\ ^1\text{S}_0^\text{e}\text{'}) = a_{31}\tilde{\Psi}^0(1\text{s}^2\ ^1\text{S}_0^\text{e}) + a_{32}\tilde{\Psi}^0(1\text{s}2\text{s}\ ^1\text{S}_0^\text{e}) + A_{33}\tilde{\Psi}^0(2\text{s}^2\ ^1\text{S}_0^\text{e}) + a_{34}\tilde{\Psi}^0(2\text{p}^2\ ^1\text{S}_0^\text{e}) + \cdots, \quad (1.25\text{c})$$

$$\tilde{\Psi}_{\text{corr.}}(\text{`2p}^2\ ^1\text{S}_0^\text{e}\text{'}) = a_{41}\tilde{\Psi}^0(1\text{s}^2\ ^1\text{S}_0^\text{e}) + a_{42}\tilde{\Psi}^0(1\text{s}2\text{s}\ ^1\text{S}_0^\text{e}) + a_{43}\tilde{\Psi}^0(2\text{s}^2\ ^1\text{S}_0^\text{e}) + A_{44}\tilde{\Psi}^0(2\text{p}^2\ ^1\text{S}_0^\text{e}) + \cdots, \quad (1.25\text{d})$$

and so on (the capital letter mixing coefficient is used for the name of the correlated state because often, but not always, its absolute value squared turns out to be the largest value).

(iv) The magnitude of the mixing coefficient depends strongly on the selected basis functions used for the description of electron–electron interaction. Hence, any discussion of the importance of electron correlations based on a comparison of the magnitudes of mixing coefficients is valid only within the selected basis. Because the HF model (with relativistic effects it is the Dirac–Fock model) provides the best *ab initio* independent-particle approach leading to the atomic shell model, these single-particle functions are usually taken as reference or basis functions in a CI approach.

These preliminary, but important clarifications, lead us to a discussion of the correlated motion within the CI picture for both electrons in the ground state of helium. For this purpose it is illustrative to analyse the three-parameter Hylleraas function, the correlation properties of which have previously been described, in terms of CI functions. Looking only for the individual components of orbital angular momenta $\ell_1 = \ell_2$ which couple to the desired S^e state, one gets [GMM53]

$$\Phi_{\text{Hyll.}}(r_1, r_2) = 0.997535\,|\text{s}^2\ \text{S}^\text{e}\rangle + 0.069227\,|\text{p}^2\ \text{S}^\text{e}\rangle + 0.010398\,|\text{d}^2\ \text{S}^\text{e}\rangle + 0.003528\,|\text{f}^2\ \text{S}^\text{e}| + \cdots, \quad (1.26\text{a})$$

or, expanding in terms of HF functions, one obtains [GMU52]

$$\Phi_{\text{Hyll.}}(r_1, r_2) = 0.9955\,|1\text{s}^2\ \text{S}^\text{e}|_{\text{HF}} - 0.0005\,|1\text{s}2\text{s}\ \text{S}^\text{e}\rangle_{\text{HF}} - 0.0012\,|1\text{s}3\text{s}\ \text{S}^\text{e}\rangle_{\text{HF}} + 0.0223\,|2\text{p}^2\ \text{S}^\text{e}\rangle_{\text{HF}} + \cdots. \quad (1.26\text{b})$$

(Due to the decoupled spin there is no difference between HF and Hartree wavefunctions; in [GMU52] a different sign convention is applied to the $2\text{p}^2\ ^1\text{S}^\text{e}$ function.) In these expressions one can see that the values of the mixing coefficients depend on the selected expansion. In addition, one can note relatively large contributions of components $\ell \neq 0$. Concentrating on the latter aspect, for simplicity it suffices to consider only the two electron configurations 1s^2 and 2p^2, i.e.,

$$\Phi_{\text{corr.}}(\text{`1s}^2\ ^1\text{S}_0^\text{e}\text{'}) = A\,\Phi^0(1\text{s}^2\ ^1\text{S}_0^\text{e}) + a\,\Phi(2\text{p}^2\ ^1\text{S}_0^\text{e}) \quad (1.27\text{a})$$

with $A > a$ and $A, a > 0$.

The two-electron basis functions $\Phi^0(1\text{s}^2\ \text{S}^\text{e})$ and $\Phi^0(2\text{p}^2\ \text{S}^\text{e})$ can be expressed by the corresponding single-electron wavefunctions, and after some manipulations one derives (see equ. (7.107a))

$$\Phi_{\text{corr.}}(\text{`}1\text{s}^2\ \text{S}^{\text{e}}\text{'}) = A\frac{1}{4\pi}R_{1\text{s}}(r_1)R_{1\text{s}}(r_2) + a(-1)\frac{\sqrt{3}}{4\pi}R_{2\text{p}}(r_1)R_{2\text{p}}(r_2)\cos\vartheta_{12}, \quad (1.27\text{b})$$

which leads, for hydrogenic wavefunctions with $Z = 2$, to

$$\Phi_{\text{corr.}}(\text{`}1\text{s}^2\ \text{S}^{\text{e}}\text{'}) = A\,2.535\,\text{e}^{-2(r_1+r_2)} + a(-0.184)r_1r_2\,\text{e}^{-(r_1+r_2)}\cos\vartheta_{12}. \quad (1.27\text{c})$$

The first term on the right-hand side represents the uncorrelated motion of both electrons in the average potential given by the nucleus with $Z = 2$ and conforms to a spherical charge distribution (see Fig. 1.2(*a*)). In contrast, the second term is responsible for the correlated motion of the two electrons. In parallel with the discussion of the three-parameter Hylleraas function, both angular and radial correlation effects can be verified (see Section 7.5). However, there is an important difference between the correlated functions represented in equs. (1.20) and (1.27c): in the Hylleraas function, electron correlations are incorporated by a multiplication of the uncorrelated part with a correlation factor, while in the CI approach, electron correlations are accomplished by an expansion, i.e., summation, with the largest term being the uncorrelated part. This difference is responsible for the different qualities of these wavefunctions when analysed quantitatively (the Hylleraas function gives better results than the truncated CI approach; for details see Section 7.5).

1.1.3 Dynamics

The treatment of the many-body problem which requires the inclusion of electron correlations plays a role in the formulation of not only the structure of the atom, but also that of the dynamical changes caused by any kind of interaction between the atom and other particles including radiation quanta. The photon interaction has the special advantage of being known exactly. For simplicity, it will be given here in the dipole approximation (and in the length form; for details see Section 8.1). For linearly polarized light with the electric field vector oscillating along the z-direction one has

$$Op(\text{photoionization}) = e_0 \sum_j z_j, \quad (1.28\text{a})$$

where z_j is the z coordinate of the electron j. The form of the operator explains the name *dipole* approximation, because the expectation value $\langle i|Op|i\rangle$ in the ground state $|i\rangle$ gives the atomic electric dipole moment. Because the summation index runs independently over all atomic electrons, the photon operator is a single-particle operator. As a consequence, *one* photon can interact only with *one* atomic electron. This implies that only one-electron processes, also called *main* photoprocesses are possible, namely:

excitation of one electron, characterized by the discrete oscillator strength f;

ionization of one electron, characterized by the continuous oscillator strength

$df/d\varepsilon$ which is proportional to the photoionization cross section σ^+. (In the energy spectrum of ejected electrons one gets, in this case, the *main* photolines.)

However, because of the correlated motion of the electrons, many-electron processes will also occur. (Looking at the many-particle effects in this way, the photon operator is a single-particle operator and electron–electron interactions have to be incorporated explicitly into the wavefunction. It is, however, also possible to describe the combined action of the electrons as an induced field which adds to the external field of the photoprocess, i.e., the transition operator becomes modified. Generally, the influence of the electron–electron interaction can be represented by modifying the wavefunction or the operator or by modifying both the wavefunction and the operator [DLe55, CWe87].) Of all the possible processes, only the important two-electron processes restricted to electron emission will be considered here. In many cases they can be divided into two different classes (see Fig. 1.3).†

The first class comprises direct two-electron processes, called *satellite* processes, which can be classified as

ionization accompanied by simultaneous excitation (cross section σ^{+*}),

direct double ionization (cross section σ^{++})

double excitation (cross section σ^{**}).

The second class consists of processes which can be described in a two-step model by an inner-shell excitation or ionization process followed by a subsequent *non-radiative* decay, i.e.,

autoionization after one-electron excitation,‡

Auger§ decay after one-electron ionization.

The operator for the non-radiative decay is the Coulomb interaction between the electrons

$$Op(\text{Auger transition}) = \frac{1}{4\pi\varepsilon_0} \sum_{i<j} \frac{e_0^2}{r_{ij}}, \tag{1.28b}$$

i.e., an effect of electron correlation. Hence, the reader might expect these transitions to be termed *satellites*. However, the Auger lines are classified separately and are called *main* or *diagram* lines. The justification for this comes from the underlying two-step mechanism which means in the case of photon-induced Auger decay that: first, the emission of a photoelectron is followed by the

† There are other electron emission processes which lie between such well-defined limiting cases, e.g., resonance affected two-electron emission which lies between direct double photoionization and photon-induced two-step double ionization (photoelectron and Auger electron emission).

‡ For autoionization following inner-shell excitation such a classification neglects the competing and interfering process of direct photoionization leading to the same final state. It would be more appropriate to consider autoionization as a resonance feature embedded in the ionization continuum of main and satellite photoprocesses.

§ Auger decay was discovered by the French physicist P. Auger [Aug25].

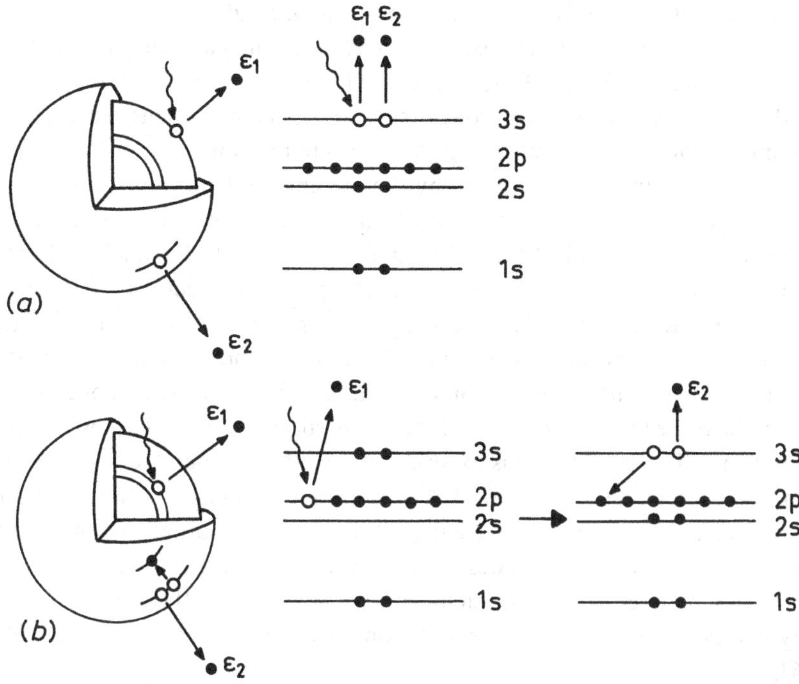

Figure 1.3 Illustration of the two classes of two-electron processes caused by photoionization using magnesium as an example, using, on the left the model-picture of Fig. 1.1 and on the right an energy-level diagram (not to scale): (*a*) direct double photoionization in the outer 3s shell; (*b*) 2p inner-shell photoionization with subsequent Auger decay where one 3s electron jumps down to fill the 2p hole and the other 3s electron is ejected into the continuum (Auger electron). The wavy line represents the incident photon (which is often omitted in such representations); electrons and holes are shown as filled and open circles, respectively; arrows indicate the movements of electrons; continuum electrons are classified according to their kinetic energy ε.

emission of an Auger electron, i.e., the intensity of a main photoprocess is transferred to an intensity of one or more main Auger processes; and second, the processes can be distinguished by their kinetic energies (see next section). Auger emission is frequently treated in this way, but there are deviations from this simple model (see [WOh76] and Section 8.3).

The interaction of a photon with an atom changes the structure of the atom, and the photon-energy-dependent change in the observables is called the *dynamics* in the photon–atom interaction. Therefore, in the present context of the study of photoionization processes using electron spectrometry and synchrotron radiation, the observations that can be made on the emitted electrons are all studies of dynamical properties. In the light of the foregoing discussion on the forces in the atom and the transition operator it can be concluded that photoprocesses in atoms provide a unique opportunity for fundamental investigations which explore the dynamics of many-body effects, because both the forces and the interaction operator are well known. In particular, since the ground state of atoms can be described rather well by taking electron correlations into account, it is the final state of the system which is of most interest. In the final state it is theoretically rather difficult to describe the partial break-up of the atom caused by electron emission: the orbitals of the remaining electrons are rearranged in response to the change in nuclear shielding produced by the missing electron (relaxation effects), the long-range Coulomb interaction will lead to important interactions between the escaping electron and the remaining electrons (and the nuclear charge), spin-orbit effects will be responsible for spin-flips of the electrons, and so on.† Therefore, a broad field of different theoretical approaches may be used to treat and explore such many-electron problems. Hand in hand with these theoretical studies, experimental investigations of many-electron phenomena and one-electron properties have been performed using electron spectrometry which is a unique and direct tool. The power of this experimental method is increased considerably by combining electron spectrometry with the unique properties of monochromatized synchrotron radiation, in particular through the tunability of the photon energies. Within the last two decades, a wealth of such theoretical and experimental studies have greatly enhanced our detailed understanding of the atomic structure and dynamics involved in photoionization (for reviews see [Sch92a, SZi92, BSh96]).

1.2 Basic measurable quantities

Single ionization of randomly oriented atoms yields a residual ion and a photoelectron with a certain kinetic energy. The quantities which can be measured for such a photoprocess are the *cross section*, the *kinetic energy*, the *angular distribution*, and the *spin polarization* of the emitted photoelectron, and the *polarization* of the ion. If the ionization process involves an inner-shell electron, a subsequent radiative (fluorescence) or non-radiative (Auger) decay may occur, and when these reaction products are observed with a certain *energy*, one can also determine the corresponding decay probability (*fluorescence/Auger yield*), the *angular distribution*, the *polarization* of the emitted photon/electron, and the *polarization* of the residual ion. Similar statements can be made for electron emission following resonance excitations. In all cases in which photoionization leads to several reaction products (the ion, one or more electrons, one or more fluorescence quanta), certain relations

† Such dynamical effects will depend on electron correlations in both the initial state and final state (which includes the correlated final ionic state and the continuum channel of the photoelectron). Within the CI picture, one therefore can distinguish

initial state configuration interaction (*ISCI*),
final ionic state configuration interaction (*FISCI*),
continuum state configuration interaction (*CSCI*),
final state configuration interaction (*FSCI*).

More details and examples of these cases will be presented in Chapter 5.

exist between the emitted particles. Coincidence measurements are then needed in order to establish the full emission pattern which follows the primary event of photon interaction, and it is possible to obtain complementary information on the photoprocess and subsequent decay process from such measurements.

The different emission products which are possible after photoionization with free atoms lead to different experimental methods being used: for example, electron spectrometry, fluorescence spectrometry, ion spectrometry and combinations of these methods are used in coincidence measurements. Here only electron spectrometry will be considered. (See Section 6.2 for some reference data relevant to electron spectrometry.) Its importance stems from the rich structure of electron spectra observed for photoprocesses in the outermost shells of atoms which is due to strong electron correlation effects, including the dominance of non-radiative decay paths. (For deep inner-shell ionizations, radiative decay dominates (see Section 2.3).) In addition, the kinetic energy of the emitted electrons allows the selection of a specific photoprocess or subsequent Auger or autoionizing transition for study.

A monochromatized photon beam of energy $h\nu$ can lead to the ejection of a photoelectron if the photon energy is larger than the ionization energy E_I. (This is often also called the *binding* energy, although strictly speaking the binding energy E_B is negative for bound electrons: $|E_B| = E_I$.) The emitted photoelectron has a kinetic energy E_{kin} given by:

$$E_{kin}(\text{photoelectron}) = E_{kin}(\text{phe}) = E_{exc} = h\nu - E_I. \tag{1.29a}$$

In this expression, the recoil energy of the ion is neglected, because practically all of the excess energy E_{exc} is carried away by the photoelectron due to the large mass ratio between the photoelectron and the ion (for a new kind of electron spectrometry based on the momentum of the recoil of the photoion see [UDM94]). Since $h\nu$ is usually known, and E_I has a definite value which depends on the ionic state produced, a specific photoprocess can be selected via the kinetic energy of the emitted photoelectrons. This is possible for the processes of single ionization (main photoionization with the partial cross section σ^+) and for ionization with simultaneous excitation (discrete photosatellites with the cross section σ^{+*}), because in these cases only one electron is emitted and, therefore, maxima appear in the intensity of detected electrons at characteristic energy positions. These structures are called photolines: *main* photolines for processes involving σ^+, and *satellite* photolines for those involving σ^{+*}. (The name satellite photo*lines* implies that these are *discrete* satellites. The continuous energy distribution from double ionization is often said to be due to *continuous satellites*.) In contrast, in the case of direct double photoionization (σ^{++}), the two ejected electrons must share the available excess energy E_{exc} which leads to a broad continuous energy distribution covering the region between zero kinetic energy and E_{exc}. Such a continuous distribution can also be measured by means of electron spectrometry, but it is more difficult to assess because the intensity is spread over this broad energy region. In addition the detection of electrons with low kinetic energy is usually difficult; scattered electrons can produce a severe disturbance, in most cases several double ionization continua are superimposed and photolines and Auger lines from other processes occur. It is then not easy to separate these processes from one another.

Within the two-step model for photoionization and subsequent Auger decay, the kinetic energy of Auger electrons for *normal* (*diagram*) transitions comes from the energy difference of the ion states before (subscript i) and after (subscript f) the Auger decay, i.e.,

$$E_{kin}(\text{Auger electron}) = E_{kin}(\text{Ae}) = E_i(\text{one-hole state}) - E_f(\text{two-hole state}). \tag{1.29b}$$

Since the energy values E_i and E_f are determined by the properties of the corresponding ions, the different kinetic energies of Auger lines allow the selection of a specific process, in this case of a certain Auger transition. Comparing equs. (1.29a) and (1.29b), one can see that photoelectrons and Auger electrons can easily be distinguished by changing the photon energy: E_{kin}(phe) varies linearly with $h\nu$, but E_{kin}(Ae) remains constant.

A similar process selection is possible for inner-shell excitation or double excitation and subsequent autoionization decay (described in the first case by a cross section of the first step, σ^*, and in the latter case by σ^{**}). These processes occur only at specific photon energies $h\nu_r$ (subscript r for resonance), and the kinetic energy of electrons from the autoionization decay is then fixed by

$$E_{kin}(\text{electron from autoionization decay}) = h\nu_r - E_f(\text{one-hole state}), \tag{1.29c}$$

where E_f is the energy of the ionic state after the autoionization transition. (A more appropriate approach is to describe inner-shell ionization and autoionization decay as resonance features embedded in the ionization continuum of main and satellite photoprocesses. Due to the 'decaying' resonance, there is a certain resonance width Γ, which appears when observing such spectra (see Section 5.1.2.1; a similar statement holds for doubly-excited states).)

In the following, the basic quantities which can be measured by means of electron spectrometry and their parametrizations will be discussed for single photoionization (for two-electron emission see Section 4.6). In Fig. 1.4 the major features relevant to the photoprocess and a possible experimental set-up are shown. For simplicity the monochromatized photon beam of energy $h\nu$ is assumed to be completely linearly polarized. The interaction of the photon with an atom will yield a certain probability for the emission of a photoelectron with a particular kinetic energy. By collecting all these emitted photoelectrons, the strength of the photon–atom interaction as expressed in the partial photoionization cross section σ^+ can be obtained (a different way of obtaining σ^+ will be given below). In contrast, the detection of electrons emitted towards the entrance slit of the electron spectrometer will give an angle-resolved signal and yield information on the spatial

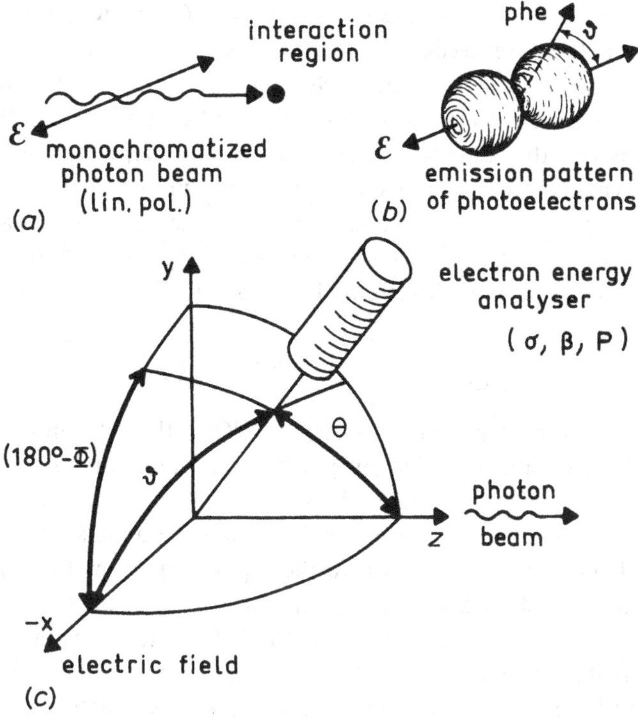

Figure 1.4 Photoionization of *ns*-electrons by linearly polarized light. (*a*) The direction of the incident photon beam and the direction of its oscillating electric field vector define a Cartesian coordinate system in the interaction region of the sample. (*b*) Photoelectron emission (phe) leads to a spatial intensity distribution that has axial symmetry around the direction of the electric field vector, $\mathscr{E}$. It is described by the dependence on the angle ϑ, and for the ejection of *ns* electrons this intensity distribution is given by $\cos^2 \vartheta$. (*c*) General arrangement for an electron energy analyser used to measure the cross section σ, the angular distribution parameter β and the spin polarization **P** of emitted electrons. The analyser is shown as a cylindrical 'black' box positioned at polar and azimuthal angles Θ and Φ given in the general x, y, z frame (note that the special angle ϑ refers to the x-axis; for details concerning the different reference axis see Section 9.1).

distribution of the photoelectrons when the spectrometer is set at different positions in space.

In general, the emission pattern of photoelectrons is not isotropic in space, but possesses a characteristic angular distribution. This can be understood because in the electric dipole approximation (Section 8.1) it is the electric field vector of the incident light which is of relevance. This field vector causes forced oscillations of the atomic electrons which can finally lead to electron emission, thus imposing a directionality on the emission process. Therefore, the electric field vector of the incident light fixes the natural reference axis against which observation of the emitted photoelectrons can be made. Since for photoionization of randomly oriented atoms no other preferred axis exists, the angular distribution depends only on the angle ϑ between the direction of the electric field vector and the direction of the ejected photoelectron. (The angles Θ and Φ are also shown in Fig. 1.4 because they are needed for the more realistic case of partially linearly polarized light as provided by monochromatized synchrotron radiation (see Section 9.1).) Because the electric field oscillates, the angles ϑ and $180° - \vartheta$ are equivalent. Hence, this angular distribution can be represented in terms of Legendre polynomials depending on even powers in $\cos\vartheta$. Within the dipole approximation, only $P_0(\cos\vartheta) = 1$ and $P_2(\cos\vartheta) = (1.5\cos^2\vartheta - 0.5)$ are possible, and the angular distribution of electrons emitted after photoionization by linearly polarized light is given by the differential cross section

$$\frac{d\sigma}{d\Omega}(\vartheta) = \frac{\sigma}{4\pi}[1 + \beta P_2(\cos\vartheta)], \tag{1.30}$$

where σ is the photoionization cross section, $d\Omega$ is the differential solid angle element in the direction specified by the polar angle ϑ, and β is the *angular distribution* or *anisotropy* parameter. This β parameter is sometimes also called the *asymmetry* parameter because in the early days the angular distribution was studied at a high photon energy where the dipole approximation breaks down and there is a forward/backward asymmetry with respect to the direction of an unpolarized photon beam. The numerical value of β determines the actual shape of the angular distribution pattern. In the special case of photoionization of an s-electron and for negligible spin–orbit effects, the β parameter has the energy-independent value $\beta = 2$. This case applies to 1s photoionization in helium, and the resulting angular distribution pattern is shown in Fig. 1.4(*b*). The pattern can be understood from the classical expectation that the electric field vector induces oscillation of the bound electron, which finally leads to electron emission along the direction of the oscillating electric field. In the general case, the β parameter varies between 2 and -1 because different amplitudes contribute to the photoionization process and interfere (see Section 2.5).

Auger electrons can also have anisotropic angular distributions (see [Meh68a, CMe74, FMS72]). These can be parametrized by the same expression as given for the photoelectrons except for some formal replacements necessary for the characterization of the Auger transition (see Section 3.5).

The spin is an inherent property of an electron. Since the photo- or Auger electrons are ejected in a certain direction in space, for an ensemble of these electrons a spin polarisation vector **P** can be defined which gives the excess of individual spin components measured in three orthogonal directions (see Section 9.2.1). In Fig. 1.5 the components of **P** are shown for a convenient decomposition into one longitudinal, P_{long}, and two transverse components, $P_{trans\parallel}$ and $P_{trans\perp}$, respectively. The measurement of these components requires an electron detector which is sensitive to spin. An example of the spectrometry of photoelectrons with spin-analysis will be described in Section 5.4.

These general remarks on the process of electron emission following photoionization lead us to a description of the experimental set-up used (spin-detection

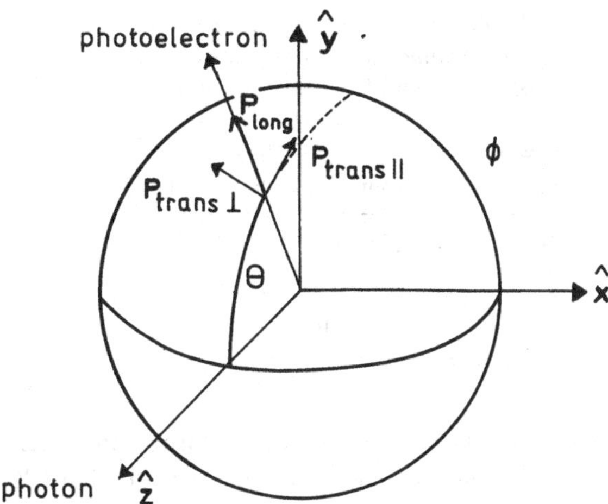

Figure 1.5 Components of the spin polarization vector **P** of ejected photoelectrons. The direction of the photoelectron is given by the polar and azimuthal angles Θ and Φ (see Fig. 1.4). For an ensemble of electrons emitted in this direction, the polarization vector **P** then points in a certain direction in space, and one possibility for representing this vector using three orthogonal components is shown in the figure: P_{long} in the direction of the photoelectron and $P_{trans\perp}$ and $P_{trans\parallel}$ both perpendicular to this direction (for the definition and measurement of these components see Section 9.2.1).

will be deferred to Section 5.4). The electron energy analyser, also called an electron *spectrometer*, is placed at a certain position in space (at angles Θ and Φ in Fig. 1.4) and receives the electrons which are emitted when photoionization occurs in the sample. The analyser can be considered as a black box, but it must have the following properties: it must accept only electrons emitted from the sample in the direction of its finite entrance slit; it must analyse the electrons with respect to their kinetic energy; and, finally, it must count the analysed electrons thus providing an intensity distribution $I(E_{kin})$. This intensity distribution represents the spectrum of ejected electrons. In particular, the intensity of a discrete photoline reflects the differential cross section $d\sigma/d\Omega$ of the respective process, as given for linearly polarized light in equ. (1.30). Therefore, measurement of the angle-dependent intensity $I(E_{kin})$ provides full information on the partial photoionization cross section σ and the angular distribution parameter β.

1.3 Properties of synchrotron radiation

The ionization energies for electron ejection from $n\ell$ orbitals in magnesium are given in Table 1.1. Since for ionization the photon energy must be larger than the ionization energy, photons with energies above the vacuum-ultraviolet region are of interest for the study of photoprocesses in atoms. (Air is opaque for photons above approximately 6 eV and, therefore, monochromators operating in the energy range above 6 eV must be evacuated. This explains why photons above 6 eV are

Table 1.1. *Ionization energies E_I of magnesium (from [BSc74, BSS76, New71, Moo71], see also [Sev79]).*

Shell $n\ell$	Ionization energy E_I [eV]
1s	1310.5
2s	96.5
2p	57.63
3s	7.646

called *vacuum*-ultraviolet photons [Boy41].) The broad and intense continuum of synchrotron radiation covers the energy range up to the hard X-ray regime and provides a unique radiation source. In the following, the properties of synchrotron radiation which are relevant to the investigation of photoprocesses by means of electron spectrometry are discussed.

A fast electron that undergoes an acceleration generates electromagnetic radiation. This phenomenon provides the basis not only for the continuous X-ray spectrum of an X-ray tube in which the electrons are decelerated when they hit the anode, but also for the emission of synchrotron radiation in which electrons in curved orbits are subject to centripetal acceleration. What is now called synchrotron radiation was first observed accidentally at the General Electric 70 MeV *synchrotron* [EGL47] where electromagnetic radiation was emitted by electrons moving in a circular orbit with highly relativistic velocities. In synchrotron radiation facilities which concentrate on the production of such radiation, the electrons are stored in a circular accelerator (called an *electron storage ring*), and synchrotron radiation is generated in the bending magnets or in special insertion devices (*wigglers/undulators*) placed into the linear sections of the storage ring. In both cases, the circulating electrons are subject to the centripetal acceleration of the Lorentz force and, therefore, emit radiation.

The properties of synchrotron radiation produced from an electron beam traversing a bending magnet are determined essentially by those of a single electron moving along a macroscopic circle. However, due to the relativistic velocity of the circulating electron, the radiation emission pattern is pushed dramatically into the forward direction (see Fig. 1.6). The opening half-angle Θ of the cone (its reciprocal value is often called γ) is given by

$$\Theta = \frac{1}{\gamma} = \frac{m_0 c^2}{E}, \tag{1.31}$$

where E is the electron energy, m_0 the rest mass of the circulating electron and c the velocity of light. For $E = 800$ MeV one obtains $\Theta = 0.64$ mrad. Thus, synchrotron radiation can be a highly collimated radiation. It should be noted that the energy lost due to the emission of synchrotron radiation is usually replenished by a radio

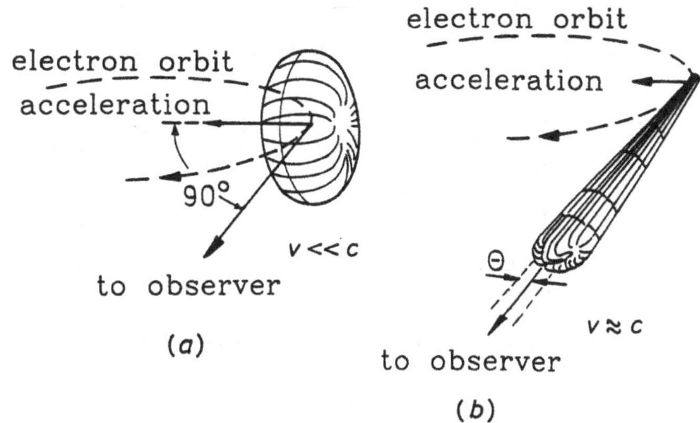

Figure 1.6 Emission pattern of an electron circulating with velocity v: (*a*) $v/c \ll 1$; (*b*) $v/c \approx 1$. From [THa56].

frequency system which accelerates the slower electrons and leads to a grouping of the electrons into 'bunches'.

The pronounced forward direction of synchrotron radiation is also called a searchlight effect. Because of this effect, an observer fixed in space can see the synchrotron light only for a short time interval Δt_{obs} which can be calculated as

$$\Delta t_{obs} = R/2\gamma^3 c, \tag{1.32a}$$

where R is the radius of the path of the circulating electron. According to Fourier analysis such a short pulse contains frequencies up to a critical value ω_c given by†

$$\omega_c = 1/\Delta t_{obs} \approx 2\gamma^3 \omega_0, \tag{1.32b}$$

where $\omega_0 = c/R$ is the fundamental angular frequency. Thus the energy spectrum of synchrotron radiation extends up to very high harmonics of ω_0 or, equivalently, to very high energies. For example, using the values $E = 800$ MeV and $R = 1.779$ m for the electron storage ring BESSY I in Berlin, one gets $\omega_0 = 1.685 \times 10^8\ \mathrm{s}^{-1}$ and $\omega_c \approx 1.293 \times 10^{18}\ \mathrm{s}^{-1}$ with the latter value corresponding to an energy of $E_c \approx 850$ eV. Due to longitudinal and transverse oscillations of the circulating electrons, the energy spectrum of emitted synchrotron radiation is a continuous distribution; an example is shown in Fig. 1.7 for the electron storage ring BESSY I. The spectral photon flux $\dot{N}_\gamma$ is presented for a mean current of 200 mA and is shown as a function of the photon wavelength λ. For increasing photon energies this flux slowly rises to its maximum value which occurs approximately at the critical energy $E_c = 850$ eV; above this value the flux rapidly decreases. As a matter of convenience, the spectral photon flux is given as the number of photons per second which reside in a wavelength interval (*bandpass*) of $\Delta\lambda = 10^{-3}\lambda$ and which enter a certain diaphragm because, if the experimenter wants to have monochromatized synchrotron radiation, this is the quantity of interest, not the overall

† ω_c is often defined as being 3/4 of the approximate value given here.

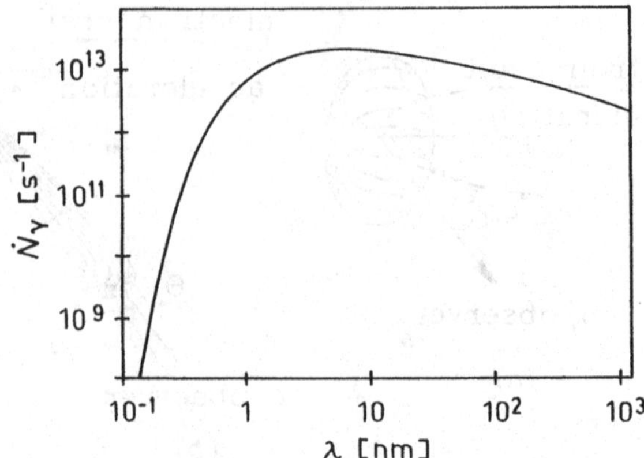

Figure 1.7 Spectral distribution of synchrotron radiation from the storage ring BESSY I (Berliner Elektronenspeicherring Gesellschaft für Synchrotronstrahlung, Germany). $\dot{N}_\gamma$ is the number of photons emitted for a beam current of 200 mA at some wavelength λ with a spectral range $\Delta\lambda = 10^{-3}\lambda$, and accepted by a given diaphragm (see text). The drop for smaller wavelengths is due to the critical value ω_c (see equ. (1.32b)) which is equivalent to $\lambda_c = 1.5$ nm. From [BES79].

intensity (see next section). In the present case the diaphram was selected to match a typical acceptance area of a monochromator; it is assumed to be $40 \times 40\ \text{mm}^2$ in size and to be placed at a distance of 7.5 m away from the electron's orbit and positioned symmetrically with respect to the plane of the storage ring. From Fig. 1.7 it can be seen that at a photon wavelength of 10 nm, which corresponds to approximately 100 eV, about 10^{13} photons/s with a bandpass of 0.1 eV enter the diaphragm and would be accepted by a typical monochromator.

In contrast to the smooth spectral distribution of radiation from a bending magnet, radiation from an undulator (and to a certain extent also from a wiggler) shows characteristic maxima. An example is shown in Fig. 1.8. These maxima are due to the periodic magnetic structure of the insertion devices (see Fig. 1.9): the transverse magnetic field with N periods of length λ_u produces an oscillatory (*wiggling* or *undulating*) path for the electrons which is similar to the response generated by $2N$ bending magnets. A characteristic parameter K for these devices relates the maximum deflection angle α of the electron path (Fig. 1.9) to the natural opening angle $\Theta = 1/\gamma$ of the searchlight cone (Fig. 1.6) and is given by

$$K = \frac{\alpha}{1/\gamma} = 93.4\, B_0 \lambda_u, \tag{1.33}$$

where B_0 is the peak field (in teslas) and λ_u is the magnetic period (in metres). Depending on the K-value, the searchlight cones from individual wiggles can contribute *separately* (in this case where $K \gg 1$ the device is called a wiggler), or they can act *coherently* (in this case where $K \lesssim 1$ the device is called an undulator).

For $K \lesssim 1$ the interference between electromagnetic waves emitted by the same

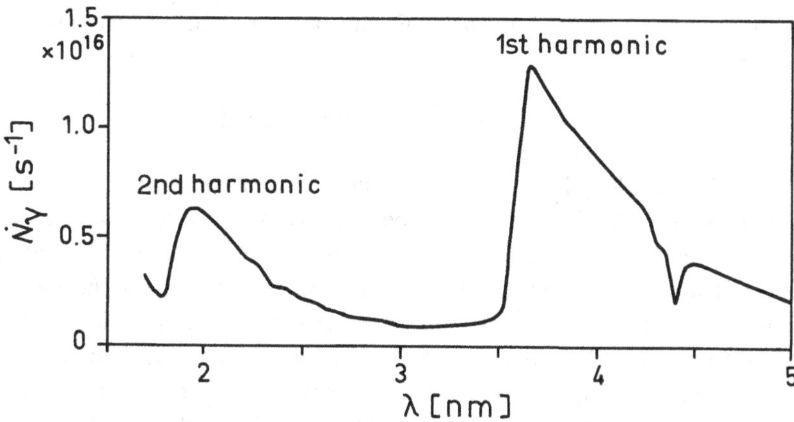

Figure 1.8 Example of the spectral flux $\dot{N}_\gamma$ of the undulator/wiggler radiation measured at the X-A1 beam line at NSLS (National Synchrotron Light Source, Brookhaven National Laboratory, USA) with the undulator parameters $K = 1.50$, $\lambda_u = 8$ cm, $N = 35$, for a 500 mA beam current, with a 0.1% bandpass and a solid angle of 1 mrad^2. The values are corrected for the beamline/monochromator efficiency and the photodiode detector response; the dip at 4.4 nm is an artifact due to carbon contamination of the optical elements. From [BRA89]. (Reproduced with permission from *Review of Scientific Instruments.*)

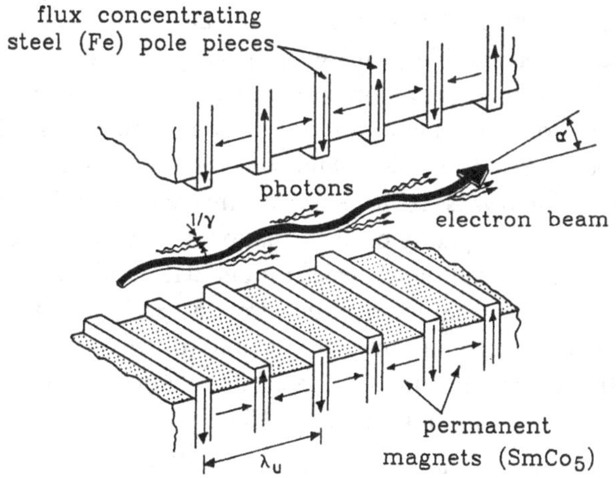

Figure 1.9 Hybrid transverse wiggler/undulator with permanent magnets and iron poles providing a wiggler/undulator magnetic field with period λ_u. The electron beam traverses this field wiggling/undulating in a plane perpendicular to the field. The maximum deflection angle α of the electron beam and the photon emission angle $1/\gamma$ at the maximum bending of the wiggles of the electron beam are shown. From [KHA84]. (Reproduced with permission from the American Institute of Physics.)

electron at different wiggles yields a redistribution of the spatial and spectral intensity. The condition for constructive interference follows from the proper timing for the movement of the electron and its emission of light with wavelength λ: the time taken by the electron to travel through one undulator period λ_u minus the time taken by light to travel this distance must equal $n\lambda/c$, where n is an integer. For radiation emitted at angles Θ and Ψ with respect to the undulator axis, one gets

$$\lambda_n = \frac{1}{n}\frac{\lambda_u}{2\gamma^2}\left[1 + \frac{K^2}{2} + \gamma^2(\Theta^2 + \Psi^2)\right], \tag{1.34}$$

where λ_n is the wavelength of the nth harmonic radiation ($n = 1$ is called the *fundamental* radiation); this Θ should not be confused with the cone angle $\Theta = 1/\gamma$ used above, it refers here to the radiation emitted in the plane of the wiggles, and Ψ is perpendicular to this plane. The harmonics are characteristic of the spectrum of observed undulator radiation. This is demonstrated in Fig. 1.8 where the first and second harmonics dominate the spectrum and provide a very high spectral flux. Hence, it is advantageous to work at the energy of such harmonics and to adapt the value of λ_n for a given undulator to the experimental needs. This is done usually by changing the gap between the permanent magnets of the insertion device, which, in turn, changes the field strength B_0, the parameter K, and thus λ_n.

From Fig. 1.8 it follows that at the first harmonic at 3.7 nm, i.e., at 335 eV, and for a beam current of 500 mA, the measured intensity of undulator radiation amounts to more than 10^{16} photons/s with a bandpass of 0.1% and in the confined solid angle of this radiation (0.25 mrad and 0.06 mrad in horizontal and vertical directions, respectively). If this number is compared with that obtained with a bending magnet, a remarkable increase in intensity by a factor of 1000 can be expected.† In addition, and even more importantly, for experiments which require the optimal conditions, the quality of undulator radiation is much better if the smallness of the source size, the tightness of the angular confinement of the radiation, and the narrowness of the bandpass are considered. The quantity which characterizes this overall quality of the radiation source is the *spectral brilliance* B (sometimes called the brightness), which for undulator radiation is several orders of magnitude larger than that for radiation from a bending magnet. The spectral brilliance takes into account both the spatial distribution of the electrons circulating in the storage ring and the spatial distribution of the emission process itself, and is defined by (see [KEF83, KPW83])

$$B = \frac{(\text{spectral flux into } 0.1\% \text{ bandpass/s}) \cdot 100 \text{ mA ring current}}{(\text{source area/mm}^2) \cdot (\text{solid angle/mrad}^2)}. \tag{1.35}$$

According to Liouville's theorem (see Section 10.3.2) the spectral brilliance B cannot be increased further by any optical system, except at the expense of total flux. The brilliance is therefore an important quantity for the design of not only electron storage rings, but also beam lines with their attached monochromators and experimental equipment (see Sections 1.4 and 1.5).

† For a comparison of the intensities achievable after monochromatization, one has, however, to take into account that – depending on the design – monochromators can accept an appreciable amount of radiation in the horizontal direction. For radiation from a bending magnet this increases the accepted radiation proportionally, but this is not so for undulator radiation due to its strong confinement in the vertical *and* horizontal directions.

A further important property of synchrotron radiation concerns its polarization characteristics. The radiation is completely polarized, and the kind of polarization depends on the direction of the circulating electron beam as well as on the direction of photon emission. In order to understand these polarization properties, it is useful to recall the result for the emission of electromagnetic radiation from an electron moving with non-relativistic velocity in a circle: the electric field vector follows the same shape and orientation as the projection of the electron's path onto a plane perpendicular to the observation direction.

This result can be transferred to the relativistic case and leads to the following polarization properties of synchrotron radiation produced in a bending magnet.† If the electrons circulate in the storage ring in a clockwise direction, an observer viewing the radiation from above/below the plane of the storage ring will see right/left elliptically polarized radiation; an observer viewing in the plane will see linearly polarized radiation. Here the optical definition for circular polarization is used: clockwise rotation of the electric field vector for an observer looking towards the light source is called right circularly polarized light (for polarization described by the helicity system of light see Section 9.2.2). In order to formulate the general case of elliptical polarization, one usually quotes the intensity components $I_{\parallel}$ and $I_{\perp}$ with respect to the plane of the storage ring, but one has to remember that between these components a fixed phase difference exists. In Fig. 1.10 a sketch of the polarization of emitted synchrotron radiation is shown. Each radiation component in a particular direction, called a *wavelet* [New66], represents completely polarized light, i.e., a pure state. However, the summation of different wavelets at the focal point of the monochromatized light and, hence, in the interaction region of the experiment, yields a mixed state. (This is just the opposite of light emission from an ordinary light bulb: unpolarized light, described as a mixed state, is emitted, and a pure state of polarized light has to be produced using suitable polarization filters.) Such a mixed state, with a pure state as a special case, is described best by the Stokes vector ([Sto52], see Section 9.2)

$$\mathbf{S}_{\text{Stokes}} = \begin{pmatrix} I \\ Q \\ U \\ V \end{pmatrix} = I \begin{pmatrix} 1 \\ S_1 \\ S_2 \\ S_3 \end{pmatrix}. \tag{1.36a}$$

The four components of the Stokes vector are determined by four measurements which refer to three different basis systems ($\hat{\mathbf{e}}_x$, $\hat{\mathbf{e}}_y$ and $\hat{\mathbf{e}}_1$, $\hat{\mathbf{e}}_2$ and $\hat{\mathbf{e}}_r$, $\hat{\mathbf{e}}_\ell$, respectively; for more details see Section 9.2):

† Undulator radiation also has specific polarization properties, e.g., the fundamental radiation of a plane undulator as shown in Fig. 1.9 is completely linearly polarized, an undulator with helical magnetic structure produces circularly polarized light, and two crossed plane undulators with a dispersive section between them are capable of producing optional polarization which depends on the phase shift introduced by the dispersive element.

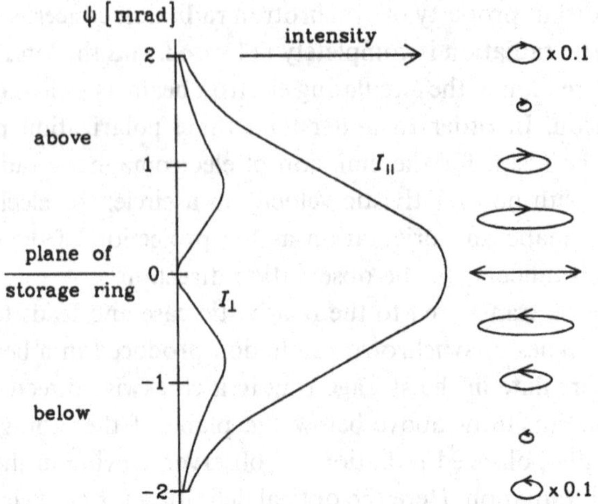

Figure 1.10 The intensity and polarization of synchrotron radiation emitted from a bending magnet. The electrons are assumed to circulate in the storage ring in the clockwise direction when viewed from above. The intensity components for linear polarization of the radiation with the electric field vector parallel or perpendicular to the plane of the storage ring are indicated by $I_\parallel$ and $I_\perp$, respectively. These values depend on the angle ψ between the direction of observation and the plane of the storage ring. Polarization ellipses are given at some selected ψ-directions, and the rotation of the electric field vector as seen by an observer looking towards the circulating electrons is indicated.

(i) Measurement of the total intensity I using a polarization-insensitive detector. The same result could be obtained by adding the intensities measured with the polarization-sensitive detectors described below. Hence, one gets

$$I = I_x + I_y = I_1 + I_2 = I_r + I_\ell. \tag{1.36b}$$

(ii) Measurement of the excess intensity using a detector which is sensitive to the linear polarization along the two orthogonal axes $\hat{\mathbf{e}}_x$ and $\hat{\mathbf{e}}_y$. One obtains

$$Q = I_x - I_y = IS_1. \tag{1.36c}$$

(iii) Measurement of the excess intensity using a detector which is sensitive to linear polarization along two orthogonal axes oriented at 45° to the right (called $\hat{\mathbf{e}}_1$ and $\hat{\mathbf{e}}_2$) with respect to the previous ones. This gives

$$U = I_1 - I_2 = IS_2. \tag{1.36d}$$

(Essentially, this measurement determines the tilt angle of the polarization ellipse (see below).)

(iv) Measurement of the excess intensity using a detector which is sensitive to right- or left-circular polarization (basis states $\hat{\mathbf{e}}_r$ and $\hat{\mathbf{e}}_\ell$ in the optical definition for clockwise and anticlockwise rotation of the respective unit vectors). This yields

$$V = I_r - I_\ell = IS_3. \tag{1.36e}$$

The components S_i are called Stokes *parameters*. They can vary between $+1$ and -1. For example, $S_1 = +1$ and $S_1 = -1$ describe completely linearly polarized light with the electric field vector oscillating along the x- and y-directions, respectively, and $S_3 = +1$ describes completely right-circularly polarized light. The components S_1, S_2, and S_3 also define P, the *degree* of polarization:

$$P = \sqrt{(S_1^2 + S_2^2 + S_3^2)}, \tag{1.37a}$$

and for this quantity the following important results hold:

$$\left.\begin{array}{ll} P = 1 & \text{is equivalent to a completely polarized beam (pure state);} \\ P < 1 & \text{is equivalent to a partially polarized beam (mixed state);} \\ P = 0 & \text{describes an unpolarized beam.} \end{array}\right\} \tag{1.37b}$$

Examples of these polarization properties, in particular for the monochromatized light in the interaction region of the experiment, will be given in the next section.

The electron bunches which circulate in the electron storage ring lead to a pulsed time structure of the emitted light which is important for some experiments (see Sections 5.7 and 10.1). The time interval Δt between two successive light flashes depends on the number of electron bunches in the ring and the repetition time of one bunch. In the single-bunch mode, Δt follows from the circumference of the electron orbit and the velocity of the electrons. At the small electron storage ring BESSY I, one gets $\Delta t = 208$ ns, and for the large storage ring DORIS in Hamburg $\Delta t = 960$ ns.

In addition to the properties of synchrotron radiation discussed so far, the fact that the spectral flux and the polarization can be calculated exactly offers the possibility of using synchrotron radiation for radiometric purposes, setting absolute radiation standards and providing the means of calibration of detectors (for references see [Sch92a]). Finally, synchrotron radiation from electron storage rings offers a clean (high-vacuum environment, approximately 10^{-9} mbar) and stable radiation source with an appreciable lifetime (the light intensity decreases because of collisions between the circulating electrons and the residual gas in the storage ring).

1.4 Monochromatization of synchrotron radiation

For experimental studies of photoprocesses on atoms by means of electron spectrometry, a particular photon energy has to be selected so that a specific photoprocess can be studied via the kinetic energy of the created photoelectrons. This energy value is taken from the broad energy range of the synchrotron radiation continuum or from the harmonic radiation of an undulator. Because there are no known optics which are transparent in the energy range between vacuum-ultraviolet and soft X-ray radiation (lithium fluoride has the lowest known transmission limit at 1040 Å, i.e., it transmits up to 11.9 eV), *reflection* grating

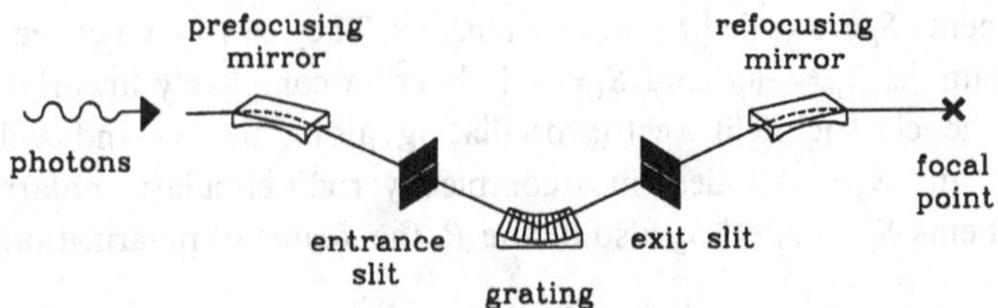

Figure 1.11 Set-up for a toroidal grating monochromator (TGM). The vital part is the grating shown in the middle of the picture. The direction of the grooves is indicated, and the grating can be rotated around an axis which coincides with the groove at its centre. Depending on the actual position of the grating, the corresponding contributions of diffracted light are imaged by the grating from the entrance slit to the exit slit.

monochromators are routinely used.† *Normal* incidence instruments operate between 5 and 35 eV, and *grazing* incidence instruments operate above 50 eV (although some grazing instruments can be used down to 5 eV). For example, the lay-out of a toroidal grating monochromator is shown schematically in Fig. 1.11. The first mirror (the prefocusing mirror) accepts, depending on its size, a certain amount of synchrotron radiation from the storage ring and focuses the path of the accepted electron beam onto the entrance slit of the monochromator. (It is also possible to work without a prefocusing mirror; in this case the path of the accepted electron beam acts as the 'entrance slit'.) The light transmitted through the entrance slit diverges and illuminates the grating. The grating diffracts and, due to its toroidal shape, focuses light of the desired wavelength onto the exit slit of the monochromator (the dispersion slit‡). After this exit slit, the monochromatized light diverges again. It might illuminate a second mirror (the refocusing mirror) which then, in turn, focuses the monochromatized light onto a spot (the focus) where experiments can be performed.

Toroidal or elliptical surfaces are used as imaging devices because they allow a correction of imaging errors. For Fresnel diffraction on a toroidal grating, one obtains the same equation as for the case of Frauenhofer diffraction on a plane grating (see [Beu45]), i.e., one has

$$n\lambda = g(\sin\alpha + \sin\beta), \tag{1.38a}$$

where g is the grating constant (the distance between successive grooves), n is the diffracted order, γ the diffracted wavelength, α the angle of incidence and β the diffraction angle ($\beta < 0$ in the present case because incident and diffracted rays are on different sides of the grating normal).

† There are many ways of achieving monochromatization (reflection and transmission gratings, Bragg reflection on crystals or multilayers, zone plates), and many different designs have been worked out (see references in [Sch92a]).

‡ The exit slit shown in Fig. 1.11 is characterized by its length and width. Since the grating disperses the 'white' synchrotron radiation into a plane perpendicular to that of the electron storage ring, the slit width in this direction is the relevant one and determines the wavelength range of transmitted, i.e., monochromatized synchrotron radiation.

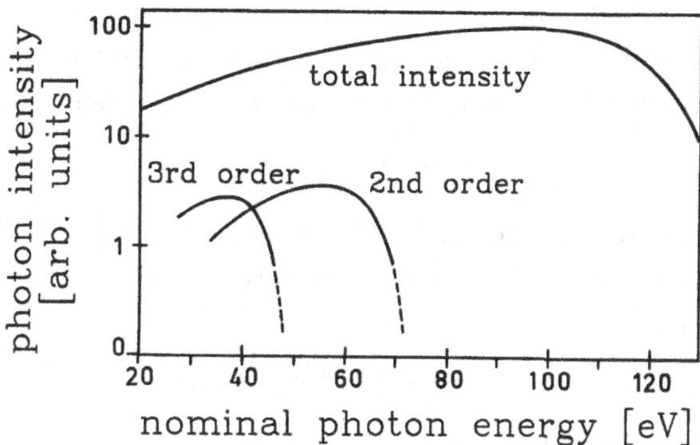

Figure 1.12 Photon flux from the TGM4 monochromator at BESSY I. The total intensity contains monochromatized light of all orders. For photon energies above 80 eV practically all light is of first order, and this photon energy is identical to the nominal value set on the monochromator. For lower photon energies, second- and third-order light also contributes at the given monochromator settings, and these individual portions are shown in the figure at the nominal photon energy given by the first-order setting of the monochromator. From [Sch92a].

In order to tune the wavelength within the working range of a grating monochromator, it has been shown by ray-tracing calculations that it suffices to rotate only the grating, and this property has made toroidal grating monochromators rather popular (albeit with only medium resolution). A small rotation of the grating by x couples the entrance and diffraction angles according to

$$\alpha = \Theta + x \quad \text{and} \quad -\beta = -(\Theta - x) \quad \text{with } \alpha + |\beta| = 2\Theta = \text{const}, \tag{1.38b}$$

where 2Θ is the angle between the incoming and diffracted photon beams, which must be constant because the entrance and exit slits of the monochromator are generally at fixed positions in space. Using equ. (1.38b) one derives the grating equation:

$$n\lambda = 2g \cos\Theta \sin x. \tag{1.38c}$$

For a grating with 950 lines/mm and $2\Theta = 146°$, one can therefore cover the wavelength range between 500 and 100 Å (between 24.8 and 124.0 eV) by changing the angle x from 4.7° to 0.9°.

An example of synchrotron radiation flux generated in a bending magnet, monochromatized by a toroidal grating monochromator, and refocused into the interaction region of an experiment is shown in Fig. 1.12. Good photon flux can be obtained in the energy range between 20 and 130 eV (the operating range of the monochromator). The upper limit follows from the 'cutoff' effect in the total reflectivity: for a given angle of incidence called α_c there is a sudden decrease of reflectivity if the wavelength gets smaller than a characteristic value λ_c or, equivalently, for photon energies higher than the corresponding value E_c ($E_c \approx$ 120 eV in the present case). For the photon energies of interest in the present context, many reflecting materials have an index of refraction n slightly smaller than unity and an index of absorption k small compared to unity Hence, neglecting the absorption, these materials show the phenomenon of total reflection if the angle of incidence is larger than a critical angle α_c which follows from

$$\sin \alpha_c = n. \tag{1.39a}$$

The index of refraction depends not only on the wavelength λ of the incident radiation, but also on the quality of the reflecting surface. Using a typical value, one has approximately

$$n \approx 1 - 3.5 \times 10^{-6}\,\text{Å}^{-2}(\lambda/\text{Å})^2, \tag{1.39b}$$

and combination of equs. (1.39a) and (1.39b) leads to

$$\lambda_c/\text{Å} \approx 380 \sin \alpha_c^{\text{g.a.}}, \tag{1.39c}$$

where $\alpha_c^{\text{g.a.}}$ is the grazing angle of incidence, $\alpha_c^{\text{g.a.}} = 90° - \alpha_c$. Hence, for the example given with $\alpha_c = \Theta = 73°$, one gets $\lambda_c \approx 111$ Å, i.e., all wavelengths longer than λ_c will be reflected with high efficiency, or, in other words, the grating will allow an operating range up to approximately 120 eV. For details see [Sam67].

The decrease of intensity towards lower energies has two causes. First, there is a change in the radiation cone delivered by the electron storage ring. Towards lower photon energies this cone increases in the direction perpendicular to the plane of the circulating electrons, thus exceeding the acceptance angle of the monochromator. Second, the grooves of the grating have a special shape (a 'blazed' grating) which produces an efficiency maximum roughly in the middle of the operation range.

Absolute measurements of the photon flux are difficult to perform. In the example shown in Fig. 1.12 the photon flux was obtained from a measurement of the photoelectron current from a clean gold foil under the impact of monochromatized radiation. (The gold foil is usually given a negative potential to drive the photoelectrons away.) Due to calibration difficulties, the total flux in Fig. 1.12 is given in arbitrary units. However, the maximum value obtained at 100 eV, with a beam current of 200 mA, and for 1 mm wide monochromator entrance and exit dispersion slits, is approximately 5×10^{12} photons/s with a bandpass of 0.8 eV. If this value is compared with 8×10^{13} photons/s with the same bandpass which enter the monochromator (see above), a total efficiency of approximately 6% is obtained for the toroidal grating monochromator with pre- and refocusing mirrors. By closing the dispersion slits to 0.05 mm, a much better resolution can be achieved; e.g., at 100 eV and for full illumination of the grating, the resolution is improved by a factor of 8 because the bandpass goes down to approximately 100 meV, but this is at the expense of a drastic decrease in intensity [TBH91]. However, photoelectron spectrometry on main photolines (large cross sections) is still possible, and the bandpass of specially designed monochromators can be as low as 1 meV below 22 eV, and up to 40 meV towards 300 eV photon energy [KCW92]. With the 'third' generation electron storage rings, dedicated to deliver

Table 1.2. *Characteristics of some discrete line sources.*

Source	Photon energy [eV]	Linewidth [eV]	Intensity at the sample [photons/s]
He I	21.217	0.003	$1 \times 10^{13(a)}$
He II	40.813	0.017	1×10^{11}
Y Mζ	132.3	0.450	3×10^{11}
Al Kα (fixed anode)	1486	0.830	1×10^{12}
Al Kα (rotating anode, monochromatized)	1486	0.165	3×10^{12}

From [FGW74], but (*a*) from [SHe83].

undulator radiation, and with specifically adapted monochromators, a further reduction of the bandpass by a factor of 5 is envisaged.

For a comparison of photon intensities and bandpasses of monochromatized synchrotron radiation with vacuum-ultraviolet or X-ray light from other sources, typical values for some discrete line sources are given in Table 1.2. It should, however, be noted that the great advantage of monochromatized synchrotron radiation lies in its tunability over a broad energy range and its good polarization characteristics. Furthermore, synchrotron radiation from wigglers and undulators yields a further gain in intensity as well as a remarkable improvement in resolution and in polarization.

In addition to the spectral light of interest, a monochromator always delivers a certain amount of higher-order light and stray light. The relative intensities of higher-order light are also shown in Fig. 1.12. The intensities of second- and third-order light are plotted against the *nominal* first-order energy setting of the monochromator; i.e., if the monochromator is set at 40 eV, one gets relative intensities of 0.88, 0.05 and 0.07 for 40 eV, 80 eV and 120 eV photons, respectively. In some cases these disturbances by photons of higher energy can be suppressed by inserting suitable absorption foils into the path of the light beam. Here one takes advantage of the low absorption coefficient μ at energies below an inner-shell absorption edge and the high values above this edge. Fixing the desired first-order light to an energy value below such an absorption edge, the small μ value leads to high transmission for the first-order light with the unwanted higher-order light being effectively suppressed.

Stray light contributions are extremely difficult to assess because several possible origins normally produce a broad continuous spectral distribution dN_{stray}/dE which covers the whole working range of the monochromator, and only crude estimates for the stray light can be given (see, for example, [KSK89]).

Another important aspect of electron spectrometry of free atoms using synchrotron radiation concerns the polarization of the monochromatized light.

As was demonstrated in Fig. 1.10, each wavelet of synchrotron radiation emitted in a certain direction represents completely polarized radiation, i.e., a pure state. However, in order to get a high photon intensity in the interaction region of the experiment, the individual wavelets are refocused after their passage through the monochromator. Hence, in the interaction region the photon beam is in a mixed state. Such a mixed state can again be described by the Stokes parameters, but due to the individual processes of reflection and diffraction in the monochromator, it is no longer an easy task to predict the values of the Stokes parameters. Because the angular distribution and the spin-polarization of the emitted photoelectrons depend on the Stokes parameters S_1, S_2 and S_3 in the interaction region, these quantities must be measured separately (see below).

In order to illustrate the mixed state, an example with five sample wavelets will be discussed in detail. Each wavelet is represented by its components α_x and α_y in the Cartesian basis (optical definition, see Section 9.2.2). If the polarization vector is described by a polarization ellipse with major and minor axes $a = \cos \gamma$ and $b = \sin |\gamma|$, by a tilt angle λ of this ellipse against a fixed coordinate frame (see Fig. 1.15), and by the direction of rotation of the electric field vector indicated by the sign of γ, the components α_x and α_y follow from

$$\alpha_x = \cos \gamma \cos \lambda + \mathrm{i} \sin \gamma \sin \lambda, \quad \alpha_y = \cos \gamma \sin \lambda - \mathrm{i} \sin \gamma \cos \lambda \tag{1.40a}$$

with

$$\tan |\gamma| = b/a \tag{1.40b}$$

and

$$\left.\begin{array}{l} \gamma > 0 \text{ for right elliptical polarization,} \\ \gamma < 0 \text{ for left elliptical polarization.} \end{array}\right\} \tag{1.40c}$$

Its polarization vector $\mathbf{P}$ is then given by

$$\mathbf{P} = \alpha_x \hat{\mathbf{e}}_x + \alpha_y \hat{\mathbf{e}}_y. \tag{1.41}$$

Knowing the individual polarization vectors $\mathbf{P}(j)$ of wavelets j, i.e., the individual $\alpha_x(j)$ and $\alpha_y(j)$ coefficients, one gets for the Stokes parameters of an ensemble (see Section 9.2.2):

$$I = \sum_j I(j) = \sum_j (|\alpha_x(j)|^2 + |\alpha_y(j)|^2), \tag{1.42a}$$

$$Q = S_1 I = \sum_j Q(j) = \sum_j (|\alpha_x(j)|^2 - |\alpha_y(j)|^2), \tag{1.42b}$$

$$U = S_2 I = \sum_j U(j) = \sum_j (\alpha_x(j)\alpha_y^*(j) + \alpha_x^*(j)\alpha_y(j)), \tag{1.42c}$$

$$V = S_3 I = \sum_j V(j) = -\mathrm{i} \sum_j (\alpha_x(j)\alpha_y^*(j) - \alpha_x^*(j)\alpha_y(j)). \tag{1.42d}$$

At this point a comment is needed. The Stokes parameters $I(j)$, $Q(j)$, $U(j)$, and $V(j)$ which are not yet normalized with regard to the overall intensity (I) do not contain the initial phases of the individual wavelets. This is the reason why these quantities are well suited for specifying a *mixed* state of wavelets with random and different phases and possibly different polarizations.

Table 1.3. *Polarization parameters for five sample wavelets (optical convention). For a detailed description see main text.*

	Polarization quantities (in the $\hat{\mathbf{e}}_x$, $\hat{\mathbf{e}}_y$ basis) for the radiation emitted from the storage ring					
Wavelet	$\{\alpha_x(j), \alpha_y(j)\}$	$Q(j)$	$U(j)$	$V(j)$	$I(j)$	$P(j)$
1	{0.906, −0.424i}	0.641	0	0.768	1	1
2	{0.974, −0.228i}	0.897	0	0.444	1	1
3	{1, 0}	1	0	0	1	1
4	{0.974, 0.228i}	0.897	0	−0.444	1	1
5	{0.906, 0.424i}	0.641	0	−0.768	1	1
Sum for all wavelets:		4.076	0	0	5	

	Polarization quantities (in the $\hat{\mathbf{e}}_x$, $\hat{\mathbf{e}}_y$ basis) for the monochromatized radiation (arbitrary changes are assumed)					
Wavelet	$\{\alpha_x(j), \alpha_y(j)\}$	$Q(j)$	$U(j)$	$V(j)$	$I(j)$	$P(j)$
1	{0.849, −0.400i}	0.561	0	0.679	0.880	1
2	{0.900, −0.2i + 0.05}	0.768	0.090	0.360	0.853	1
3	{0.950, 0}	0.903	0	0	0.903	1
4	{0.908, 0.210i}	0.780	0	−0.381	0.869	1
5	{0.85 + 0.1i, 0.39i}	0.598	0.078	−0.671	0.902	1
Sum of all wavelets:		3.610	0.168	−0.013	4.407	

The relations above are then applied to the five sample wavelets which represent an ensemble of photons. The data in the upper part of Table 1.3 simulate the emission of synchrotron radiation, and that in the lower part monochromatized light in the target region. For simplicity, all wavelets of the emitted radiation are assumed to have the same intensity. Each wavelet describes a different kind of fully polarized light ($P = 1$), where wavelets 1 and 2 are right elliptically polarized, wavelets 4 and 5 are left elliptically polarized, and wavelet 3 is linearly polarized. If an ideal imaging device focuses these wavelets without any change in the polarization character or intensity into the interaction region of the experiment, each photon wavelet has a certain probability of interacting with an atom, i.e., the atoms are confronted with an incoherent mixture of wavelets with different polarization properties, and the individual interactions add to the observable result. According to the above formula one gets

$$\mathbf{S}_{\text{Stokes}} = \begin{pmatrix} I \\ Q \\ U \\ V \end{pmatrix} = I \begin{pmatrix} 1 \\ S_1 \\ S_2 \\ S_3 \end{pmatrix} = 5 \begin{pmatrix} 1 \\ 0.815 \\ 0 \\ 0 \end{pmatrix} \quad \text{and} \quad P = 0.815 < 1. \qquad (1.43a)$$

This result reflects exactly what may be expected intuitively:

(i) The atoms are confronted with a mixed polarization state of the beam, because $P < 1$.

(ii) The beam contains the Stokes parameter S_1 (or Q) only, because the contributions from right and left circular polarization expressed in the $V(j)$ components of wavelets 1 and 5, and 2 and 4 cancel. As a result, there is no excess of circular polarization as represented by the Stokes parameter S_3.

(iii) The resulting light is partially linearly polarized light, because equ. (1.43a) can also be expressed as

$$\mathbf{S}_{\text{Stokes}} = 4.076\begin{pmatrix}1\\1\\0\\0\end{pmatrix} + 0.924\begin{pmatrix}1\\0\\0\\0\end{pmatrix}. \tag{1.43b}$$

This formulation shows that the monochromatized light can be described as being composed of two contributions: completely linearly polarized light ($S_1 = +1$) with a weight of 4.076, and completely unpolarized light with a weight of 0.924. The representations given in equs. (1.43a) and (1.43b) are equivalent and both will be used when discussing angular distributions following photoionization by arbitrarily polarized light (see Section 9.1).

In a more realistic treatment of the monochromatized light, one has to take into account that the individual wavelets are affected by the optical elements in the monochromator before they reach the focal region of the monochromator. This means the wavelets are attenuated and their polarization ellipses can change. To demonstrate this effect, some arbitrary changes are assumed and listed in the lower part of Table 1.3. Each wavelet is attenuated, the ratio of $\alpha_x(j)$ to $\alpha_y(j)$ is changed, and a tilting of the polarization ellipse is indicated by the presence of the imaginary parts in wavelets 2 and 5. It can be seen that each wavelet is still completely polarized. However, for the mixed polarization state of the beam in the interaction region, one now gets

$$\mathbf{S}_{\text{Stokes}} = 4.407\begin{pmatrix}1\\0.819\\0.038\\-0.029\end{pmatrix} \quad \text{and} \quad P = 0.820 < 1. \tag{1.44a}$$

Again, this result reflects what may be expected:

(i) The total intensity I is reduced from 5 to 4.407.

(ii) The atoms are subject to a mixed polarization state of the beam, because $P < 1$.

(iii) Although S_1 is still the largest Stokes parameter, the parameters S_2 and S_3 also contribute because of the small differences which occur in the reflection of wavelets which originate from above and below the plane of the storage ring. These differences prevent the compensation of the $U(j)$ and $V(j)$ components seen in the former example.

(iv) The polarization of the light can be described as partial polarization, containing completely polarized contributions related to S_1, S_2, and S_3 and completely unpolarized light:

$$\mathbf{S}_{\text{Stokes}} = 3.609\begin{pmatrix}1\\1\\0\\0\end{pmatrix} + 0.167\begin{pmatrix}1\\0\\1\\0\end{pmatrix} + 0.128\begin{pmatrix}1\\0\\0\\-1\end{pmatrix} + 0.503\begin{pmatrix}1\\0\\0\\0\end{pmatrix}. \quad (1.44b)$$

Due to the mixed polarization of monochromatized synchrotron radiation, the angle dependence of photoelectron emission as expressed in equ. (1.30) for completely linearly polarized light requires modification. This is considered in detail in Section 9.1, but the implication for the corresponding appropriate experimental set-up is treated in the next section.

1.5 Basic aspects of electron spectrometry with synchrotron radiation

When performing electron spectrometry with monochromatized synchrotron radiation, one has to consider the dependences of the observables on the parameters relevant to the process. The essential parameters are:

(i) the properties of the incident light, namely, its energy, intensity, and polarization;

(ii) the target for which the atom–photon interaction is to be studied (theoretical and experimental interest lies in the differential cross section of photoionization);

(iii) the angle settings of the electron spectrometer describing its position in space (geometrical aspect), including possible effects of the finite entrance slit;

(iv) the properties of the spectrometer itself, i.e., its acceptance solid angle, transmission, and resolution, and the efficiency of the electron detector.

The properties of monochromatized synchrotron radiation have been discussed in detail in the previous section and the characteristic features of electrostatic spectrometers will be discussed in detail in Chapter 4, with examples of photoionization processes in certain atoms and specific questions of interest presented in Chapter 5. Therefore, the following discussion is restricted to basic aspects of electron spectrometry with monochromatized synchrotron radiation, in particular to some of the fundamental properties of electron spectrometers and to the special polarization properties of this radiation which require appropriate experimental set-ups for angle-resolved electron spectrometry (without spin-analysis; for the determination of spin-polarization see Section 5.4).

The analyser accepts, transmits, analyses and possibly focuses electrons of a certain energy emitted by the sample. An important measure for the analyser's capability to meet these requirements is comprised in the *luminosity* L of the analyser. The luminosity is a measure of both the accepted source volume ΔV and

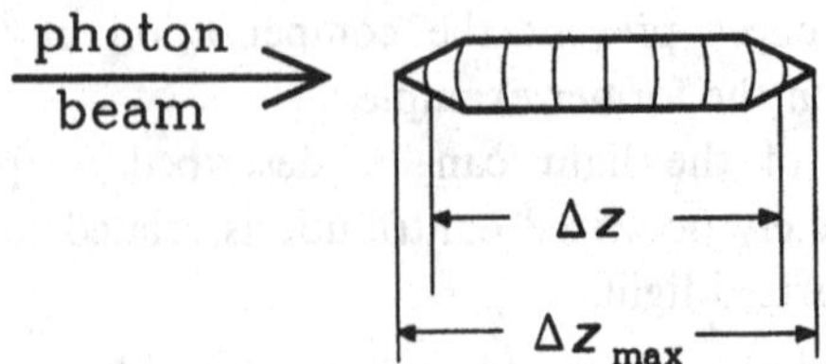

Figure 1.13 Definition of the source volume of an electrostatic energy analyser. In the case shown, the diameter of the source volume is determined by the diameter of the photon beam, and the restricted length Δz_{max} by diaphragms in the electron spectrometer which prevent the acceptance of electrons from regions outside of Δz_{max}. The average length, Δz, usually identified with 'the' length is also indicated.

the average analyser transmission T, because one has (see equ. (4.16))

$$L = \Delta V\, T. \tag{1.45}$$

As the name implies, the *accepted* source volume ΔV is that part of the more general photon–atom interaction region from which electrons are accepted by the analyser. Usually, photoionization occurs along the whole path along which the photons travel in the experimental chamber, but the ionization strengths differ, because they depend on the actual gas density. An ideal situation would be to confine photoionization almost exclusively to an interaction region filled with target gas (see Sections 5.5.1 and 10.7). This, however, can be achieved only approximately, and one has to distinguish the *actual* ionization region from those parts of it from which electrons are *accepted* by the electron spectrometer. (Note that the volume ΔV_{seen} seen by the electron spectrometer can be smaller than, equal to or even larger than the actual source ΔV_{actual}. The accepted source volume ΔV is introduced here as $\Delta V = \Delta V_{seen} \cap V_{actual}$.) Such an *accepted* source volume ΔV is shown schematically in Fig. 1.13 for the special case in which the diameter of the source volume is equal to the diameter of the incident photon beam. Such a situation occurs if the analyser images the actual source volume into its focal plane. This special property can be realized at the expense of a diameter-dependent analyser resolution. For angle-dependent measurements, where the electron spectrometer has to be rotated around the source volume, such a spectrometer design has the advantage that the registered intensities do not depend on a small misalignment between this rotation axis and the middle position of the actual source volume.

The analyser transmission T describes the fraction of electrons, out of all those created in ΔV and emitted isotropically, that will be accepted and transmitted by the electron spectrometer for optimum conditions, i.e., for electrons with kinetic energy E^0_{kin} and a matching pass energy E^0_{pass} of the analyser. Assuming that it is only the finite entrance slit (solid angle) of the analyser which plays a role in the analyser's acceptance (i.e., the number of electrons will not be reduced by other slits or by the size of the electron detector), T becomes a purely geometrical quantity. Its value follows if the acceptance solid angle is averaged over all points of the source volume ΔV and divided by 4π. In the more general case the luminosity L is calculated by a ray-tracing procedure, and this approach also provides information on T and ΔV.

Both T and ΔV are important instrumental parameters. Together with the detection efficiency ε of the electron detector they determine the *intensity* of registered electrons, $I_{reg}(E^0_{kin})$:

$$I_{reg}(E^0_{kin}) \sim \Delta V\, T\varepsilon, \tag{1.46}$$

where E^0_{kin} is the kinetic energy of these electrons. Since the spectrometer analyses the accepted electrons according to their kinetic energy, it contains an energy-dispersive element which selects a certain *pass* energy E_{pass} of the analyser. Electrons whose kinetic energy matches the pass energy will then be transmitted while other electrons will be rejected. In electrostatic spectrometers, the pass energy is determined by setting a certain voltage, called the *spectrometer voltage* U^0_{sp} on the deflection plate(s) of the analyser. The pass energy is then proportional to this spectrometer voltage, i.e.,

$$E^0_{kin} = fU^0_{sp}. \tag{1.47}$$

The proportionality factor f is called the *spectrometer factor*. Its value is fixed by the geometry and focusing properties of the analyser. Hence, for a given spectrometer, one can use the kinetic energy or alternatively the spectrometer voltage.

In a more detailed description of the analyser transmission, one has to distinguish the *nominal* values of voltages or energies, indicated, for convenience, by a superscript zero, from *actual* values (without a superscript), because maximum transmission occurs only for matched nominal values U^0_{sp} and E^0_{kin}. In the more general case of arbitrary values, one has to consider the full response function of the analyser. This instrumental function is shown schematically in Fig. 1.14.

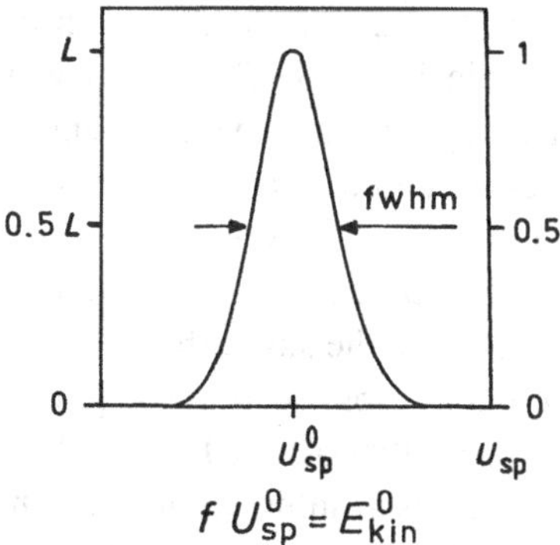

Figure 1.14 Instrumental response function of an electrostatic energy analyser shown as a function of the spectrometer voltage U_{sp}. Maximum transmission is achieved at the nominal voltage U^0_{sp}, and this maximum value is equal to the luminosity L (left-hand scale) or set to unity (right-hand scale), respectively. For values other than U^0_{sp} the response function decreases, and the characteristic fwhm value is indicated. For the relation $fU^0_{sp} = E^0_{kin}$ see the main text.

Maximum transmission occurs for $U_{sp} = U^0_{sp}$, or, alternatively, for E^0_{kin}. The maximum value of the response function gives the luminosity value L (left-hand scale in the figure). Since for practical applications it is convenient to treat the luminosity L separately, the instrumental function is usually normalized with its maximum set to unity (right-hand scale in the figure), and it is this latter function which is called the *spectrometer function* $F_{sp}(E^0_{kin}, E_{pass} = fU_{sp})$. This function can be interpreted in two ways: first as the response of the spectrometer to incident monochromatic electrons with energy E^0_{kin}, changing the spectrometer voltage U_{sp} (pass energy E_{pass}):

$$F_{sp}(E^0_{kin}, E_{pass} = fU_{sp}); \tag{1.48a}$$

second, as the response of the spectrometer at a fixed spectrometer voltage U^0_{sp} to a continuum of incident electrons with energies E_{kin} (constant intensity distribution):

$$F_{sp}(E_{kin}, E^0_{pass} = fU^0_{sp}). \tag{1.48b}$$

As well as the general shape of the spectrometer function, a simpler but also important characteristic feature of an electron spectrometer is its full-width-at-half-maximum (*fwhm*) value ΔE_{sp} (see Fig. 1.14). This quantity determines the instrumental *resolution* given usually as the relative value R:

$$R = \Delta E_{sp}/E^0_{kin} \tag{1.49a}$$

or as the *resolving power*

$$R^{-1} = E^0_{kin}/\Delta E_{sp}. \tag{1.49b}$$

The electron spectrometer is placed at a certain position in space, and it accepts only electrons which are emitted in the direction of its entrance slit (see Fig. 1.4). Since, in general, the emission of photoelectrons and Auger electrons is non-isotropic in space, the intensity I_{exp} observed with an electron spectrometer will then depend on the geometrical set-up. These geometrical aspects will now be treated in detail. In Fig. 1.15 a coordinate system is shown which is centred in the middle of the interaction region and defined by the direction $\hat{\mathbf{z}}$ of the photon beam and two other orthogonal directions, one of which is in the plane of the electron storage ring ($\hat{\mathbf{x}}$-axis). The direction for the emission of photoelectrons is then described by the polar angle Θ and the aximuthal angle Φ (see also Fig. 1.4).† With these definitions, the differential cross section for photoionization for randomly oriented atoms, partially elliptically polarized incident light and within the dipole approximation of the photon–atom interaction is given by ([Sch73, SSt75], see

† In the discussion it has been assumed that the electron spectrometer accepts an infinitesimally small solid angle. In this case the individual angles for electron emission (Θ, Φ) coincide with the angles which characterize the entrance slit, i.e., the setting of the electron spectrometer (Θ^0, Φ^0). Quite often, however, a large acceptance solid angle is needed in order to get a sufficient intensity of registered electrons. One is then faced with the problem of solid-angle corrections (a theoretical treatment of this is given in Section 10.5 and an example in Section 4.6.3).

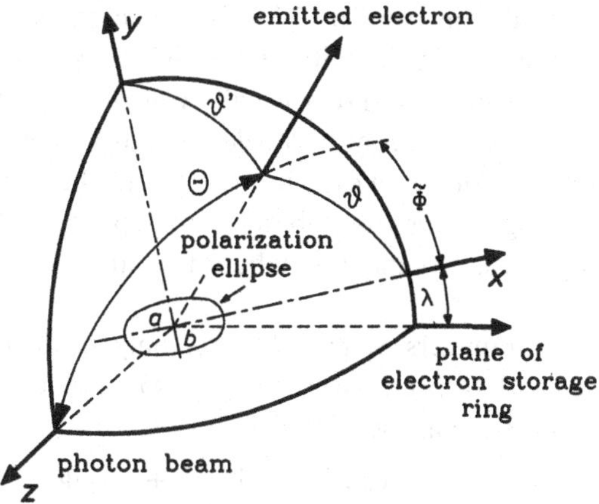

Figure 1.15 Tilted collision frame at the sample. The photon beam direction defines the z-axis; the x- and y-axes are aligned with the major (a) and minor (b) axes of the polarization ellipse which lies in the plane perpendicular to the direction of the photon beam. λ is the tilt angle between the x-axis and the plane of the storage ring. The direction of the emitted electron is described by the polar and azimuthal angles Θ and $\tilde{\Phi}$ measured in the tilted coordinate frame. From [Sch92a].

also Section 9.1)[†]

$$\frac{d\sigma}{d\Omega}(\Theta, \Phi) = \frac{\sigma}{4\pi}\left\{1 - \frac{\beta}{2}\left[P_2(\cos\Theta) - \frac{3}{2}A\sin^2\Theta\right]\right\} \tag{1.50a}$$

with

$$A = S_1 \cos 2\Phi + S_2 \sin 2\Phi, \tag{1.50b}$$

where S_1 and S_2 are the Stokes parameters defined in equs. (1.42). (The same formula applies for non-coincident emission of Auger electrons.) The dynamical parameters are given by the cross section σ and the angular distribution parameter β, the geometrical aspects are contained in the dependences on the angles Θ and Φ.

Some remarks are necessary in connection with this formula:

(i) The total cross section σ defined by

$$\sigma = \int \frac{d\sigma}{d\Omega} d\Omega \tag{1.51}$$

follows from this expression: integration over the full solid angle yields in equ. (1.50a) for the term '1' within the braces the value 4π which cancels with $1/4\pi$ in front of the braces, and the integration over the angle-dependent terms in the braces gives zero. It should be emphasized that σ is independent of the polarization of the incident light, i.e., interactions with any kind of polarization

[†] Note that this equation follows after two steps in which the Stokes parameters for one wavelet and an ensemble of wavelets are used: first, the differential cross section is calculated for one wavelet which is completely elliptically polarized and, second, the presence of differently polarized wavelets in the incident light is taken into account using equ. (1.42).

contribute with equal probability. This is a consequence of the random distribution of the atoms in the initial state. Strictly speaking, σ is the cross section for a selected process and is usually called the *partial* cross section.

(ii) The angle-dependent emission of photoelectrons is connected with the light polarization through the Stokes parameters. In the present case of no spin-analysis, the angular distribution does not depend on the Stokes parameter S_3 describing circular polarization, but on S_1 and S_2 which refer to linear polarization.

(iii) Since the Stokes parameters S_1 and S_2 differ only in their reference frames which are rotated with respect to each other by 45°, a special coordinate frame must exist in which one has (see Fig. 1.15)

$$A = \tilde{S}_1 \cos 2\tilde{\Phi} \quad \text{with} \quad \tilde{\Phi} + \lambda = \Phi. \tag{1.52a}$$

From the condition that both expressions for A have to be equal, one derives

$$\tilde{S}_1 = \sqrt{(S_1^2 + S_2^2)}, \quad \tilde{S}_2 = 0 \quad \text{and} \quad \tan 2\lambda = S_2/S_1. \tag{1.52b}$$

It is convenient to make these substitutions, i.e, to replace the general Stokes parameters S_1 and S_2 by $\tilde{S}_1$ and a tilt angle λ (which implies $\tilde{S}_2 = 0$). $\tilde{S}_1$ is then defined as the greatest possible value of the Stokes parameter for linear polarization or, alternatively, by the excess of linear polarization intensity along the major (a) and minor (b) axes of the polarization ellipse, i.e.,

$$\tilde{S}_1 = \frac{I_a - I_b}{I_a + I_b}. \tag{1.52c}$$

The *tilt* angle λ describes how the polarization ellipse is tilted against the former reference system (the plane of the storage ring). With these quantities equs. (1.50) can be replaced by the simpler form

$$\frac{d\sigma}{d\Omega}(\Theta, \tilde{\Phi}) = \frac{\sigma}{4\pi}\left(1 - \frac{\beta}{2}[P_2(\cos\Theta) - \tfrac{3}{2} A \sin^2\Theta]\right) \tag{1.53a}$$

with

$$A = \tilde{S}_1 \cos 2\tilde{\Phi} \quad \text{and} \quad \tilde{\Phi} = \Phi - \lambda. \tag{1.53b}$$

In the following, the consequences of equs. (1.53) for two conventional experimental set-ups will be discussed.

To measure relative partial cross sections σ, it is advantageous to avoid the dependences on the polarization quantities $\tilde{S}_1$ and λ and on the angular distribution parameter β. This can be achieved with the experimental set-up shown in Fig. 1.16 where a cylindrical mirror analyser CMA (see Section 4.2) is used which accepts all electrons around the Φ-direction. This corresponds to a Φ-integration in the above formula and yields

$$\int_0^{2\pi} \frac{d\sigma}{d\Omega}(\Theta, \tilde{\Phi})\, d\Phi = \frac{\sigma}{4\pi}\left[1 - \frac{\beta}{2} P_2(\cos\Theta)\right] 2\pi, \tag{1.54a}$$

i.e., all dependences on the light polarization disappear. It is possible now also to

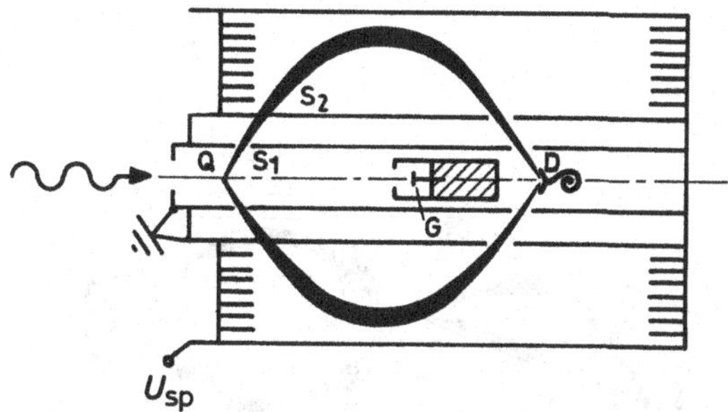

Figure 1.16 Example of an experimental set-up for electron spectrometry with synchrotron radiation which is well suited to measurements of relative intensities. The CMA is aligned with its symmetry axis coinciding with the direction of the photon beam and accepts electrons in a cone of $\vartheta_0 = 54.7°$. According to equ. (1.54c) the intensity of observed electrons is then proportional to σ, independent of the light polarization (and the angular distribution of the emitted electrons). The light flux is measured with a gold-foil detector (G). The paths of electrons from the source volume (Q) which enter the analyser through a slit system (S_1 and S_2) bend in the electric field of the analyser due to the applied spectrometer voltage U_{sp}. By proper matching of U_{sp} and the selected kinetic energy E_{kin} of the electrons, focusing of these electrons into a small opening in front of the electron detector D is achieved (this opening is not shown in the figure). Since at a given spectrometer voltage electrons of other energies are not focused into this small opening, this slit (called the *dispersion slit*) ensures the detection of electrons with the selected energy only. From [WAD77].

eliminate the remaining dependence on β by choosing as the entrance angle Θ of the electrons into the spectrometer the *magic angle* Θ_m to be 54.7° which is determined by

$$P_2(\cos\Theta_m) = \tfrac{3}{2}\cos^2\Theta_m - \tfrac{1}{2} = 0. \tag{1.54b}$$

Under these conditions the set-up of Fig. 1.16 provides an intensity I_{exp} of detected electrons which is related directly to the partial cross section σ:

$$I_{exp} \sim \int_0^{2\pi} \frac{d\sigma}{d\Omega}(\Theta_m, \tilde{\Phi})\, d\Phi = \frac{\sigma}{2}. \tag{1.54c}$$

If it is desired to measure the angular distribution parameter β, the experimental set-up of Fig. 1.17 can be used. A rotation of the sector-analyser around the photon beam direction keeps $\Theta = 90°$, but changes the angle $\tilde{\Phi}$ or, equivalently, Φ. This set-up has the advantage that the analyser always views the same source volume, independent of the angle Φ. The angle-dependent intensity I_{exp} of detected electrons, equ. (1.53), then reduces to

$$I_{exp} \sim \frac{d\sigma}{d\Omega}(\Theta = 90°, \Phi) = \frac{\sigma}{4\pi}\left\{1 + \frac{\beta}{4}[1 + 3\tilde{S}_1 \cos 2(\Phi - \lambda)]\right\}. \tag{1.55a}$$

From this relation it can be seen that the determination of β is possible, provided that the polarization parameter $\tilde{S}_1$ does not vanish and that $\tilde{S}_1$ and λ are known. These parameters can be obtained from a calibration measurement, i.e., an angular distribution measurement for a target with a known and non-vanishing β (see

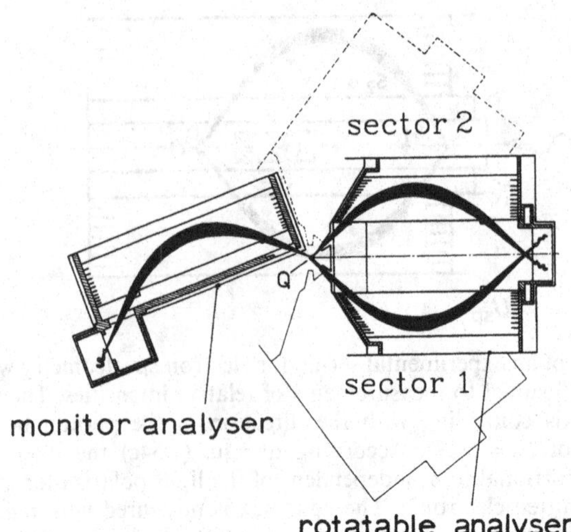

Figure 1.17 An experimental set-up for electron spectrometry with synchrotron radiation which is well suited to angle-resolved measurements. A double-sector analyser and a monitor analyser are placed in a plane perpendicular to the direction of the photon beam and view the source volume Q. The double-sector analyser can be rotated around the direction of the photon beam thus changing the angle $\tilde{\Phi}$ between the setting of the analyser and the electric field vector of linearly polarized incident photons. In this way an angle-dependent intensity as described by equ. (1.55a) can be recorded. The monitor analyser is at a fixed position in space and is used to provide a reference signal against which the signals from the rotatable analyser can be normalized. For all three analysers the trajectories of accepted electrons are indicated by the black areas which go from the source volume Q to the respective channeltron detectors. Reprinted from *Nucl. Instr. Meth.*, **A260**, Derenbach *et al.*, 258 (1987) with kind permission of Elsevier Science—NL, Sara Burgerhartstraat 25, 1055 KV Amsterdam, The Netherlands.

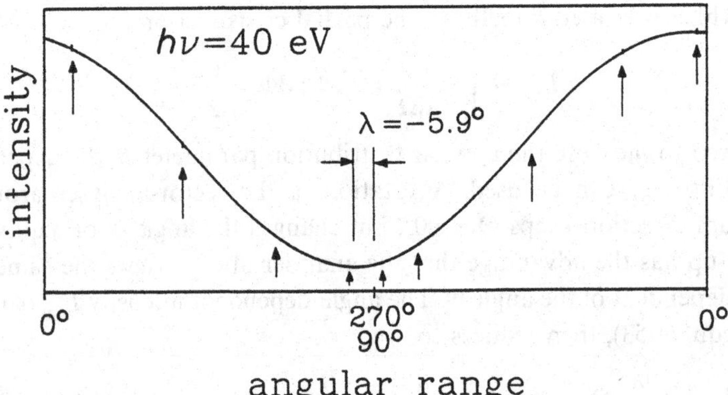

Figure 1.18 Example of the determination of the Stokes parameter $\tilde{S}_1$ and tilt angle λ measured at 40 eV photon energy. It is based on an angular distribution measurement of 1s-photoelectrons in helium for which equ. (1.55a) holds with $\beta = 2$. The arrows point to the measured values (with error bars, some of which are too small to be visible); the solid curve is a fit of the experimental data to this equation from which $\tilde{S}_1 = 0.819(3)$ and $\lambda = -5.9(9)^\circ$ can be extracted.

Section 6.2). An example of such a calibration measurement is shown in Fig. 1.18 where 1s photoionization in helium with the energy-independent β value of $\beta = 2$ was selected. The angle-dependent intensity varies between a maximum and a minimum value. Within these limits, at a certain angle Φ_{qm}, the intensity is proportional to the relative partial cross section. The above relation requires for this case that

$$1 + 3\tilde{S}_1 \cos 2(\Phi_{qm} - \lambda) = 0. \tag{1.55b}$$

At this *quasi-magic* angle Φ_{qm} one gets for the intensity I_{exp} of detected electrons

$$I_{exp} \sim \frac{d\sigma}{d\Omega}(\Theta = 90°, \Phi_{qm}) = \frac{\sigma}{4\pi}. \tag{1.55c}$$

Therefore, with angle-resolved measurements one can determine both σ and β. (For determining σ this method has the disadvantage of losing electron intensity as compared to angle-integrated measurements.)

2
Photoelectron spectrum of neon

2.1 Theoretical formulation of the photoprocess

In order to facilitate the theoretical treatment, photoionization of a 1s-electron in neon will be discussed in detail for the case of linearly polarized light, where the emitted photoelectron is observed without spin-analysis, and within the dipole approximation. Neon has the electron configuration $1s^2 2s^2 2p^6$ which gives a $^1S_0^e$ state (closed-shell atom). In principle the photon operator acts on all electrons which are energetically accessible for ionization. However, because of the energy-analysis of the ejected photoelectrons a distinction between photoprocesses in the individual shells can be made which allows a partitioning into incoherent sums belonging to individual shells only. In the present case the part with the two 1s-electrons is of relevance. If the quantization axis coincides with the direction of the electric field vector of the linearly polarized light, the interaction operator is then given in the dipole approximation by (see equ. (1.28a) and Section 8.1; atomic units)

$$Op = z_1 + z_2, \tag{2.1}$$

where z_j is the z coordinate for one of the two 1s electrons. The wavefunctions for the initial and final states of the photoprocess are assumed to be Slater determinantal wavefunctions (uncorrelated wavefunctions), i.e.,

$$|J_i M_i\rangle = \{1s0^+, 1s0^-, 2s0^+, 2s0^-, 2p1^+, 2p1^-, 2p0^+, 2p0^-, 2p{-1}^+, 2p{-1}^-\}, \tag{2.2a}$$

$$|J_I M_I, \boldsymbol{\kappa} m_s^{(-)}\rangle = \{1s0^{M_I}, 2s0^+, 2s0^-, 2p1^+, 2p1^-, 2p0^+, 2p0^-, 2p-1^+, 2p-1^-, \psi^{(-)}_{\boldsymbol{\kappa} m_s}\}, \tag{2.2b}$$

where J_I and M_I are the angular momentum and its projection number, respectively, of the final ionic state. Since this state is a 1s hole-state, J_I is equal to 1/2 and M_I is equal to the spin-projection number of the 1s orbital (cf. first orbital in the determinantal wavefunction). $\psi^{(-)}_{\boldsymbol{\kappa} m_s}$ is the single-particle function of the electron ejected into the continuum. The electron has the wave vector $\boldsymbol{\kappa}$, the energy $\varepsilon = \kappa^2/2$ (in atomic units), the direction $\hat{\boldsymbol{\kappa}} = (\vartheta, \varphi)$, and the spin-projection quantum number m_s. (It is common practice in theoretical expressions to use ε to describe the kinetic energy E_{kin} of an emitted electron. Both symbols will be used in the text. In order to indicate against which axis the direction $\hat{\boldsymbol{\kappa}}$ of the photoelectron is measured, the angles (ϑ, φ) are used if the reference axis is the direction of the electric field vector, and the angles (Θ, Φ) if this axis is the direction of the incident light.) The superscript (−) indicates that the boundary condition for the process of electron *emission* has been taken into account (for details see Section 7.1.3).

The differential cross section then follows from (see Section 8.1):

$$\frac{d\sigma}{d\Omega}(\vartheta, \varphi) = 4\pi^2 \alpha E_{ph} \kappa \sum_{M_I} \sum_{m_s} |M_{fi}(M_I, m_s)|^2. \tag{2.3}$$

The fine-structure constant α indicates that first-order perturbation theory has been applied; the linear dependence on the photon energy E_{ph} is due to the length form of the dipole operator used in equ. (2.1), and the wavenumber κ compensates the $1/\kappa$ which appears if the absolute squared value of the continuum wavefunction is used (see equ. (7.29)). The summations over the magnetic quantum numbers M_I of the photoion and m_s of the photoelectron's spin are necessary because no observation is made with respect to these substates. Due to the closed-shell structure of the initial state with $J_i = 0$ and $M_i = 0$, the averaging over the magnetic quantum numbers M_i simply yields unity and is omitted.

The matrix element is understood to be 'on-the-energy-shell', i.e., the energy ε of the photoelectron has to be calculated according to equ. (1.29a). Due to the different binding energies of electrons ejected from different shells of the atom, it is therefore possible to restrict the calculation of the matrix element to the selected process: in the present example to photoionization in the 1s shell only. As a consequence, the matrix element factorizes into two contributions, a matrix element for the two electrons in the 1s shell where one electron takes part in the photon interaction, and an *overlap* matrix element for the other electrons which do not take part in the photon interaction (*passive* electrons). The overlap matrix element is given by

$$\begin{aligned}\langle\text{overlap}\rangle = \langle\{&2s0^+, 2s0^-, 2p1^+, 2p1^-, 2p0^+, 2p0^-, 2p{-}1^+, 2p{-}1^-\} \\ &|\{2s0^+, 2s0^-, 2p1^+, 2p1^-, 2p0^+, 2p0^-, 2p{-}1^+, 2p{-}1^-\}\rangle.\end{aligned} \tag{2.4a}$$

In the *frozen* atomic structure approximation, where the same orbitals are used in the initial and final states, this overlap matrix element yields unity. Hence, one obtains for the remaining matrix element

$$\begin{aligned} M_{fi}(M_I, m_s) &= \langle\{1s0M_I, \boldsymbol{\kappa} m_s\}|z_1 + z_2|\{1s0^+, 1s0^-\}\rangle \\ &= \tfrac{1}{2}[\langle\varphi_{1s0M_I}(1)\psi_{\boldsymbol{\kappa} m_s}(2)|z_1 + z_2|\varphi_{1s0^+}(1)\varphi_{1s0^-}(2)\rangle \\ &\quad - \langle\varphi_{1s0M_I}(1)\psi_{\boldsymbol{\kappa} m_s}(2)|z_1 + z_2|\varphi_{1s0^+}(2)\varphi_{1s0^-}(1)\rangle \\ &\quad - \langle\varphi_{1s0M_I}(2)\psi_{\boldsymbol{\kappa} m_s}(1)|z_1 + z_2|\varphi_{1s0^+}(1)\varphi_{1s0^-}(2)\rangle \\ &\quad + \langle\varphi_{1s0M_I}(2)\psi_{\boldsymbol{\kappa} m_s}(1)|z_1 + z_2|\varphi_{1s0^+}(2)\varphi_{1s0^-}(1)\rangle]. \end{aligned} \tag{2.4b}$$

Renaming the electron numbers shows that the first and fourth terms in the matrix element $M_{fi}(M_I, m_s)$ and also the second and third terms are identical, and they can be combined. As a next step one calculates the action of the photon operator on the single-particle wavefunctions. Omitting for simplicity the wavefunction symbols φ and ψ, one gets

$$\begin{aligned} M_{\mathrm{fi}}(M_{\mathrm{I}}, m_s) = {} & \langle 1\mathrm{s}0 M_{\mathrm{I}}|z|1\mathrm{s}0^+\rangle\langle \boldsymbol{\kappa} m_s|1\mathrm{s}0^-\rangle \\ & + \langle 1\mathrm{s}0 M_{\mathrm{I}}|1\mathrm{s}0^+\rangle\langle \boldsymbol{\kappa} m_s|z|1\mathrm{s}0^-\rangle \\ & - \langle 1\mathrm{s}0 M_{\mathrm{I}}|z|1\mathrm{s}0^-\rangle\langle \boldsymbol{\kappa} m_s|1\mathrm{s}0^+\rangle \\ & - \langle 1\mathrm{s}0 M_{\mathrm{I}}|1\mathrm{s}0^-\rangle\langle \boldsymbol{\kappa} m_s|z|1\mathrm{s}0^+\rangle. \end{aligned} \tag{2.4c}$$

As will be demonstrated below, the dipole operator z requires a change in the orbital angular momentum. Since such a change does not occur in the first and third terms of the matrix element $M_{\mathrm{fi}}(M_{\mathrm{I}}^{\mathrm{c}}, m_s)$, one derives

$$\begin{aligned} M_{\mathrm{fi}}(M_{\mathrm{I}}, m_s) = {} & \langle 1\mathrm{s}0 M_{\mathrm{I}}|1\mathrm{s}0^+\rangle\langle \boldsymbol{\kappa} m_s|z|1\mathrm{s}0^-\rangle \\ & - \langle 1\mathrm{s}0 M_{\mathrm{I}}|1\mathrm{s}0^-\rangle\langle \boldsymbol{\kappa} m_s|z|1\mathrm{s}0^+\rangle. \end{aligned} \tag{2.4d}$$

The single-electron matrix elements for the passive and the active electrons contain different projections of the electron spin. Since neither the photon operator nor the 'unity operator' (in the overlap matrix element) acts on the spin, the quantum numbers M_{I} and m_s are fixed by the corresponding spin of the formerly bound 1s-electron. This yields

$$M_{\mathrm{fi}}(M_{\mathrm{I}}, m_s) = \langle 1\mathrm{s}0 \mid 1\mathrm{s}0\rangle\langle \boldsymbol{\kappa}|z|1\mathrm{s}0\rangle\{\delta_{M_{\mathrm{I}},1/2}\,\delta_{m_s,-1/2} - \delta_{M_{\mathrm{I}},-1/2}\,\delta_{m_s,1/2}\}. \tag{2.4e}$$

(It is also possible to derive this result by incorporating the condition (2.5) below from the beginning by using the factorization of a two-electron function into a symmetrical spatial function and an antisymmetrical spin function, see equ. (1.16).) The expression in the braces indicates that the two electrons in the final state have opposite spins, i.e., the photoprocess reaches a singlet final state. This can be easily understood, because in LS-coupling spin–orbit effects are absent, and the photon operator does not act on the spin. Therefore, the selection rule

$$\Delta S = 0, \ \Delta M_S = 0 \tag{2.5}$$

is satisfied.

In a next step the absolute squared value of the matrix element has to be evaluated together with the summation over the magnetic quantum numbers M_{I} and m_s. This gives

$$\begin{aligned} & \sum_{M_{\mathrm{I}}}\sum_{m_s} |M_{\mathrm{fi}}(M_{\mathrm{I}}, m_s)|^2 \\ & \quad = |\langle 1\mathrm{s}0 \mid 1\mathrm{s}0\rangle|^2\, |\langle \boldsymbol{\kappa}^{(-)}|z|1\mathrm{s}0\rangle|^2 \sum_{M_{\mathrm{I}}} (|\delta_{M_{\mathrm{I}},-1/2}|^2 + |\delta_{M_{\mathrm{I}},1/2}|^2) \\ & \quad = 2\,|\langle 1\mathrm{s}0 \mid 1\mathrm{s}0\rangle|^2\, |\langle \boldsymbol{\kappa}^{(-)}|z|1\mathrm{s}0\rangle|^2. \end{aligned} \tag{2.6}$$

This result can be interpreted very well; it contains:

(i) the factor 2 which is equal to the number of equivalent electrons in the photoprocess (in neon, one has two 1s-electrons);
(ii) an overlap matrix element for the second 1s-electron which stays in its orbital (it is a passive electron like the ones in the other shells);

(iii) the dipole matrix element for the active 1s-electron which participates in the photon interaction.

The dipole matrix element can be evaluated further and yields, using the expansion of the continuum function $\psi^{(-)}_{\boldsymbol{\kappa}m_s}$ into partial waves as given in equ. (7.28b):

$$d^{(-)}_{\mathrm{fi}} = \langle \boldsymbol{\kappa}^{(-)}|z|1s0\rangle$$
$$= \frac{1}{\sqrt{\kappa}} \sum_{\ell m} (-\mathrm{i})^{\ell}\, \mathrm{e}^{\mathrm{i}(\sigma_\ell + \delta_\ell)}\, Y_{\ell m}(\hat{\boldsymbol{\kappa}}) \langle R_{\varepsilon\ell}(r)\, Y_{\ell m}(\hat{\mathbf{r}})|z|R_{1s}(r)\, Y_{00}(\hat{\mathbf{r}})\rangle. \tag{2.7a}$$

(The spinor function $\kappa^{m_s}_{1/2}$ has already been used in the derivation of equ. (2.4e); the magnetic quantum number of the orbital angular momentum ℓ is written as m. This matrix element together with the overlap contribution $\langle 1s0\,|\,1s0\rangle$ is proportional to the more general channel matrix element D_γ whose standard form is given in equ. (8.35). In the present case this channel is given in *LSJ*-coupling by $\gamma = 1s\varepsilon p\,{}^1P^o_1$.) Three important features are introduced in this expansion:

(i) two phases, δ_ℓ from the short-range atomic potential and σ_ℓ from the long-range Coulomb potential (the symbol σ_ℓ used for the Coulomb phase should not be confused with the symbol $\sigma_{n\ell}$ used for a partial cross section);
(ii) the angle dependence of photoelectron emission, contained in $Y_{\ell m}(\hat{\boldsymbol{\kappa}}) = Y_{\ell m}(\vartheta, \varphi)$;
(iii) the spatial parts $R_{\varepsilon\ell}(r)$ and $Y_{\ell m}(\hat{\mathbf{r}}) = Y_{\ell m}(\vartheta', \varphi')$ of the photoelectron's wavefunction.

Because

$$z = r\cos\vartheta' = \sqrt{\left(\frac{4\pi}{3}\right)}\; r\; Y_{10}(\vartheta', \varphi'), \quad Y_{00}(\vartheta', \varphi') = \frac{1}{\sqrt{(4\pi)}}, \tag{2.8}$$

the spatial integration leads to

$$d^{(-)}_{\mathrm{fi}} = \frac{1}{\sqrt{\kappa}} \sum_{\ell m} (-\mathrm{i})^{\ell}\, \mathrm{e}^{\mathrm{i}(\sigma_\ell + \delta_\ell)}\, Y_{\ell m}(\hat{\boldsymbol{\kappa}}) \frac{1}{\sqrt{3}} \langle R_{\varepsilon\ell}(r)|r|R_{1s}(r)\rangle \int Y^*_{\ell m}(\hat{\mathbf{r}})\, Y_{10}(\hat{\mathbf{r}})\, \mathrm{d}\Omega. \tag{2.7b}$$

For the last integral, the orthogonality condition of the spherical harmonics requires $\ell = 1$ and $m = 0$ for a non-vanishing value. This is again a selection rule which means that of all orbital angular momenta ℓ in the partial-wave expansion of the photoelectron's wavefunction $\psi^{(-)}_{\boldsymbol{\kappa}m_s}$ only $\ell = 1$ contributes, if the photoprocess occurs with an s-electron. For the general case of a formerly bound active electron with principal quantum number n and orbital angular momentum ℓ_{i} being ejected by the photon interaction one obtains

$$\left.\begin{array}{l} n\ell_{\mathrm{i}} \to \varepsilon\ell \qquad \text{with} \qquad \ell = \begin{cases} \ell_{\mathrm{i}} \pm 1 & \text{for } \ell_{\mathrm{i}} \neq 0 \\ 1 & \text{for } \ell_{\mathrm{i}} = 0 \end{cases} \\ m = m_{\mathrm{i}} \text{ (for linearly polarized light).} \end{array}\right\} \tag{2.9a}$$

(Note that m is an abbreviation for m_ℓ.) It can be seen that in general there exist two photoionization channels which differ in the orbital angular momentum ℓ of the photoelectron and can interfere. Equ. (2.9a) is the dipole selection rule for the orbital angular momentum, obtained here in the single-particle model, i.e., for ℓ and m instead of L and M_L, respectively. In the general case of LS-coupling, equ. (2.9a) has to be replaced by

$$\left.\begin{aligned} &\Delta L = L_f - L_i = \pm 1 \text{ or } 0, \text{ but } L = 0 \rightarrow L = 0 \text{ forbidden}, \\ &\Delta M_L = M_{L_f} - M_{L_i} = 0 \text{ for linearly polarized light (with the electric field} \\ &\qquad\qquad \text{vector defining the quantization axis)}, \end{aligned}\right\} \quad (2.9b)$$

and for LSJ- or jjJ-coupling by

$$\left.\begin{aligned} &\Delta J = J_f - J_i = \pm 1 \text{ or } 0, \text{ but } J = 0 \rightarrow J = 0 \text{ forbidden}, \\ &\Delta M_J = M_{J_f} - M_{J_i} = 0 \text{ for linearly polarized light}, \end{aligned}\right\} \quad (2.9c)$$

where these angular momenta refer to the initial state of the atom and the complete final state which includes the ion and the photoelectron.

In the context of selection rules for photoionization it must be added that the dipole operator requires a change of the parity from the initial to the final state of the photoprocess, i.e.,

$$\Delta\pi = \pi_i \pi_f = (-1) \quad (2.10)$$

In equ. (2.7b) the matrix element $\langle R_{\varepsilon\ell}(r)|r|R_{1s}(r)\rangle$ is to be understood to be an integration over the r coordinate. Hence it is called the *radial dipole integral.* In the present case only one such integral is possible and is given by:

$$\begin{aligned} R_{\varepsilon p, 1s} &= \langle R_{\varepsilon p}(r)|r|R_{1s}(r)\rangle \\ &= \int_0^\infty R_{\varepsilon p}(r) r R_{1s}(r) r^2 \, dr = \int_0^\infty P_{\varepsilon p}(r) r P_{1s}(r) \, dr. \end{aligned} \quad (2.11)$$

The last form can be considered as an overlap integral for the radial functions $P_{1s}(r)$ and $P_{\varepsilon p}(r)$, modified by r. Such radial functions are shown in Fig. 2.1 for the case of 1s photoionization in neon. $P_{1s}(r)$ has no nodes and appreciable amplitude only in the spatial region which agrees with the average radius of the 1s orbital. In contrast, the functions $P_{\varepsilon p}(r)$ extend up to infinity. At large distances $rP_{\varepsilon p}(r)$ approaches a sinusoidal oscillation with the wavelength $\lambda = 2\pi/\sqrt{(2\varepsilon)}$, in atomic units. In the atomic region $P_{\varepsilon p}(r)$ is strongly modified by the special shape of the effective potential (atomic short-range potential and centrifugal potential). Knowing these radial functions, the basic energy dependence of the radial dipole integral $R_{\varepsilon p, 1s}$ which depends both on $P_{\varepsilon p}(r)$ and $P_{1s}(r)$, and on r, can be understood: the weighting with r is responsible for the largest value being obtained in the present case at threshold ($\varepsilon = 0$). For increasing photon energies the kinetic energy ε also becomes larger, and the first negative lobe of the radial function $P_{\varepsilon p}(r)$ moves into the spatial region of the bound 1s orbital. Hence, the integration over the product of $P_{1s}(r)$ and $P_{\varepsilon p}(r)$, modified by the factor r, becomes smaller. In particular, for very high photon energies the oscillatory structure of the continuum function $P_{\varepsilon p}(r)$ makes the integrand smaller and smaller, finally approaching zero. Hence, the cross section for 1s photoionization starts at threshold with a high value and decreases for increasing photon energies. In the absorption spectroscopy of X-rays

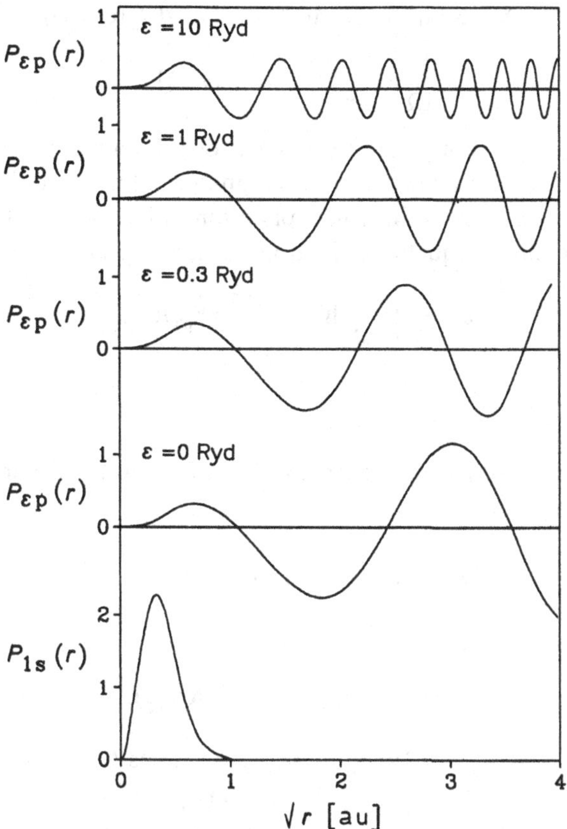

Figure 2.1 Plot of the radial functions $P_{1s}(r)$ and $P_{\varepsilon p}(r)$ for the bound 1s and a photoionized εp electron in neon as functions of the square root of the radial coordinate r (all quantities are in atomic units). The parameter ε indicates the kinetic energy of the continuum electron. The radial functions have been calculated for $P_{1s}(r)$ and $P_{\varepsilon p}(r)$ using the HF potentials of the electron configurations $1s^2 2s^2 2p^6$ and $1s 2s^2 2p^6$, respectively. From [Sch95].

these typical onsets of the elements are called K-absorption *edges*.

Collecting all contributions for the dipole matrix element $d_{fi}^{(-)}$ of the active electron, one gets

$$d_{fi}^{(-)} = \frac{1}{\sqrt{\kappa}}(-i)\, e^{i(\sigma_{\varepsilon p}+\delta_{\varepsilon p})}\, Y_{10}(\vartheta, \varphi)\, \frac{1}{\sqrt{3}}\, R_{\varepsilon p, 1s} \tag{2.12}$$

and, finally,

$$\frac{d\sigma}{d\Omega} = 4\pi^2 \alpha E_{ph} \kappa 2\, |d_{fi}^{(-)}|^2$$
$$= 2\pi\alpha E_{ph} R^2_{\varepsilon p, 1s} \cos^2 \vartheta. \tag{2.13a}$$

The result can be brought into the standard form (see equ. (1.30)) which separates the dynamics (comprised in the angular distribution parameter β) from the geometry (given by the second Legendre polynomial $P_2(\cos\vartheta)$):

$$\frac{d\sigma}{d\Omega} = \frac{\sigma}{4\pi}[1 + \beta P_2(\cos\vartheta)], \tag{2.13b}$$

but the partial cross section σ and the angular distribution parameter β can now be traced back to radial integrals from the photon interaction. (Relative phases also play a role for cases other than photoionization of s-electrons, see equ. (2.15b).) Representing all quantities in atomic units, one has

$$\sigma_{1s} = \int \frac{d\sigma}{d\Omega} d\Omega_\kappa = \frac{8\pi^2}{3}\alpha E_{ph} R^2_{\varepsilon p,1s} \tag{2.13c}$$

and

$$\beta_{1s} = 2. \tag{2.13d}$$

In a similar way, the other partial cross sections for photoionization in neon can be calculated. One obtains:

$$\sigma_{2s} = \frac{8\pi^2}{3}\alpha E_{ph} R^2_{\varepsilon p,2s}, \tag{2.14a}$$

$$\beta_{2s} = 2, \tag{2.14b}$$

$$\sigma_{2p} = \frac{8\pi^2}{3}\alpha E_{ph}(R^2_{\varepsilon s,2p} + 2R^2_{\varepsilon d,2p}), \tag{2.15a}$$

$$\beta_{2p} = \frac{2R^2_{\varepsilon d,2p} - 4R_{\varepsilon d,2p}R_{\varepsilon s,2p}\cos\Delta}{R^2_{\varepsilon s,2p} + 2R^2_{\varepsilon d,2p}}, \tag{2.15b}$$

with $\Delta = \sigma_{\varepsilon d} + \delta_{\varepsilon d} - \sigma_{\varepsilon s} - \delta_{\varepsilon s}$. Obviously, the dipole selection rule for the change of the orbital angular momentum in the case of 2p photoionization allows two partial waves εd and εs for the photoelectron. Both contributions can interfere in the expression for the angular distribution parameter which contains the phase difference Δ of the channels (for more details see below and Section 8.2).

2.2 Analysis of line position

For photoionization of a 1s-electron in neon, incident photons must have an energy higher than the binding energy of 870.2 eV. Such high energies can be provided by characteristic radiation from an X-ray tube. Therefore, before the advent of synchrotron radiation many pioneering studies for neon were performed with such radiation. Because the results from these studies are good examples of detailed investigations, they are included here even though the main topic is electron spectrometry with synchrotron radiation.

Due to relativistic effects important for the inner shells of atoms, the deep $n\ell$ shells split in energy according to the j value from the spin–orbit coupling: $\mathbf{j} = \boldsymbol{\ell} + \mathbf{1/2}$. Therefore, deep inner shells are classified by their $n\ell j$ values. Frequently, in X-ray emission and Auger electron spectrometry, the shell index

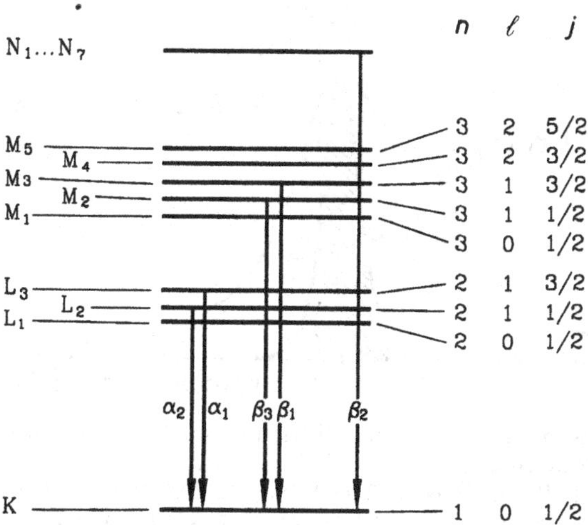

Figure 2.2 Demonstration of the two equivalent nomenclatures used for the description of inner-shell levels and X-ray transitions (also Auger transitions, see below). The vertical direction is regarded as the energy axis (but is not to scale here). On the left-hand side is given the notation which is frequently used in inner-shell spectroscopy, on the right-hand side the corresponding single-orbital quantum numbers with n, ℓ and j being, respectively, the principal quantum number, the orbital angular momentum and the total angular momentum which includes the spin of the electron. Also shown are the main X-ray transitions with their spectroscopic notation (for a more complete plot which includes satellite transitions see [Urc79]).

$n = 1, 2, 3, \ldots$ is replaced by the symbols K, L, M, ..., and the subshells ℓj are characterized by numbers (the KLM ... nomenclature was introduced by C. G. Barkla [Bar11]; see [INo74]). Both nomenclatures are illustrated with the help of the term level diagram shown in Fig. 2.2. Some of the main transitions for X-ray decay following K-shell ionization are also included in this figure; they are characterized by α_1, α_2, β_1, β_3, and β_2. In the present case of photoionization in neon with characteristic radiation from an X-ray tube, Al Kα or, alternatively, Mg Kα radiation is of interest. These contain the two components Kα_1 and Kα_2 which can also be monochromatized in order to decrease the linewidth of the X-ray radiation (see Fig. 2.3) and to reject further lines (X-ray satellites, see below).

A compilation of photoelectron spectra of rare gases obtained with characteristic X-ray radiation is shown in Fig. 2.4. (K. Siegbahn (Uppsala, Sweden) was awarded a Nobel prize in 1981 for his work on the development of very precise spectroscopic methods which allowed such detailed studies.) First, the kinetic energy of ejected photoelectrons is measured and plotted on the x-axis, but using the energy relation of photoelectrons, $E_{kin} = h\nu - E_I$ (equ. (1.29a)), and knowing the photon energy, this axis is then redrawn to become an ionization energy (E_I) axis (called in the figure the binding energy). The most obvious features in these spectra are discrete structures which occur at certain E_I values. These peaks are called *photolines*, and their E_I values are related to specific ionization processes with electrons from $n\ell$

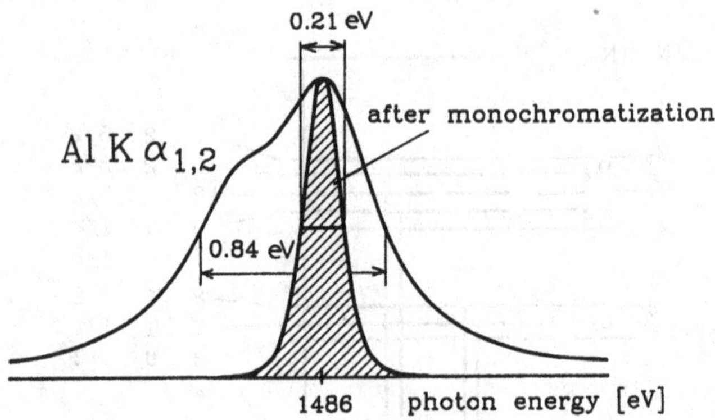

Figure 2.3 Comparison of the spectral distributions of normal (broad pattern) and monochromatized (shaded area) Al Kα radiation; for details see main text. Reprinted from *J. Electron Spectrosc. Relat. Phenom.*, **2**, Gelius *et al.*, 405 (1973) with kind permission of Elsevier Science, NL, Sara Burgerhartstraat 25, 1055 KV Amsterdam, The Netherlands.

orbitals as indicated. In all the spectra one can recognize the shell structure of the individual atoms. For helium with the electron configuration $1s^2$ only one photoline appears, which is at the 1s ionization energy of 24.6 eV. For neon with the electron configuration $1s^22s^22p^6$ three photolines are seen which correspond to ionization in the 1s, 2s and 2p shells, respectively. (The small features accompanying each photoline at lower energy values are due to the presence of Kα', Kα_3 and Kα_4 components in the characteristic X-ray radiation.) The corresponding ionization energies are listed in Table 2.1. Proceeding in this way to the other rare gases one notices that, due to their increasing atomic number, the ionization energies of subshells separate more and more, and it becomes possible to resolve even the fine-structures by electron spectrometry, as shown by the $2p_{1/2}$ and $2p_{3/2}$ photolines of argon. However, for xenon the photoline from the $4p_{1/2}$ orbital is missing. This peculiarity is due to phenomena outside the single-particle model, and is a clear manifestation for the importance of electron–electron interactions in photoionization [WOh76]. As explained in Section 1.1, the interpretation of such anomalies, including also the presence of satellite lines, and the dynamical changes in these features for varying photon energy, are the topics of main interest in present research activities.

2.3 Analysis of linewidth

In addition to the different energy positions of the photolines shown in Fig. 2.4, the different heights of these lines are also a pronounced property of such spectra. It seems natural to use this height as a measure of the intensity of a photoline and, hence, for the cross section of a specific $n\ell$ orbital at the given photon energy. However, as will be seen in the next section, the appropriate quantity for the intensity is the *area* under a photoline (more correctly the dispersion corrected

Table 2.1. *Ionization energies in neon (from [PNS82] and [Moo71]; for some atomic ionization energies, see [Sev79]).*

Ionization energy	E_I
1s	870.2
2s	48.48
2p	21.60

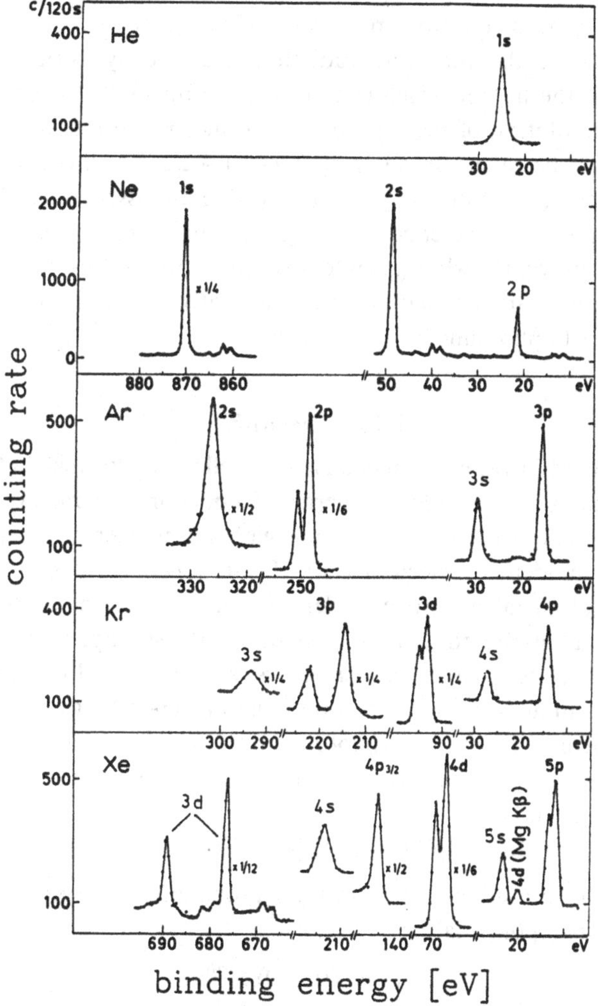

Figure 2.4 Photoelectron spectra of rare gases obtained with monochromatized Mg Kα radiation. The spectra have been plotted as functions of the ionization energy (here called the binding energy) by using equ. (1.29a) with $h\nu = 1254$ eV. A detailed description of the features observed is given in the main text. Reprinted from Siegbahn *et al.*, *ESCA applied to free molecules* (1969) with kind permission from Elsevier Science – NL, Sara Burgerhartstraat 25, 1055 KV Amsterdam, The Netherlands.

area, see equ. (2.39b)), and not its height. Since the area depends strongly on the specific lineshape, this aspect has to be treated first. A complete discussion requires a full analysis of this lineshape. However, very often attention is only paid to the linewidth because this quantity can be extracted easily from the spectra by taking the *fwhm*. In the following, this measured linewidth will be discussed taking into account three contributions† and assuming certain standard lineshapes (see Section 10.4):‡ (i) the linewidth of the incoming radiation (bandpass B_m of the monochromator); (ii) the levelwidth $\Gamma_{n\ell j}$ of the hole-state produced by the ionization of an $n\ell j$ electron; and (iii) the instrumental width ΔE_{sp} from the finite energy resolution of the electron spectrometer.

The bandpass of the incoming radiation has already been considered in connection with the monochromatization of synchrotron radiation, Section 1.4, and the finite resolution of the electron spectrometer, introduced in Section 1.5 (equ. (1.49)), will be taken up again in Section 4.2.2. Therefore, only the level width $\Gamma_{n\ell j}$ and the explanation of its origin will be discussed here. Then how the three contributions, characterized by B_m, $\Gamma_{n\ell j}$ and ΔE_{sp}, form the observed photoline with its width fwhm_{exp}, by convolution procedures will be discussed. Finally, the results are applied to the quantitative analysis of the linewidth obtained for the 1s photoline in neon.

2.3.1 Level width

Any state produced by the ejection of an inner-shell $n\ell j$-electron has a finite *lifetime* since it can decay to states of lower energy. (To be more general, every state with some excess energy will decay to a state with lower energy, except when the transition is forbidden by selection rules. These selection rules follow from the transition operators taken into consideration: in the present context these are electric dipole radiation and non-radiative decay. If these operators do not allow a transition, the state is called *metastable*, because it has a long lifetime which, however, is not infinitely long due to other and higher-order transition operators (e.g., spin–orbit interactions, two-photon transitions, and so on).) If there is no return to the initial state, the rate of change for the number $N(t)$ of atoms in the excited state is given by

$$dN/dt = \dot{N}(t) = -N(t)P_{n\ell j}, \tag{2.16a}$$

where $P_{n\ell j}$ is the *transition rate* (transition probability per second). Hence, the decay follows an exponential law

$$N(t) = N(t=0)\,e^{-t/\tau_{n\ell j}} \tag{2.16b}$$

† It is obvious that neighbouring lines which are not resolved in the experiment broaden the observed linewidth. For example, the two fine-structure components corresponding to $np_{1/2}$ and $np_{3/2}$ ionization in the outermost shell of the rare gases are not resolved in the spectra shown in Fig. 2.4, except for xenon.
‡ In addition to the lineshapes of Gaussian and Lorentzian functions, one ought to consider the *Beutler–Fano* lineshape [Beu35, Fan35, Fan61] and the lineshape due to *post-collision* interaction. For the first of these see [Sch92a], for the second see Section 5.5.

with the time constant (mean lifetime) $\tau_{n\ell j}$, and with the characteristic properties related by

$$P_{n\ell j} = 1/\tau_{n\ell j}. \tag{2.16c}$$

The two decay mechanisms of interest in the following discussion are radiative (i.e., fluorescence) and non-radiative (i.e., Auger or autoionization) transitions. For the decaying state the following ansatz is used for the time-dependent wavefunction $\Psi_{n\ell j}(t)$, starting at $t = 0$:

$$\Psi_{n\ell j}(t) = \Psi_{n\ell j}(0)\, \mathrm{e}^{-\mathrm{i}E_{n\ell j}t}\, \mathrm{e}^{-t/2\tau_{n\ell j}}, \tag{2.17a}$$

i.e.,

$$|\Psi_{n\ell j}(t)|^2 = |\Psi_{n\ell j}(0)_{\text{stationary}}|^2\, \mathrm{e}^{-t/\tau_{n\ell j}}, \tag{2.17b}$$

which shows that there is a stationary state of energy $E_{n\ell j}$, but the probability of finding the system in this state decreases exponentially with the mean lifetime $\tau_{n\ell j}$. Related to this decay is an uncertainty in the energy $E_{n\ell j}$. This energy distribution follows from the Fourier transformation

$$g_{n\ell j}(E) = \frac{1}{\sqrt{(2\pi)}} \int \mathrm{d}t\, \Psi_{n\ell j}(t)\, \mathrm{e}^{\mathrm{i}Et}. \tag{2.18a}$$

Integration between the limits $0 \le t \le \infty$ gives

$$g_{n\ell j}(E) = \frac{\Psi(0)}{\sqrt{(2\pi)}} \frac{\mathrm{i}}{(E_{n\ell j} - E) + \mathrm{i}\Gamma_{n\ell j}/2}, \tag{2.18b}$$

where the transition rate has been replaced by the natural level width $\Gamma_{n\ell j}$ (see below). The probability $F(E, E_{n\ell j})$ of finding, in the presence of the decaying state at $E_{n\ell j}$, an energy value in the interval between E and $(E + \mathrm{d}E)$ follows from this Fourier amplitude as

$$F(E, E_{n\ell j}) = \text{const}\,|g_{n\ell j}(E)|^2. \tag{2.19a}$$

If $F(E, E_{n\ell j})$, integrated over all energies E, is normalized to unit area, i.e.,

$$\int_{-\infty}^{+\infty} F(E, E_{n\ell j})\, \mathrm{d}E = 1, \tag{2.19b}$$

one obtains for $F(E, E_{n\ell j})$ the so-called *Lorentzian* distribution function

$$L(E, E_{n\ell j}) = \frac{\Gamma_{n\ell j}}{2\pi} \frac{1}{(E_{n\ell j} - E)^2 + (\Gamma_{n\ell j}/2)^2}. \tag{2.19c}$$

This result describes quantitatively the energy distribution of the decaying $n\ell j$ hole-state. The function is symmetric in E around $E_{n\ell j}$. For $E = E_{n\ell j}$ it has a maximum, and its fwhm value is given by $\Gamma_{n\ell j}$ which is called the *natural* or *inherent* level width because it originates from the decaying hole-state which is inherent to the atom. As an example, a compilation of level widths $\Gamma_{n\ell}$ in neon is given in Table 2.2. Because of the replacement made in the derivation of equ. (2.18b) for $\tau_{n\ell}$, one has (in atomic units)

$$\Gamma_{n\ell j}\tau_{n\ell j} = 1 \tag{2.20}$$

Table 2.2. *Natural level widths $\Gamma_{n\ell}$ in neon (see, for example, [Kra79]).*

Shell $n\ell$	$\Gamma_{n\ell}$ [eV]
1s	0.27[(a)]
2s	<0.1
2p	0 (there is no decay possible)

[(a)] This value from [SMB76] replaces the 0.23 eV given before [GSS74].

or, if combined with equ. (2.16c)

$$\Gamma_{n\ell j} = P_{n\ell j}. \tag{2.21}$$

(If equ. (2.20) is not expressed in atomic units, one gets

$$\Gamma_{n\ell j}\tau_{n\ell j} = \hbar, \tag{2.22a}$$

and the result follows from the Heisenberg uncertainty principle [Hei27]

$$\Delta E\ \Delta t \geq \hbar, \tag{2.22b}$$

if the lower limit of uncertainty is used and the associations $\Delta E = \Gamma_{n\ell j}$ and $\Delta t = \tau_{n\ell j}$ are made.)

With $\Gamma_{n\ell}$ only the *total* decay rate or, equivalently, the *total* level width of an inner-shell hole-state has been considered so far. In general, the system has different decay *branches.* In many cases these branches can be classified as radiative (fluorescence) or non-radiative (Auger or autoionizing) transitions, and even further, by specifying within each group individual decay branches to different final ionic states. (Combinations of radiative and non-radiative transitions are also possible in which a photon is emitted and simultaneously an electron is excited/ejected. These processes are termed *radiative Auger decay* (see [Abe75]).) As a result, the total transition rate $P_{n\ell}$ and, hence, the *total* level width $\Gamma_{n\ell}$, is composed of sums over *partial* values:

$$P_{n\ell} = \sum_f P_{n\ell}(\text{radiative; decay to } f) + \sum_{f'} P_{n\ell}(\text{non-rad.; decay to } f'), \tag{2.23a}$$

$$\Gamma_{n\ell} = \sum_f \Gamma_{n\ell}(\text{radiative; decay to } f) + \sum_{f'} \Gamma_{n\ell}(\text{non-rad.; decay to } f'). \tag{2.23b}$$

Before these partial quantities are discussed further, an important comment has to be made; unlike the partial transition rates, the partial level widths have no direct physical meaning, because even for a selected decay branch it is always the *total* level width which determines the natural energy broadening. The partial level width is only a measure of the partial transition rate. Both aspects can be inferred from the Lorentzian distribution attached to a selected decay branch, e.g., Auger decay, which is given by

$$L(E, E_{n\ell})|_{\text{Auger}} = \frac{\Gamma_{n\ell}(\text{Auger})}{2\pi}\,\frac{1}{(E - E_{n\ell})^2 + (\Gamma_{n\ell}(\text{total})/2)^2}. \tag{2.24}$$

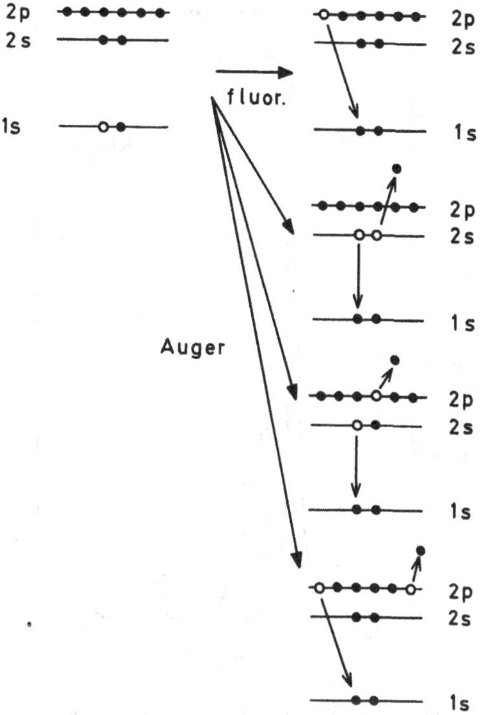

Figure 2.5 The main decay branches for fluorescence and Auger decay following 1s ionization in neon. The initial and final states are characterized by single-orbital energy level diagrams (not to scale) showing their occupation with electrons (filled circles) or holes (open circles). The initial 1s hole-state is shown on the left-hand side, different possibilities leading to some different final states are shown on the right-hand side. The change of occupation is indicated by arrows pointing to the electron which moved leaving a hole behind.

The dominant decay branches which result from 1s ionization in neon are shown in Fig. 2.5. The fluorescence decay concerns the dipole transitions 2p → 1s, shown in the upper row. In the X-ray nomenclature they are called $K{-}L_{2,3}$ transitions where the dash is used to separate the initial and final hole-states. These transitions are also called $K\alpha_{1,2}$ transitions (see Fig. 2.2). As can be seen in the experimental fluorescence spectrum shown in Fig. 2.6, in addition to these $K{-}L_{2,3}$ *main* transitions there are other lines called *satellites*. These satellites result from $KL{-}L^2$ and even $KL^2{-}L^3$ transitions where the dash again separates initial and final hole-states, e.g., $KL{-}L^2$ radiative transitions start with two holes, one in the K-shell, and one in the L-shell, and they end with two holes in the L-shell.

The dominant non-radiative branches for the filling of the created 1s-hole in neon are the K–LL Auger transitions. As can be seen in Fig. 2.5 there are different ways for the two holes to be distributed in the final state in the L-subshells L_1, $L_{2,3}$, and the corresponding Auger transitions can be grouped into $K{-}L_1L_1$,

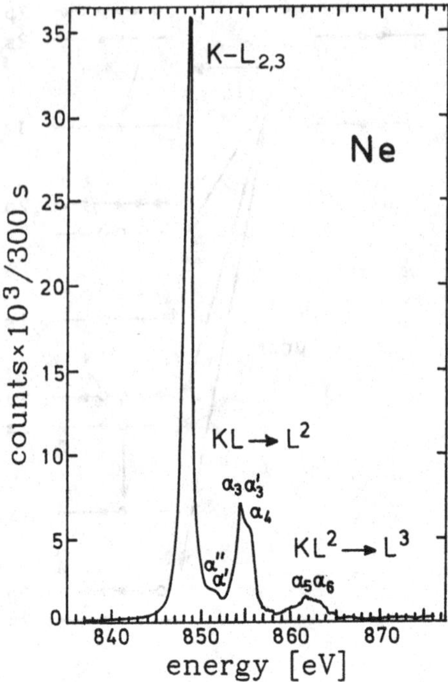

Figure 2.6 Fluorescence spectrum following 1s ionization in neon. The most intense line represents the unresolved doublet of main lines, $K–L_{2,3}$, the structures with lower intensity are due to accompanying satellite transitions where the initial state contains one ($KL \rightarrow L^2$) or even two ($KL^2 \rightarrow L^3$) additional holes in the L-shell. From [Kes73].

$K–L_1L_{2,3}$, and $K–L_{2,3}L_{2,3}$ transitions. These *normal* or *diagram* Auger transitions will be discussed in detail in Chapter 3, and there it will be shown that in addition *satellites* again appear. The existence of satellites becomes important if the 1s level width Γ_{1s} of neon is determined quantitatively, because all decay processes originating from the 1s initial hole-state have to be taken into account. In the case under consideration these are the main K–L fluorescence radiation and the main K–LL Auger lines, as well as K–LL fluorescence satellites (*radiative* Auger decay) and K–LLL Auger satellites (*double* Auger decay).

After these preliminaries, the partition into radiative (subscript R) and non-radiative (subscript A for Auger) transitions for the 1s-hole decay in neon can be given in terms of intensities I_R and I_A or partial transition rates P_R and P_A by introducing a fluorescence *yield* ω_R and an Auger *yield* ω_A defined by

$$\omega_R(1s) = \frac{I_R(1s)}{I_R(1s) + I_A(1s)} = \frac{P_R(1s)}{P(1s)} = \frac{P_R(K\text{–}L) + P_R(K\text{–}LL)}{P(1s)}, \tag{2.25a}$$

$$\omega_A(1s) = \frac{I_A(1s)}{I_R(1s) + I_A(1s)} = \frac{P_A(1s)}{P(1s)} = \frac{P_A(K\text{–}LL) + P_A(K\text{–}LLL)}{P(1s)} \tag{2.25b}$$

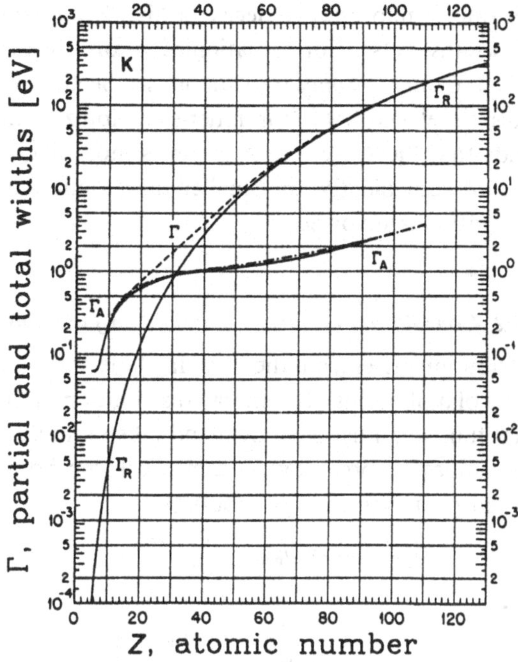

Figure 2.7 Theoretical level widths for K-shell ionization as a function of the atomic number. The total level width Γ is the sum of two contributions that come from radiative (fluorescence) decay, Γ_R, and non-radiative (Auger) decay, Γ_A. From [Kra79].

with

$$P(1s) = P_R(1s) + P_A(1s), \tag{2.25c}$$

or

$$\Gamma(1s) = \Gamma_R(1s) + \Gamma_A(1s), \tag{2.25d}$$

and

$$\omega_R + \omega_A = 1. \tag{2.25e}$$

Some radiative and non-radiative partial widths of K-shell ionization are shown in Fig. 2.7. In the case of 1s photoionization in neon, one has a rather small fluorescence yield, proposed to be $\omega_R(1s) = 0.018$, therefore the Auger effect with its yield $\omega_A = 0.982$ greatly dominates the decay routes. In general, it can be seen from Fig. 2.7 that for atomic numbers smaller than 30 the Auger effect dominates, but above this value fluorescence radiation (characteristic X-ray radiation) is the main decay branch. This behaviour can be understood qualitatively by calculating the radiative and non-radiative transition rates with hydrogenic wavefunctions. One gets [BTH53]

$$P_R(1s) = aZ^4, \tag{2.26}$$

$$P_A(1s) = b, \tag{2.27}$$

where Z is the atomic number and a and b are constants. (Note that according to equ. (2.21) the high rate of fluorescence decay in high-Z elements is responsible for their high level widths $\Gamma(1s)$. For radiationless transitions starting in shells other than the K-shell, *Coster–Kronig* transitions (see Section 3.1.1) also lead to high transition rates (level widths).) For inner-shell ionizations other than in the K-shell, one finds that the Auger effect becomes more likely the smaller the transition energy is, or, equivalently, the smaller the atomic number is, and the more outer-shell ionization is involved.

2.3.2 Convolution of energy distribution functions

The Lorentzian distribution arises from the decaying hole-state, but for an analysis of the observed shape of a photoline, the energy distribution of the incoming light and the spectrometer function must be known and taken into account. In general, the latter functions cannot be presented in a closed form; however, quite often they are approximated by a Gaussian distribution:

$$G(E, E_0) = G_{max}\, e^{-(E-E_0)^2/\delta^2}, \tag{2.28a}$$

and

$$\delta = \frac{1}{2\sqrt{(\ln 2)}}\,\text{fwhm} \tag{2.28b}$$

and

$$\int_{-\infty}^{+\infty} G(E, E_0)\, dE = G_{max}\sqrt{\pi}\,\delta = 1.06 G_{max}\text{fwhm}. \tag{2.28c}$$

If the Gaussian function is normalized at its maximum to unity, $G_{max} = 1$, one gets

$$\int_{-\infty}^{+\infty} G(E, E_0)|_{max=1}\, dE = 1.06\ \text{fwhm} \approx \text{fwhm}, \tag{2.28d}$$

while if it is normalized to unit area, one gets

$$\int_{-\infty}^{+\infty} G(E, E_0)\, dE = 1 \qquad \text{for} \qquad G_{max} = (\sqrt{\pi}\,\delta)^{-1}; \tag{2.28e}$$

both forms of normalization are used in the following. In Fig. 2.8 a comparison is made between a Gaussian and a Lorentzian lineshape with the same height and width. The Lorentzian profile is characterized by its broad feet. A Voigt profile, which results from the convolution of a Lorentzian and a Gaussian profile, is also shown.

The combined effect of the individual energy distribution functions which are of relevance for the photoionization process and for the detection of photoelectrons, can now be discussed. As a first step, the photoionization process alone, i.e., without any detection device, is discussed with the help of Fig. 2.9. The y-axis represents an energy scale with respect to the ground state, and different states of neon are plotted along the x-direction. For photons with sufficient energy $h\nu$ and with a distribution function $G_B(E_{ph}, E^0_{ph})$, one reaches the continuum above the 1s

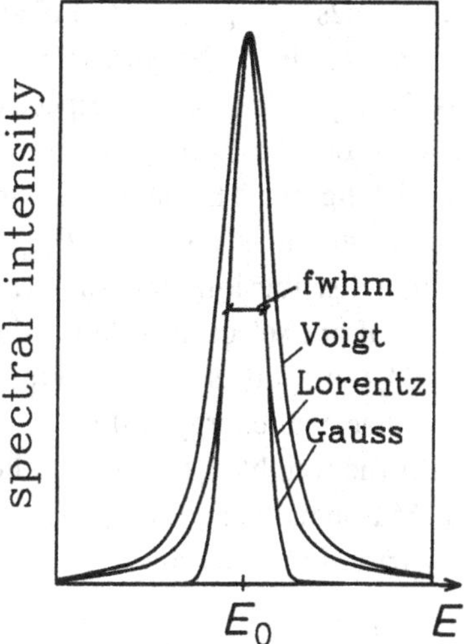

Figure 2.8 Several energy-distribution functions, all normalized to the same height. The Lorentzian and Gaussian distributions are shown for equal fwhm values. The Voigt profile results from the convolution of the shown Lorentzian and Gaussian functions.

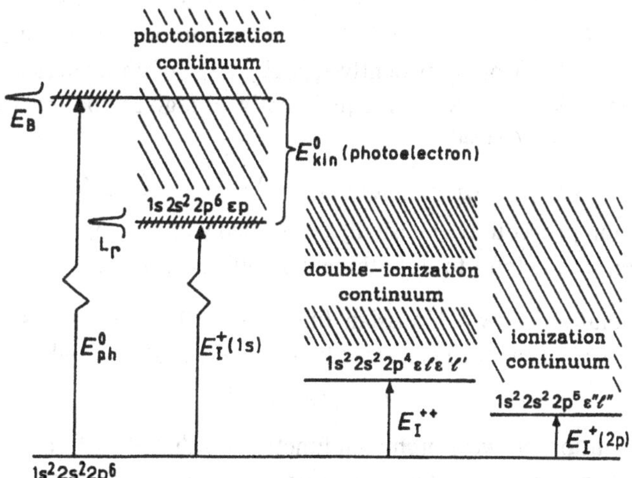

Figure 2.9 Illustration of energies relevant for 1s photoionization of neon and subsequent Auger or fluorescence decay. Energies are measured in the vertical direction and refer to the energy of the ground state $1s^2 2s^2 2p^6$; however, the individual energy positions are not to scale. Single and double ionization continua reached by photoionization ($1s2s^2 2p^6 \varepsilon p$) and subsequent fluorescence decay ($1s^2 2s^2 2p^5 \varepsilon'' \ell''$) or by 1s photoionization and subsequent Auger decay ($1s^2 2s^2 2p^4 \varepsilon\ell\varepsilon'\ell'$) are indicated by the regions with shading tilted to the left; the corresponding ionization energies are given by $E_I(\ldots)$. The finite bandpass of the incident photons of nominal energy E^0_{ph} and the finite level width Γ of the 1s hole-state, characterized by the corresponding Gaussian and Lorentzian distributions, G_B and L_Γ, respectively, are the regions with shading tilted to the right.

hole-state which is indicated as $1s2s^22p^6\varepsilon p$. This 1s hole-state may decay via an Auger or a fluorescence transition into the $1s^22s^22p^4\varepsilon\ell\varepsilon'\ell'$ double-ionization continuum or into the $1s^22s^22p^5\varepsilon''p''$ continuum, respectively. Hence, the energy of the 1s hole-state is smeared out according to the Lorentzian distribution $L_\Gamma(E_\ell, E_I^+)$, as indicated in the figure. The nominal kinetic energy of photoelectrons created in the source region is given by $E^0_{kin} = E^0_{ph} - E_I^+$, but other kinetic energies E_{kin} occur with a distribution function $F_s(E_{kin}, E^0_{kin})$. This distribution function can be worked out if one selects a certain value $E_{kin} = E_{ph} - E_\ell$ with weight $G_{ph}(E_{ph} = E_{kin} + E_\ell, E^0_{ph})$ and takes into account the spread of different level energies E_ℓ: first for a selected energy value E_ℓ by a multiplication of $G_{ph}(E_{ph} = E_{kin} + E_\ell, E^0_{ph})$ with the weight $L_\Gamma(E_\ell, E_I^+)$, followed by an integration over all possible E_ℓ values. Mathematically this procedure is described by the *convolution* product of the distribution functions:

$$F_s(E_{kin}, E^0_{kin}) = G_B \otimes L_\Gamma$$

$$= \int G_B(E_{kin} + E_\ell, E^0_{ph})\, L_\Gamma(E_\ell, E_I^+)\, dE_\ell. \tag{2.29}$$

(Note that such convolutions are commutative, associative, and distributive.) If these photoelectrons, created with the distribution function $F_s(E_{kin}, E^0_{kin})$ in the source region, are detected with an energy analyser, a further convolution is necessary to account for the instrumental resolution as described by the spectrometer functions $G_{sp}(E_{kin}, E_{pass})$, equ. (1.48). The convolution yields the distribution function $F_{exp}(E_{pass}, E^0_{kin})$, or equivalently $F_{exp}(U_{sp}, U^0_{sp})$, of photoelectrons observed at a preselected pass-energy E_{pass} or, equivalently, at a given spectrometer voltage U_{sp}. One has (see equ. (10.54b))

$$F_{exp}(E_{pass}, E^0_{kin}) = F_{exp}(U_{sp}, U^0_{sp}) = G_{sp} \otimes L_B \otimes G_\Gamma. \tag{2.30}$$

Such convolution procedures are usually performed with a computer. For practical work, however, it is helpful to note three convolution results:

(i) The convolution of two Gaussian functions with fwhm values B_1 and B_2, respectively, yields again a Gaussian function, the width B of which is given by

$$B = \sqrt{(B_1^2 + B_2^2)}. \tag{2.31a}$$

(ii) The convolution of two Lorentzian functions with widths Γ_1 and Γ_2, respectively, yields again a Lorentzian function, the width Γ of which is given by

$$\Gamma = \Gamma_1 + \Gamma_2. \tag{2.31b}$$

(iii) The convolution of a Lorentzian with a Gaussian function (or vice versa) yields a Voigt profile which cannot be presented analytically in a closed form. An example of a Voigt profile is shown in Fig. 2.8. It results from the convolution of the two other distribution functions shown in the figure.

From the general discussion of the width of photolines, the observed width $(\text{fwhm})_{exp}$ in neon can now be related to individual contributions. The main interest

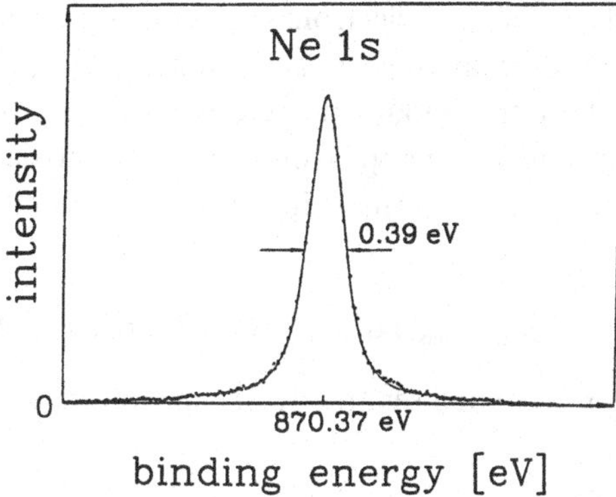

Figure 2.10 Neon 1s photoline obtained with monochromatized Al Kα radiation. The solid line is a fit to the experimental data given by the points. The observed fwhm value of 0.39 eV is indicated. From this value the level width $\Gamma(1s) = 0.27(2)$ eV can be obtained (see text). For $\Gamma(1s)$ see [SMB76]; for the ionization (binding) energy see also [PNS82]. Reprinted from *J. Electron Spectrosc. Relat. Phenom.* **2**, Gelius *et al.* 405 (1973) with kind permission of Elsevier Science – NL, Sara Burgerhartstraat 25, 1055 KV Amsterdam, The Netherlands.

lies in the extraction of a value for the level width $\Gamma(1s)$. In Fig. 2.10 the experimental data (points) for the neon 1s photoline obtained with monochromatized Al Kα radiation are shown together with the result from a convolution procedure (solid line). The necessary instrumental energy distribution functions for the monochromatized light (see Fig. 2.3) and for the electron spectrometer were determined by calculation and by experiment on the performance of these devices. For the Lorentzian function different values of the unknown level width $\Gamma(1s)$ were assumed, and the agreement of the convolution result with the experimental spectrum was then checked. The best agreement, shown in Fig. 2.10, implies a value of $\Gamma(1s) = 0.27(2)$ eV. This value will be needed in Chapter 3 to obtain absolute values of the neon K–LL Auger transition probabilities.

2.4 Analysis of line intensity

As asserted in the previous section, the height of the photolines shown in Fig. 2.4 does not provide the correct measure of the intensity of a photoline. It will now be demonstrated that the appropriate measure for intensities is the area A under the line, recorded within a certain time interval, at a given intensity of the incident light, and corrected for the energy dispersion of the electron spectrometer. This quantity, called the *dispersion corrected area* A_D, then depends in a transparent way on the photoionization cross section σ and on other experimental parameters. In order to derive this relation, the photoionization process which occurs in a finite source volume has to be considered, and the convolution procedures described above have to be included. In order to facilitate the formulation, it has to be assumed that certain requirements are met. These concern:

(i) The energy-analysis of the electrons is performed with an electrostatic deflection analyser, by changing the deflection voltage (spectrometer voltage U_{sp}, see equ. (1.47)). (Other modes of operation are described in Section 4.2.4.) The spectrometer function is approximated by a Gaussian function which is used in two different forms: normalizing at the maximum to the spectrometer transmission T, one has

$$\int G_{sp}(E_{kin}, E_{pass})\, dE_{kin} = 1.06\, T\, \Delta E_{sp} \approx T\, \Delta E_{sp}; \tag{2.32a}$$

if this transmission is treated separately, the maximum value is unity (see Fig. 1.14), and one has

$$\int \tilde{G}_{sp}(E_{kin}, E_{pass})\, dE_{kin} = 1.06\, \Delta E_{sp} \approx \Delta E_{sp}. \tag{2.32b}$$

(ii) The energy distribution of the monochromatized photon beam incident on the sample is described by a Gaussian function, $G_B(E_{ph}, E^0_{ph})$, with the fwhm given by the bandpass E_B, normalized to unit area

$$\int G_B(E_{ph}, E^0_{ph})\, dE_{ph} = 1, \tag{2.33}$$

i.e., the total number N_{ph} of incident photons/s is treated separately.

(iii) The density n_v of the target gas is uniform.

(iv) The cross section σ is assumed to be constant in the energy range relevant for the selected photoionization process. (This condition is not fulfilled for a resonance which, therefore, needs a slightly different treatment.) Its angle-dependent terms in equ. (1.50) are neglected, i.e.,

$$\frac{d\sigma}{d\Omega} = \frac{\sigma}{4\pi}(1 + \cdots). \tag{2.34}$$

(The incorporation of angle-dependent effects leads to attenuation factors for the angle functions attached to the angular distribution parameter β (see Section 10.5).)

(v) A possible inherent level width Γ of the photoionized state is described by a Lorentzian function.

(vi) The source volume ΔV accepted by the analyser is defined by the diameter of the photon beam (cross section q)† and a length Δz along the direction of the photon beam (see Fig. 1.13). The luminosity L of the electron spectrometer introduced in equ. (1.45) can then be expressed as

$$L = q\, \Delta z\, T, \tag{2.35}$$

where the average transmission T is related to the average solid angle Ω accepted by the analyser by

$$T = \Omega/4\pi. \tag{2.36}$$

(vii) The efficiency ε of the electron detector depends only on the mean kinetic energy E^0_{kin} of the impinging electrons.

† This prerequisite can be adapted easily to other situations.

When these requirements are met, the intensity $I_{exp}(E^0_{kin}, E_{pass} = fU_{sp})$ of a photoline with nominal energy E^0_{kin}, obtained by scanning the spectrometer voltage U_{sp} in equal steps and for equal time intervals across the adapted value U^0_{sp} follows from equ. (2.30) to be

$$I_{exp}(E^0_{kin}, fU_{sp}) = N_{ph} n_v \sigma \, \Delta z \, T\varepsilon \, \tilde{G}_{sp} \otimes L_\Gamma \otimes G_B. \tag{2.37}$$

As demonstrated in Section 10.4.2, the convolution procedure leads to a simple and impressive result if an integration is performed over all spectrometer voltages (for the corresponding treatment of the experimental data see below). One derives (see also [WTW77])

$$A_{exp}(E^0_{kin}) = \int I_{exp}(E^0_{kin}, fU_{sp}) f \, dU_{sp} = N_{ph} n_v \sigma \, \Delta z \, T\varepsilon \, \Delta E_{sp}. \tag{2.38}$$

Note that it is the fwhm value of the spectrometer function, ΔE_{sp}, which appears in the formula and not the value ΔE_{exp} attached to the observed photoline, as might have been expected naively (for the role of ΔE_{exp} see below). Hence, it is convenient to introduce the constant relative resolution $R = (\Delta E_{sp}/E^0_{kin})$ of electrostatic deflection analysers introduced in equ. (1.49) which leads to the *dispersion corrected* area A_D:†

$$A_D = A_{exp}(E^0_{kin})/E^0_{kin},$$

$$A_D = \int I_{exp}(E^0_{kin}, fU_{sp}) f \, dU_{sp}/E^0_{kin} = N_{ph} n_v \sigma \, \Delta z \, T\varepsilon \, R. \tag{2.39a}$$

Experimentally, the dispersion corrected area is obtained from the area of the photoline plotted on an energy scale (fU_{sp}), divided by the nominal kinetic energy E^0_{kin}:

$$A_D = \sum_{\text{photoline}} I_{exp}(E^0_{kin}, fU_{sp}) f \, \Delta U_{sp}/E^0_{kin}, \tag{2.39b}$$

where the photoline is scanned in steps ΔU_{sp} of the spectrometer voltage and for equal time intervals. Hence, A_D directly follows from the experimental data, but due to equ. (2.39a) it is also directly related to the photoionization cross section σ and other relevant experimental and apparatus parameters. Equs. (2.39) provide the basis for all quantitative determinations of relative intensities of photolines within a given spectrum of ejected electrons, because for any line (i) one has

$$A_D(i) = \text{const } \sigma(i). \tag{2.40}$$

Based on this relation, the relative *partial* cross sections $\sigma(i)$ for photoionization can be determined. Using known values for the *total* absorption cross section σ_{tot}, these relative values can then be placed on an absolute scale, because in the energy range of interest the total absorption cross section is identical to the total

† In this final equation, the factor 1.06 in equ. (2.28d) has been approximated to unity because for the electron spectrometer shown in Fig. 1.17 ray-tracing calculations lead to the result that the spectrometer function deviates slightly from the Gaussian shape (see Fig. 5.25) and gives the value $k = 1.00$ instead of $k = 1.06$ as required by an exact Gaussian function.

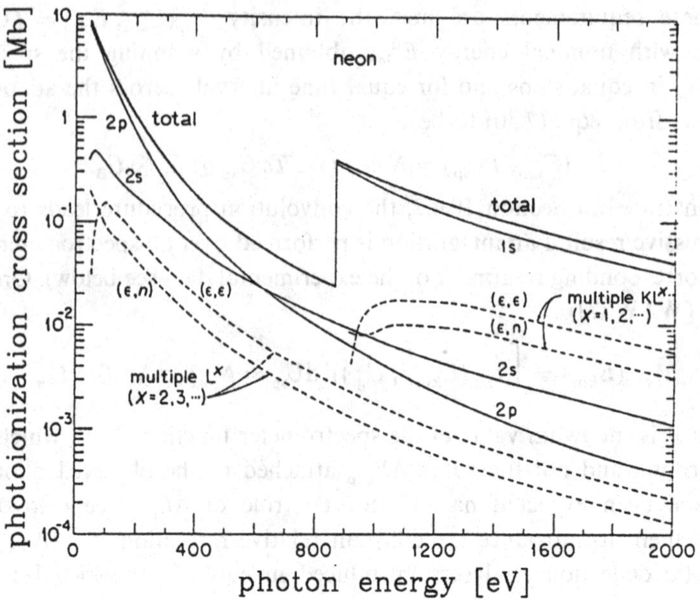

Figure 2.11 Total and partial cross sections for photoionization processes in neon; for details see main text. From [WKr74].

ionization cross section, and one has

$$\sigma_{\text{tot, abs.}} = \sigma_{\text{tot, ioniz.}} = \sum_{\text{all processes } i} \sigma(i). \tag{2.41}$$

Some partial photoionization cross sections, derived in this way for neon, are shown in Fig. 2.11 as a function of photon energy. The uppermost curve is the total absorption cross section. At the onset of the ionization thresholds for the ejection of 1s, 2s and 2p electrons this quantity shows the corresponding absorption edges (see the discussion related to equ. (2.11)). The partition of the total cross section into partial contributions $\sigma(i)$ clearly demonstrates that the dominant features are due to main photoionization processes described by the partial cross sections $\sigma^+(1s)$, $\sigma^+(2s)$, and $\sigma^+(2p)$ which correspond to the ejection of a 1s, a 2s, and a 2p electron, respectively. However, in addition, satellite transitions from multiple photoionization processes are also present. If these are related to a K-shell ionization process, they are called in Fig. 2.11 'multiple KL^X' where the symbol KL^X indicates that one electron from the K-shell and X electrons from the L-shell have been released by the photon interaction. Similarly, 'multiple L^X' stands for processes where X electrons from the L-shell are ejected. Furthermore, these two groups of multiple processes are classified with respect to ionization accompanied by excitation, (ε, n), or double ionization, (ε, ε). If one compares in Fig. 2.11 the magnitude of the partial cross sections for 2p, 2s and 1s photoionization at 1253.6 eV photon energy (Mg Kα radiation) and takes into account the different angular distribution parameters and the dispersion effect of the electron spectrometer, one gets the intensity ratios as observed in the spectrum of Fig. 2.4.

Finally, an estimate of the counting rate $I_{exp}(\max)$ at the maximum of the photoline can be derived. Approximating the measured photoline with a Gaussian distribution of fwhm $= \Delta E_{exp}$ one derives from equ. (2.28c)

$$A_{exp}(E^0_{kin}) \approx I_{exp}(\max)\,\Delta E_{exp}. \tag{2.42a}$$

If $A_{exp}(E^0_{kin})$ from equ. (2.38) is inserted, one obtains

$$I_{exp}(\max) \approx N_{ph} n_v \sigma\, T\, \Delta z\, \varepsilon \frac{\Delta E_{sp}}{\Delta E_{exp}}, \tag{2.42b}$$

where ΔE_{exp} may be approximated by

$$\Delta E_{exp} \approx \sqrt{[(B, \Gamma)^2 + \Delta E^2_{sp}]}, \tag{2.42c}$$

where (B, Γ) represents the fwhm value from the combined influences of B and Γ. This last relation clearly shows that the observed height of a photoline depends on the conditions of the experimental resolution and is, therefore, not well suited as a measure of intensities (for other operation modes of electron spectrometers see Section 4.2.4).

2.5 Analysis of angular distribution

The angular distribution of photoelectrons is described by the angular distribution parameter β (and the polarization properties of the incident light). For 1s, 2s, and 2p photoionization in neon, theoretical expressions for the β values have been presented in equs. (2.13), (2.14), and (2.15). In particular, $\beta(1s)$ and $\beta(2s)$ have the energy-independent value 2, and the corresponding photoelectron angular distribution for linearly polarized light was sketched in Fig. 1.4(*b*). The more interesting case of $\beta(2p)$ will be discussed here. Within the LS-coupling model two dipole-allowed channels contribute, $2p \to \varepsilon s$ and $2p \to \varepsilon d$, and in the expression for the β parameter they interfere. Reproducing equ. (2.15b) one has

$$\beta(2p) = \frac{2R^2_{\varepsilon d, 2p} - 4R_{\varepsilon d, 2p} R_{\varepsilon s, 2p} \cos(\Delta_{\varepsilon d} - \Delta_{\varepsilon s})}{R^2_{\varepsilon s, 2p} + 2R^2_{\varepsilon d, 2p}}. \tag{2.43}$$

The individual quantities needed in this expression for $\beta(2p)$ of neon are shown in Fig. 2.12. They have been obtained from a calculation based on the Herman–Skillman potential. The resulting $\beta(2p)$ parameter and, for convenience, also the partial cross section $\sigma(2p)$ are plotted in Fig. 2.13. The cross section has a slight maximum which is due to the maximum in the $R_{\varepsilon d, 2p}$ radial integral, and there the large value can be traced back to a favourable matching of the lobes in the $P_{2p}(r)$ and $P_{\varepsilon d}(r)$ radial wavefunctions (modified by the dipole operator r). As was discussed in connection with 1s photoionization (Fig. 2.1), for higher photon energies the continuum functions $P_{\varepsilon s}(r)$ and $P_{\varepsilon d}(r)$ oscillate more and more, thus reducing the dipole-modified overlap with the initial orbital function $P_{2p}(r)$. Hence,

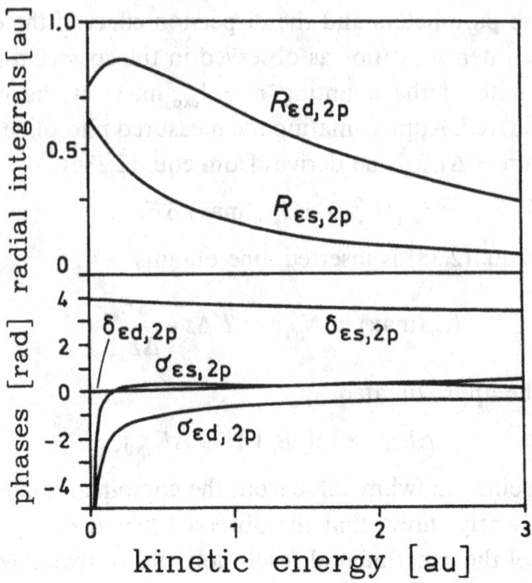

Figure 2.12 Radial integrals and phases for 2p photoionization in neon as functions of the kinetic energy of the photoelectron. The radial integrals $R_{\varepsilon d,2p}$ and $R_{\varepsilon s,2p}$ and the corresponding phases refer to the photoionization channels 2p → εd and 2p → εs, respectively. Instead of the total phase $\Delta_{\varepsilon\ell,2p}$ the individual contributions are shown (equ. (7.27)), namely the Coulomb phases $\sigma_{\varepsilon d,2p}$ and $\sigma_{\varepsilon s,2p}$, and the phases $\delta_{\varepsilon d,2p}$ and $\delta_{\varepsilon s,2p}$ from the short-range atomic potential. The data have been calculated using the Herman–Skillman potential with Latter correction [HSk63; Lat55]; the values are taken from [DSa73].

the 2p cross section also decreases with photon energy. This energy dependence of the radial integrals $R_{\varepsilon s,2p}$ and $R_{\varepsilon d,2p}$ also plays a role in the β(2p) parameter. However, as can be seen in equ. (2.43), these radial integrals occur in the numerator and in the denominator, and this leads to some cancellation of the energy dependences. A detailed analysis shows that the photon-energy dependence of the angular distribution parameter β(2p) in neon seen in Fig. 2.13 reflects the interference between the contributing partial waves and is due to the energy dependence of the relative phase $(\Delta_{\varepsilon d} - \Delta_{\varepsilon s})$.

A comparison between experimental and theoretical values for the β(2p) parameter in neon is shown in Fig. 2.14. (The corresponding comparison between experimental and theoretical values for the partial cross section σ(2p) of the 2p main photoline is omitted because most calculations give only σ(2p, main) + σ(2p, satellites), see Section 5.2.) The experimental data are given by the solid curve surrounded by a hatched area which takes into account the error bars. Theoretical results from advanced photoionization theories (many-body perturbation theory, R-matrix theory, and random-phase approximation) are represented by the other lines, and they are in close agreement with the experimental data (for details see [Sch86]). The theoretical β(2p) data of Fig. 2.13 are also close to the experimental values, except in the threshold region.

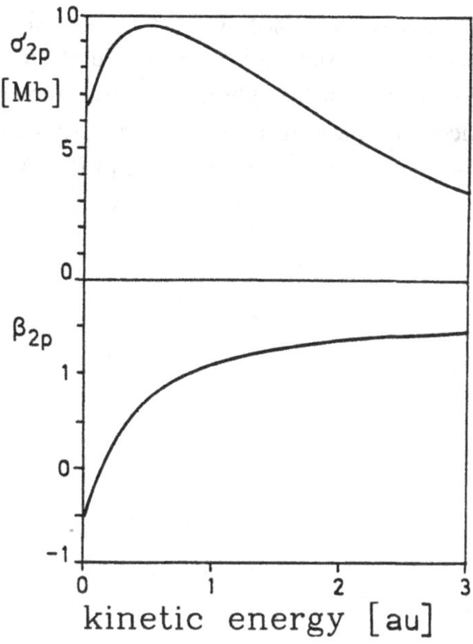

Figure 2.13 Theoretical values for the partial cross section σ_{2p} and angular distribution parameter β_{2p} of 2p photoionization in neon as functions of the kinetic energy of the photoelectron. The data have been calculated using the values in Fig. 2.12 (see [Man72]).

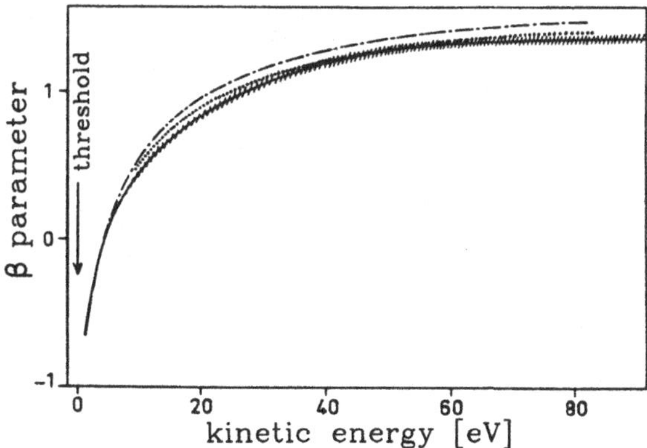

Figure 2.14 Compilation of data for the β parameter of 2p photoionization in neon as a function of the kinetic energy of the photoelectron. The solid line surrounded by the hatched area represents the experimental values including an error of ± 0.03. The other curves come from theoretical calculations (see main text and [Sch86]). From [Sch86].

Finally, a comment will be made concerning the relative simplicity of 2p photoionization in neon. Atoms other than neon have more interesting features in the observables which could be due to the photoionization of an electron with higher ℓ value, to potential barrier effects, to stronger influences from electron–electron interactions, to the appearance of Cooper [Coo62] and correlation minima, to increased importance of spin–orbit effects, and so on (see references in [Sch92a]). The necessarily more elaborate formulation for $np^{-1}\,{}^2P_J$ photoionization is presented in Section 8.2 where it is also shown which simplifying approximations are needed in order to derive the cross section σ and the β parameter in the forms given here in equs. (2.15) and (2.43).

3

Auger spectrum of neon following 1s ionization

3.1 Description of the K–LL Auger spectrum

Inner-shell ionization is accompanied by subsequent radiative and non-radiative decay. In the context of electron spectrometry, the non-radiative or Auger decay is of special interest, because the emitted Auger electron can be detected. After some remarks on the general description and classification of Auger transitions following 1s ionization in neon, the calculation of K–LL Auger transition rates and the formulation of intermediate coupling in the final ionic state of the K–LL Auger transition will be addressed. This information then provides the basis for a detailed analysis of the experimental K–LL Auger spectrum of neon which is organized similarly to the previous discussion of photoelectrons: namely, with respect to line positions, linewidths, line intensities, and angular distributions.

3.1.1 General aspects

In addition to the photoelectron lines, other discrete structures appear in the electron spectrum of neon if the photon energy is higher than the threshold for 1s ionization. These lines are due to radiationless transitions called *Auger* transitions [Aug25]; the 1s-hole created by photoionization is filled by a subsequent two-electron transition induced by the Coulomb interaction between the electrons. This interaction causes one outer-shell electron to jump down, filling the 1s-hole, simultaneously ejecting another outer-shell electron, the Auger electron, into the continuum. This process has been sketched schematically in Figs. 1.3 and 2.5. As has been discussed in connection with the latter figure, $K-L_1L_1$, $K-L_1L_{2,3}$, and $K-L_{2,3}L_{2,3}$ Auger transitions can be distinguished, and these all belong to the *main* (or *diagram*) Auger transitions denoted by the symbol K–LL. (For radiationless transitions starting in shells other than the K-shell, the possibility exists that one or even both of the two holes in the final ionic state belong to the same principal shell as the initial vacancy. These special transitions are termed *Coster–Kronig* [CKr35] and *super* Coster–Kronig [McG72] transitions, respectively. Examples are the $L_1-L_{2,3}M_1$ Coster–Kronig transitions in magnesium, and the $N_{2,3}-N_{4,5}N_{4,5}$ super Coster–Kronig transitions in xenon.)

Because the kinetic energies of photoelectrons depend on the photon energy, equ. (1.29a), but the kinetic energies of Auger electrons are independent of the photon energy, equ. (1.29b), the processes can be easily distinguished in an experimental spectrum of ejected electrons if the kinetic energies are observed at

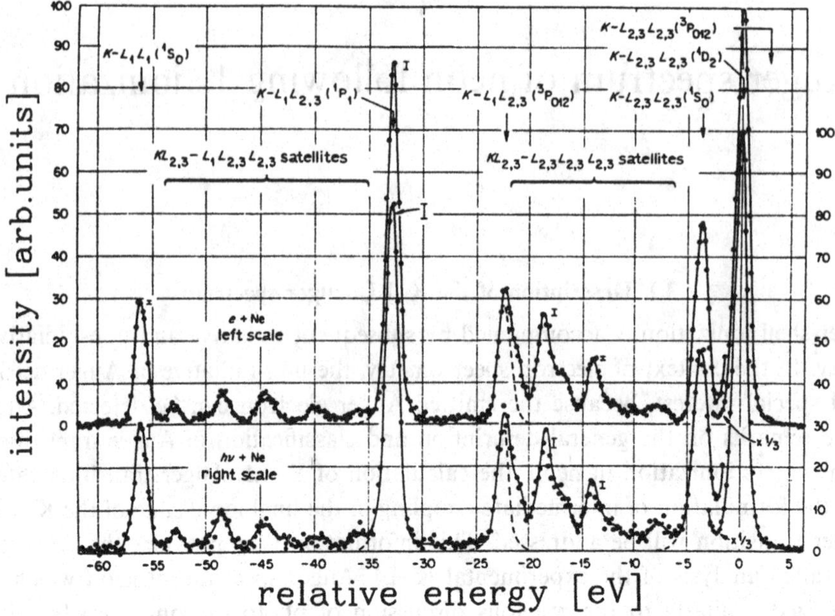

Figure 3.1 Comparison of neon K Auger spectra induced by the impact of 3.2 keV electrons (upper trace) and 1.5 keV photons (lower trace). The zero on the relative energy scale corresponds to 804.5 eV kinetic energy. The small symbols close to the peak maxima indicate the error bars at peak intensity. The main Auger lines are classified as $K\text{–}LL(^{2S+1}L_J)$; the peaks collectively called 'satellites' are predominantly due to Auger transitions from $KL_{2,3}$ photosatellites. For details see main text. Reprinted from *Phys. Lett.*, **31A**, Krause *et al.*, 81 (1970) with kind permission of Elsevier Science – NL, Sara Burgerhartstraat 25, 1055 KV Amsterdam, The Netherlands.

different photon energies: photoelectron lines move with the photon energy, but Auger lines remain stationary. This statement, however, is correct only if thc two-step model applies in which the first step of inner-shell ionization and the second step of Auger decay can be treated separately, a condition which is well fulfilled for 1s ionization in neon with subsequent K–LL Auger decay (provided that the photon energy is high enough, see Section 5.5). (For the two-step model see Section 8.3; for a striking example of a strong deviation see [WOh76]; for post-collision interaction see Sections 4.5.5 and 5.5.) Within the two-step model, the Auger spectrum is also independent of the ionization mechanism. Therefore, Auger spectra caused by photon or electron impact are equivalent, as demonstrated in Fig. 3.1, and most of the results for the K-shell Auger spectrum of neon described below are based on Auger spectra obtained by electron impact. One should, however, keep in mind that Auger electron spectrometry using synchrotron radiation has several advantages over electron impact studies and that the full equivalence of Auger spectra following electron or photon impact fails for a more detailed comparison. In this context the following points need to be mentioned:

(i) In the case of electron impact, the energy transfer to the atom covers a broad energy range (in practice between zero and the energy of the primary electron beam). As a result, not only 1s ionization processes are possible, but also 1s → $n\ell$ excitations. The autoionizing decay of such excited states produces additional structures in the electron spectrum (Auger satellites of category B in the nomenclature of [KCM71]). In contrast, electron spectrometry with synchrotron radiation allows separate studies of excitation and ionization processes if the photon energy is selected appropriately.

(ii) Related to the broad range of energy transfers in electron impact are large ranges of kinetic energies of the scattered and primarily ejected electrons. Hence, in non-coincidence experiments the electron spectrometer will record an electron spectrum in which the Auger lines are superimposed on a rather large, smooth 'background' of scattered and primarily ejected electrons. In contrast, Auger spectra induced by photoionization have a much smaller 'background' (continuous multiple ionizations) and, in addition, the spectrum of photoelectrons provides valuable direct information about the initial state of the Auger process.

(iii) The effects of postcollision interaction are different for photon and electron impact ionization (see Sections 4.5.5 and 5.5). This leads to differences in the main and satellite Auger lines as observed in both cases.

(iv) At lower energies of the incident photon or electron, in particular, the production mechanism for satellites may depend on the nature of the incoming particle.

Even though the detailed discussion in the next section will be restricted to the K–LL *main* Auger transitions, information about Auger *satellites* is also necessary, because otherwise a correct description of the main processes is not possible. Since Auger transitions follow inner-shell ionization, in the present case induced by photoionization, Auger satellites can in lowest order be due to photosatellites (two-electron processes during photoionization), to double Auger transitions (three-electron processes during Auger decay), or to a combination. The first two cases are demonstrated with the help of Fig. 3.2. Spectra of ejected electrons are shown in the centre of this figure, at the top for the total spectrum of photoelectrons, at the bottom (at a different energy range) for the total spectrum of Auger electrons, and in parts (*a*)–(*e*) for the individual constituents of the Auger spectrum (typical main and satellite Auger lines). Alongside these individual Auger spectra are shown pictures for the change of individual electron orbitals (placed at the right- and left-hand sides, respectively).

Looking at the total photoelectron spectrum at the top of Fig. 3.2, one sees the huge main photoline from 1s photoionization, some much smaller lines (discrete satellites with electron configurations $1s2s^22p^5np$ and others), and a continuum (continuous satellite processes with electron configurations $1s2s^22p^5$). Following first the Auger decays from the intense 1s photoline, the possible decay branches

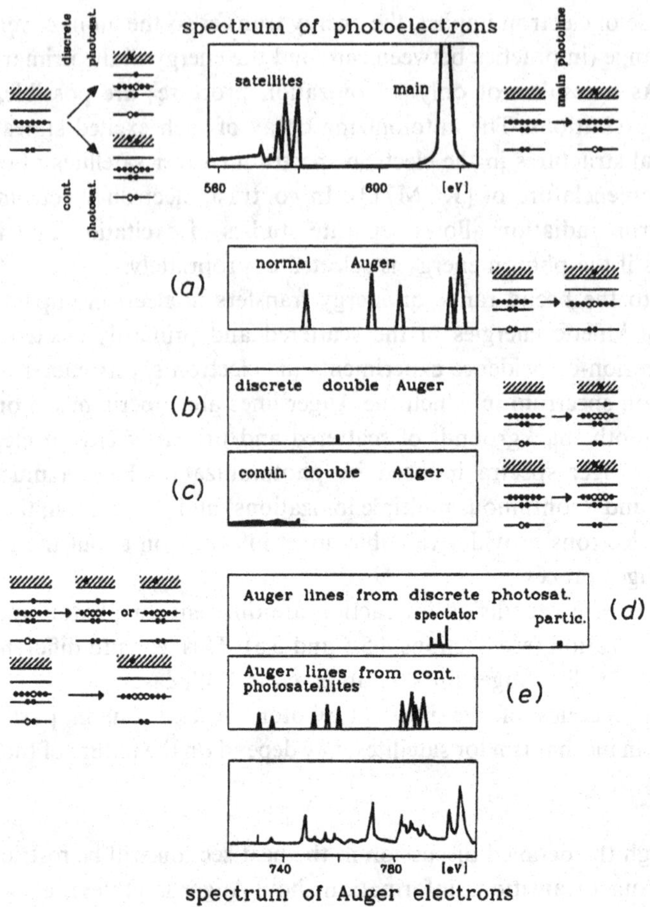

Figure 3.2 Illustration of the connection between inner-shell photoionization and subsequent Auger decay of neon. The photoelectron spectrum is shown in the centre at the top, the spectrum of Auger electrons in the centre at the bottom. Between these spectra individual components of the Auger spectrum are shown (parts (*a*)–(*e*)) which together add up to the complete Auger spectrum at the bottom. The intensities of all spectra are given on a logarithmic scale (the participator Auger decay from discrete photosatellites is magnified by a factor of 10). Only a selection of all the possible lines is included, because some lines have not yet been identified (e.g., lines from discrete double Auger decay). Alongside the electron spectra are shown the energy-level diagrams (not to scale) for electrons in the 1s, 2s, 2p shells, in an excited orbital or in the continuum (shaded area), sometimes the 2s and 2p levels are combined. These orbitals are occupied by electrons (filled circles) or holes (open circles) in order to show the changes due to photoionization and Auger decay. A detailed explanation for these transitions is given in the main text.

are listed on the right-hand side of the figure. If the 1s-hole is filled by an ordinary K–LL Auger transition, electrons from the 2s and/or 2p shells are involved, and their possible states give in LS-coupling the five normal Auger lines (*main* or *diagram* lines) shown in Fig. 3.2(*a*). In addition, the 1s hole-state can also decay by *double* Auger transitions where three electrons are involved and electron correlations play an essential role. Two classes can be distinguished, *discrete* double Auger decay in which the third electron is excited to an unoccupied orbital, K–LLL* transitions (Fig. 3.2(*b*)), and *continuous* double Auger decay in which this electron goes into the continuum, K–LLL transitions (Fig. 3.2(*c*)).

The discrete and continuous photosatellites at the top of Fig. 3.2, which accompany K-shell ionization, lead to KL* and KL initial-state configurations for the subsequent Auger decay. The decay of KL* can follow either of two paths: the excited electron can stay in a bound orbital when the 1s-hole is filled (*spectator* transition†) leading to KL*–LLL* Auger satellites, or it can take part in the transition (*participator* or *involved* transition) leading to KL*–LL Auger satellites. Both possibilities are shown in Fig. 3.2(*d*). Finally, the KL initial-state configuration with one hole in each of the K- and L-shells leads to the intense KL–LLL Auger transitions shown in Fig. 3.2(*e*).

The result of all the individual processes is the total Auger spectrum shown at the bottom of Fig. 3.2. It can be seen that even for such a simple case as 1s photoionization in neon the resulting Auger spectrum is very complicated, and about 90 lines have been analysed [KCM71, ATW90]. In this work the following classifications are made:

category A: normal Auger lines arising from 1s-electron ionization;

categories Bα and Bβ: satellite lines arising from 1s $\rightarrow n\ell$ electron excitations with subsequent spectator or participator transition, respectively;

category Cα and Cβ: KL*–LLL* Auger transitions arising from 1s, 2s $\rightarrow \infty$, $n\ell$ and 1s, 2p $\rightarrow \infty$, $n\ell$ two-electron processes of ionization and excitation with subsequent Auger decay where the excited electron is involved or acts as spectator, respectively;

category D: KL–LLL Auger transitions arising from 1s, 2s $\rightarrow \infty, \infty$ and 1s, 2p $\rightarrow \infty, \infty$ two-electron ionization processes with subsequent Auger decay.

In this section only the normal K–LL Auger transitions will be considered further.

3.1.2 Calculation of transition rates

Within the two-step model, one can say that the intermediate photoionized state is the initial state for the Auger transition. For the K–LL spectrum of neon this initial state is described by $1s2s^22p^6\ {}^2S^e_{1/2}$. For the final state the possible electron configurations of the ion were shown in Fig. 2.5. Within the *LS*-coupling scheme which applies well to neon, these electron configurations yield the following final

† Also called *resonance Auger* or *autoionization spectator* transition; the 'spectator' electron may also change its orbital, $n\ell \rightarrow n'\ell'$, mostly with $\ell' = \ell$ (*shake-modified* spectator transitions); see Section 5.1.2.1.

states:

$$1s^22p^6\ {}^1S^e \tag{3.1a}$$

$$1s^22s2p^5\ {}^1P^o \quad \text{or} \quad 1s^22s2p^5\ {}^3P^o \tag{3.1b}$$

$$1s^22s^22p^4\ {}^{2S+1}L^o \tag{3.1c}$$

In equ. (3.1c) the possible final ionic states arising from the different couplings of angular momenta of the four 2p-electrons cannot immediately be discerned. Therefore, this case will be discussed in more detail. (A rather simple description is given here. For a more powerful method, see Section 7.2.) First, the coupling of four 2p-electrons is replaced by the equivalent coupling of the two missing 2p-electrons. This reduction of the four-electron problem to a two-electron problem (more correctly to a two-hole problem) is possible, because the angular momenta **L** and **S** of the four-electron states must be equal to those of the two-electron states necessary to reproduce the ${}^1S_0^e$ state of the closed-shell six-electron system. Second, one seeks all possibilities for the determinantal wavefunctions of two 2p-electrons, taking into account the different combinations for the individual magnetic quantum numbers m_ℓ and m_s which are allowed by the Pauli principle. These possibilities give the Slater determinantal wavefunctions listed in Table 3.1 together with the resulting values of $M_L = (m_\ell(1) + m_\ell(2))$ and $M_S = (m_s(1) + m_s(2))$. (In this context it should be noted that the Pauli principle prohibits the two 2p-electrons, which are called *equivalent* electrons, from coupling to ${}^3S^e$, ${}^3D^e$ and ${}^1P^e$ states. These states, however, can occur for two *non*-equivalent electrons, $npn'p$ with $n \neq n'$, because then all combinations from the couplings $\boldsymbol{\ell}_1 + \boldsymbol{\ell}_2 \to \mathbf{L}$ and $\mathbf{s}_1 + \mathbf{s}_2 \to \mathbf{S}$ are allowed which give $L = 0, 1, 2$ and $S = 0, 1$ (see

Table 3.1. *Slater determinantal wavefunctions for the 2p²-electron configuration* (*see* [*Sla60*]).

Number	Slater wavefunction	M_L	M_S
$\tilde{\Phi}_1$	$\{2p\,1^+, 2p\,1^-\}$	2	0
$\tilde{\Phi}_2$	$\{2p\,1^+, 2p\,0^+\}$	1	1
$\tilde{\Phi}_3$	$\{2p\,1^+, 2p\,0^-\}$	1	0
$\tilde{\Phi}_4$	$\{2p\,1^+, 2p-1^+\}$	0	1
$\tilde{\Phi}_5$	$\{2p\,1^+, 2p-1^-\}$	0	0
$\tilde{\Phi}_6$	$\{2p\,1^-, 2p\,0^+\}$	1	0
$\tilde{\Phi}_7$	$\{2p\,1^-, 2p\,0^-\}$	1	−1
$\tilde{\Phi}_8$	$\{2p\,1^-, 2p-1^+\}$	0	0
$\tilde{\Phi}_9$	$\{2p\,1^-, 2p-1^-\}$	0	−1
$\tilde{\Phi}_{10}$	$\{2p\,0^+, 2p\,0^-\}$	0	0
$\tilde{\Phi}_{11}$	$\{2p\,0^+, 2p-1^+\}$	−1	1
$\tilde{\Phi}_{12}$	$\{2p\,0^+, 2p-1^-\}$	−1	0
$\tilde{\Phi}_{13}$	$\{2p\,0^-, 2p-1^+\}$	−1	0
$\tilde{\Phi}_{14}$	$\{2p\,0^-, 2p-1^-\}$	−1	−1
$\tilde{\Phi}_{15}$	$\{2p-1^+, 2p-1^-\}$	−2	0

Section 7.2).) These wavefunctions are not yet eigenfunctions of the angular momentum and spin operators, **L** and **S**, respectively. Therefore, in a third step these eigenfunctions have to be constructed as linear combinations of the determinantal functions of Table 3.1. Some rules for such angular momentum couplings are given in Section 7.2. However, in the present example it is possible to derive the necessary information without going into detail, by looking at the M_L and M_S values of the wavefunctions in Table 3.1 (see [Sla60]). By considering the wavefunctions $\tilde{\Phi}_1$ and $\tilde{\Phi}_{15}$, one can conclude that a $^1D^e$ state must exist. Similarly, from the wavefunctions $\tilde{\Phi}_2$ and $\tilde{\Phi}_{14}$, the $^3P^e$ state can be inferred. A ^1D state has five eigenfunctions (for $M_L = 2, 1, 0, -1, -2$ with $M_S = 0$), and a ^3P state has nine eigenfunctions (for the corresponding M_L and M_S combinations), i.e., these two states already cover 14 eigenfunctions. Since 15 determinantal functions can yield only 15 eigenfunctions, only one eigenfunction is left which must then be a ^1S state. These 15 eigenfunctions which belong to the $^1D^e$, $^3P^e$, and $^1S^e$ final ionic states of the $2p^2$-electron configurations are shown in Table 3.2. (If atoms heavier than neon are considered, the *LS* eigenfunctions in Table 3.2 have to be augmented to *LSJ*-coupled functions (see equ (7.41)) or even to wavefunctions describing intermediate coupling (see below).)

After this excursion one returns to the neon K–LL Auger transitions and their final ionic states, given in equ. (3.1). Within the *LS*-coupling scheme one expects the following six transitions which are classified according to X-ray nomenclature (see Fig. 2.2):

$$\text{K–L}_1\text{L}_1\ ^1\text{S}^\text{e}; \tag{3.2a}$$

$$\text{K–L}_1\text{L}_{2,3}\ ^1\text{P}^\text{o}; \tag{3.2b}$$

$$\text{K–L}_1\text{L}_{2,3}\ ^3\text{P}^\text{o}; \tag{3.2c}$$

$$\text{K–L}_{2,3}\text{L}_{2,3}\ ^1\text{S}^\text{e}; \tag{3.2d}$$

$$\text{K–L}_{2,3}\text{L}_{2,3}\ ^1\text{D}^\text{e}; \tag{3.2e}$$

$$\text{K–L}_{2,3}\text{L}_{2,3}\ ^3\text{P}^\text{e}. \tag{3.2f}$$

The Auger transition is caused by the Coulomb interaction between the electrons, i.e., by the operator of equ. (1.28b) which is given in atomic units by:

$$Op = \sum_{i<j} \frac{1}{r_{ij}}. \tag{3.3}$$

(Using the basis functions which follow from the approximate Hamiltonian H^0 of equ. (1.3), it is the residual interaction $H - H^0$ which causes the Auger transitions. This operator, however, reduces to the Coulomb interaction if more than one electron changes its orbital.) Within the *LS*-coupling scheme this transition operator requires the following selection rules

$$\Delta L = \Delta S = \Delta M_L = \Delta M_S = 0 \tag{3.4a}$$

and no change of parity, i.e.,

$$\Delta\pi = \pi_i \pi_f = (+1). \tag{3.4b}$$

Table 3.2. *Angular momentum eigenfunctions in LS-coupling for the electron configuration $2p^2$. From J. C. Slater, Quantum theory of atomic structure (1960), with the kind permission of J. F. Slater and The McGraw-Hill Companies.*

$$\tilde{\Psi}(2p^2\ {}^1D\ M_L = 2\ M_S = 0) = \{2p\,1^+, 2p1^-\}$$

$$\tilde{\Psi}(2p^2\ {}^1D\ M_L = 1\ M_S = 0) = \frac{1}{\sqrt{2}}(\{2p\,1^+, 2p\,0^-) - \{2p\,1^-, 2p\,0^+\})$$

$$\tilde{\Psi}(2p^2\ {}^1D\ M_L = 0\ M_S = 0) = \frac{1}{\sqrt{6}}(2\{2p\,0^+, 2p\,0^-\} + \{2p\,1^+, 2p\,-1^-\} - \{2p\,1^-, 2p\,-1^+\})$$

$$\tilde{\Psi}(2p^2\ {}^1D\ M_L = -1\ M_S = 0) = \frac{1}{\sqrt{2}}(\{2p\,0^+, 2p\,-1^-\} - \{2p\,0^-, 2p\,-1^+\})$$

$$\tilde{\Psi}(2p^2\ {}^1D\ M_L = -2\ M_S = 0) = \{2p\,-1^+, 2p\,-1^-\}$$

$$\tilde{\Psi}(2p^2\ {}^3P\ M_L = 1\ M_S = 1) = \{2p\,1^+, 2p\,0^+\}$$

$$\tilde{\Psi}(2p^2\ {}^3P\ M_L = 1\ M_S = 0) = \frac{1}{\sqrt{2}}(\{2p\,1^-, 2p\,0^+\} + \{2p\,1^+, 2p\,0^-\})$$

$$\tilde{\Psi}(2p^2\ {}^3P\ M_L = 1\ M_S = -1) = \{2p\,1^-, 2p\,0^-\}$$

$$\tilde{\Psi}(2p^2\ {}^3P\ M_L = 0\ M_S = 1) = \{2p\,1^+, 2p\,-1^+\})$$

$$\tilde{\Psi}(2p^2\ {}^3P\ M_L = 0\ M_S = 0) = \frac{1}{\sqrt{2}}(\{2p\,1^-, 2p\,-1^+\} + \{2p\,1^+, 2p\,-1^-\})$$

$$\tilde{\Psi}(2p^2\ {}^3P\ M_L = 0\ M_S = -1) = \{2p\,1^-, 2p\,-1^-\}$$

$$\tilde{\Psi}(2p^2\ {}^3P\ M_L = -1\ M_S = 1) = \{2p\,0^+, 2p\,-1^+\}$$

$$\tilde{\Psi}(2p^2\ {}^3P\ M_L = -1\ M_S = 0) = \frac{1}{\sqrt{2}}(\{2p\,0^-, 2p\,-1^+\} + \{2p\,0^+, 2p\,-1^-\}))$$

$$\tilde{\Psi}(2p^2\ {}^3P\ M_L = -1\ M_S = -1) = \{2p\,0^-, 2p\,-1^-\}$$

$$\tilde{\Psi}(2p^2\ {}^1S\ M_L = 0\ M_S = 0) = \frac{1}{\sqrt{3}}(\{2p\,1^+, 2p\,-1^-\} - (2p\,1^-, 2p\,-1^+\} - \{2p\,0^+, 2p\,0^-\})$$

The selection rules have to be fulfilled for the transition from the $1s2s^22p^6\ {}^2S^e$ initial state to the possible final states. Thus, the final state contains one of the final ionic states listed in Table 3.2 and the wavefunction $\psi^{(-)}_{\kappa m_s}$ for the emitted Auger electron in its partial wave expansion (see equ. (7.28b)). Due to the selection rules, only a few ℓ values from the partial wave expansion contribute. In the present case there is only one possibility which will be characterized by $\varepsilon\ell$. Therefore, one gets for the neon Auger transitions in *LS*-coupling

$$1s2s^22p^6\ {}^2S^e \rightarrow 1s^2\ \ 2p^6({}^1S^e)\varepsilon s\ {}^2S^e \text{ is allowed,} \tag{3.5a}$$

$$1s2s^22p^6\ {}^2S^e \rightarrow 1s^22s2p^5({}^1P^o)\varepsilon p\ {}^2S^e \text{ is allowed,} \tag{3.5b}$$

$$1s2s^22p^6\ {}^2S^e \rightarrow 1s^22s2p^5({}^3P^o)\varepsilon p\ {}^2S^e \text{ is allowed,} \tag{3.5c}$$

$$1s2s^22p^6\ {}^2S^e \rightarrow 1s^22s^22p^4({}^1S^e)\varepsilon s\ {}^2S^e \text{ is allowed,} \tag{3.5d}$$

$$1s2s^22p^6\ {}^2S^e \rightarrow 1s^22s^22p^4({}^1D^e)\varepsilon d\ {}^2S^e \text{ is allowed,} \tag{3.5e}$$

$$1s2s^22p^6\ {}^2S^e \rightarrow 1s^22s^22p^4({}^3P^e)\varepsilon\ell\ {}^2S^e \text{ is forbidden.} \tag{3.5f}$$

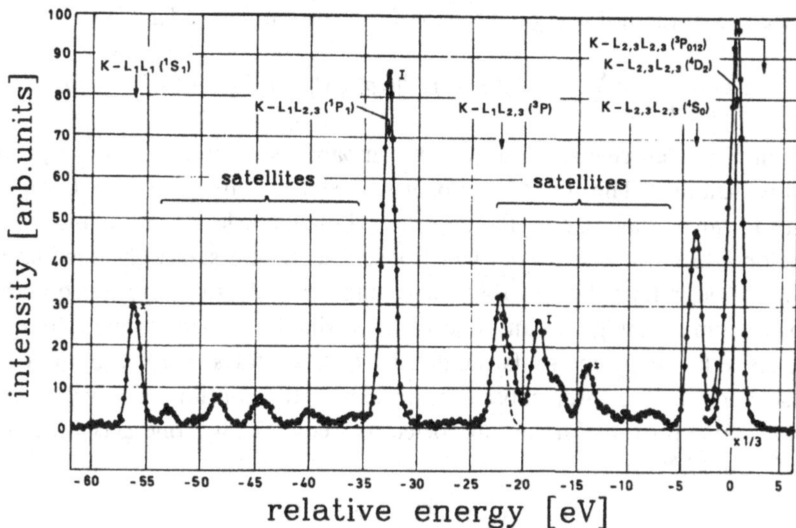

Figure 3.3 The K Auger spectrum of neon (shown in Fig. 3.1) caused by electron impact. See also [KMe66, KCM71, ATW90]. Reprinted from *Phys. Lett.*, **31A**, Krause *et al.*, 81 (1970) with kind permission from Elsevier Science – NL, Sara Burgerhartstraat 25, 1055 KV Amsterdam, The Netherlands.

The last transition is forbidden because the demands from the angular momentum coupling and the parity requirement are mutually exclusive: the coupling of the orbital angular momenta requires the vector addition $\mathbf{L} + \boldsymbol{\ell} = 0$ with $L = 1$ and hence also $\ell = 1$; on the other hand, the parity selection rule requires ℓ = even, and both conditions cannot be fulfilled simultaneously. Therefore, only five transitions are expected for the K–LL Auger spectrum in neon, and these can be identified in Fig. 3.3.

The theoretical expression for the transition probability will be evaluated for the simplest one of these Auger transitions, K–L_1L_1 $^1S_0^e$. The transition operator Op of equ. (3.3) connects the wavefunctions of the initial and final states, $|J_i M_i\rangle$ and $|J_f M_f, \boldsymbol{\kappa} m_s^{(-)}\rangle$ given by

$$|J_i M_i\rangle = \{1s0^{M_i}, 2s0^+, 2s0^-, 2p1^+, 2p1^-, 2p0^+, 2p0^-, 2p-1^+, 2p-1^-\}, \quad (3.6a)$$

$$|J_f M_f, \boldsymbol{\kappa} m_s^{(-)}\rangle = \{1s0^+, 1s0^-, 2p1^+, 2p1^-, 2p0^+, 2p0^-, 2p-1^+, 2p-1^-, \psi^{(-)}_{\boldsymbol{\kappa} m_s}\}, \quad (3.6b)$$

where J_i and M_i stand for the angular momentum and its projection of the initial state of the Auger decay, respectively. Since this state is a 1s hole-state, J_i is equal to 1/2, and M_i is equal to the spin projection number of the initial 1s orbital (cf. the first orbital in the determinantal wavefunction). $\psi^{(-)}_{\boldsymbol{\kappa} m_s}$ represents the single-particle function of the Auger electron ejected into the continuum (cf. the corresponding function of a photoelectron in equ. (2.2b)).

According to Fermi's golden rule, one derives for the transition rate $P(\hat{\boldsymbol{\kappa}})$ which depends on the direction $\hat{\boldsymbol{\kappa}}$ of the emitted Auger electron (measured against a preselected axis) the following expression (in atomic units) [Wen27]

$$P(\hat{\boldsymbol{\kappa}}) = 2\pi \frac{1}{2J_i + 1} \sum_{M_i} \sum_{M_f} \sum_{m_s} |\langle J_f M_f, \boldsymbol{\kappa} m_s^{(-)} | Op | J_i M_i \rangle|^2 \delta(E_I^+ - E_I^{++} - \varepsilon)\rho. \quad (3.7a)$$

(For Fermi's golden rule see equ. (8.11); compare also equ. (2.3) for the process of photoionization.) The summation over M_i together with the factor $1/(2J_i + 1)$ accounts for an averaging over the unobserved magnetic levels in the initial state, because one does not know from which level the system starts; the summations over M_f and m_s reflect the fact that these magnetic substrates are not observed, and the system might go to any one of them (in the present example one has $M_f = 0$). The factor ρ represents the density of final states in the continuum. Its value depends on the form chosen to normalize the continuum function $\psi_{\boldsymbol{\kappa} m_s}^{(-)}$. Using the same form as for the photoelectron, equ. (7.28b), one gets (in atomic units, cf. equ. (7.28g))

$$\rho = \kappa^2 \, d\kappa \, d\Omega_\kappa = \kappa \, d\varepsilon \, d\Omega_\kappa, \quad (3.8)$$

where the direction $\hat{\boldsymbol{\kappa}}$ of the emitted Auger electron is described by the polar and azimuthal angles measured against the preselected quantization axis. The δ-function ensures energy conservation (cf equ. (1.29b)). It can be eliminated by an integration over the energy parameter ε. (It is implicitly understood that the occurrence of a δ-function is always accompanied by an integration over the continuous parameter in the δ-function.) This integration fixes the kinetic energy ε of the Auger electron to $\varepsilon = E_I^+ - E_{II}^{++}$, and its wavenumber κ to $\kappa = \sqrt{(2\varepsilon)}$, and one describes the remaining matrix element as 'on-the-energy-shell'. With these requirements equ. (3.7a) can be replaced by

$$P(\hat{\boldsymbol{\kappa}}) = 2\pi \frac{1}{2J_i + 1} \kappa \sum_{M_i} \sum_{M_f} \sum_{m_s} |\langle J_f M_f, \boldsymbol{\kappa} m_s^{(-)} | Op | J_i M_i \rangle|^2 \, d\Omega_\kappa. \quad (3.7b)$$

Because the Coulomb operator is a two-particle operator, the transition matrix element M_{fi} is non-zero only for cases in which at most two orbitals differ in the initial- and final-state wavefunctions. For normal Auger transitions it will turn out that these are just the electron orbitals used to characterize the Auger transition, including the Auger electron itself. To show this for the K–L_1L_1 $^1S_0^e$ transition one starts with the matrix element

$$M_{fi} = \langle J_f M_f, \boldsymbol{\kappa} m_s^{(-)} | \sum_{k<\ell} \frac{1}{r_{k\ell}} | J_i M_i \rangle. \quad (3.9a)$$

For a calculation of this matrix element one first changes the order of orbitals in such a way that the two *different* orbitals in the determinantal wavefunctions are at the same positions. Since in the expansion of the continuum function into partial waves, equ. (3.5a), only εs is allowed, one gets

$$M_{fi} = \langle \{1s0^+, 1s0^-, \psi(\varepsilon s, m_s), 2p1^+, 2p1^-, 2p0^+, 2p0^-, 2p-1^+, 2p-1\} | Op | \{1s0^{M_i}, 2s0^+, 2s0^-, 2p1^+, 2p1^-, 2p0^+, 2p0^-, 2p-1^+, 2p-1^-\} \rangle. \quad (3.9b)$$

From this matrix element it can be seen: first, that the 2p orbitals on either side of the matrix element are the same; second, that two 2s orbitals present on the right-hand side are absent on the left-hand side; and third, that on the left-hand side two orbitals are present, $\psi(\varepsilon s, m_s)$ and one of the $1s0^{m_s'}$- orbitals, which are absent on the right-hand side. (One of the spin-orbitals, $1s0^+$ or $1s0^-$, on the left-hand side must coincide with $1s0^{M_i}$ on the right-hand side.) Therefore, exactly two different orbitals remain on each side of this matrix element. They are connected by the Coulomb operator and determine the value of the matrix element. As with photoionization, the two electrons relevant for the Auger transition are called *active* electrons while the remaining electrons which lead to an overlap integral are termed *passive* electrons. Hence, the calculation yields

$$\begin{aligned} M_{fi} &= \langle\{1s0^-, \psi(\varepsilon s, m_s)\}|1/r_{12}|\{2s0^+, 2s0^-\}\rangle\langle\text{overlap1}\rangle\, \delta_{M_i,+1/2} \\ &\quad - \langle\{1s0^+, \psi(\varepsilon s, m_s)\}|1/r_{12}|\{2s0^+, 2s0^-\}\rangle\langle\text{overlap2}\rangle\, \delta_{M_i,-1/2} \end{aligned} \tag{3.9c}$$

(note that for the second term a change in the order of the two 1s-electrons on the left-hand side leads to a minus sign) with the overlap integrals

$$\begin{aligned} \langle\text{overlap 1 or 2}\rangle &= \langle 1s0^- \text{ or } 1s0^+, 2p1^+, 2p1^-, 2p0^+, 2p0^-, 2p-1^+, 2p-1^- | 1s0^- \\ &\quad \text{or } 1s0^+, 2p1^+, 2p1^-, 2p0^+, 2p0^-, 2p-1^+, 2p-1^-\rangle. \end{aligned} \tag{3.9d}$$

This result shows that the original matrix element containing the orbitals of all electrons factorizes into a two-electron Coulomb matrix element for the active electrons and an overlap matrix element for the passive electrons. Within the frozen atomic structure approximation, the overlap factors yield unity because the same orbitals are used for the passive electrons in the initial and final states. Considering now the Coulomb matrix element, one uses the fact that the Coulomb operator does not act on the spin. Therefore, the m_s value in the wavefunction of the Auger electron is fixed, and one treats the matrix element M_{fi} as

$$M_{fi} = M_1 + M_2 \tag{3.9e}$$

with

$$M_1 = \langle\{1s0^-, \psi(\varepsilon s0^+)\}|1/r_{12}|\{2s0^+, 2s0^-\}\rangle\, \delta_{m_s,+1/2}\, \delta_{M_i,+1/2} \tag{3.10a}$$

and

$$M_2 = \langle\{1s0^+, \psi(\varepsilon s0^-)\}|1/r_{12}|\{2s0^+, 2s0^-\}\rangle\, \delta_{m_s,-1/2}\, \delta_{M_i,-1/2}. \tag{3.11a}$$

The first expression yields

$$\begin{aligned} M_1 = \tfrac{1}{2}[&\langle\varphi_{1s0^-}(1)\psi_{\varepsilon s0^+}(2)|1/r_{12}|\varphi_{2s0^+}(1)\varphi_{2s0^-}(2)\rangle \\ &- \langle\varphi_{1s0^-}(1)\psi_{\varepsilon s0^+}(2)|1/r_{12}|\varphi_{2s0^+}(2)\varphi_{2s0^-}(1)\rangle \\ &- \langle\varphi_{1s0^-}(2)\psi_{\varepsilon s0^+}(1)|1/r_{12}|\varphi_{2s0^+}(1)\varphi_{2s0^-}(2)\rangle \\ &+ \langle\varphi_{1s0^-}(2)\psi_{\varepsilon s0^+}(1)|1/r_{12}|\varphi_{2s0^+}(2)\varphi_{2s0^-}(1)\rangle]\, \delta_{m_s,+1/2}\, \delta_{M_i,+1/2}. \end{aligned} \tag{3.10b}$$

Because the spin value of an individual electron also remains unaffected by the Coulomb interaction, the first and fourth terms vanish, and one gets

$$\begin{aligned} M_1 = -\tfrac{1}{2}[&\langle\varphi_{1s0}(1)\psi_{\varepsilon s0}(2)|1/r_{12}|\varphi_{2s0}(1)\varphi_{2s0}(2)\rangle \\ &+ \langle\varphi_{1s0}(2)\psi_{\varepsilon s0}(1)|1/r_{12}|\varphi_{2s0}(2)\varphi_{2s0}(1)\rangle]\, \delta_{m_s,+1/2}\, \delta_{M_i,+1/2}. \end{aligned} \tag{3.10c}$$

In these remaining matrix elements the numbers 1 and 2 for the two electrons can be interchanged without altering the value, and one obtains

$$M_1 = -\langle \varphi_{1s0}(1)\psi_{\varepsilon s0}(2)|1/r_{12}|\varphi_{2s0}(1)\varphi_{2s0}(2)\rangle\, \delta_{m_s,+1/2}\, \delta_{M_i,+1/2}. \quad (3.10d)$$

Using equ. (7.28b) without the spin function, one can now introduce the wavefunction $\psi_{\varepsilon s0}$

$$\psi_{\varepsilon s0}(\mathbf{r}, \boldsymbol{\kappa}^{(-)}) = \frac{1}{\sqrt{\kappa}}\, \mathrm{e}^{-\mathrm{i}(\sigma_{\varepsilon s}+\delta_{\varepsilon s})}\, Y_{00}(\hat{\boldsymbol{\kappa}})\, R_{\varepsilon s}(r)\, Y_{00}(\hat{\mathbf{r}}), \quad (3.12)$$

and perform the integration over $\mathbf{r}_1$ and $\mathbf{r}_2$. In this way one finally gets

$$M_1 = \frac{(-1)}{\sqrt{\kappa}}\, \mathrm{e}^{+\mathrm{i}(\sigma_{\varepsilon s}+\delta_{\varepsilon s})}\, Y_{00}(\hat{\boldsymbol{\kappa}})\, R^0(1s\varepsilon s, 2s2s)\, \delta_{m_s,+1/2}\, \delta_{M_i,+1/2} \quad (3.10e)$$

with the Coulomb integral (Slater integral; see equ. (7.60a))

$$R^0(1s\varepsilon s, 2s2s) = \int_0^\infty R_{1s}(r_1) R_{\varepsilon s}(r_2) \gamma_0(r_1, r_2) R_{2s}(r_1) R_{2s}(r_2) r_1^2 r_2^2\, \mathrm{d}r_1\, \mathrm{d}r_2 \quad (3.13a)$$

with (see equ. (7.58))

$$\gamma_0(r_1, r_2) = 1/\max(r_1, r_2). \quad (3.13b)$$

A similar calculation yields

$$M_2 = \frac{(-1)}{\sqrt{\kappa}}\, \mathrm{e}^{+\mathrm{i}(\sigma_{\varepsilon s}+\delta_{\varepsilon s})}\, Y_{00}(\hat{\boldsymbol{\kappa}})\, R^0(1s\varepsilon s, 2s2s)\, \delta_{m_s,-1/2}\, \delta_{M_i,-1/2}. \quad (3.11b)$$

From the expressions for M_1 and M_2 one can see immediately that the Auger electrons of this transition do not possess an anisotropic angular distribution, because $Y_{00}(\hat{\boldsymbol{\kappa}})$ is a constant. (The same result follows for the other K–LL transitions: it is a consequence of the vanishing alignment of the initial state for the Auger decay, see below equ. (3.30c).) The total transition rate P then follows from

$$P = \int P(\hat{\boldsymbol{\kappa}})\, \mathrm{d}\Omega_\kappa \quad (3.14a)$$

to

$$P = \pi\, R^0(1s\varepsilon s, 2s2s)^2\, \tfrac{1}{2} \sum_{M_i} \sum_{m_s} |\delta_{M_i,+1/2}\, \delta_{m_s,+1/2} + \delta_{M_i,-1/2}\, \delta_{m_s,-1/2}|^2 \int |Y_{00}(\hat{\boldsymbol{\kappa}})|^2\, \mathrm{d}\Omega_\kappa \quad (3.14b)$$

and yields

$$P(\mathrm{K{-}L_1L_1}\ {}^1\mathrm{S^e}) = 2\pi\, R^0(1s\varepsilon s, 2s2s)^2. \quad (3.15a)$$

Similarly, one can calculate the other transition rates and get

$$P(\mathrm{K{-}L_1L_{2,3}}\ {}^1\mathrm{P^o}) = 2\pi\, \tfrac{3}{2}\, [R^0(1s\varepsilon p, 2s2p) + \tfrac{1}{3} R^1(1s\varepsilon p, 2p2s)]^2, \quad (3.15b)$$

$$P(\mathrm{K{-}L_1L_{2,3}}\ {}^3\mathrm{P^o}) = 2\pi\, \tfrac{9}{2}\, [R^0(1s\varepsilon p, 2s2p) - \tfrac{1}{3} R^1(1s\varepsilon p, 2p2s)]^2, \quad (3.15c)$$

(within the *LSJ*-coupling scheme one obtains

$$P(\mathrm{K{-}L_1L_{2,3}}\ {}^3\mathrm{P}^o_J) = 2\pi \frac{2J+1}{2} [R^0(1s\varepsilon p, 2s2p) - \tfrac{1}{3}R^1(1s\varepsilon p, 2p2s)]^2 \quad (3.15c')$$

$$P(\mathrm{K{-}L_{2,3}L_{2,3}}\ {}^1\mathrm{S}^e) = 2\pi\, \tfrac{1}{3}\, R^1(1s\varepsilon s, 2p2p)^2, \quad (3.15d)$$

$$P(\mathrm{K{-}L_{2,3}L_{2,3}}\ {}^1\mathrm{D}^e) = 2\pi\, \tfrac{2}{3}\, R^1(1s\varepsilon d, 2p2p)^2, \quad (3.15e)$$

$$P(\mathrm{K{-}L_{2,3}L_{2,3}}\ {}^3\mathrm{P}^e) = 0. \quad (3.15f)$$

A detailed discussion of these transition rates involves the intensity of observed Auger lines and will be discussed below.

3.1.3 Intermediate coupling

The relations given in equs. (3.15) hold for pure *LS*-coupling as realized in neon. For the more general case one has to consider intermediate coupling in the final ionic state because the different electron configurations with equal *J* values and the same parity will mix (see equ. (1.24)). From the possible final ionic states described in equs. (3.1), one finds that $\mathrm{L_1L_{2,3}}\ {}^1\mathrm{P}^o_1$ can mix with $\mathrm{L_1L_{2,3}}\ {}^3\mathrm{P}^o_1$, $\mathrm{L_{2,3}L_{2,3}}\ {}^1\mathrm{D}^e_2$ with $\mathrm{L_{2,3}L_{2,3}}\ {}^3\mathrm{P}^e_2$, and $\mathrm{L_{2,3}L_{2,3}}\ {}^1\mathrm{S}^e_0$ with $\mathrm{L_{2,3}L_{2,3}}\ {}^3\mathrm{P}^e_0$. Hence, one gets for the wavefunctions in intermediate coupling the ansatz†

$$\Psi_{\mathrm{I}}(\mathrm{L_1L_{2,3}}\ \text{`}{}^1\mathrm{P}^o_1\text{'}) = A_1\Psi(\mathrm{L_1L_{2,3}}\ {}^1\mathrm{P}^o_1) + b_1\Psi(\mathrm{L_1L_{2,3}}\ {}^3\mathrm{P}^o_1), \quad (3.16a)$$

$$\Psi_{\mathrm{II}}(\mathrm{L_1L_{2,3}}\ \text{`}{}^3\mathrm{P}^o_1\text{'}) = a_2\Psi(\mathrm{L_1L_{2,3}}\ {}^1\mathrm{P}^o_1) + B_2\Psi(\mathrm{L_1L_{2,3}}\ {}^3\mathrm{P}^o_1) \quad (3.16b)$$

and similarly

$$\Psi_{\mathrm{I}}(\mathrm{L_{2,3}L_{2,3}}\ \text{`}{}^1\mathrm{D}^e_2\text{'}) = A'_1\Psi(\mathrm{L_{2,3}L_{2,3}}\ {}^1\mathrm{D}^e_2) + b'_1\Psi(\mathrm{L_{2,3}L_{2,3}}\ {}^3\mathrm{P}^e_2), \quad (3.17a)$$

$$\Psi_{\mathrm{II}}(\mathrm{L_{2,3}L_{2,3}}\ \text{`}{}^3\mathrm{P}^e_2\text{'}) = a'_2\Psi(\mathrm{L_{2,3}L_{2,3}}\ {}^1\mathrm{D}^e_2) + B'_2\Psi(\mathrm{L_{2,3}L_{2,3}}\ {}^3\mathrm{P}^e_2) \quad (3.17b)$$

and‡

$$\Psi_{\mathrm{I}}(\mathrm{L_{2,3}L_{2,3}}\ \text{`}{}^1\mathrm{S}^e_0\text{'}) = A''_1\Psi(\mathrm{L_{2,3}L_{2,3}}\ {}^1\mathrm{S}^e_0) + b''_1\Psi(\mathrm{L_{2,3}L_{2,3}}\ {}^3\mathrm{P}^e_0), \quad (3.18a)$$

$$\Psi_{\mathrm{II}}(\mathrm{L_{2,3}L_{2,3}}\ \text{`}{}^3\mathrm{P}^e_0\text{'}) = a''_2\Psi(\mathrm{L_{2,3}L_{2,3}}\ {}^1\mathrm{S}^e_0) + B''_2\Psi(\mathrm{L_{2,3}L_{2,3}}\ {}^3\mathrm{P}^e_0). \quad (3.18b)$$

If these final states from the intermediate coupling case are used in the calculation of transition probabilities, one obtains not only a splitting of the Auger transition energies according to their *J* values, but also the result that $\mathrm{K{-}L_{2,3}L_{2,3}}$ transitions forbidden in pure *LS*-coupling are allowed for $J = 0$ and $J = 2$. Hence, nine transitions are possible in intermediate coupling, instead of the only five transitions in pure *LS*-coupling [ABu58, Asa63]. (In the limit of pure *jjJ*-coupling these nine

† For the explicit calculation of the mixing coefficients see Section 7.4.3 where the case sp ${}^1\mathrm{P}^o_1$ mixed with sp ${}^3\mathrm{P}^o_1$ is treated in detail.

‡ For these states one should also include configuration interaction with the $\mathrm{L_1L_1}\ {}^1\mathrm{S}^e_0$ state (see [Asa65]).

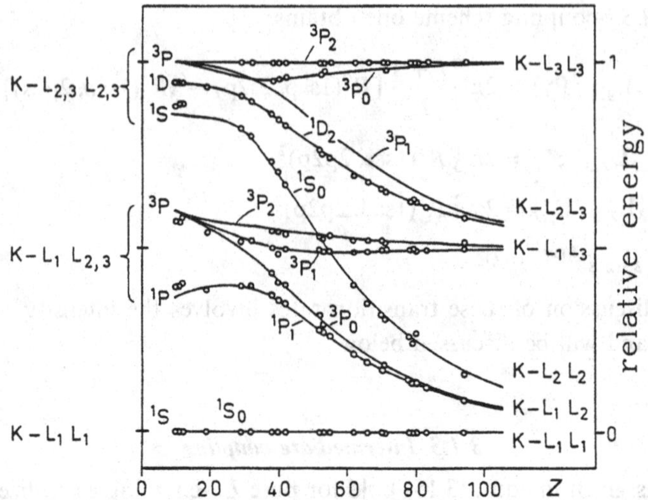

Figure 3.4 Relative energies of K–LL Auger transitions as functions of the atomic number. The experimental values shown as open circles have been adjusted so that the energies K–L_1L_1 and K–$L_{2,3}$ $^3P_2$ fall on the theoretical solid lines. The *LS*-coupling limit is shown on the left-hand side, and the *jj*-coupling limit on the right. The states K–L_1L_1 $^1S_0$, K–$L_1L_{2,3}$ $^3P_2$, K–$L_1L_{2,3}$ $^3P_0$, and K–$L_{2,3}L_{2,3}$ $^3P_1$ are pure and can be described unambiguously in either coupling limit, but all other states are mixed and must be represented by intermediate coupling. From [Meh78] using the data of [SNF67].

transitions reduce to six, see Fig. 3.4.) These transition rates are given by

$$P(\text{K–L}_1\text{L}_1\ {}^1\text{S}^e_0) = \text{same as in equ. (3.15a)}, \tag{3.19a}$$

$$P(\text{K–L}_1\text{L}_{2,3}\ \text{‘}{}^1\text{P}^o_1\text{’}) = |A_1|^2 P(\text{K–L}_1\text{L}_{2,3}\ {}^1\text{P}^o_1) + |b_1|^2 P(\text{K–L}_1\text{L}_{2,3}\ {}^3\text{P}^o_1), \tag{3.19b}$$

$$P(\text{K–L}_1\text{L}_{2,3}\ {}^3\text{P}^o_0) = \text{same as in equ. (3.15c}')\text{ with } J = 0, \tag{3.19c}$$

$$P(\text{K–L}_1\text{L}_{2,3}\ \text{‘}{}^3\text{P}^o_1\text{’}) = |a_2|^2 P(\text{K–L}_1\text{L}_{2,3}\ {}^3\text{P}^o_1) + |B_2|^2 P(\text{K–L}_1\text{L}_{2,3}\ {}^1\text{P}^o_1), \tag{3.19d}$$

$$P(\text{K–L}_1\text{L}_{2,3}\ \text{‘}{}^3\text{P}^o_2\text{’}) = \text{same as in equ. (3.15c}')\text{ with } J = 2, \tag{3.19e}$$

$$P(\text{K–L}_{2,3}\text{L}_{2,3}\ \text{‘}{}^1\text{S}^e_0\text{’}) = |A''_1|^2 P(\text{K–L}_{2,3}\text{L}_{2,3}\ {}^1\text{S}^e_0), \tag{3.19f}$$

$$P(\text{K–L}_{2,3}\text{L}_{2,3}\ \text{‘}{}^1\text{D}^e_2\text{’}) = |A'_1|^2 P(\text{K–L}_{2,3}\text{L}_{2,3}\ {}^1\text{D}^e_2), \tag{3.19g}$$

$$P(\text{K–L}_{2,3}\text{L}_{2,3}\ \text{‘}{}^3\text{P}^e_0\text{’}) = |a''_2|^2 P(\text{K–L}_{2,3}\text{L}_{2,3}\ {}^1\text{S}^e_0), \tag{3.19h}$$

$$P(\text{K–L}_{2,3}\text{L}_{2,3}\ \text{‘}{}^3\text{P}^e_2\text{’}) = |a'_2|^2 P(\text{K–L}_{2,3}\text{L}_{2,3}\ {}^1\text{D}^e_2), \tag{3.19i}$$

$$P(\text{K–L}_{2,3}\text{L}_{2,3}\ {}^3\text{P}^e_1) = 0. \tag{3.19j}$$

Such intermediate coupling is realized for K–LL Auger transitions in elements with medium Z values. For atoms with very high Z the extreme *jjJ*-coupling case is approached. A sketch of the relative energies of such allowed and forbidden K–LL Auger transitions, covering the whole range from *LS*- to *jj*-coupling, is given in Fig. 3.4. The solid lines represent theoretically expected values, the open circles are experimental data. Good agreement can be seen, and the large changes introduced by the different coupling schemes are verified.

Table 3.3. *Energy values relevant for K–LL Auger transitions in neon.*

Transition	E(final)/eV optical data	E(Auger)/eV experiment	E(Auger)/eV HF theory
$K\text{-}L_1L_1\ ^1S^e$		748.1	747.0
$K\text{-}L_1L_{2,3}\ ^1P^o$	98.4	771.7	770.9
$K\text{-}L_1L_{2,3}\ ^3P^o$	87.9	782.2	783.0
$K\text{-}L_{2,3}L_{2,3}\ ^1S^e$	64.4	800.6	801.0
$K\text{-}L_{2,3}L_{2,3}\ ^1D^e$	65.7	804.3[(a)]	806.0

Experimental values are from [ATW90], see also [KMe66, KCM71]; [(a)] this line has been used as the reference; with $E_I(1s)$ + 870.2 eV from Table 2.1 and the optical data it should be positioned at 804.5 eV. For comparison the results from Hartree–Fock calculations performed separately for the two initial and final ion-states are also shown (see [Kel75]).

3.2 Analysis of line position

The position of an Auger line is fixed by its kinetic energy which is the energy difference between the initial and final ionic states, equ. (1.29b). Two energy values determine the third value. Since in some cases the energy for a final ionic state is known from optical spectroscopy (see [Moo71]), the measurement of the kinetic energy of Auger electrons allows the determination of ionization energies for inner-shell electrons. (Ionization energies can also be measured directly via the kinetic energy of the corresponding photoelectrons, equ. (1.29a).) This procedure has been applied using the method of electron impact ionization to determine many ionization energies E_I. (For the influence of postcollision interaction on the energies of Auger and photoelectrons see Sections 4.5.5 and 5.5.) Experimental and theoretical energy values for the K–LL Auger transitions in neon are given in Table 3.3.

3.3 Analysis of linewidth

In analogy to the discussion of Fig. 2.9 describing the energy relations and energy widths relevant for 1s photoionization in neon, the observed Auger lines have a certain width which comes from natural and instrumental sources. Since the Auger decay proceeds from a specific initial ionic state to a specific final ionic state with both states having inherent *level* widths, Γ_i and Γ_f, respectively, the natural *line*width Γ_{fi} of an Auger transition will depend on both. Because the energy distributions attached to the initial and final states of the Auger transition are Lorentzian, the resulting natural linewidth Γ_{fi} of the transition is

$$\Gamma_{fi} = \Gamma_f + \Gamma_i. \tag{3.20a}$$

In the special case of a K–LL Auger transition this gives

$$\Gamma_{K-LL} = \Gamma_K + \Gamma_{LL}. \tag{3.20b}$$

In addition, the observed width of an Auger line is also affected by the spectrometer resolution. However, the bandpass of the incident radiation which produces the initial state for the Auger decay does not play a role, unlike in the case of the width of an observed photoline. (This statement only holds for the two-step model of inner-shell ionization and subsequent Auger decay. In the vicinity of the inner-shell ionization threshold it significantly fails due to postcollision interaction (Section 5.5) and the resonant Raman Auger effect (Section 5.1.2.1).) Hence, Auger transitions often appear in the spectra of ejected electrons as lines much sharper than the corresponding photolines.

In order to estimate the inherent linewidth of K–LL transitions in neon, one needs, according to equ. (3.20b) the value $\Gamma_K = 0.27$ eV discussed earlier (see Table 2.2), and information on Γ_{LL}. This latter quantity is fixed by the probabilities of decay of the corresponding two-hole states which are given in the present case by L_1L_1, $L_1L_{2,3}$, and $L_{2,3}L_{2,3}$. Therefore, different $\Gamma_{L_1L_1}$, $\Gamma_{L_1L_{2,3}}$, and $\Gamma_{L_{2,3}L_{2,3}}$ values exist. For neon, for example, one has $\Gamma_{L_{2,3}L_{2,3}} = 0$, because no further decay is possible, but $\Gamma_{L_1L_{2,3}}$ and $\Gamma_{L_1L_1}$ do not vanish because the L_1 hole can be filled by a radiative decay, i.e., $1s^22s2p^5 \rightarrow 1s^22s^22p^4$ and $1s^22p^6 \rightarrow 1s^22s2p^5 \rightarrow 1s^22s^22p^4$, respectively. Within the single-particle model, the rate for radiative decay or, equivalently, the partial width Γ_{L_1} for the corresponding electron jump $2p \rightarrow 2s$ can be calculated. One gets ([BSa57], used in [McG69]; atomic units)

$$\Gamma_{L_1} = 7.76 \times 10^{-7} N_{2p} f_{2p\rightarrow 2s}\, \Delta E^2, \tag{3.21}$$

where N_{2p} is the number of 2p-electrons initially in the 2p shell, $f_{2p\rightarrow 2s}$ is the oscillator strength per electron for the $2p \rightarrow 2s$ transition, and ΔE is the $2p \rightarrow 2s$ transition energy. With this relation, a rough estimate for Γ_{L_1} can be obtained, assuming: for the oscillator strength the upper boundary, $f_{2p\rightarrow 2s} \approx 1$; for ΔE the difference of the 2s and 2p binding energies (see Fig. 2.4), $\Delta E \approx 30$ eV, and $N_{2p} = 5$. These data give $\Gamma_{L_1} < 0.15$ meV. This small value leads to small values for $\Gamma_{L_1L_1}$ and $\Gamma_{L_1L_{2,3}}$ also which, therefore, can be neglected with respect to $\Gamma_K = 270$ meV. Hence, the linewidth of all K–LL Auger transitions in neon reduces to the level width Γ_K.

3.4 Analysis of line intensity

From the Auger spectrum of Fig. 3.3 relative intensities of Auger lines can be evaluated following the same method as outlined for the photoelectrons in Section 2.4. However, a slight modification is necessary to relate the observed intensity of a photoinduced Auger line to the experimental parameters. Within the two-step model of photoionization and subsequent Auger decay this modification just reflects these two steps: the photoionization process depends on the photon intensity N_{ph} and the photoionization cross section σ_{ph}, and the Auger decay brings in the partial Auger yield $\omega_A(\text{K–LL}\ {}^{2S+1}L_J)$ which describes the fraction of 1s-holes that decay into this specific branch. This partial Auger yield is defined in

Table 3.4. *Relative intensities for K–LL Auger lines in neon (see also [MSV68, KCM71]).*

Auger transition	Exper. relative intensity [ATW90]
$K\text{–}L_1L_1\ ^1S^e$	10.1
$K\text{–}L_1L_{2,3}\ ^1P^o$	28.3
$K\text{–}L_1L_{2,3}\ ^3P^o$	10.3
$K\text{–}L_{2,3}L_{2,3}\ ^1S^e$	15.6
$K\text{–}L_{2,3}L_{2,3}\ ^1D^e$	100.0

analogy to the total Auger yield (equ. (2.25b)) as

$$\omega_A(\text{K–LL}\ ^{2S+1}L_J) = P_A(\text{K–LL}\ ^{2S+1}L_J)/P(1s). \tag{3.22}$$

Hence, one gets for the intensity $I_{exp}(E^0_{kin}, fU_{sp})$ of K–LL $^{2S+1}L_J$ Auger electrons with nominal energy E^0_{kin} recorded at different settings of the spectrometer voltage U_{sp} and per unit of time (see Section 10.4.2 and equ. (2.37))

$$I_{exp}(E^0_{kin}, fU_{sp}) = N_{ph} n_v \sigma_{ph}\ \omega_A(\text{K–LL}\ ^{2S+1}L_J)\ \Delta z\ T\varepsilon\ L_i \otimes \tilde{G}_{sp} \tag{3.23}$$

and, similarly, for the dispersion corrected area of an Auger line (see equs. (2.39))

$$A_D = N_{ph} n_v \sigma_{ph}\ \omega_A(\text{K–LL}\ ^{2S+1}L_J)\ \Delta z\ T\varepsilon\ R. \tag{3.24}$$

Dispersion corrected areas for the normal K–LL Auger transitions in neon obtained from Auger spectra like the one shown in Fig. 3.3 are listed in Table 3.4 as relative intensities $I_{rel}(i)$ with i an abbreviation for the selected transition K–LL $^{2S+1}L_J$:

$$I_{rel}(i) \sim A_D(i). \tag{3.25a}$$

These intensities are proportional to the partial Auger transition rates P_A(K–LL $^{2S+1}L_J) = P_A(i)$

$$I_{rel}(i) \sim P_A(i). \tag{3.25b}$$

Equ. (3.25b) then leads to

$$I_{rel}(i) = \frac{I(i)}{\sum_{\text{K–LL}} I(i)} = \frac{P_A(i)}{P_A(\text{all K–LL})} = \frac{\Gamma_A(i)}{\Gamma_A(\text{all K–LL})}. \tag{3.26}$$

This equation directly relates the relative intensities $I_{rel}(i)$ to the absolute values of the partial Auger rates $P_A(i)$, provided P_A(all K–LL), or equivalently Γ_A(all K–LL), is known. Following [MSV68, Meh85], this latter quantity can be derived from experimental data as follows: $\Gamma(1s)$ is known to be $\Gamma(1s) = 0.27(2)$ eV (Table 2.2). Taking into account the different branches for the decay of this 1s hole-state, one has (see equ. (2.25d))

$$\Gamma(1s) = \Gamma_A(\text{all K–LL}) + \Gamma_A(\text{all K–LLL}) + \Gamma_R(\text{all K–L}) + \Gamma_R(\text{all K–LL}). \tag{3.27}$$

Using the ratio $r = \Gamma_A(\text{all K–LLL})/\Gamma_A(\text{all K–LL})$, measured to be $r = 0.081(8)$

Table 3.5. *Absolute transition rates $P_A(i)$ for K–LL Auger lines in neon (see [Meh78, Meh85] and [TAM92]).*

Auger transition	Transition rate [10^{-4} au] exp. [ATW90]	theor. [1]	[2]	[3]	[4]	[5]	[6a]	[6b]
$K-L_1L_1\ {}^1S$	5.5	9.5	4.9	4.5	5.1	5.9	6.6	4.8
$K-L_1L_{2,3}\ {}^1P$	15.5	20.3	13.7	15.0	14.4	18.1	18.8	14.4
$K-L_1L_{2,3}\ {}^3P$	5.6	7.9	4.9	7.0	5.6	8.2	9.8	5.4
$K-L_{2,3}L_{2,3}\ {}^1S$	8.6	4.6	7.7	8.3	8.0	8.7	9.0	7.0
$K-L_{2,3}L_{2,3}\ {}^1D$	54.8	56.8	49.3	55.0	52.8	56.8	58.9	47.9
	90 ± 8	99.1	80.5	89.8	85.9	97.7	103.1	79.5

[1] Hartree–Fock calculation [Kel75]; [2] calculation with many-body perturbation theory [Kel75]; [3] calculations which include electron correlation and relaxation in the final ionic state ([HAG78]; the data given refer to basis orbitals which are chosen to take some account of the nature of the doubly ionized residual ions); [4] calculation in the close-coupling approximation [Pet82]; [5] relativistic close-coupling approach [Bru87]; [6] multichannel multiconfiguration Dirac–Fock calculation with two different basis sets: [6a] optimized with respect to the initial state of the ion, [6b] optimized with respect to the final state of the ion [TAM92].

[CKr65], and the fluorescence yield $\omega_R(1s) = (\Gamma_R(\text{all K–L}) + \Gamma_R(\text{all K–LL}))/\Gamma(1s)$, measured to be $\omega_R(1s) = 0.025(15)$ (mean value of two data sets [BTH53, FJH59; listed in [FJM66]), one derives from equ. (3.27) together with $\Gamma(1s) = 0.27(2)$ meV the value $\Gamma_A(\text{all K–LL}) = P_A(\text{all K–LL}) = 90(8) \times 10^{-4}$ au. With this number the transition rates $P_A(i)$ for all individual Auger transitions can now be calculated using equ. (3.26). Some K–LL experimental and theoretical Auger transition rates in neon are shown in Table 3.5. The theoretical data come from different approaches: calculation [1] is an independent-particle calculation using Hartree–Fock orbitals, but calculations [2]–[6] also consider effects from the correlated motion of the electrons. It can be seen that these latter theories are in fair agreement with the experimental data.

3.5 Analysis of angular distribution

Auger electrons can also possess an angular distribution (see the discussion in connection with equ. (1.30)). For Auger transitions induced by arbitrary polarized light this angular distribution is parametrized by an expression similar to equ. (1.53) for photoelectrons:

$$\left.\frac{dI}{d\Omega}(\Theta, \tilde{\Phi})\right|_{\text{Auger}} = \frac{\sigma_{ph}\omega_A}{4\pi}\left\{1 - \frac{\beta_A}{2}\left[P_2(\cos\Theta) - \tfrac{3}{2}\tilde{S}_1 \cos 2\tilde{\Phi} \sin^2\Theta\right]\right\}, \quad (3.28)$$

where the index A refers to the Auger transition. In this expression, the polar and azimuthal angles Θ and $\tilde{\Phi}$ are defined in the tilted coordinate system of Fig. 1.15 where the quantization axis z points in the direction of the incident photon beam, the Stokes parameter $\tilde{S}_1$ describes the excess linear polarization intensity along the major and minor axes of the polarization ellipse (equ. (1.52c)), and β_A is the angular distribution parameter of emitted Auger electrons.

As a consequence of the two-step model for photoionization and subsequent Auger decay, there is a factorization of all observables. For the intensity, the product between the photoionization cross section σ_{ph} and the Auger yield ω_A has already been discussed in connection with equ. (3.22), and for the angular distribution parameter β_A one has [BKa77]

$$\beta_A = (-2)\mathscr{A}_{20}\,\alpha_2. \tag{3.29}$$

(A factor (-2) is introduced here in order to use published α_2-coefficients [BKa77] which were evaluated for incident unpolarized light on the basis of an angular distribution function $I = 1 + \beta_A P_2(\cos\Theta)$ instead of $I = 1 - 0.5\beta_A P_2(\cos\Theta)$ as implied here.) The factor $\mathscr{A}_{20}$ is due to the first step of photoionization. It is called the *alignment* parameter, because it describes the alignment of the ion after the photoprocess. (This parameter is one component of the more general alignment tensor $\mathscr{A}_{2q}$ (see Section 8.5).) It is this quantity through which photoionization and subsequent Auger decay are linked together and, due to its importance, $\mathscr{A}_{20}$ will be treated below in more detail. The second factor in the expression for β_A, α_2, contains the properties of the second step, i.e., of the Auger transition, which are those of the Coulomb matrix elements including their relative phases. Hence, it is called the *Auger decay parameter*. In cases where the selection rules of the Auger transition allow only one partial wave for the Auger electron, α_2 becomes a simple numerical factor only (compare the similar situation with $\beta = 2$ for photoionization of an s-electron (equ. (2.13d))), and in other cases α_2 provides complementary information on the Auger decay to the Auger yield ω_A.

The alignment parameter $\mathscr{A}_{20}$ is related to the populations $a(JM_J)$ of the photoionized substrates JM_J classified with respect to the angular momentum J and its projection M_J onto a preselected quantization axis. In general, an ensemble of atoms in the state J can have different populations $a(JM_J)$ for the corresponding substates M_J, leading to *isotropy* or *polarization* where the latter can be split into *alignment* and *orientation* [Blu81]. (Note that sometimes, particularly in nuclear physics, the meanings of the words polarized and oriented are interchanged (see [Gro52]).) While isotropy is characterized by equal populations for all M_J, i.e., $a(JM_J) = \text{const}$, the aligned state describes a deviation from isotropy characterized by equal populations of $-M_J$ and $+M_J$, i.e.,

$$\text{alignment: } a(JM_J) \neq \text{const}, \quad \text{and} \quad a(J-M_J) = a(J+M_J), \tag{3.30a}$$

and the oriented state describes a deviation from isotropy characterized by populations which decrease/increase with M, i.e.,

$$\text{orientation: } a(JM_J) \neq \text{const}, \quad \text{and} \quad a(JM_J) \gtrless a(JM_J-1). \tag{3.31a}$$

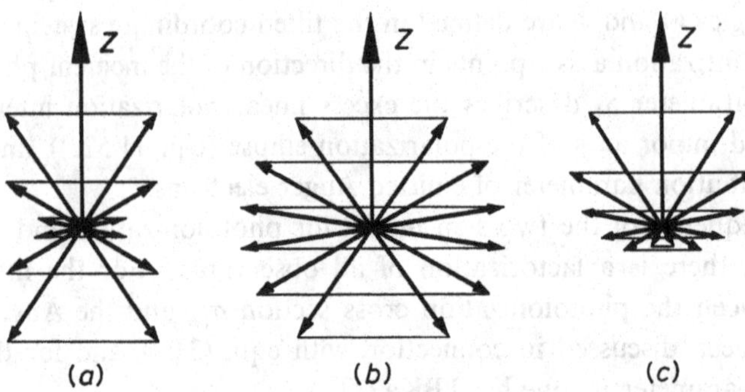

Figure 3.5 Characterization of different polarizations of an axially symmetric system JM_J for the special example $J = 5/2$. According to the vector model the projections $M_J = 5/2$, $3/2$, $1/2$, $-1/2$, $-3/2$ and $-5/2$ are shown as tilted arrows precessing around the z-axis (in order to indicate this precession, the arrows have been drawn to both the left and right). The length of the arrows is used as a measure for the number of particles in the corresponding magnetic state, thus giving information about polarization. The polarizations shown are: (*a*) an aligned, (*b*) an isotropic, and (*c*) an oriented system.

This situation is sketched in Fig. 3.5 where different polarizations are shown for $J = 5/2$. The system has axial symmetry around the preselected quantization axis (z-axis). The M_J-associated directions are indicated by the precessing angular momentum J, and the populations $a(JM_J)$ are characterized by the length of these precessing angular momenta. In this way one obtains a spatial view of (*a*) an aligned, (*b*) an isotropic, and (*c*) an oriented state.

Formally, alignment and orientation follow from the general symmetry properties of statistical tensors $\rho_{k\kappa}$ introduced in Section 8.4. Spherical symmetry leads to $k = \kappa = 0$, axial symmetry to $\kappa = 0$, and alignment requires $k =$ even, and orientation $k =$ odd. Since the dipole approximation in the photoionization process restricts k to $k \leq 2$, a photoionized axially symmetric state can only have ρ_{00}, ρ_{10}, and ρ_{20}; ρ_{00} describes isotropy, and the alignment is given by [BKa77]

$$\mathscr{A}_{20} = \rho_{20}/\rho_{00}, \tag{3.30b}$$

and the orientation by

$$\mathscr{O}_{10} = \rho_{10}/\rho_{00}. \tag{3.31b}$$

From the definitions for alignment and orientation, given in equs. (3.30a) and (3.31a), it follows that alignment can occur only for $J > 1/2$, and orientation only for $J > 0$, i.e.,

$$\mathscr{A}_{20}(J \leq 1/2) = 0 \tag{3.30c}$$

and

$$\mathscr{O}_{10}(J = 0) = 0. \tag{3.31c}$$

The non-vanishing components then follow from

$$\mathscr{A}_{20}(J) = \sqrt{\left[\frac{5}{(2J+3)(J+1)J(2J-1)}\right]} \sum_{M_J} [3M_J^2 - J(J+1)]\, a(JM_J) \tag{3.30d}$$

and

$$\mathcal{O}_{10}(J) = \sqrt{\left[\frac{3}{J(J+1)}\right]} \sum_{M_J} M_J\, a(JM_J), \tag{3.31d}$$

with normalized population probabilities $a(JM_J)$:

$$\sum_{M_J} a(JM_J) = 1. \tag{3.32}$$

For photoionization with unpolarized or linearly polarized light the intermediate state for the Auger decay can only be aligned and not oriented, provided the photoelectron is not observed. (In coincidence experiments between the photoelectron and the subsequent Auger electron one needs not only the alignment parameter $\mathcal{A}_{20}$, but also the full alignment tensor $\mathcal{A}_{2\kappa}$, and similarly for circular polarized light the full orientation tensor $\mathcal{O}_{1\kappa}$, see Section 8.5.2.) This alignment property follows from the fact that the only observable of the intermediate photoionized state J is its projection quantum number M_J measured against the preselected reference axis. For linearly polarized light this reference axis is conveniently chosen to coincide with the direction of the electric field vector of the incident light. Since this field vector oscillates, no preference for a positive or negative direction of the quantization axis exists and it is clear that $a(JM_J) = a(J\ -M_J)$ must hold. For unpolarized light the convenient reference axis coincides with the propagation direction of the incident light. In this case the alignment condition cannot be seen immediately, but parity conservation can be used to verify it. (Of course, the argument of parity conservation also applies for demonstrating $a(JM_J) = a(J\ -M_J)$ for the case of linearly polarized incident light.) Very generally, the parity operation Π introduced in equ. (1.13a) can be replaced by two other operations, a reflection onto an arbitrarily selected plane and a 180° rotation around an axis perpendicular to that plane:

$$\Pi = \text{(mirror reflection) and (180° rotation)}. \tag{3.33}$$

If parity conservation is applied to a physical process, it requires that the actual process and its mirror image must have the same probability. (For these pictures of a physical process the supplementary 180° rotation is of no relevance.) In the present context the actual process is represented by the propagation axis (z-axis) of the incident light and the precessing angular momentum J of the photoionized state with projection M_J. Placing a mirror perpendicular to the selected z-axis, the direction of this axis is changed (a property of a *polar vector*) while the angular momentum keeps its direction of precession (a property of an *axial vector*). Hence, the actual process and its mirror image differ by the sign of M_J, and parity conservation then demands $a(JM_J) = a(J\ -M_J)$. Alternatively, one can place the mirror parallel and at a distance to the z-axis. Then in the mirror image the z-axis is unchanged, but the angular momentum changes its direction of precession, and the same conclusion is reached.

In contrast to the case of states reached by unpolarized or linearly polarized light, orientation (and alignment) is possible for circular polarization of the incident light. Since the characteristic property of circular polarized light is the rotation of the electric field vector around the direction of propagation, parity conservation only leads to the requirement $a(JM_J)_{\text{left circ.}} = a(J\,-M_J)_{\text{right circ.}}$, and vice versa. This, however, permits $a(JM_J)_{\text{left circ.}} \neq a(J\,-M_J)_{\text{left circ.}}$ and $a(JM_J)_{\text{right circ.}} \neq a(J\,-M_J)_{\text{right circ.}}$, and according to equ. (3.31a) this property allows orientation. (Orientation plays a role if the spin polarization of the emitted Auger electrons is observed.)

According to equ. (3.30c) the alignment parameter $\mathscr{A}_{20}(J)$ is zero for $J = 0$ or $J = 1/2$. Hence, Auger transitions which start from such an intermediate ionic state have an isotropic angular distribution, and this is the situation for all K–LL Auger transitions in neon. (See equ. (3.12) which only contains $Y_{00}(\hat{\mathbf{\kappa}}) = 1/\sqrt{(4\pi)}$. Note, however, that K–LL Auger transitions in open-shell atoms can have anisotropic angular distributions [LFr94].) This means that the previous extensive discussion on the alignment is not applicable to the neon K–LL Auger transitions. It was, however, included here because of its general importance. A simple, but rather instructive, example for a non-vanishing angular distribution of Auger electrons is the L_3–M_1M_1 Auger transition in magnesium. The corresponding experiment and its interpretation are described in detail in Section 5.2, and a theoretical discussion of alignment and orientation as well as of the angle-dependent detection of Auger electrons in coincidence with the primary ejected photoelectron is presented in Sections 4.6.1 and 8.5, respectively.

Part B

Experimental aspects and recent examples

4
Experimental aspects

In Part A the aims and the potential of electron spectrometry of free atoms have been discussed for the example of photoionization and subsequent Auger decay in neon. Now the apparatus details and the basic features of the technique of electron spectrometry will be considered. The discussion is restricted to electrostatic deflection analysers. However, the properties discussed can easily be adapted and transferred to other kinds of electron energy analysers.

4.1 The dispersive element of electron energy analysers

In order to analyse electrons with respect to their kinetic energy, an energy *dispersive* element is needed which separates the electrons according to their energy. For this purpose one of the following properties of moving electrons can be used (non-relativistic electrons only are considered):

(i) The *time-of-flight* that the electrons need to travel a given distance d,

$$t = d/v; \tag{4.1a}$$

because for non-relativistic velocities v one has

$$v = \sqrt{(2E_{kin}/m_0)}. \tag{4.1b}$$

The prerequisite for such a measurement is information about the starting time. This can be provided by the pulsed time structure of the primary photon beam (single-bunch mode for operating the electron storage ring), or by a reaction product of the photoprocess which is in coincidence with the emitted electron and can be another electron or a photon. More details on time-of-flight electron spectrometry are given in Section 10.1.

(ii) The deflection of the electrons in a homogeneous *magnetic field B*: because an electron moving in a plane perpendicular to the magnetic field describes a circular path with a radius of curvature ρ, given by

$$\rho = \frac{m_0 v}{e_0 B}, \tag{4.2}$$

i.e., for a given field strength the radius depends on the momentum of the electrons, and *momentum dispersive* instruments can be developed which are based on this relation. (Magnetic field analysers are used frequently in β-ray spectrometry (see, for example, [Sie66]), because with these at high electron energies there is no problem with high voltages as is the case for electrostatic

electrostatic analysers. On the other hand, stray fields at the edges of the experimental field region strongly influence the electron path. The necessary compensation of such edge effects [Her40] is more difficult to achieve for magnetic fields than for electrical fields. Therefore, electric field analysers are mostly used for medium and lower kinetic energies of the electrons.)

(iii) The deflection of the electrons in an *electric field*: because electrons which travel a distance d against an electrical field $\mathscr{E}$ lose energy E

$$E = e_0 \mathscr{E} d = e_0 U, \tag{4.3}$$

where U is the potential difference. Hence, *energy dispersive* instruments can be developed which are based on this relation. There are two classes of electric field analysers: *retardation* field analysers (for an example see [EDH80]) where the electrons enter the field region parallel to the field vector, and *deflection* analysers where the electrons enter the field region at a certain angle to the field vector.

These three possibilities for an energy dispersive element can be used singly or in combination. In the following, only electrostatic deflection analysers will be considered further (see also [Sev72, GvW83, Lec87, RTr90]). For a novel and powerful technique of high-resolution momentum spectroscopy see [UDM94].

In Fig. 4.1 four electrostatic deflection analysers are shown. They are classified according to the geometrical part that determines the electric field inside the

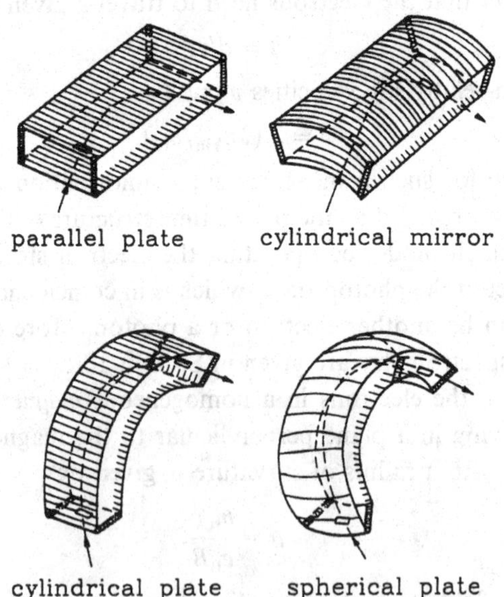

Figure 4.1 Electrostatic deflection analysers with finite entrance (and exit) slits (*sector analysers*). The diagrams show the field-defining elements which give these analysers their names, the entrance and exit slits, and one principal trajectory which represents the mean trajectory of electrons from the source through the analyser towards the detector. From [Sto83].

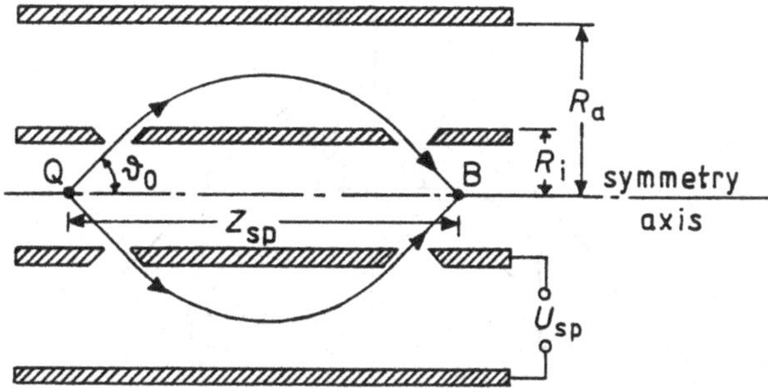

Figure 4.2 Cross-cut through a CMA which accepts electrons into a cone set by the average polar angle ϑ_0 with angular spread $\pm\Delta\vartheta$ (not shown). Since the analyser accepts the full 2π range of the azimuthal angle it is called a *2π-analyser*. The diagram shows the principal ray from the source Q to the image B. The characteristic dimensions are the radii of the inner and outer cylinders, R_i and R_a, and the total distance z_{sp} between Q and B. The spectrometer voltage U_{sp} is applied between the outer and inner cylinders and produces the electric field in the analyser.

analyser. In all cases there is only a small entrance (and exit) slit through which electrons are accepted (and released). For an electron source located at a certain distance from this entrance slit, only a small angular range is accepted, and these analysers are called *sector-analysers*. (For an example of a sector cylindrical mirror analyser see Fig. 1.17. It is also possible to use focusing properties in which the source is placed directly at the analyser entrance. For such an example see Fig. 5.35.) In some cases it is possible to enlarge the entrance and exit slits such that the analyser accepts azimuthal angles between 0 and 2π, and these analysers are called *2π-analysers*. Examples are the cylindrical mirror analyser (CMA, see Fig. 4.2 and also Fig. 1.16) and the toroidal analyser (see Fig. 4.3).

4.2 Characteristic features of electrostatic analysers

The characteristic features of electrostatic analysers will be demonstrated using a CMA, in either a 2π-analyser or a sector analyser form, as an example. These features concern: the principal trajectory (principal ray) and the focusing properties; the energy dispersion and energy resolution; the accepted solid angle; spectrometer transmission; spectrometer function and luminosity; and finally different modes of operation. Specific design criteria are given in Section 10.2, and typical features of a sector CMA which are characteristic for its use in combination with synchrotron radiation can be found in [DFM87].

4.2.1 Principal trajectory and focusing properties

Usually, electrostatic deflection analysers contain two regions through which the electrons pass, a field region and a field-free region. Very often the field-free region

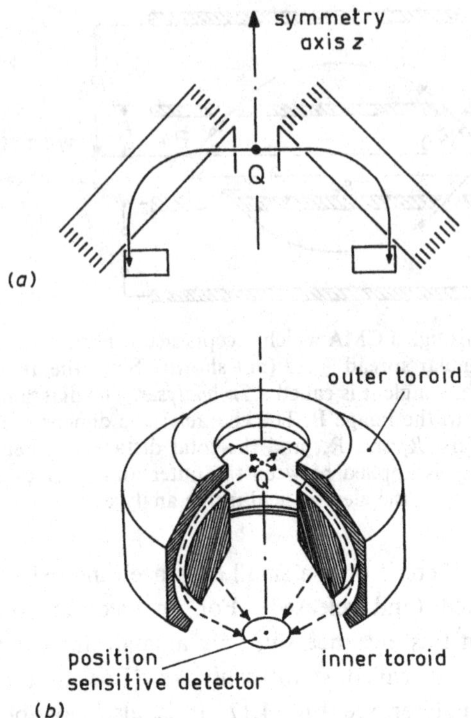

Figure 4.3 Two examples of a 2π-spectrometer collecting electrons within a plane. (*a*) Suggested set-up with full rotational symmetry around the symmetry axis z. The electric field is supplied by conical bodies, and the detector collects simultaneously all electrons emitted from the source Q into the plane perpendicular to the z-axis (for directional information of electron emission using a discretised detector, see below). From [Kuy68]. (*b*) Spatial view of a toroidal analyser. The outer and inner toroids which again possess axial symmetry provide the electric field. The detector is a position-sensitive detector (see Section 4.3.2) which records directional information about the electron emission. Hence, the angle dependence of electron emission from the source Q into the plane perpendicular to the symmetry axis is preserved. See also [EBM81, Hue93]. Part (*b*) reprinted from *Nucl. Instr. Meth.* **B12**, Toffoletto *et al*, 282 (1985) with kind permission of Elsevier Science – NL, Sara Burgerhartstraat 25, 1085 KV Amsterdam, The Netherlands.

is further divided into two parts, one before and one after the field region, because then a symmetrical arrangement can be used (see Fig. 4.2), but this is not a necessary requirement. In the field-free regions the electrons travel along straight lines. In the field region, they are deflected according to their kinetic energy E^0_{kin} by the action of the electric field. This action depends on the type of electric field used (the geometry of the analyser) and on the spectrometer voltage U^0_{sp} applied to the deflection plate of the analyser. In the example of a CMA shown in Fig. 4.2 the negative spectrometer voltage is on the outer cylinder, and the inner cylinder is kept at ground potential. Among the possible electron paths through the analyser, one can be singled out which will be called the *principal trajectory.* This trajectory is shown by the solid lines in Figs. 4.2 and 4.4: it starts in the middle

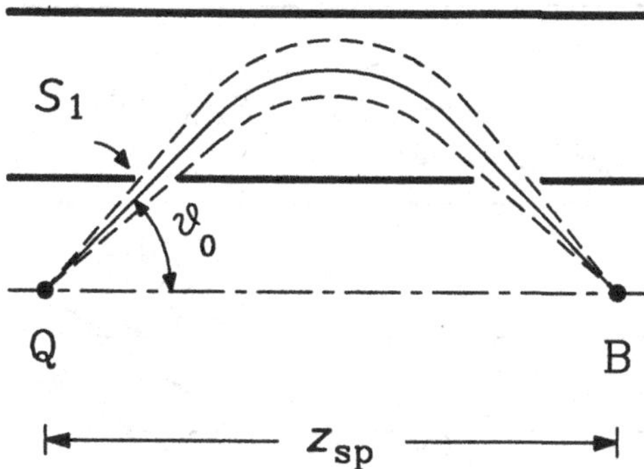

Figure 4.4 Imaging property of a CMA as introduced in Fig. 4.2. The diagram shows the principal ray (solid curve with cone angle ϑ_0) together with a finite bundle of electrons (dashed curves) accepted by the entrance slit S_1. All electrons have the same kinetic energy E^0_{kin}, and the spectrometer voltage U^0_{sp} is selected such that these electrons are imaged from the source point Q to the focal point B.

of the electron source (Q), enters the field region at an angle ϑ_0, goes through the analyser field and ends at the image point (B) where the dispersion slit of the detector can be placed (see below). The two dashed trajectories shown in Fig. 4.4 lie either side of this principal ray. They represent electron emission into a finite angular range $\Delta\vartheta$. (A finite angular range $\Delta\varphi$ in the φ-direction also has to be considered. However, for a CMA with full 2π symmetry and focusing onto its symmetry axis, the full range of $\Delta\varphi = 2\pi$ is accepted and needs no further treatment.) Due to the focusing properties of the analyser, these dashed rays intersect the principal trajectory at the *focal point* where they form the image B of the source. Hence, the principal trajectory is used to represent a bunch of electrons emitted into a finite solid angle and focused in the focal plane. Electrons of such a principal trajectory obey the relation (see equ. (1.47))

$$E^0_{kin} = fU^0_{sp}, \tag{4.4}$$

where the spectrometer factor f is fixed by the geometry and the focusing properties of the analyser design. For example, for a CMA with $\vartheta_0 = 42.3°$ and axis-to-axis focusing (see Fig. 4.2) one gets (see Section 10.2) $f = 1.310/\ln(R_a/R_i)$, measured in eV/V. (The ln dependence comes from the cylindrical electrostatic potential φ in the analyser given by $\varphi(r) = (V_i - V_a)\ \ln(r/R_i)/\ln(R_a/R_i)$.)

If the voltage U^0_{sp} is kept fixed, but electrons, which have energies different from E^0_{kin} are considered, one has different principal trajectories. As shown in Fig. 4.5 they reach different image points which define a *focal plane*, and the spatial dispersion $\Delta z'$ for electrons with energy deviation ΔE allows the desired energy analysis. Two methods can be applied:

(i) A simple rejection of electrons with energies outside the range $E^0_{kin} \pm \Delta E$ can be achieved by placing a *dispersion slit* of width $\Delta z'$ in the focal plane. (The other dimension of the dispersion slit, i.e., its 'length', must also be considered.

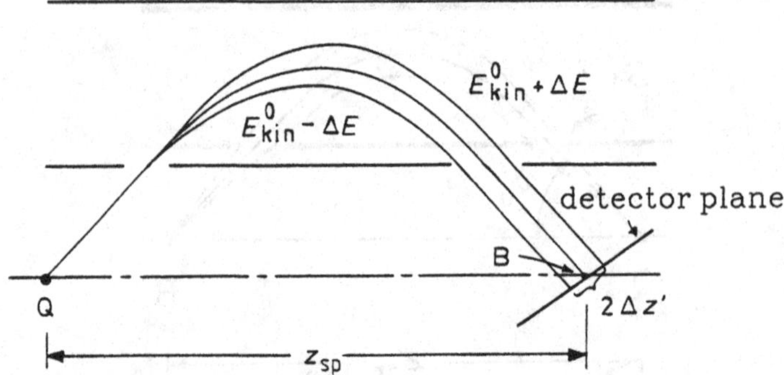

Figure 4.5 Principal trajectories of electrons with different kinetic energies in an electrostatic deflection analyser. The centre of the detector plane coincides with the focal point B of the principal ray at a U^0_{sp} corresponding to the E^0_{kin}. Principal rays, which also start in the source volume Q, but differ in their kinetic energy by $\pm\Delta E_{kin}$, reach the detector plane at different positions, thus illustrating the energy dispersion of the analyser. In the case shown the detector plane is aligned perpendicular to the principal rays; it should be noted that the actual focal plane of a CMA is inclined in a clockwise direction with respect to this plane.

It depends on the geometry of the analyser (sector or 2π) and is related to the dimensions of the acceptance source volume.) The transmitted electrons are then detected by a common channeltron detector placed behind this dispersion slit.

(ii) Electrons with different energies E^0_{kin} can be collected simultaneously using a position-sensitive detector positioned in the focal (or detector) plane.

Both types of detection device will be discussed in Section 4.3.

In order to fix the principal trajectory, appropriate focusing properties of the analyser have to be selected. These properties follow from the treatment of the analyser as an electron optical device which images the source Q at the analyser entrance to the image B at the analyser exit, and a detailed treatment for a point source is given in Section 10.2. Here only the main results necessary for a basic understanding will be reproduced.

When considering the focusing properties of an electron spectrometer, a point source is assumed which emits electrons into the directions (ϑ, φ) where ϑ and φ are the polar and azimuthal angles in the spectrometer frame (see Fig. 4.2). Obviously, the 2π-CMA with its rotational symmetry in φ possesses ideal focusing for all φ angles, but in the ϑ-direction focusing can be achieved only for a limited angular range $\pm\Delta\vartheta$ around a certain ϑ_0 value. The appropriate ϑ_0 value can be found if the distance z_{sp} between the source and the image is analysed with respect to its dependence on the angular spreads $\pm\Delta\vartheta$ around a still unknown, but preselected, ϑ_0 value. A Taylor expansion yields (see equ. (10.20))

$$\Delta z_{sp} = \sum_{n>0} \frac{1}{n!} \Delta\vartheta^n \frac{\partial^n}{\partial\vartheta^n} z_{sp}(\vartheta)\bigg|_{\vartheta_0} . \tag{4.5}$$

The expansion coefficients describe the degradation of the image due to the angular

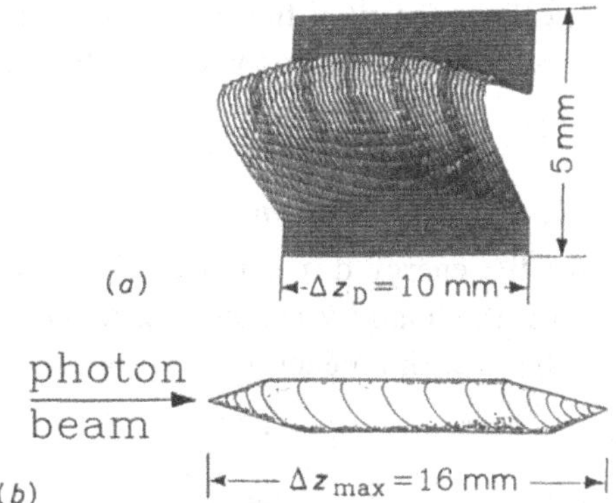

Figure 4.6 Imaging property of a sector CMA. (*a*) Intensity distribution of electrons starting in the finite source volume and reaching the detector plane at different positions (the detector plane is fixed at the focal point of the principal trajectory, point B in Fig. 4.5, and aligned perpendicular to the principal trajectory). (*b*) Source volume produced by a photon beam of 2 mm diameter (see Fig. 1.13). Reprinted from *Nucl. Inst. Meth.* A, **260**, Derenbach *et al.*, 258 (1987) with kind permission of Elsevier Science – NL, Sara Burgerhartstraat 25, 1055 KV Amsterdam, The Netherlands.

spread. If the first partial derivative is zero, one has *first-order* focusing (in ϑ). For the CMA this can be achieved for any desired angle ϑ_0. Hence, a special angle ϑ_0 exists where the second derivative also vanishes. For axis-to-axis focusing, this second-order focusing (in ϑ) occurs at $\vartheta_0 = 42.3°$. The non-vanishing third partial derivative taken at this angle then yields for a point source the contribution of the angular spread $\pm\Delta\vartheta$ to the finite size of the image. With this information one gets for a CMA with axis-to-axis focusing the compact results: in the φ-direction it has focusing in all orders, in the ϑ-direction it has focusing in first order for all ϑ_0 values, and in second order for $\vartheta_0 = 42.3°$.

For a finite source, the focusing properties may change. They have to be determined by a ray-tracing procedure which takes into account individual electron trajectories from individual points of the source volume. As an example, the image of a finite source is shown in Fig. 4.6 for a sector CMA in which the direction of the photon beam is perpendicular to the symmetry axis of the CMA (see Fig. 1.17). One can verify two observations, one concerning the size and the other the shape of the image as compared to that of the source. In the direction perpendicular to the photon beam (2 mm diameter) the image is slightly larger. In the direction along the photon beam, the restricted width in the focal plane ($\Delta z_D = 10$ mm in the example under discussion which is determined by the size of the position-sensitive detector) in turn fixes the length Δz in the source volume. (For the definition of Δz see below. Note that the photon beam and the symmetry axis of the CMA can be in the same direction (Fig. 1.16) or perpendicular to each other (Fig. 1.17). Therefore, different axes are used for characterizing the length of the source volume which is always along the direction of the photon beam.) The curved shape of the image results from the cylindrical symmetry of the analyser and the cylindrical symmetry of the source volume which, however, do not match, because they refer here to orthogonal directions.

4.2.2 Energy dispersion and energy resolution

With the help of Fig. 4.5 the energy dispersion D of an electrostatic deflection analyser can be introduced: the quantity D normalized against E^0_{kin} describes the spatial spread $\Delta z'$ of electrons with an energy deviation $\Delta E = E^0_{kin} - E_{kin}$, i.e.,

$$D = E^0_{kin}\,\frac{\Delta z'}{\Delta E}. \tag{4.6}$$

Hence, the energy dispersion contains information about the ability of the analyser to sort electrons of different kinetic energies in space and, therefore, it is an important chacteristic property of the energy resolution of an analyser. It depends essentially on the size of the analyser, but also on the focusing properties chosen. For example, for a CMA analyser with axis-to-axis focusing one gets $D = 5.60\ R_i$ where R_i is the radius of the inner cylinder of the analyser.

The definition of the energy dispersion given in equ. (4.6) on the basis of Fig. 4.5 goes back directly to the operational principle of an analyser whereby electrons enter the spectrometer (set at U^0_{sp}) with different energies $E^0_{kin} \pm \Delta E$. To show this, Fig. 4.5 will be interpreted slightly differently, because a similar picture also applies to a monochromatic beam of electrons (E^0_{kin}) which enters the spectrometer at slightly different spectrometer voltages $U^0_{sp} \mp \Delta U$. (An electron trajectory $E^0_{kin} + \Delta E$ with U^0_{sp} fixed corresponds to an electron trajectory of E^0_{kin} at $U^0_{sp} - \Delta U$, and vice versa. Compare in this context the two ways in which the spectrometer function introduced in equ. (1.48) may be interpreted.) Tuning the spectrometer voltage then moves the image B in the focal plane. The size of the image, B, will depend on the size of the source, Q, and on the non-vanishing terms in the Taylor expansion for the angular spread $\pm\Delta\vartheta$, and for a sector analyser also those for $\pm\Delta\varphi$. Therefore, the image B may be written as $B(\mathrm{Q}, \Delta\vartheta, \Delta\varphi)$. If the analyser is equipped with a dispersion slit (width S) in front of the channeltron detector, a situation arises which is demonstrated in Fig. 4.7: variation of the spectrometer voltage moves the image B across the dispersion slit which leads to a specific transmission function $T(U_{sp})$. Comparing the upper and lower parts of Fig. 4.7, it can be seen that the distance $\Delta z' = B(Q, \Delta\vartheta, \Delta\varphi) + S$, measured in the focal plane, just corresponds to a change of the spectrometer voltage U^0_{sp} of $\pm\Delta U^{basis}_{sp}$.

If this result is put into the dispersion formula, one gets

$$\left.\frac{\Delta E_{sp}}{E^0_{kin}}\right|_{basis} = \left.\frac{\Delta U_{sp}}{U^0_{sp}}\right|_{basis} = \frac{S + B(Q, \Delta\vartheta, \Delta\varphi)}{D}. \tag{4.7a}$$

The subscript *basis* indicates that the basis width of the transmission function $T(U_{sp})$ is taken. This basis width is approximately twice the fwhm value which is

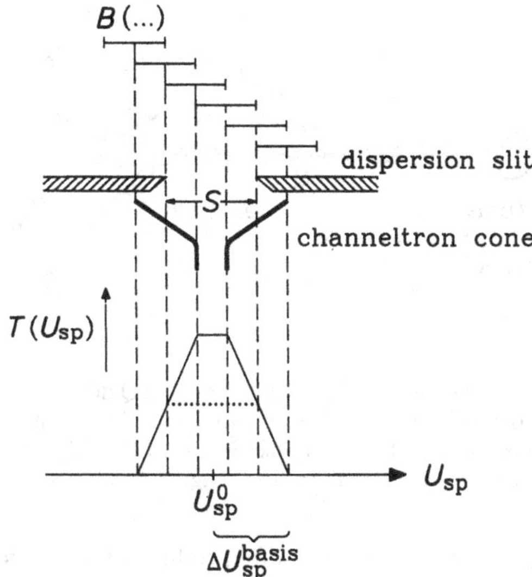

Figure 4.7 Explanation of the transmission of an electrostatic analyser. The upper part shows how different spectrometer voltages U_{sp} (see scale at the bottom) move the image $B(\ldots)$ across the width S of the dispersion slit. It should be noted that in reality all images lie in the plane of the dispersion slit, but for clarity they are shown in different planes. In the lower part the resulting transmission function $T(U_{sp})$ is plotted (later $T(U_{sp})$ will be written as $T(U_{sp}^0, U_{sp})$ in order to express its dependence on both voltages). The dashed vertical lines help to establish relations between the quantities in the upper and lower parts, respectively. At U_{sp}^0 the image $B(\ldots)$ fits well in the width S of the slit, and maximum transmission is achieved. However, outside $U_{sp}^0 \pm \Delta U_{sp}^{basis}$ the image $B(\ldots)$ cannot pass the slit and, hence, the transmission is zero. Combining the lower and upper parts of the figure, it can be seen that the basis width $2\Delta U_{sp}^{basis}$ on the voltage scale corresponds to a range $(S + B)$ in the dispersion plane.

called ΔE_{sp}, i.e.,

$$\Delta E_{sp}^{basis} \approx 2\Delta E_{sp}, \tag{4.8}$$

and hence one derives for the relative resolution R

$$R = \left(\frac{\Delta E}{E}\right)_{sp} \approx \frac{1}{2}\frac{\Delta E_{sp}}{E_{kin}^0}\bigg|_{basis} \approx \frac{1}{2}\frac{S + B(Q, \Delta\vartheta, \Delta\varphi)}{D}. \tag{4.7b}$$

From this equation several important conclusions can be drawn concerning a good energy resolution, i.e., a small ΔE_{sp}:

(i) The energy dispersion D has to be large in comparison to the source size Q and the slit width S. As was discussed above, D is determined essentially by the dimensions of the analyser. The dimensions, however, are limited by the general size of the apparatus and increasing disturbances by field inhomogeneities (see Sections 4.5.2 and 4.5.3).

(ii) The source size Q has to be small, because $B(Q, \Delta\vartheta, \Delta\varphi) \approx Q + B(\Delta\vartheta, \Delta\varphi)$. In many cases, the simple ratio Q/D limits the resolution. In order to improve the resolution in these cases, the actual source volume produced by the diameter

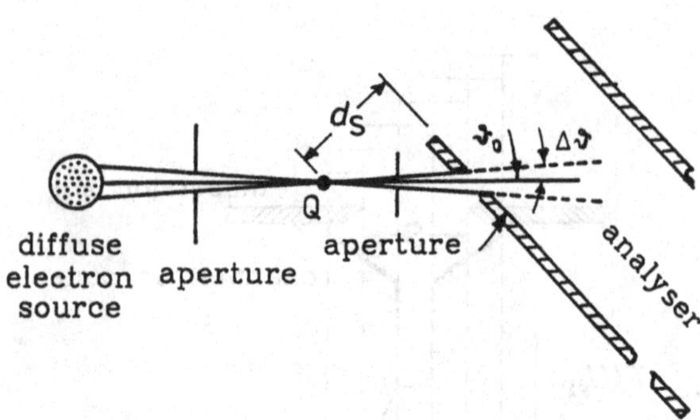

Figure 4.8 Schematic representation of an effective source Q defined by two apertures which collimate the diffuse original source. The analyser then has to accept only the smaller effective source Q placed at a vertical distance d_S. From [Ris72], see also Fig. 1.16 where S_1 and S_2 act as apertures to restrict Δz, the length of the acceptance source volume.

of the incoming beam (Fig. 1.13) is often reduced by an additional slit system which defines an *effective* source of smaller size Q_{eff}. This is shown in Fig. 4.8. (If a lens system is installed in front of the electron spectrometer, it is also possible to place such a diaphragm at the focal plane of the lens (see, for example, [MBB94]).) The improvement in the resolution, however, is accompanied by a reduction of the acceptance source volume and, therefore, of the counting rate. In addition, angle-resolved measurements may become difficult, because, depending on the angle setting of the analyser, different parts of the actual source volume are seen and, hence, an inhomogeneous density distribution in the source volume and/or a misalignment of the analyser can produce severe angle-dependent disturbances.

(iii) As can be seen from Fig. 4.7, the slit width S itself contributes to the resolution and controls the transmission function $T(U_{sp})$. These aspects are demonstrated quantitatively in Fig. 4.9 for a specially selected analyser: for S larger than $B(Q, \Delta\vartheta, \Delta\varphi)$, shown in Fig. 4.9(*c*), one gives away good energy resolution; for S smaller than $B(Q, \Delta\vartheta, \Delta\varphi)$, one loses transmission (Fig. 4.9(*a*)). An optimum slit width S follows for

$$S = B(Q, \Delta\vartheta, \Delta\varphi) \tag{4.9}$$

(Fig. 4.9(*b*)).

(iv) The angular spreads $\pm\Delta\vartheta$ and $\pm\Delta\varphi$ also contribute to the resolution. Their influence is due to the focusing properties of the analyser. It should be remembered that for a realistic estimation of their contribution a ray-tracing calculation has to be performed which takes into account the finite source size (point source results are not sufficient).

(v) Equ. (4.7) reflects the important feature of electrostatic deflection analysers: that the relative resolution $R = (\Delta E_{sp}/E^0_{kin})$ is constant, i.e., independent of the kinetic energy of the analysed electrons. This result has already been used in the discussion of the dispersion correction for the recorded counting rates

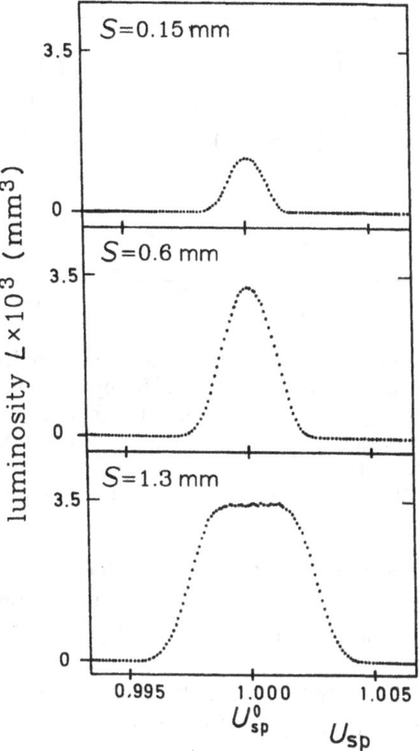

Figure 4.9 Calculated luminosity functions $L(U_{sp}^0, U_{sp})$ for a sector CMA as described in [DFM87]. The parameter S gives the width of the dispersion slit measured perpendicular to the principal ray (the length of this slit is 8 mm). The data refer to a photon beam of 0.5 mm diameter which results in an image width B of approximately 0.6 mm. Therefore, the three cases shown refer to the situations: (*a*) $B \gg S$, (*b*) $B \approx S$, (*c*) $B \ll S$. From [Ehr91].

(equs. (2.39) and (3.24)). The reason for this property can be understood better now, because with the help of Figs. 4.5 and 4.7 it was shown how an electrostatic deflection analyser operates. Applying this principle to two electrons created in the source volume with same intensity, but different kinetic energies E_{kin}^0, one gets the corresponding response functions (spectrometer functions) as shown in Fig. 4.10. Both functions are centred at their spectrometer voltages U_{sp}^0, and they have the same height which is the result of the energy-independent transmission function of the spectrometer but they differ in width. This difference can be explained by the fact that for higher kinetic energies the corresponding principal trajectories become less bent, and a larger range, ΔU_{sp}, of the spectrometer voltage is necessary in order to move the image B across the dispersion slit S. The dispersion correction is then just a compensation of this effect, because the observed peak areas are divided by U_{sp}^0 (or E_{kin}^0).

(vi) The constant resolution of the electrostatic deflection analyser means a small

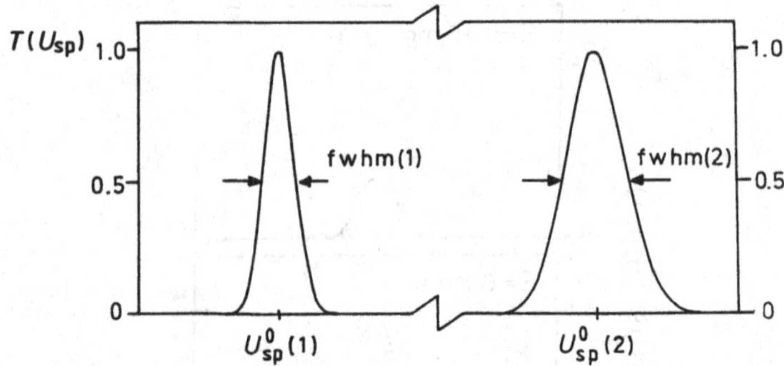

Figure 4.10 Transmission functions $T(U_{sp})$ of an electrostatic deflection spectrometer for electrons with two different kinetic energies which differ by a factor of 2: $E^0_{kin}(2) = 2E^0_{kin}(1)$ with $E^0_{kin}(1) = fU^0_{sp}(1)$ and $E^0_{kin}(2) = fU^0_{sp}(2)$. Because these transmission functions are plotted on the same abscissa scale, the voltage range needed to produce the transmission function with $E^0_{kin}(2)$ is twice as large as that needed for $E^0_{kin}(1)$. As a consequence, the spectrometer function at $U^0_{sp}(2)$ is twice as broad as that at $U^0_{sp}(1)$, with fwhm(2) = 2 fwhm(1).

ΔE_{sp} for low kinetic energies. Therefore, one can improve the resolution by retarding the electrons of higher kinetic energy before they enter the analyser. This point will be considered in Sections 4.2.4 and 4.4.2.1.

The following example has been selected to give an idea of the spectrometer resolution (see [DFM87]). The sector CMA chosen has $R_i = 60$ mm, $R_a = 135$ mm and $\vartheta_0 = 42.8°$, $\Delta\vartheta = \pm 3°$, $\Delta\varphi = \pm 10°$, a slight off-axis focusing and the energy dispersion $D = 6.115R_i$. For a photon beam diameter of 1 mm ($Q = 1$ mm) and a dispersion slit $S = 1$ mm (perpendicular to the direction of the principal ray) the resolution is

$$\left(\frac{\Delta E}{E}\right)_{sp} \approx \frac{1}{2}\left\{\frac{B + S + B(\Delta\varphi) + B(\Delta\vartheta)}{D}\right\}$$
$$\approx 0.0031 + \frac{1}{2}\frac{B(\Delta\varphi)}{D}. \tag{4.10}$$

(For a point source $B(\pm\Delta\vartheta)$ can be calculated from the first non-vanishing term of the Taylor expansion in equ. (4.5); for the present example it is the third derivative, and one gets $B(\pm\Delta\vartheta) = 0.3$ mm.) Ray-tracing calculations which include not only the influence of $B(\Delta\varphi)$, but also of the non-point-like source on $B(\Delta\vartheta)$ give $(\Delta E/E)_{sp} = 0.0042$.

4.2.3 Acceptance solid angle, spectrometer transmission, spectrometer function and luminosity

The electron spectrometer accepts (and transmits) only a fraction of the many electrons created at different points **r** in the actual source volume and emitted into the full space. This is due to the finite *acceptance* solid angle Ω_{acc} of the

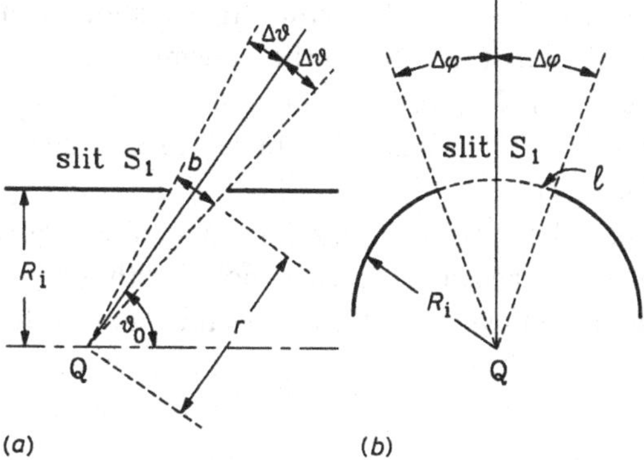

Figure 4.11 Definition of the solid angle Ω_0 of a point source Q accepted by the entrance slit S_1 of a sector CMA. Two cross-cuts are shown: (*a*) for a plane containing the symmetry axis of the analyser, (*b*) for a plane perpendicular to this axis. The principal ray starting at Q is shown together with the angular spreads from the finite acceptances in $\pm\Delta\vartheta$ and $\pm\Delta\varphi$. If expressed in spherical coordinates, the slit S_1 has a width $b = r2|\pm\Delta\vartheta|$ and a length $\ell = R_i 2|\pm\Delta\varphi| = r \sin\vartheta_0\, 2|\pm\Delta\varphi|$ as indicated. Their product $b\ell$ is the area of a sphere with radius r. If divided by r^2, one obtains the solid angle $\Omega_0 = \sin\vartheta_0\, 2|\pm\Delta\vartheta| 2|\pm\Delta\varphi|$.

spectrometer. However, Ω_{acc} is difficult to determine and instead Ω_0 the solid angle which accepts all trajectories which start at the middle of the source volume ($r = 0$) and enter the analyser's entrance slit is often used: according to Fig. 4.11 this solid angle Ω_0 is given by

$$\Omega_0 = \iint_{\text{opening}} \sin\vartheta \, d\vartheta \, d\varphi. \tag{4.11a}$$

and for the relatively small slits of the sector-analyser one gets

$$\Omega_0 = \sin\vartheta_0\, 2\,|\pm\Delta\vartheta|\, 2\,|\pm\Delta\varphi|, \tag{4.11b}$$

or, with an entrance slit of width $b = r2|\pm\Delta\vartheta|$ and length $\ell = R_i 2|\pm\Delta\varphi|$, and with a slit area $A_\perp$ on a sphere of radius r, this solid angle is given by

$$\Omega_0 = \frac{A_\perp}{r^2}. \tag{4.11c}$$

It should be noted that the solid angle Ω_0 is not automatically the same as the acceptance solid angle $\Omega_{acc}(\mathbf{r} = 0)$ of a point source. This only occurs if there is no further degradation due to additional slits and/or the size of the electron detector. In the following, such additional reductions will be assumed to be absent.

For a point source, the transmission T_0 of the electron spectrometer is then defined by

$$T_0 = \frac{\Omega_{acc}(\mathbf{r} = 0)}{4\pi}\, \varepsilon_g, \tag{4.12}$$

where ε_g takes care of the electron transmission through the meshes which usually separate the field and field-free regions. The transmission T_0 given is defined for the specific spectrometer voltage U^0_{sp}. The actual transmission at other spectrometer voltages is then described by the shape of the transmission function $T(U_{sp})$ which, therefore, also depends on U^0_{sp} and is expressed as $T(U^0_{sp}, U_{sp})$. (Except for the different normalization, $T(U^0_{sp}, U_{sp})$ is identical to the spectrometer function $F_{sp}(U^0_{sp}, U_{sp})$, see Fig. 1.14.)

For a finite source volume the situation for the analyser transmission becomes much more complicated, because for each point within the source volume a different transmission $T(\mathbf{r})$ exists, and the relevant quantity to be considered is the *luminosity function* $L(U^0_{sp}, U_{sp})$

$$L(U^0_{sp}, U_{sp}) = \int_{\substack{\text{source}\\ \text{volume}}} T(\mathbf{r}, U^0_{sp}, U_{sp})\, d\mathbf{r}. \tag{4.13}$$

(Equivalently, the luminosity function can be expressed as $L(E^0_{kin} = fU^0_{sp}, E_{pass} = fU_{pass})$.) The term *luminosity* is borrowed from optical radiometry† where it describes the flux-accepting and transmitting capabilities of an optical system for extended planar source and image areas. In the present context it is generalized for the non-planar, but three-dimensional, source volume. $L(U^0_{sp}, U_{sp})$ describes the response of the electron spectrometer to electrons of $U^0_{sp} = E^0_{kin}/f$ at the voltage setting U_{sp}. The luminosity $L(U^0_{sp}, U_{sp})$ has to be calculated by a ray-tracing procedure for the given analyser geometry (including all slits), for a selected diameter of the photon beam (and possibly a certain spatial distribution of this beam), for electrons emitted isotropically with kinetic energy E^0_{kin}, and for a certain setting of the spectrometer voltage U_{sp}. In this ray-tracing calculation a source region larger than the actual source volume is partitioned into finite volume elements $\Delta V'$, and from each element the isotropic electron emission into the whole 4π solid angle is also discretized. This gives a finite number of starting events with trajectories characterized by polar and azimuthal angles ϑ and φ, and separated by $\Delta\vartheta$ and $\Delta\varphi$. By following the individual trajectories, it is then checked whether such an electron reaches the detector or not. The luminosity L is finally obtained by a summation over all trajectories that reach the detector, multiplied by the individual weighting factor $\sin\vartheta\,\Delta\vartheta\,\Delta\varphi$ and the volume element $\Delta V'$. Repeating the calculation for different spectrometer voltages U_{sp}, one then obtains in $L(U^0_{sp}, U_{sp})$ the full response of the analyser. For practical reasons, however, it is convenient to distinguish different portions of the luminosity function $L(U^0_{sp}, U_{sp})$. In a first step the luminosity, L, and the spectrometer function $F_{sp}(U^0_{sp}, U_{sp})$ are introduced by defining (see Fig. 1.14)

$$L = L(U^0_{sp}, U^0_{sp}) \tag{4.14a}$$

and

$$F_{sp}(U^0_{sp}, U_{sp}) = L(U^0_{sp}, U_{sp})/L(U^0_{sp}, U^0_{sp}). \tag{4.15}$$

† See [HRi70, Hed71] and the discussion in Section 10.3.2; the name étendue is also used.

In a second step the luminosity

$$L = \int_{\text{source volume}} T(\mathbf{r})\,\mathrm{d}\mathbf{r}\Big|_{\text{at } U^0_{\text{sp}}} \tag{4.14b}$$

is factorized into the transmission T and the source volume ΔV, and the latter can be factorized by introducing the length Δz along the photon beam direction and the area q of the photon beam (see equ. (1.45)):

$$L = T\,\Delta V = T\,\Delta z\, q. \tag{4.16}$$

There is some freedom to choose T and ΔV (or Δz) in different ways which all yield the same L, because L is the only quantity which is uniquely determined by the ray-tracing procedure (for a preselected q value). However, when L is calculated, one also can determine which points within the actual source region contribute to the spectrometer response. Hence, an appropriate value for Δz can be selected, e.g., the value which delivers 90% of the intensity accepted by the spectrometer, and this Δz value then fixes the corresponding value for T. Because of their transparent interpretation the quantities T and Δz are frequently used, but it should be kept in mind that it is only their product (together with q), i.e., the luminosity L, which is the fundamental quantity.

Because of their great importance, approximate values for the luminosity L (for transmission T and length Δz of the source volume) will be derived from simple arguments. It will be assumed that the dispersion slit has a width S comparable to or larger than the width B of the image. This has the advantage that the diameter of the photon beam (area q) does not play an essential role; only the dependence on the z coordinate along the photon beam is of relevance. The finite value of Δz arises from the slit(s) in the electron spectrometer which reject trajectories for electrons starting from outside Δz. Therefore, the transmission $T(z)$ must approach zero at the edges of Δz. These limits and the transmission T_0 defined for a point source (equ. (4.12)) establish the $T(z)$ function approximately as demonstrated in Fig. 4.12. The average T value must then lie between the rectangular and triangular shapes. Making use of the nearly one-to-one imaging properties of the analyser (see Fig. 4.6), it can be seen that the essential contributions to the transmission come from a source region Δz the length of which is approximately equal to the Δz_{D} of the detector, and for these source points one will have approximately full transmission T_0. Therefore, one can make the approximation

$$L \approx T_0\, \Delta z_{\text{D}}\, q. \tag{4.17}$$

For the sector CMA described in [DFM87] one has $\vartheta_0 = 42.8°$, $\Delta\vartheta = \pm 3°$, $\Delta\varphi = \pm 10°$, which gives the transmission T_0 (with $\varepsilon_{\text{g}} = 1$) $\approx \Omega_0/4\pi \approx 2.0 \times 10^{-3}$. Then using $\Delta z_{\text{D}} = 8$ mm and $q = 0.2\ \text{mm}^2$ (0.5 mm diameter), one gets an estimated value of $L \approx 3.1 \times 10^{-3}\ \text{mm}^3$ which is close to the result from the ray-tracing calculation, $L = 3.5 \times 10^{-3}\ \text{mm}^3$ ($S \geq B$; Fig. 4.9(*c*)).

In the discussion it was assumed that the width S of the dispersion slit of the

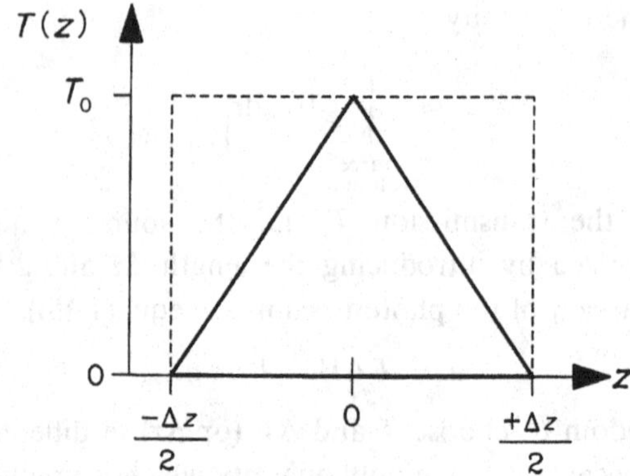

Figure 4.12 Two extreme cases of transmission functions $T(z)$ of an electrostatic analyser plotted for different points z of a linear source: triangular and rectangular shapes are shown by the solid and dashed lines, respectively. Δz is the length of the accepted source volume, T_0 the transmission obtained for electrons from the centre of the source (see equ. (4.12)).

electron spectrometer was large enough to accept in this dimension the imaged source volume. In practice, however, it may be difficult to fulfil or control this condition. For example, in the experimental set-up shown in Fig. 1.17 where the analyser is rotated around the direction of the photon beam, a non-circular photon beam would lead to different widths d of the source volume viewed at different angles. Therefore, ray-tracing calculations explore how the luminosity L depends for fixed width S of the dispersion slit on the diameter d of the photon beam. Results for a sector CMA are shown in Fig. 4.13; in (*b*) for the luminosity L divided by the cross section q of the photon beam and, for comparison, in (*a*) for the relative resolution $(\Delta E/E)_{sp}$. As may be expected from Fig. 4.7, the luminosity L starts to decrease, and the relative fwhm value $(\Delta E/E)_{sp}$ starts to increase, when d becomes larger than S ($S = 0.73$ mm in this example). However, the product of the two quantities divided by the area q of the photon beam is constant: in this example $L\,(\Delta E/E)_{sp}/q = 5.2 \times 10^{-5}$ mm.† The property

$$\frac{L}{q}\left(\frac{\Delta E}{E}\right)_{sp} = \left[\begin{array}{l}\text{const (independent of the photon beam diameter and}\\ \text{of the kinetic energy of the electrons)}\end{array}\right] \tag{4.18}$$

of an electrostatic analyser with a given geometry and slits again underlines that the dispersion-corrected area $A_D(i)$ of an electron line is the proper quantity for extracting relative intensities, because it is the dispersion-corrected area which is proportional to the quantity defined in equ. (4.18) and found to be constant. One has (see equs. (2.39) and (2.40))

$$A_D(i) = N_{ph}\,\sigma(i)\,n_v\,\frac{L}{q}\left(\frac{\Delta E}{E}\right)_{sp}\varepsilon_g\varepsilon, \tag{4.19}$$

† This value differs from the one quoted in [DFM87] because of a numerical mistake in the ray-tracing program.

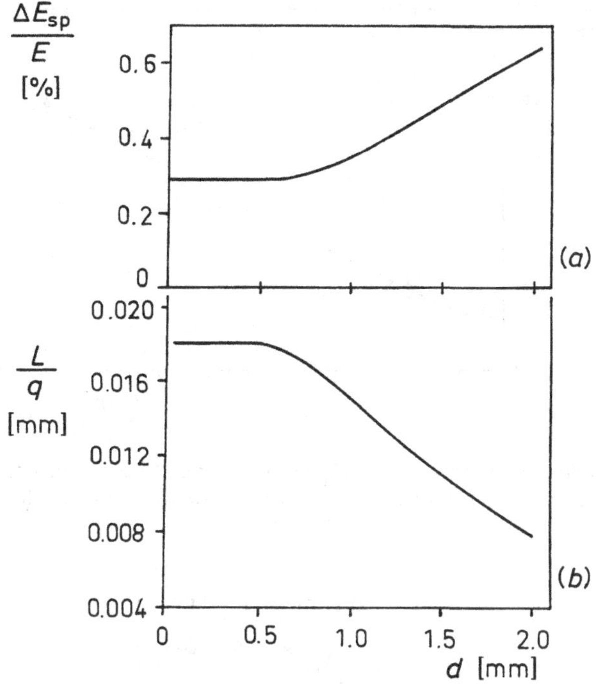

Figure 4.13 Dependence of (*a*) the relative energy resolution $\Delta E_{sp}/E$ and (*b*) the luminosity L (divided by the cross section q of the photon beam) on the photon beam diameter d. The data refer to the sector CMA described in [DFM87] with a dispersion slit which is 0.73 mm wide (measured perpendicular to the principal trajectory) and 8.0 mm long. From [Ehr91].

i.e., at a given photon energy all parameters are the same except for the different cross sections $\sigma(i)$ which belong to different photolines.

4.2.4 Operation modes

The different modes in which electrostatic deflection analysers can be operated will now be discussed, with some comments on their correct performance.

The first mode allows the recording of *energy distribution curves* (the EDC mode, also called the *photoelectron spectrometry* (PES) mode) by changing, for a fixed photon energy $h\nu$, the spectrometer voltage U_{sp} (see Figs. 2.4 and 3.1). In this mode, the spectrometer transmission T and the length Δz along the photon beam, as well as the relative spectrometer resolution $R = (\Delta E/E_{kin})_{sp}$ are kept constant. This last property requires the application of the dispersion correction to the measured electron spectrum before relative intensities can be extracted.

Generally, inhomogeneities in electric fields and in residual magnetic fields disturb the regular performance of the spectrometer by changing the resolution and transmission. Such effects become extremely important at lower kinetic energies. Figs. 4.14 and 4.15 show examples of changing the resolution and transmission,

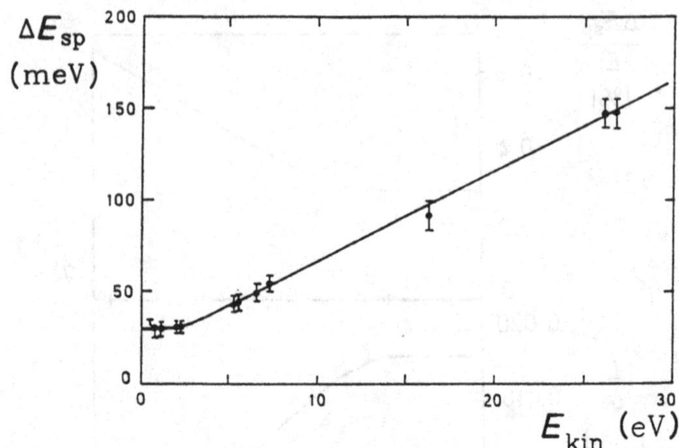

Figure 4.14 Experimental fwhm values ΔE_{sp} for a sector CMA as a function of the kinetic energy of the electrons. The solid curve is drawn to guide the eye through the data. From [Der84].

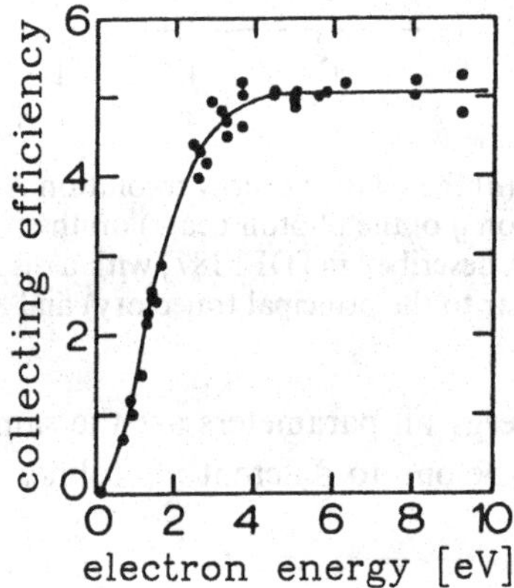

Figure 4.15 Relative transmission T (called here the 'collecting efficiency') of a CMA as a function of the kinetic energy of the electrons. The solid curve guides the eye through the experimental data. For energies higher than 4 eV the transmission is constant, below this value it drops to zero (the *cut-off* effect due to disturbances in the performance of the analyser caused by residual magnetic fields and inhomogeneities of electrical fields, see Sections 4.5.2 and 4.5.3). Reprinted from *J. Electron Spectrosc. Relat. Phenom.*, **15**, Samson, 257 (1979) with kind permission of Elsevier Science – NL, Sara Burgerhartstraat 25, 1055 KV Amsterdam, The Netherlands.

respectively. From Fig. 4.14 it can be seen that for E_{kin} approaching zero, the width ΔE_{sp} approaches a finite value which limits the ultimate resolution and violates $(\Delta E/E_{kin})_{sp}$ = const. (The finite value of 30 meV shown in Fig. 4.14 does not imply a general resolution limit. For example, for a hemispherical analyser with 200 mm mean radius and a resolution-defining slit of 0.2 mm, an instrumental width ΔE_{sp} better than 2.7 meV has been found at 2 eV pass energy [MBB94].)

Fig. 4.15 shows how for small kinetic energies the transmission T drops to zero. This is called the *cut-off* effect of the analyser transmission.

A second mode of operation which also allows EDC curves to be recorded uses a *constant pass energy* E_{pass} for the electrons transmitted through the analyser. Since the photoprocess yields electrons with energies different from E_{pass}, a system (usually a lens) installed at the entrance to the electron analyser changes the electron energies by acceleration/retardation from E_{kin} to E_{pass}. In order to obtain the energy distribution of ejected electrons for this mode, the retardation voltage U_{ret} is ramped while the spectrometer voltage U_{sp} which is superimposed on U_{ret} is kept constant. If E_{kin} is larger than E_{pass}, this mode of operation brings the advantage of improved experimental linewidth, because one has

$$\Delta E_{sp}(\text{at } E_{pass}) < \Delta E_{sp}(\text{at } E_{kin}) \quad \text{for} \quad E_{pass} < E_{kin}. \tag{4.20}$$

For the dispersion corrected area, $A_D = F(E_{pass})/E_{pass}$, of a photoline one gets in this case

$$A_D = \left[N_{ph} n_v \sigma \, \Delta z \; T(E_{pass}) \, \varepsilon(E_{pass}) \frac{\Delta E_{pass}}{E_{pass}} \right] T_{ret}(E_{kin}, E_{pass}). \tag{4.21}$$

The quantities in the square brackets are just the ones known from equ. (4.19) with equ. (4.16). Due to the fixed value of E_{pass} it can be seen that for the evaluation of relative intensities the dispersion correction can be omitted. However, the transmission factor $T_{ret}(E_{kin}, E_{pass})$ which describes the change of transmission caused by the retardation becomes very important in this case, see Fig. 4.16. It has to be determined experimentally, and in ideal cases it can be estimated on the basis of Liouville's theorem for optical systems (see Section 10.3.2). In the example shown in Fig. 4.16 the essential action of the retardation field is to change the brightness B in one dimension. (One has a one-dimensional problem because the lens produces focusing of the line source in one dimension only (for details see [GSa75]).) Following equ. (10.47) one gets (subscripts ℓ and r denote quantities before and after retardation)

$$\left. \frac{B_r}{B_\ell} \right|_{1-\dim} = \sqrt{\left(\frac{E_{pass}}{E_{kin}} \right)}, \tag{4.22a}$$

and from equ. (10.45b)

$$T_{ret}(E_{kin}, E_{pass}) \sim I_r \sim B_r \Big|_{1-\dim}. \tag{4.22b}$$

Since the brightness B_ℓ is fixed by the source strength, the combination of equs. (4.22a) and (4.22b) leads to

$$T_{ret}(E_{kin}, E_{pass}) \sim \sqrt{(E_{pass}/E_{kin})}, \tag{4.22c}$$

and this dependence describes rather well the collection efficiency observed in Fig. 4.16 when normalized to unity at $E_{kin} = E_{pass}$.

Disturbances due to inhomogeneities in electric fields and to residual magnetic fields on electrons of low kinetic energy also occur in this mode of operation. They

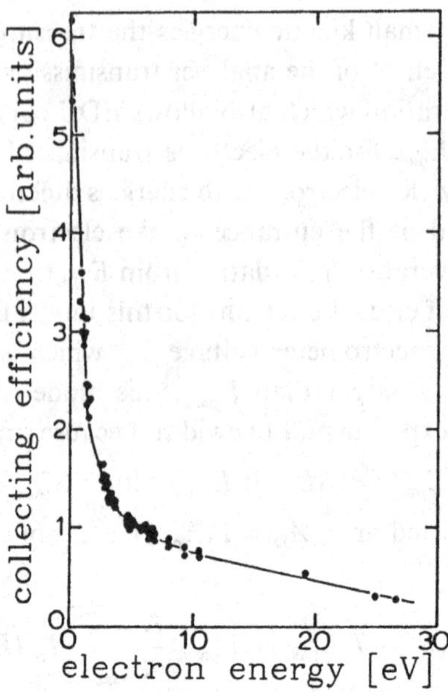

Figure 4.16 Relative transmission function T_{ret} (called here the 'collecting efficiency') of a CMA with constant pass energy (E_{pass} = 3 eV) as a function of the kinetic energy of the electrons. For E_{kin} > 3 eV the electrons are retarded, for E_{kin} < 3 eV they are accelerated, before they enter the analyser with the required 3 eV pass energy. The solid curve is drawn to guide the eye through the experimental data (for its explanation see the main text). Reprinted from *J. Electron Spectrosc. Relat. Phenom.*, **15**, Samson, 257 (1979) with kind permission of Elsevier Science – NL, Sara Burgerhartstraat 25, 1055 KV Amsterdam, The Netherlands.

cause a degradation of analyser transmission $T(E_{pass})$ which, however, may be partly compensated by an increase in the collection efficiency of electrons with $E_{kin} < E_{pass}$ (effect of Δz). On the other hand, such disturbances are responsible for the finite value of ΔE_{sp} for $E_{kin} \rightarrow 0$, and this sets a lower bound to practical values of E_{pass}: pass energies which are too small do not improve the resolution further, they only degrade the transmission.

The discussion so far has concentrated on EDCs taken at a fixed photon energy $h\nu$. Often, only one specific line in the spectrum of ejected electrons and how this changes as a function of the photon energy that is of interest. For this purpose, EDCs have to be taken in discrete and closely spaced steps for many photon energies, and the selected line analysed each time (see Section 5.1, in particular the discussion in connection with Fig. 5.5). However, a faster approach is to make use of a third mode of operation called the *constant-ionic-state* (CIS) mode. Here the photon energy $h\nu$ is swept synchronously with the spectrometer voltage U_{sp}, (alternatively, the acceleration/retardation voltage U_{ret} of a spectrometer with an entrance lens can be varied with the photon energy) and one follows the maximum of a selected photoline and explores its intensity variations. The photoline is selected by the corresponding ionization energy E_I which is a quantity of the *ionic* state, hence the name for this mode. (Originally, this CIS mode was called *constant-initial-state* energy spectroscopy, but a better name is *constant-ionic-state* spectroscopy [PGG77].) With a constant spectrometer pass energy the peak height is proportional to the intensity of the photoline, so the CIS mode has the two advantages that the resolution of the CIS spectrum is independent of the spectrometer resolution† and the intensity only depends on the wavelength-dependent monochromator function. However, precautions are necessary to ensure that during such a scan no other lines (e.g., Auger lines) and/or a changing background affect the intensity information from the preselected photoline.

Finally, a fourth mode of operation can also be useful, the *constant-final-state* (CFS) mode. Here the photon energy $h\nu$ is again varied, but the spectrometer voltage U_{sp} (or alternatively, the acceleration/retardation voltage U_{ret} of a spectrometer with an entrance lens) is kept constant. In this way one always observes electrons with the same kinetic energy E_{kin}: e.g., the photon-energy dependence of a preselected Auger transition. Due to its similarity to the CIS mode, this mode has the same advantage of independence of spectrometer resolution in the E_{pass} mode, but precautions are again necessary to ensure that no other lines (in this case photolines) or a changing background intensity affect the results.

4.3 Electron detectors (channeltron/channelplate)

A critical part of the electron spectrometer is the detector which registers the energy-analysed electrons. Channeltrons or channelplates are very convenient detectors, and they will now be discussed with respect to their performance characteristics, the use of channelplates as position-sensitive detectors and their detection efficiencies.

4.3.1 Performance characteristics

The names of both detectors reflect that these devices are channels which act as continuous dynode electron multipliers. If there is one channel, it is called a *channeltron*‡ (channeltron electron multiplier, CEM), if many microchannels are used to form a plate it is called a *microchannel electron multiplier plate* (in short a *microchannelplate*, MCP, or *channelplate*), see Fig. 4.17. A comprehensive description of these devices is given in [Wiz79].

A channeltron consists of one channel (about 1 mm in diameter, and 50 mm in length) with a special shape (a curved channel with often a cone at the entrance)

† Of course, a structure not well resolved in the EDC mode will remain not well resolved also in this CIS-mode.

‡ The word channeltron is a registered trademark of the Galileo Electro-Optics Corporation.

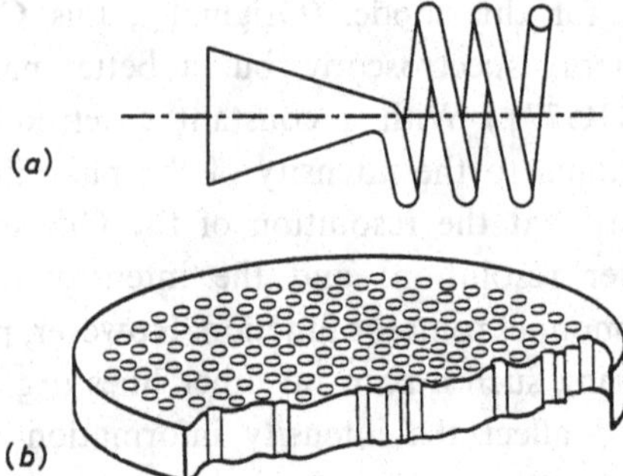

Figure 4.17 (*a*) Channeltron with entrance cone (from [Gal77]) and (*b*) channelplate, cutaway view; (*b*) is reprinted from *Nucl. Instr. Meth.*, **162**, Wiza, 587 (1979) with kind permission of Elsevier Science – NL, Sara Burgehartstraat 25, 1055 KV Amsterdam, The Netherlands.

while a channelplate consists of a large number (10^4–10^7) of small channels (about 10 μm in diameter, 0.5–1 mm in length) which have a small bias angle (~8°) to the input surface of the formed plate. The channel surface is covered with a semiconducting material which possesses good characteristics for the emission of secondary electrons. (Electrical contact to the channel(s) at the entrance and exit is provided by the deposition of a metallic coating.) Therefore, each channel can be considered to be an electron multiplier with continuous dynodes which act as their own dynode resistor chain. If a positive voltage is applied at the channel exit, an electron which hits the surface at the entrance can generate one or more secondary electrons which are accelerated in the field along the channel axis and strike the opposite side of the channel, generating additional secondaries and so on, see Fig. 4.18. As a result, a single electron at the detector entrance has a certain probability ε (the detection efficiency) for starting an avalanche of electrons. The charge cloud produced can be collected at the detector exit in a cup connected to the anode of the channeltron or on an addition anode behind the exit of the channelplate. Frequently, two or even more plates are stacked one after the other with opposite bias angles in order to improve the performance (see below); this mounting is called the *Chevron* configuration (Chevron is a trade mark), see Fig. 4.18.

The electron avalanche leads to a charging of the collector's capacitance C_D, and this process stops when the total charge Q of the avalanche has been deposited (see Fig. 4.19). The collection time t_{coll} is equal to the transit time that the electrons need for their passage through the channel. It is approximately 100 ns for a channeltron and 1 ns for a channelplate. Subsequent processes add more and more charges to the collector. Hence, in order to allow the counting of single events, discharging of the collector is provided through a load resistor R_a. Fig. 4.20 shows the equivalent circuit for the channeltron/channelplate as a current source $i(t)$ together with the inherent collector capacitance C_D and external electronic

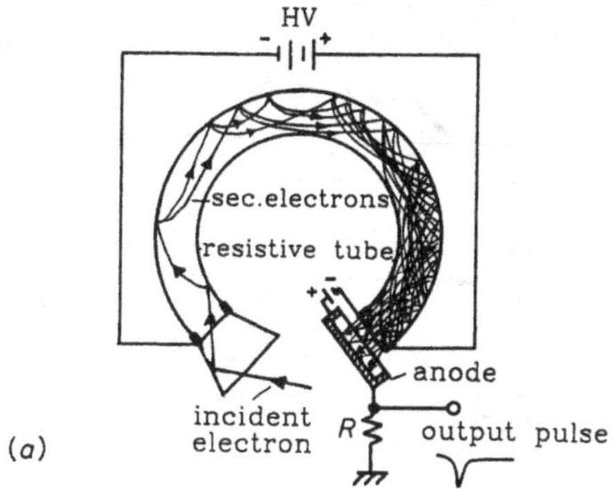

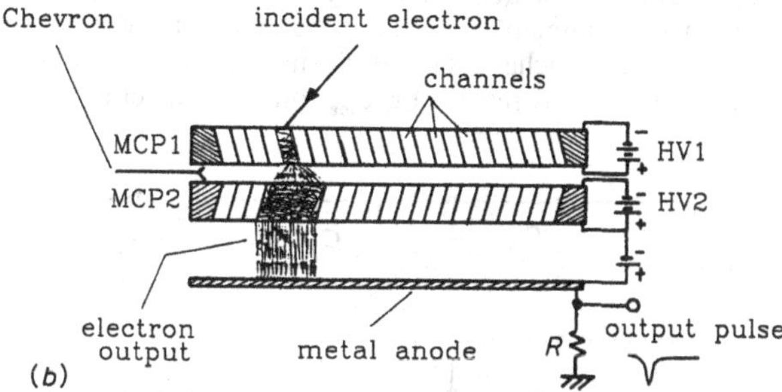

Figure 4.18 Illustration for the operation of (*a*) a channeltron (reprinted from Granneman and Van der Wiel, *Handbook on Synchrotron Radiation* (1983) with kind permission from Elsevier Science – NL, Sara Burgerhartstraat 25, 1055 KV Amsterdam, The Netherlands) and (*b*) a channelplate (reprinted from *Nucl. Instr. Meth.* **162**, Wiza, 587 (1979) with kind permission from Elsevier Science – NL, Sara Burgerhartstraat 25, 1055 KV Amsterdam, The Netherlands). Two channelplates, MCP1 and MCP2, are stacked together changing their bias angle as shown in the figure (Chevron mounting). An electron hitting the inner wall of the channeltron (similarly for the first channelplate) leads by iterative emission of secondary electrons to an avalanche of electrons. The electron avalanche is collected on the anode. The high voltage power supply and the extraction of the output pulse on the load resistor R are also shown.

elements (load resistor R_a, coupling capacitor C_K, and preamplifier Amp). The maximum amplitude U_{max} of the voltage pulse sent to the amplifier is given by†

$$U_{max} = Q/C_D. \tag{4.23a}$$

† These values will also depend on the input impedance of the preamplifier with resistive and capacitive components R_i and C_i which then modify R_a and C_D to R_{tot} and C_{tot}. For $R_{tot}C_{tot} \gg t_{coll}$ the induced charge will remain on C_D and be transformed to the voltage as given in equs. (4.23); this is called the *voltage mode.* For $R_{tot}C_{tot} \ll t_{coll}$ the detector capacitance will discharge faster than charge is induced on it by the collection process, and the current into the input of the preamplifier is then equal to the signal current in the detector; this is called the *current mode.* These items and the derivation of a good timing signal by appropriate charge-, voltage- or current-sensitive preamplifiers as well as the correct termination using a cable between the detector and the preamplifer are discussed in detail in [Spi82].

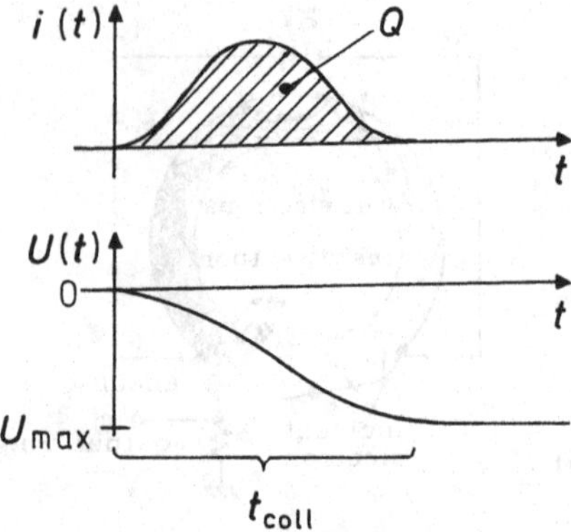

Figure 4.19 Time-dependent current $i(t)$ and voltage $U(t)$ signal produced by an electron avalanche in the channeltron/channelplate. The shaded area of the current pulse represents the total charge Q of the avalanche collected during the time t_{coll} on the detector's capacitance, thus producing U_{max} on this capacitor.

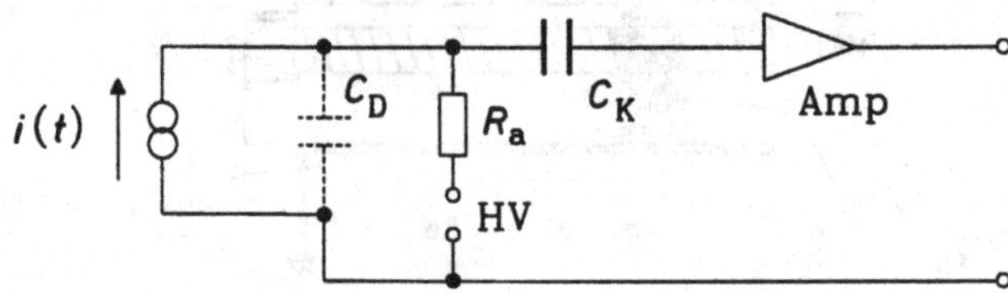

Figure 4.20 Equivalent circuit for the channeltron/channelplate as the current source $i(t)$ shown together with the main components of the external circuit. C_D is the capacitance of the detector which is depicted by the dotted lines in order to indicate that – in contrast to actual electronic components – it is a property inherent to the detector; HV is a high voltage source, R_a the load resistor, C_K the high voltage coupling capacitor and Amp the preamplifier.

Typically, one has

$$U_{max}\ (\text{channeltron}) = \frac{e_0 \times 10^8}{20\ \text{pF}} = 0.8\ \text{V}, \tag{4.23b}$$

$$U_{max}\ (\text{set of 2 plates}) = \frac{e_0 \times 10^7}{60\ \text{pF}} = 0.03\ \text{V}. \tag{4.23c}$$

From the function of a channeltron/channelplate detector it is obvious that high gains are desirable. However, ion feedback and space charge effects limit the gain: with increasing charge of the electron avalanche, electron impact ionization with molecules of the residual gas or molecules desorbed under electron bombardment from the channel surface occurs more frequently. The ions produced are then accelerated towards the channel input. If such an ion hits the surface at the channel entrance, it may release an electron which again can start an avalanche of practically the same size, i.e., it causes after-pulses.

For straight channel multipliers, the onset of ion feedback sets an upper limit to the operation gain of approximately 10^4. At such a gain, the pulse height distribution of registered events approximates an exponential function with a negative exponent. In order to achieve higher gains that produce a more suitable pulse height distribution (with a maximum; see below), ion feedback must be avoided. To do this, advantage is taken of the different paths of ions and electrons in the same electric field and of electronic signal discrimination: the different masses of electrons and ions lead the particles to have different paths in the same electric field; the paths of the ions are less bent than those of the electrons. A curved channel structure means that the ions hit the surface before they reach the channel entrance. This results in small pulses of such disturbing processes which can be rejected by pulse height discrimination. In a channeltron, ion feedback can be suppressed completely by use of an appropriate geometrical shape. However, in channelplates with straight channels this cannot be achieved so easily (for channelplates with curved channels see, for example, [Tim81]). Therefore, channelplates are used almost exclusively in a sandwich or Chevron mounting of two channelplates which approximately simulates a curved channel by a proper adjustment of the corresponding bias angles (see Fig. 4.18).

If ion feedback is suppressed, the gain is still limited by space charge effects. At high enough gains, the large space charge of the electron cloud shields the electric field across the channel. As a consequence, the secondary electrons are accelerated less, i.e., they recover less kinetic energy in the weakened electric field, which results in a lower rate of production of further secondaries. Saturation occurs when the coefficient δ for secondary emission approaches unity. For a channeltron, saturation effects occur at a gain of approximately 10^8 (see Fig. 4.21), and the channeltron is operated in the saturation region because it provides large voltage

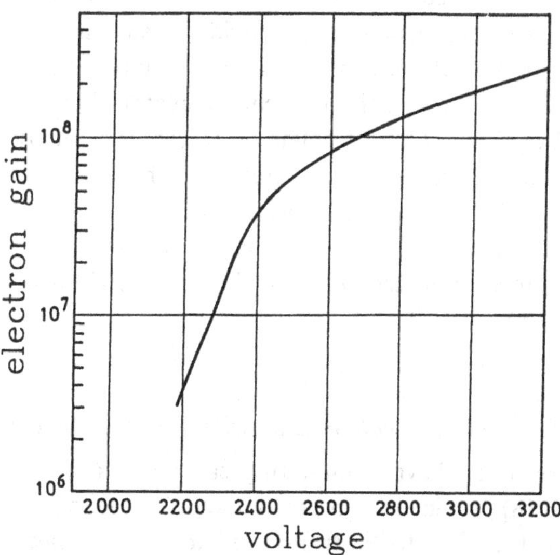

Figure 4.21 Typical gain of a channeltron as a function of the applied voltage. The change in the initial slope is due to saturation effects in the channeltron. From [Gal77].

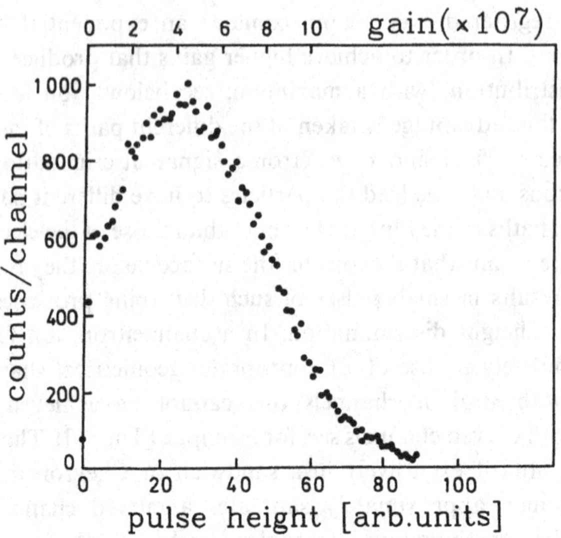

Figure 4.22 Distribution of the pulse height of detector impulses for a set of two channelplates in a Chevron mounting. The pulse height given at the lower abscissa in arbitrary units is transformed on the upper abscissa to the average gain values of the corresponding electron avalanches. The voltage drop across each plate is 1 kV; no extra voltage is applied between the plates. Reprinted from *Nucl. Inst. Meth.* **162**, Wiza 587 (1979) with kind permission from Elsevier Science – NL, Sara Burgerhartstraat 25, 1055 KV Amsterdam, The Netherlands.

pulses of nearly equal amplitudes. For a single channelplate with its smaller channel diameter, space charge saturation would require a gain of approximately 10^6 which cannot be reached because of the lower limit set by ion feedback. As mentioned above, a Chevron mounting provides a sufficiently large directional change to suppress ion feedback from the second channelplate to the entrance of the first channelplate. However, the two channelplates allow the electron cloud leaving a channel of the first channelplate to spread over many channels of the second channelplate. Depending on this spreading, which can be controlled by the distance and voltage between the channelplates, the total gain for the Chevron mounting of two channelplates is approximately 10^7, and because of saturation effects in many channels of the second plate, the pulse height distribution becomes peaked (see Fig. 4.22).

4.3.2 A channelplate as a position-sensitive detector

Two channelplates in a Chevron mounting can be used as a *position-sensitive* detector (see Fig. 4.18) provided the collector anode is able to preserve the position information which then has to be transferred to suitable electronics. Position-sensitive detectors are used in electron spectrometry mainly for four different applications:

(i) Simultaneous recording of an angular distribution pattern for electrons of the same kinetic energy. An example of this is the toroidal analyser shown in Fig. 4.3 where the channelplate detector is placed off the focal plane of the spectrometer and has a collector anode of an appropriate shape in order to detect angular-dependent intensities.

(ii) Simultaneous recording of a certain portion of the electron spectrum. For this purpose, the dispersion slit of the electron analyser has to be removed and replaced by the channelplate detector which is aligned in the focal plane. (Often the channelplate is aligned perpendicular to the incident electrons in order to have a high detection efficiency and to minimize secondary peaks (see Fig. 4.5).) Again, an appropriate collector anode shape is needed in order to read out the whole intensity distribution.

(iii) Improvement of timing properties in coincidence experiments (see Fig. 4.50).

(iv) In the previous cases the position-sensitive detector was used in conjunction with an electron spectrometer. Recent developments add another very interesting application of a position-sensitive detector, its operation as a charged-particle-imaging device, which allows the direct and simultaneous observation of the energy and angular distribution of electrons emitted from a point source without a spectrometer (for details see [HDy94, UDM94]).

Several methods have been developed for the extraction of the position information. (For a review see [Fra84, RHo86]; for an integrated circuit multi-detector see [HBC92]; for an extension to different geometries and/or to two-dimensional information see [MJL81, TCR91] with references therein.) Here only one-dimensional discrete or continuous structures of the collecting anode will be discussed. An electron entering a channelplate at a certain position can initiate an electron avalanche. Independent of the finite size, the centroid of this avalanche always contains information about the spatial position of the primary impact. Therefore, position-sensitive devices concentrate on the determination of this centroid, and it can even be helpful to spread the avalanche on the collector over a broader spatial range (see Fig. 4.18(*b*)).

In Figs. 4.23 and 4.24 two examples with discrete collector strips are shown, which differ only in their subsequent electronics. In the set-up shown in Fig. 4.23 the centroid of the electron avalanche is given by the largest signal in any of the individual electronic channels. In the set-up shown in Fig. 4.24 the total charge Q of the electron avalanche flows to both ends of the RC line. Depending on position, the fractions Q_1 and Q_2 arrive at the left- or right-hand sides, respectively, and these charges are proportional to the distances $(L - P)/L$ and P/L. Therefore, the two amplifiers yield signals U_1 and U_2:

$$U_1 = \text{const}\, Q \frac{(L-P)}{L} \quad \text{and} \quad U_2 = \text{const}\, Q \frac{P}{L}. \tag{4.24}$$

By dividing one of the signals by the sum of both, one gets the desired dependence on the position P/L. The same concept lies behind the resistive strip anode

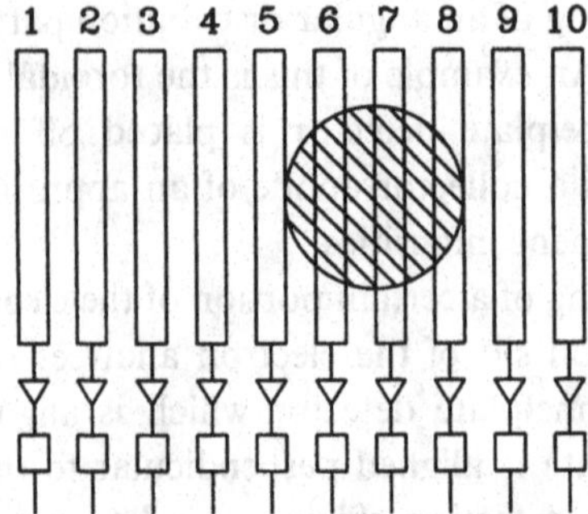

Figure 4.23 Extraction of one-dimensional position information from a channelplate detector by using a discrete multianode with individual electronics. The ten strips indicate the discrete multianode which is in a Chevron mounting behind the second channelplate (not shown in this figure). The shaded circular area represents an electron avalanche incident on the multianode, hitting three strips in the example shown. Each strip is connected to its own pre- and main amplifier indicated by triangles and rectangles, respectively. In the example shown the electronic circuit no. 7 provides the highest output signal indicating that the centroid of this electron avalanche was close to the position of strip no. 7. From [Wac85].

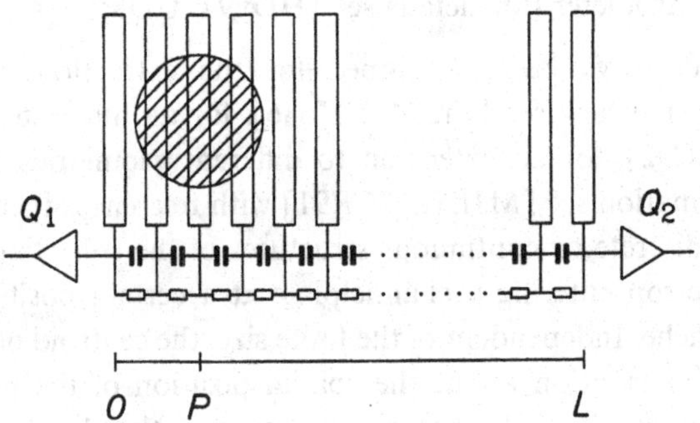

Figure 4.24 Extraction of one-dimensional position information from a channelplate detector by using a discrete multianode with an *RC* line. For an explanation of the strips and the dashed circular area see the caption of Fig. 4.23. Each strip is connected to the *RC* line indicated by the resistor and capacitor symbols. The total charge Q of the electron avalanche (shaded area) incident on the multianode flows to both ends of the *RC* line, giving Q_1 and Q_2 in amounts proportional to the distances (P, L) and (O, P), respectively. The two preamplifiers in which these charges are collected are indicated by triangles. From [Wac85].

discussed below. It is, however, also possible to replace the discrete *RC* line by a delay line using inductive and capacitive components. This introduces an artificial time delay between each individual strip which accumulates to a measurable time delay of the pulse at the end of the line. The arrival time, measured as the difference in the overall non-delayed signal or between the two ends of the delay line, can then be used to extract the desired position information (for details see [KKK87, SWi88]). There are advantages and disadvantages for all the set-ups which require

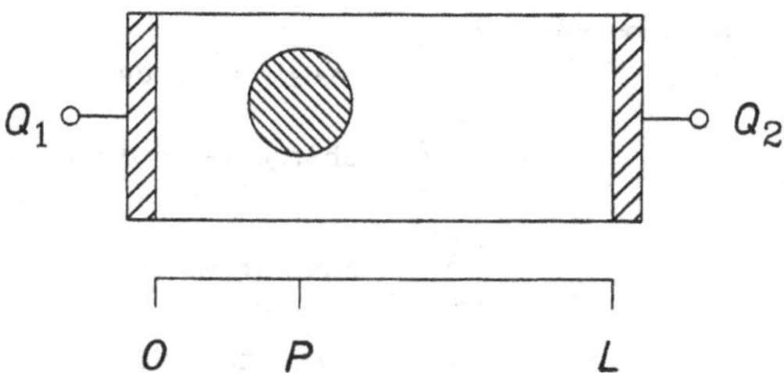

Figure 4.25 Extraction of one-dimensional position information from a channelplate detector by using a continuous resistive strip anode. If an electron avalanche (shaded circular area) hits the resistive material of the anode, the charges Q_1 and Q_2, collected on the electrical contacts (shaped rectangular areas), are proportional to the distances (P, L) and (O, P), respectively.

a detailed balancing in practical applications. The main points to consider are the number and accommodation of electronic components, cross-talk between neighbouring strips, and the processing of high counting rates. In addition, attention must be paid to the limitation of the spatial resolution caused by the width of the strips.

In Fig. 4.25 an example of a continuous collector anode (the *resistive strip* anode) is shown. This device is frequently used because of its mechanical and electronic simplicity as well as its ability to handle moderate counting rates with good spatial resolution. The anode consists of an isolating sheet (typically 30 mm long and 10 mm wide for a channelplate set with 25 mm diameter) covered with a homogeneous film of resistive material (for example, a carbon layer with typically some 10 kΩ/strip). Both ends of the sheet are goldplated for good electrical contacts to the preamplifier. A rear metal plate stabilizes the capacitance C_D of the system. The resistive strip anode acts as a diffusive RC transmission line where the total charge Q flows – depending on the location of the event – in different amounts Q_1 and Q_2 to each side of the collector anode which provides the desired information about the position P (compare the set-up of Fig. 4.24). In most applications, the heights of the respective voltage signals U_1 and U_2 are used for the analysis (see [MDG75]); however, the rise time of these signals can also be used (see [PEM74]).

As an example of deriving position information from a resistive strip anode, the main components of the necessary electronics are shown in Fig. 4.26 which is for the most part self-explanatory. The necessary information about $Q = Q_1 + Q_2$ and Q_1 or Q_2 is obtained here using the digital signals delivered by the two fast analogue-to-digital converters. It is, however, also possible to derive this information from analogue signals using analogue devices for summation and division. In either case there is a limiting resolving ability of one of the instruments used and, hence, a 'dead' time T_{dead}. Usually, an analogue divider has $T_{\text{dead}} \approx 12$ μs, but with fast analogue-to-digital converters and fast digital sum-and-divider-units a $T_{\text{dead}} \approx 3$ μs can be achieved. This shorter dead time is advantageous because

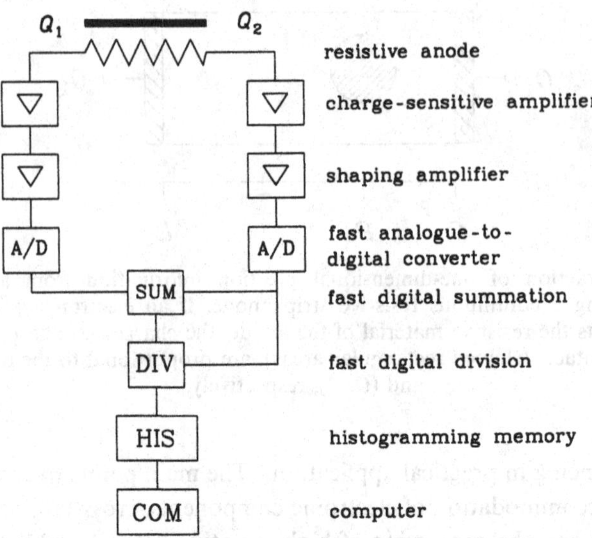

Figure 4.26 Possible electronic circuit for deriving one-dimensional position information from a position-sensitive detector with a resistive strip anode. The two charges Q_1 and Q_2 on the ends of the anode are amplified, shaped and converted to a digital signal. The mathematical operations of $\{Q = Q_1 + Q_2\}$ and $\{Q_2/Q\}$ are performed electronically, and the result is stored in a histogramming memory from which it is read into the computer. Q_2/Q carries the information first that an electron has been detected and second at which position this electron has hit the detector. From [Wac85].

it leads to smaller losses in the observed counting rate I_{obs} as compared to the true counting rate I_{true} delivered by the process.

Assuming a random distribution of true events which have a Poisson distribution, the relation between I_{obs} and I_{true} can be derived as follows. The average number n of events during the time interval T_{dead} is given by

$$n = I_{true}\, T_{dead}, \tag{4.25}$$

and the number n_{lost} of events unrecorded (lost) during the time T_{dead} follows as a summation over the Poisson probabilities $P_k(n)$ for k events, multiplied by k,

$$n_{lost} = \sum_{k=1}^{\infty} kP_k(n) = \sum_{k=1}^{\infty} k \frac{e^{-n}}{k!} n^k = n \sum_{\tilde{k}=0}^{\infty} \frac{e^{-n}}{\tilde{k}!} n^{\tilde{k}} \tag{4.26a}$$

(with $\tilde{k} = k - 1$), which gives

$$n_{lost} = n. \tag{4.26b}$$

The result states that n_{lost} is equal to the average number n of events during T_{dead} as one might have expected. Since n_{lost} belongs to *one* observed event, the total rate I_{lost} of lost events is

$$I_{lost} = I_{obs}\, n_{lost} = I_{obs}\, I_{true}\, T_{dead}, \tag{4.27}$$

provided the unrecorded counts do not themselves extend the dead time. Using

$$I_{true} = I_{obs} + I_{lost}, \tag{4.28}$$

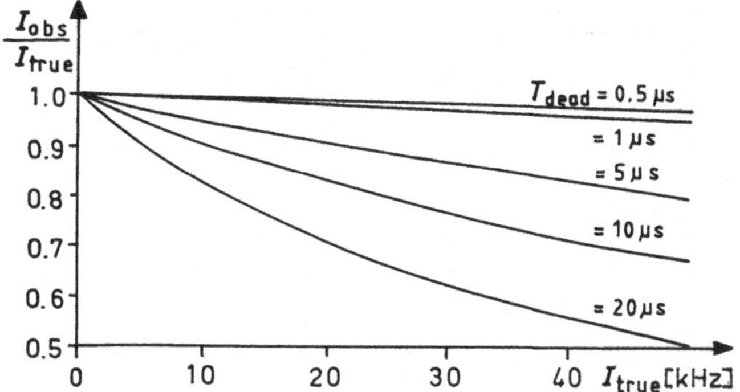

Figure 4.27 Dependence of the ratio of the observed counting rate I_{obs} to the true counting rate I_{true}, shown as a function of I_{true}, and for values of the electronic dead time T_{dead}. From [Wac85].

one derives

$$\frac{I_{obs}}{I_{true}} = \frac{1}{1 + I_{true} T_{dead}}. \tag{4.29}$$

(Using $I_{obs} = I_{true} - I_{lost}$, the equivalent form $I_{obs}/I_{true} = 1 - I_{obs} T_{dead}$ would have been obtained, and this form shows that for $I_{true} \to \infty$ the observed counting rate approaches $1/T_{dead}$.) The ratio I_{obs}/I_{true} is plotted in Fig. 4.27 for several values of the dead time T_{dead}. It can be seen that a dead time of 10 µs limits the counting rate to 10^4 Hz with an electronic detection efficiency of 90%. (In this context it should be added that discrete multianodes with separate processing electronics have the advantage of much smaller electronic dead times, and can thus handle counting rates up to 10^7 Hz.)

4.3.3 Detection efficiency

Not every electron which hits a channeltron or channelplate detector will be detected. The detection efficiency depends on the processes at the detector entrance, on the formation and passage of the electron avalanche, and on the registering of the resultant electronic signal. The electronic signal can be handled in a controlled way. In particular, for a channeltron detector operating in the saturation region, a high electronic amplification ensures that practically all electron avalanches are counted. Therefore, the detection efficiency ε is determined primarily by the processes at the detector's entrance. The incident electron must release one or more secondary electrons which then start the electron avalanche. If the detector surface has a mean coefficient δ for the emission of secondary electrons, the ejection of m secondary electrons is governed by the Poisson distribution

$$P(m) = \frac{e^{-\delta}}{m!} \delta^m. \tag{4.30}$$

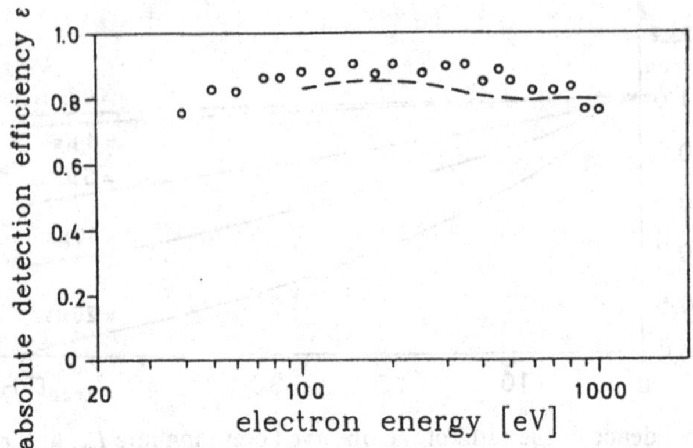

Figure 4.28 Absolute efficiency ε for the detection of electrons with different kinetic energies by a channeltron. Reprinted from Granneman and Van der Wiel in *Handbook on Synchrotron Radiation* (1983) with kind permission from Elsevier Science – NL, Sara Burgerhartstraat 25, 1055 KV Amsterdam, The Netherlands. Open circles from [Ost79]; dashed curve from [Was74]; see also [SSm91].

From this relation one gets for the probability W of the ejection of one or more secondary electrons

$$W = 1 - P(0) = 1 - e^{-\delta}. \tag{4.31}$$

Under the assumption that the ejection of one or more secondary electrons leads to an electron avalanche, the detection efficiency ε is then equal to the probability W. With $\delta(24\text{ eV}) \approx 1$ and $\delta(300\text{ eV}) \approx 3$ [GHe67] one gets $\varepsilon(24\text{ eV}) \approx 0.63$ and $\varepsilon(300\text{ eV}) \approx 0.95$. Fig. 4.28 shows the measured detection efficiency ε of electrons with different kinetic energies incident on a channeltron. It can be seen that the estimation of ε on the basis of the typical secondary emission coefficients δ given above yields the correct order of magnitude. The actual ε values depend also on the angle of incidence of the electrons (see Fig. 4.29 and also [Was74]), on the position of electron impact (strength of the electrical field) and on the quality of the channeltron surface (type of semiconducting material, pollution, age and particle damage of the surface).

The same arguments hold for the detection efficiency of a channelplate detector as for the channeltron, but in addition the ratio $r_{\text{open area}}$ of the area channel openings to the total plate area has to be included also. As a rough estimate one gets

$$\varepsilon_{\text{channelplate}} > \varepsilon_{\text{channeltron}}\, r_{\text{open area}}\, \varepsilon_{\text{electronics}}, \tag{4.32}$$

where the product on the right-hand side of the equation gives a lower limit, because electrons striking the surface at the interchannel material also produce secondaries which might reach a neighbouring microchannel and start an avalanche there, thus increasing the detection probability. With $\varepsilon_{\text{channeltron}}$ being typically 90% and an open area of 50%, one obtains $\varepsilon_{\text{channelplate}} > 0.45$; actual values for 200 eV electrons are between 0.6 and 0.9.

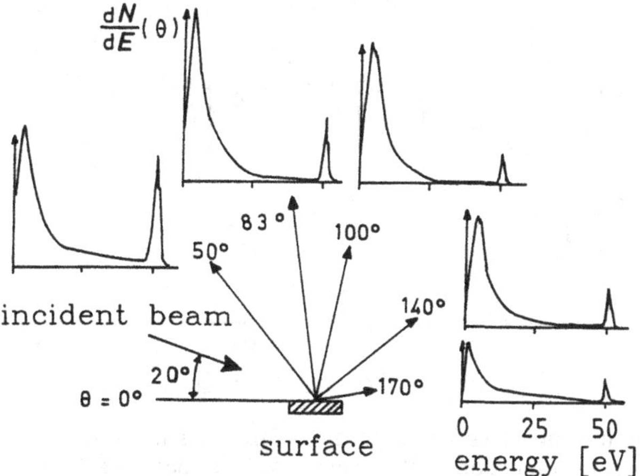

Figure 4.29 Energy–angular distributions $dN(\Theta)/dE$ of electrons emitted by a 51 eV primary electron beam hitting the surface of a channeltron material at a grazing incidence of 20°. The lengths of the arrows shown in the middle part of the figure represent for the different directions selected the corresponding total intensities of secondary electrons. The energy distributions found at the selected angles are plotted close to these directions; they all refer to the energy axis shown in the lower right-hand corner. From [GHe67].

The quantitative evaluation of relative intensities for selected photo- or Auger processes requires information about both the relative kinetic energy dependence of the analyser transmission T (see Fig. 4.15) and the accompanying detection efficiency ε of the electron detector. The *relative* magnitude for the desired product $T\varepsilon$ can be determined directly if, for example, non-coincident electron and ion spectrometry are combined: with helium as target gas, the 1s photoline is recorded as a function of the photon energy and yields the dispersion corrected area A_D (electron); see equ. (2.39):

$$A_D(\text{electron}) = N_{ph} n_v \, \Delta z_{el} \, \sigma^+_{1s}(\text{He}) \, T\varepsilon \frac{\Delta E_{sp}}{E^0_{kin}}. \tag{4.33}$$

In this expression the product $T\varepsilon$ may change with the kinetic energy. (For kinetic energies approaching zero, the relative spectrometer resolution $\Delta E_{sp}/E^0_{kin}$ will also change (see Fig. 4.14), but this change can be measured by means of electron spectrometry.) The number N^+_{ion} of He^+ ions can be recorded simultaneously by the ion spectrometer:

$$N^+_{ion} = N_{ph} n_v \, \Delta z_{ion} \, \sigma^+_{1s}(\text{He}) \left[1 + \frac{\sigma^{+*}_{1s}(\text{He})}{\sigma^+_{1s}(\text{He})} \right] T_{ion} \varepsilon_{ion}. \tag{4.34}$$

Comparing equ. (4.34) with equ. (4.33), two important differences can be noted. First, the product of ion transmission and detection efficiency, $T_{ion}\varepsilon_{ion}$, is independent of the photon energy, while $T\varepsilon$ can change with the kinetic energy of the electrons. Second, the number of singly-charged ions, N^+_{ion}, comes not only from single ionization, $\sigma^+_{1s}(\text{He})$, which leads to the 1s photoelectrons, but also from ionization and simultaneous excitation, $\sigma^{+*}_{1s}(\text{He})$. Fortunately, these other contributions can

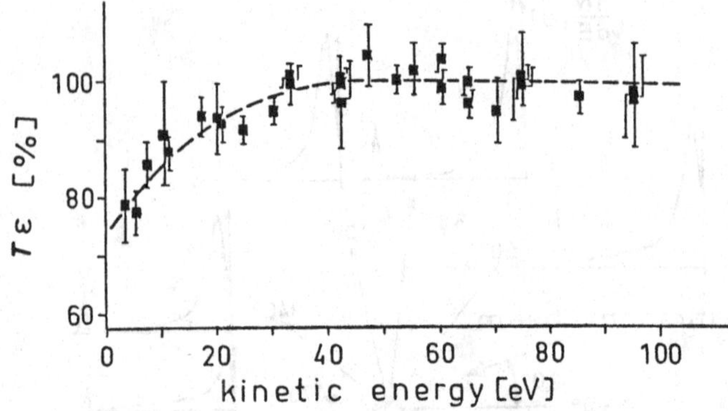

Figure 4.30 Experimental values for the product of transmission T and detection efficiency ε for a sector CMA as a function of the nominal kinetic energy of the analysed electrons (note the suppressed zero on the ordinate). The data have been normalized to 100% at high energies. All electrons were postaccelerated by a +50 V potential before hitting the cone of the channeltron detector. From [KKS89].

easily be taken into account, because for helium all partial cross sections are known [BWu95]. Therefore one can collect together all measurable and easily known quantities on one side giving

$$\frac{A_D(\text{electron})}{N^+_{\text{ion}}}\left[1+\frac{\sigma^{+*}_{1s}(\text{He})}{\sigma^+_{1s}(\text{He})}\right]=\left(\frac{\Delta z_{\text{el}}}{\Delta z_{\text{ion}}}\frac{1}{T_{\text{ion}}\varepsilon_{\text{ion}}}\right)\frac{\Delta E_{\text{sp}}}{E^0_{\text{kin}}}T\varepsilon. \tag{4.35}$$

The quantities in the brackets on the right-hand side are constant (energy-independent), and the relative instrumental resolution of the electron spectrometer can be measured. Therefore, the dependence of $T\varepsilon$ on the kinetic energy can be established on a relative scale. An example is shown in Fig. 4.30.

Absolute values for the product of transmission and detection efficiency may be obtained from equ. (4.35) if, in addition, the transmission and detection efficiencies T_{ion} and ε_{ion} of the ion analyser are known, together with the corresponding lengths of the source volumes (in suitable experimental set-ups, these are purely geometrical quantities). To obtain the product $T_{\text{ion}}\varepsilon_{\text{ion}}$, a coincidence experiment between $He^+(1s)$ photoelectrons and He^+ ions can be performed, using the detected electron as START for a time-of-flight analysis of the ions. (In order to measure electrons in coincidence with thermal ions, one needs a pulsed electric field which extracts and accelerates the ions after an electron has been detected. For details of this technique and the problems of taking into account the disturbance by accidental coincidences see [KKS92, Kos93].) In this case one has: for the counting rate $N_{\text{el}}(E_{\text{pass}})$ of photoelectrons (equ. (2.37))

$$N_{\text{el}}(E_{\text{pass}})=N_{\text{ph}}n_{\text{v}}\,\Delta z_{\text{el}}\,\sigma^+_{1s}(\text{He})\;T\varepsilon\;F(E_{\text{pass}}), \tag{4.36a}$$

where E_{pass} is the pass energy of the electron spectrometer and $F(E_{\text{pass}})$ takes care of the necessary energy convolution functions; and for the counting rate of true coincidences (see equ. (4.108))[†]

$$N_{\text{true}}(\text{el., ion}) = N_{\text{ph}} n_{\text{v}}\, \Delta z_{\text{el., ion}}\, \sigma^{+}_{1\text{s}}(\text{He})\, T\varepsilon\, F(E_{\text{pass}})\, T_{\text{ion}}\varepsilon_{\text{ion}}. \tag{4.36b}$$

This leads to

$$\frac{N_{\text{true}}(\text{el., ion})}{N_{\text{el}}(E_{\text{pass}})} = T_{\text{ion}}\varepsilon_{\text{ion}} \frac{\Delta z_{\text{el., ion}}}{\Delta z_{\text{el}}}, \tag{4.37}$$

and, with known source volumes, the desired product $T_{\text{ion}}\varepsilon_{\text{ion}}$ may be determined.

Finally, two practical aspects for working with a channeltron or channelplate detector are considered. Firstly, from the energy dependence of the detection efficiency ε shown in Fig. 4.28 it can be seen, and from the mechanism of secondary emission it can be understood, that the detection efficiency ε drops for incident electrons with low kinetic energies. Therefore, in most applications of electron spectrometry which involve low-energy electrons, the electrons are passed through a postacceleration field before they hit the detector. Secondly, a channelplate may show a non-uniform surface sensitivity (and thus also detection efficiency) which, however, is essential if it is used in electron spectrometry as a position-sensitive detector. In order to overcome this problem for the case of simultaneous recording of a certain portion of an electron spectrum, a *dithering mode* can be applied for the data collection (see [PTL81]): in this the dispersed electron spectrum is swept across the channelplate so that each spectral element spends an equal amount of time on each part of the channelplate surface used.

4.4 Electrostatic lenses

For some electron spectrometry applications it is desirable to install a lens system at the entrance and/or the exit of the electron spectrometer. Some examples are: (i) when adapting the original kinetic energy of the electrons to the pass energy of the spectrometer (and/or to a certain value before the electrons hit the detector), (ii) when transforming the solid angle and/or the size of the source volume accepted by the spectrometer (and/or similar transformations between the spectrometer and the electron detector), (iii) to create a larger distance between the reaction zone and the spectrometer (and/or between the spectrometer and the electron detector), (iv) when combinations of these possibilities are needed. In order to understand the performance characteristics of a lens system, some basic properties will first be recalled (for details see the monographs [Gla70, ElK70, Gri72, Dah73, GvW83, Wol90]). Some examples of the different uses of electrostatic lenses in conjunction with electron spectrometers will then be given.

† The complications described in equ. (4.108) by the symbol f(e.c.) are absent for ion–electron coincidences where one has $f(\text{e.c.}) = 1$.

4.4.1 Basic properties

The passage of electrons or other particles with charge q and mass m through an electrostatic lens system is governed by their motion under the action of the electric field. In the case considered here, cylindrical symmetry around the optical axis (z-axis) and paraxial rays will be assumed. Of the cylindrical coordinates only the transverse radial coordinate ρ and the distance coordinate z are of relevance, and the electrostatic potential of the lens is given by $\varphi(\rho, z)$. As shown in Section 10.3.1, in the paraxial approximation the potential $\varphi(\rho, z)$ is fully determined by the potential $\Phi(z) = \varphi(\rho = 0, z)$ at the symmetry axis. Hence, the equations of motion and the fundamental differential equation of an electrostatic lens depend only on this potential. The fundamental lens equation is given by (see equ. (10.38))

$$\frac{\mathrm{d}^2\rho}{\mathrm{d}z^2} + \frac{\tilde{\Phi}'(z)}{2\tilde{\Phi}(z)}\frac{\mathrm{d}\rho}{\mathrm{d}z} + \frac{\tilde{\Phi}''(z)}{4\tilde{\Phi}(z)}\rho = 0. \tag{4.38}$$

This establishes a connection between the coordinate ρ and the potential $\tilde{\Phi}(z)$ with its derivatives in z (for the meaning of the tilde on the potential $\tilde{\Phi}(z)$, which differs from $\Phi(z)$, see below). Several important factors can be deduced from this differential equation:

(i) The charge-to-mass ratio q/m does not appear, i.e., the trajectory is the same for all particles with a constant q/m value provided they enter the lens with the same initial energy. (In this context it should be noted that $\tilde{\Phi}(z)$ depends not only on the actual potential $\Phi(z)$, but also on the kinetic energy and charge on the incident particle (see equ. (10.35)). Therefore, the sign of $\Phi(z)$ changes with the charge of the incident particle.)

(ii) The equation is homogeneous in ρ, hence, if $\rho = \rho_1(z)$ is a solution, then $\rho = c\rho_1(z)$, where c is a constant, is also a solution. This means a lens focusing particles which follow the trajectory $\rho_1(z)$ also focuses particles following other paraxial trajectories given by $c\rho_1(z)$.

(iii) The equation is homogeneous in $\tilde{\Phi}(z)$, so a proportional increase of the voltages on all electrodes of the lens leaves the trajectories unchanged.

(iv) A change in the dimensions of the whole system (in ρ and z) corresponds to a change of the unit of length. Since the potentials keep their form, a geometrical scaling leads to a corresponding scaling in the dimensions of the trajectories.

(v) The potential $\tilde{\Phi}(z)$ differs from the actual potential $\Phi(z)$ by U_e (for electrons; see equ. (10.35c)):

$$\tilde{\Phi}(z) = \Phi(z) + U_e, \tag{4.39}$$

where U_e is the kinetic energy E_{kin} of the electrons before they enter the lens system, divided by e_0. As a consequence of this normalization, all voltages $\tilde{V}_j$ found as solutions of the fundamental lens equations are related to the voltages V_j^a actually applied to these electrodes by

$$\tilde{V}_j = V_j^a + E_{kin}^0/e_0. \tag{4.40}$$

For example, if voltages $\tilde{V}_1 = 0$ V and $\tilde{V}_2 = 100$ V are calculated and electrons with $E^0_{\text{kin}} = 10$ eV are considered, the voltages $V^a_1 = -10$ V and $V^a_2 = 90$ V must be set for the corresponding electrodes.

In order to derive the full properties of a lens system, the fundamental lens equation has to be solved for individual rays subject to the influence of the given potential $\tilde{\Phi}(z)$. However, without going into the details for such a procedure, some aspects, which are important for axial-symmetrical electrostatic lenses with paraxial rays, can be listed by using results from light optics:

(i) The overall action of several lenses which are combined in series can be characterized by the properties of one lens only.† For such an overall lens a *left-* and *right-hand* region can be distinguished, each of which is field-free, but carries different potentials $\tilde{\Phi}_\ell$ and $\tilde{\Phi}_r$, respectively; one can identify these two different sides with the *object* and *image* regions, respectively.‡

(ii) The action of a lens is fully described by simple geometrical constructions if two *principal planes* at positions P_ℓ and P_r and two *focal lengths*, f_ℓ and f_r, extending to the positions F_ℓ and F_r are given. It should, however, be remembered that the *actual* rays do not follow the *asymptotic* rays used in the geometrical construction. Three rules for such a geometrical construction will be recapitulated with the help of Fig. 4.31; the first two rules help to determine the position and size of the image while the third rule allows the asymptotic trajectory to be traced for an arbitrary direction of the particle. First, Fig. 4.31(*a*) is used to define the characteristic parameters along the optical axis. If a reference point S somewhere in the middle of the lens is fixed, the points defining the principal planes, P_ℓ and P_r, and the focal planes, F_ℓ and F_r, can be marked, which then give the focal distances f_ℓ and f_r. Second (Fig. 4.31(*b*)), for an object y_ℓ placed on the left-hand side in the region of the potential $\tilde{\Phi}_\ell$ its image y_r on the right-hand side at the potential $\tilde{\Phi}_r$ can be found by selecting two straight asymptotic trajectories. The two rays which are chosen for convenience are the ray which enters the lens parallel to the optical axis and leaves the lens intersecting the focal point F_r on the image side, and the ray which does the reverse of this, i.e., it enters the lens through the focal point F_ℓ and leaves it parallel to the optical axis. (Note that the incident ray is assumed to proceed up to its principal plane P_ℓ where it is projected back to the principal plane P_r before its direction is changed.) Third, rays entering the lens at an arbitrary angle α_ℓ to the optical axis (Fig. 4.31(*c*); left-hand side) will proceed on the right-hand side through a point X in the

† A very elegant method exists for the combination of individual optical systems by using a matrix formulation. Each optical system is described by a 2×2 matrix which acts on the transverse components ρ and $\rho' = \tan\alpha$ of the trajectories before they enter the system and transforms these into new components after passage through the system. This formulation, however, is beyond the scope of the present book (see, for example, [Sep67, Gri72, Dah73, GvW83]).

‡ Although the field still exists mathematically up to infinity, in practice the field of the lens is zero at a distance larger than a few lens diameters.

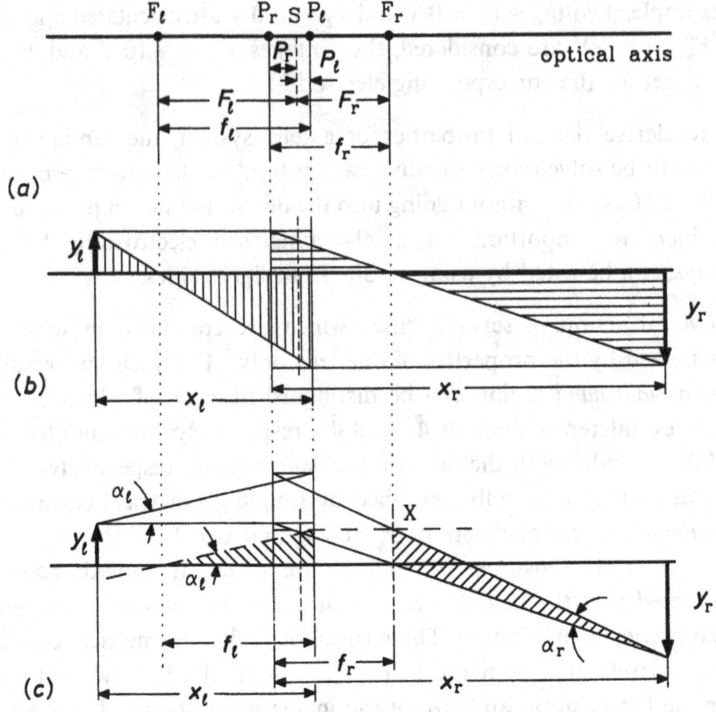

Figure 4.31 Data for the characterization of an electrostatic lens. (*a*) Positions of the focal and principal planes (left-hand and right-hand sides are indicated by the subscripts 'ℓ' and 'r', respectively) and their distances (optical sign conventions are disregarded, i.e., the distances are described only by their lengths). (*b*) Geometrical construction applied to image the arrow y_ℓ by means of characteristic asymptotic trajectories. (*c*) Geometrical construction for an asymptotic ray with a pencil angle α_ℓ. The shaded areas are needed for the derivation of the linear and angular magnification factors of the lens. For details see main text.

focal plane whose distance x from the optical axis can be found by following a parallel incident ray (with α_ℓ) that crosses the (left) focal plane at the optical axis at the point F_ℓ and continues on the right-hand side parallel to the optical axis and with a distance x.

(iii) The solution of the fundamental differential equation of an electrostatic lens yields for the ratio of the two focal distances f_ℓ and f_r located in the regions of potentials $\tilde{\Phi}_\ell$ and $\tilde{\Phi}_r$, the so-called *Helmholtz–Lagrange relation*, given by

$$f_r/f_\ell = \sqrt{(\tilde{\Phi}_r/\tilde{\Phi}_\ell)}. \tag{4.41}$$

If this equation is compared with the result from light optics where glass bodies are used, one notices that the square-root of the potentials $\tilde{\Phi}_\ell$ and $\tilde{\Phi}_r$ takes over the role of the refraction indices n_ℓ and n_r. The relation states that there are different focal lengths in an immersion lens in which the asymptotic potentials $\tilde{\Phi}_\ell$ and $\tilde{\Phi}_r$ or, correspondingly, the kinetic energies of the electrons entering and leaving the lens are different:

$$E^{\ell}_{\mathrm{kin}} = E^{0}_{\mathrm{kin}} + e_0 \Phi^{\mathrm{a}}_{\ell} = e_0 \tilde{\Phi}_{\ell}, \tag{4.42a}$$

$$E^{\mathrm{r}}_{\mathrm{kin}} = E^{0}_{\mathrm{kin}} + e_0 \Phi^{\mathrm{a}}_{\mathrm{r}} = e_0 \tilde{\Phi}_{\mathrm{r}}. \tag{4.42b}$$

(iv) Further interesting properties of a lens are the *linear magnification* M_{lin} defined by

$$M_{\mathrm{lin}} = y_{\mathrm{r}}/y_{\ell} \tag{4.43a}$$

and the *angular magnification* M_{α} defined by

$$M_{\alpha} = \tan \alpha_{\mathrm{r}}/\tan \alpha_{\ell}. \tag{4.44a}$$

Using the similarity of the triangles shown in Fig. 4.31 (shaded vertically and horizontally), one obtains for the linear magnification the equivalent expression

$$M_{\mathrm{lin}} = \frac{f_{\ell}}{x_{\ell} - f_{\ell}} = \frac{x_{\mathrm{r}} - f_{\mathrm{r}}}{f_{\mathrm{r}}} \tag{4.43b}$$

from which the *Newton relation*

$$(x_{\ell} - f_{\ell})(x_{\mathrm{r}} - f_{\mathrm{r}}) = f_{\ell} f_{\mathrm{r}} \tag{4.45a}$$

or the *Gaussian relation*

$$\frac{f_{\ell}}{x_{\ell}} + \frac{f_{\mathrm{r}}}{x_{\mathrm{r}}} = 1 \tag{4.45b}$$

can be obtained. By writing this last equation slightly differently as

$$(x_{\ell} - f_{\ell})x_{\mathrm{r}} = x_{\ell} f_{\mathrm{r}}, \tag{4.45c}$$

one gets

$$M_{\mathrm{lin}} = \frac{f_{\ell} x_{\mathrm{r}}}{f_{\mathrm{r}} x_{\ell}}. \tag{4.43c}$$

Similarly, from the shaded triangles in Fig. 4.31(*c*) in the paraxial approximation, i.e., for small y and α values one obtains for the angular magnification the equivalent form

$$M_{\alpha} = \frac{f_{\ell}}{x_{\mathrm{r}} - f_{\mathrm{r}}}. \tag{4.44b}$$

Multiplying equ. (4.43b) by equ. (4.44b) and noting equ. (4.41) gives the important *Helmholtz–Lagrange relation*

$$M_{\mathrm{lin}} M_{\alpha} = f_{\ell}/f_{\mathrm{r}} = \sqrt{(\tilde{\Phi}_{\ell}/\tilde{\Phi}_{\mathrm{r}})}. \tag{4.46}$$

A detailed discussion of this equation is presented in Section 10.3.2.

The requirements of the Helmholtz–Lagrange relation and further constraints on the properties of a lens (for example, a constant distance d between the object and the image which is of great practical importance) do not allow an arbitrary choice of values for the magnification and/or acceleration/retardation of the electrons. Hence, in order to obtain freedom in the selection of certain properties, a lens system must have a certain number of elements. For example, if the electron energy is to be varied, keeping the distance d constant, at least two free parameters are needed, and this implies that two (or even more) variable voltage ratios must exist, i.e., such a lens must have at least three elements carrying the three voltages. In general the following rules apply:

for a three-element lens system with variable acceleration/retardation, the distance d can be kept constant, but M_α and/or M_{lin} change; and

for a four-element lens system with variable acceleration/retardation, the distance d can be kept constant together with one of the magnifications (M_α or M_{lin}) while the other one changes.

Due to the Helmholtz–Lagrange relation, no greater flexibility in the parameters determining the image is possible. However, lenses with even more elements than four can have other advantages, such as lower aberration or a more extended operation range. (See, for example, the 'movable' electrostatic lens in [Rea83].)

In principle, the potential $\Phi(z)$ can be produced by apertures as well as by cylinders, and both systems are frequently used. (The performance of an actual lens system with several elements may be more reliable if cylinders are used instead of apertures, because the field actions between the cylinders are more isolated.) Of all possible geometrical arrangements, a standardization is frequently applied by choosing certain values for the diameter D of the apertures/cylinders, the thickness G and the distance A between the individual elements. Usually, all these quantities are given in units of D, because all characteristic lens parameters scale with D. Extensive tables for electrostatic lenses can be found in [HRe76]; a ray-tracing program [DDA90] with which the trajectories of individual rays can be calculated and displayed is very helpful.

4.4.2 Selected examples

In order to demonstrate the convenient properties of lenses combined with an electrostatic energy analyser, three examples will be discussed: first the important case in which an acceleration/retardation lens is placed in front of the analyser which allows an operation mode with a constant pass energy in the analyser; second the case in which increased acceptance angles are used, which is important for coincidence experiments; and third the case in which 0 eV electrons are handled and analysed with high resolution and high acceptance.

4.4.2.1 *Lens system for acceleration/retardation*

An experimental set-up using a lens system for the acceleration/retardation of the electrons in conjunction with an electrostatic analyser is shown in Fig. 4.32. The electron spectrometer is a spherical analyser with a mean radius R_0 of 101.6 mm combined with a three-aperture zoom lens at the entrance and the exit. Only the purpose and the characteristic properties of the entrance lens will be

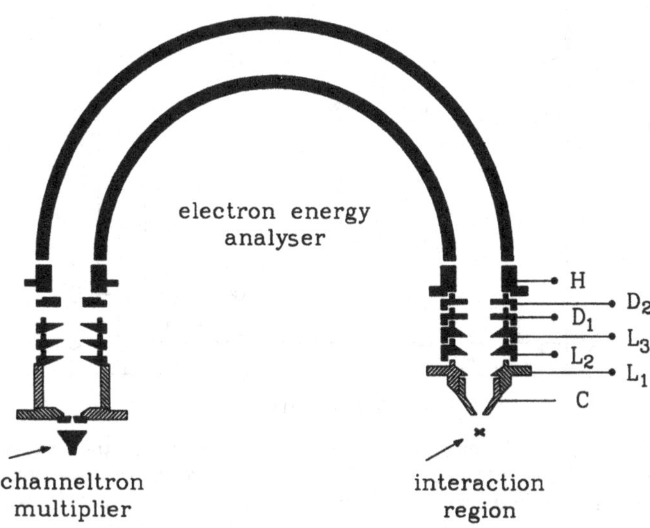

Figure 4.32 Schematic drawing of an electron spectrometer with acceleration/retardation of the electrons by a three-aperture zoom lens placed at the entrance and exit of a spherical electron spectrometer. The elements of the entrance lens are given by L_1 (with entrance cone C), L_2, and L_3; D_1 and D_2 are x–y deflectors for steering the beam; and H is a Herzog diaphragm for the compensation of edge effects (see [Her40]). Reprinted from *Nucl. Instr. Meth.* **222**, Parr *et al.*, 221 (1984) with kind permission of Elsevier Science – NL, Sara Burgerhartstraat 25, 1055 KV Amsterdam, The Netherlands.

discussed here. As the word *zoom* lens implies, the lens system is used to allow acceleration/ retardation of the electrons before they enter the analyser, keeping the distance d between the source and image constant. Hence, energy analysis can be performed with a constant pass energy, E_{pass}, in the spectrometer and a constant experimental resolution, ΔE_{sp} (see Section 4.2.4). Because two parameters, the change of electron energy and the constant distance d between the source and the image, have to be controlled by the experimenter, a lens system with at least three elements is needed. In the present example, a three-aperture lens system with $A/D = 1$, $P = Q = 5\,A$ and $D = 9.2$ mm was selected (D is the opening of the aperture, A the distance between two apertures, and P and Q the object and image distances, each measured with respect to the centre aperture).

In Fig. 4.33 the relevant potentials and energies are shown for the case of acceleration. Because the electrode L_1 is grounded, and L_3 is at the potential needed to accelerate/retard the incident electrons to the same pass energy in the spectrometer, one derives from equ. (4.40) the following relations for the voltages

$$\left.\begin{aligned}\tilde{V}_1 &= V_1^{\text{a}} + E_{\text{kin}}/e_0 = E_{\text{kin}}/e_0 \quad \text{because } V_1^{\text{a}} = 0,\\ \tilde{V}_2 &= V_2^{\text{a}} + E_{\text{kin}}/e_0,\\ \tilde{V}_3 &= V_3^{\text{a}} + E_{\text{kin}}/e_0 = E_{\text{pass}}/e_0,\end{aligned}\right\} \tag{4.47a}$$

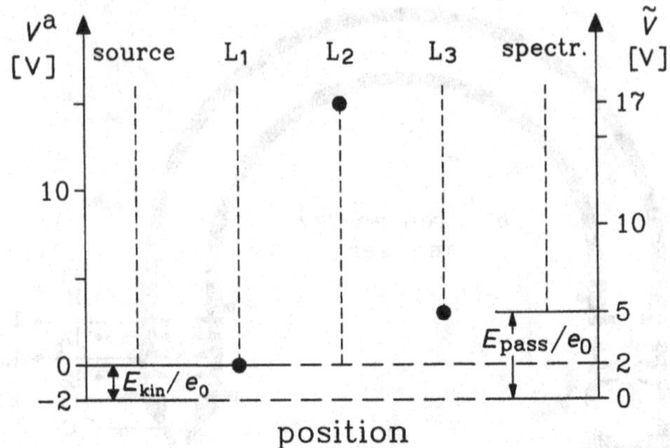

Figure 4.33 Definition of the energies and potentials relevant for the three-aperture lens shown in Fig. 4.32 at the entrance of an electron spectrometer. In this example electrons with $E_{kin} = 2$ eV emitted in the source volume which travel through the lens with electrodes L_1, L_2, L_3 and enter the spectrometer set at $E_{pass} = 5$ eV are shown. The potentials V_i on the electrodes can be given with respect to different reference values: using the ordinate on the left-hand side one gets the actual potentials V_i^a at the electrodes, using the ordinate on the right-hand side one gets the values $\tilde{V}_i$ used in the theoretical calculation (see equ. (4.47a)).

and one has

$$\left.\begin{aligned} \tilde{\Phi}_\ell &= \tilde{V}_1 = E_{kin}/e_0, \\ \tilde{\Phi}_r &= \tilde{V}_3 = E_{pass}/e_0. \end{aligned}\right\} \tag{4.47b}$$

Hence, the ratio $\tilde{V}_3/\tilde{V}_1$ determines the acceleration/retardation for the selected electron line and the given pass energy. For example, if electrons between 2 eV and 20 eV are to be analysed with a pass energy of 5 eV, one gets a $\tilde{V}_3/\tilde{V}_1$ of between 2.5 and 0.25. The voltage ratio $\tilde{V}_2/\tilde{V}_1$ then follows from the operation mode of the lens system and has to be calculated. For the geometry shown, the result is displayed in Fig. 4.34 as a function of different $\tilde{V}_3/\tilde{V}_1$ settings and with two (mid-object, mid-image)-distances as parameters. First it can be noted that the operation range of the lens system covers the range of $\tilde{V}_3/\tilde{V}_1$ ratios needed for the above example. Second, within this operation range there are two solutions (marked by the solid and dashed curves) which differ in the potential $\tilde{V}_2$ of the middle electrode L_2. Because a more detailed examination of this lens system leads to the result that in the case of the higher potential (solid curves) there is a practically constant linear magnification M_{lin}[†] and a lower spherical aberration error, preference is given to this mode. One then obtains the necessary voltage ratios $\tilde{V}_2/\tilde{V}_1$ from one of the solid surves in Fig. 4.34 (in the present example from the curve labelled (5, 5)). The ratio $\tilde{V}_2/\tilde{V}_1$ fixes the actual voltage V_2^a for the electrode L_2. By changing V_2^a, the energy analysis of the incident electrons can be performed for a preselected pass energy. As this operation mode has variable

[†] For this lens with (mid-object, mid-image)-parameter equal to (5, 5), and for $\tilde{V}_3/\tilde{V}_1$ between 0.1 and 3.0, one has $M_{lin} \approx 1$.

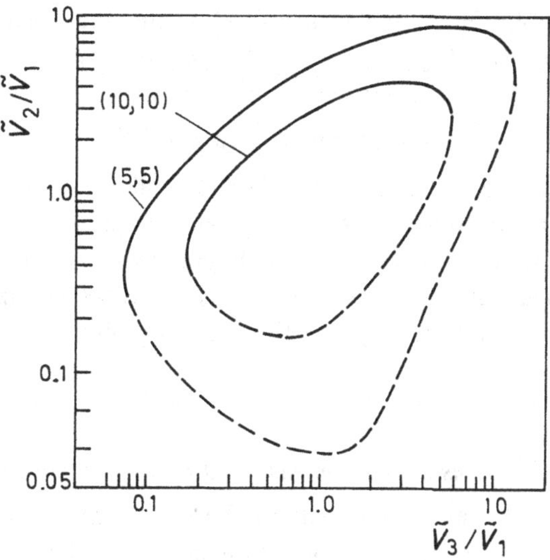

Figure 4.34 The relationship between the voltage ratios $\tilde{V}_2/\tilde{V}_1$ and $\tilde{V}_3/\tilde{V}_1$ of an asymmetric electrostatic lens with the apertures calculated for $A/D = 1$ and two values of the mid-object and mid-image distances P/D and Q/D which are labelled $(P/D, Q/D)$ on the curves. The solid and dashed curves belong to two different solutions for the potential $\tilde{V}_2$ applied to the middle electrode. From [Rea70].

acceleration/retardation and practically constant linear magnification M_{lin}, the Helmholtz–Lagrange relation requires a change of the angular magnification M_α. Therefore, it is necessary to investigate whether a change in M_α affects the experimental resolution of the electron spectrometer equipped with a lens system.

The energy resolution of the spherical analyser which has first-order focusing in $\Delta\vartheta$ and all-order focusing in $\Delta\varphi$ will be evaluated first. According to equ. (4.7b) one has for a spherical analyser

$$\left(\frac{\Delta E}{E}\right)_{\text{sp}} \approx \frac{1}{2}\left(\frac{S + B(Q)}{D} + \frac{B(\Delta\vartheta, \Delta\varphi)}{D}\right) \tag{4.48a}$$

with $D = 2R_0$ and $B(\Delta\vartheta, \Delta\varphi)/D = \Delta\hat{\vartheta}^2$ [Sev72]. Hence, the influence of finite angles $\Delta\vartheta$ on the experimental resolution can be neglected only if

$$B(\Delta\vartheta, \Delta\varphi) < S + B(Q), \tag{4.48b}$$

or, equivalently,

$$(\pm\hat{\vartheta})^2 < \frac{S + B(Q)}{2R_0}, \tag{4.48c}$$

which requires in the present case ($S \approx B \approx 1.5$ mm. $R_0 = 101.6$ mm)

$$\Delta\vartheta < \pm 7.0^\circ. \tag{4.48d}$$

Next, the modification of the angular range $\Delta\vartheta$ by the lens is considered. At the entrance of the lens, $\Delta\vartheta_\ell$ is restricted to $\Delta\vartheta_\ell \leq \pm 3^\circ$ by an aperture (cone C in Fig. 4.32). (Together with a counterpart in the image plane of the lens this aperture determines the resolution of the instrument.) Using the Helmholtz–Lagrange relation, equ. (4.46), with $E_{pass} = 5$ eV and $M_{lin} \approx 1$, for the angular range $\Delta\vartheta_r$ at the entrance of the electron spectrometer one obtains

$$\tan \Delta\vartheta_r \approx \tan 3^\circ \sqrt{(E_{kin}/5\text{ eV})}, \tag{4.48e}$$

and thus for $E_{kin} = 20$ eV and $E_{kin} = 2$ eV it follows that $\Delta\vartheta_r \approx \pm 6^\circ$ and $\Delta\vartheta_r \approx \pm 2^\circ$, respectively. Since $\Delta\vartheta_r$ is smaller than the critical value of equ. (4.48d), one can conclude that no degradation of the analyser resolution occurs if it is equipped with this lens system. This means that full advantage can be taken of the retardation system's ability to improve the energy resolution without noticeable loss of transmission.

4.4.2.2 *Lens system for increased acceptance angles*

In some electron spectrometry applications the instrumental resolution is not as important as a high efficiency, which is equivalent to a large acceptance solid angle, and possibly a large electron velocity in the spectrometer, which reduces the time spread between different electron trajectories. In particular such demands are essential for coincidence experiments with electrons (see Section 4.6). A lens system in front of the electron spectrometer can help to fulfil these requirements, and an example of such an experimental set-up is shown in Fig. 4.35. The spectrometer is a sector CMA with a relative resolution of $\Delta E_{sp}/E = 0.7\%$, a relative acceptance solid angle of $T_0 = 3.3 \times 10^{-3}$ and a luminosity of $L = 2.6 \times 10^{-3}$ mm^3 (these and the following data are valid for a photon beam diameter of

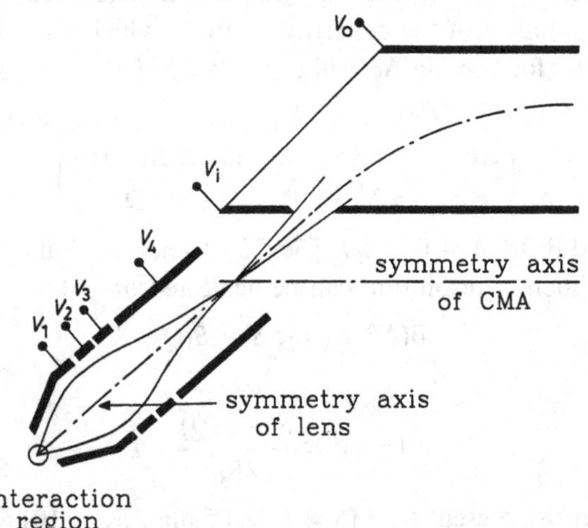

Figure 4.35 Schematic drawing of the set-up of an electron spectrometer (sector CMA) with a lens (four cylindrical electrodes) at its entrance used to increase the acceptance of electrons from the interaction region. The voltages applied to the elements of the lens are characterized by V_1–V_4, and the voltages on the inner and outer cylinders of the electron spectrometer by V_i and V_o, respectively. From [Elm92].

1 mm, and for Δz = 1 mm). If the lens system is included and operates to accelerate the electrons towards the electron spectrometer, the corresponding values for L and T_0 increase to T_0 (with lens) = 9.7×10^{-3} and L (with lens) = 7.6×10^{-3} mm^3, while the relative resolution deteriorates to $\Delta E_{sp}/E = 5.9\%$. The operational conditions of this lens system for electrons with kinetic energies around 10 eV can be characterized as follows [Elm92]:†

(i) a fixed distance d between the actual source volume and its image in the focal plane where the image then acts as the object for the electron spectrometer;
(ii) a linear magnification $M_{\text{lin}} \approx 1$ leading to a one-to-one correspondence between source and image which allows an easy comparison of the performance of the set-up with and without the lens;
(iii) an angular magnification $M_\alpha < 0.5$, which provides the desired increase in the acceptance solid angle at the source by a factor of 2, keeping the acceptance angle of the electron spectrometer constant;
(iv) an increase of the electron energy at the exit of the lens, e.g., $\tilde{\Phi}_\ell/\tilde{\Phi}_r \approx 10$, because the time spread of electron trajectories in the spectrometer is reduced (see Section 4.6.3).

Due to these requirements, the minimum number of lens elements is four, and the system shown in Fig. 4.35 consists of four cylindrical electrodes (the shape of the first electrode is modified towards the source volume). When operating this lens system, the first electrode has to be at ground potential whilst the last one is connected to the inner cylinder of the analyser. As in the example of the acceleration/retardation lens described above, the potentials on the second and third electrodes follow from the characteristic data of the lens (length A of the second and third cylinders, $A = 0.5D$; distance G between these cylinders $G = 0.1D$; distance d between the source and image $d = 4D$; diameter D of the lens elements $D = 30$ mm). For $\tilde{\Phi}_r/\tilde{\Phi}_\ell = 10$, i.e., $\tilde{V}_4/\tilde{V}_1 = 10$, one gets the required voltage ratios $\tilde{V}_3/\tilde{V}_1 = 34$ and $\tilde{V}_2/\tilde{V}_1 = 14$ [MSR83]. These data refer to a lens operating in the paraxial approximation. In the present case, however, the desired large acceptance angle of the lens (a cone of 20°) together with the desired slim form of the lens for angle-resolved measurements (due to spatial restrictions) strongly violate the approximation of paraxial rays. In addition, the geometries of the source volume, the lens, and the electron spectrometer do not match at all; the source volume has rotational symmetry around the incident beam and is perpendicular to the symmetry axis of the lens, and the image of the lens is perpendicular to the symmetry axis of the sector CMA. Therefore, the theoretical voltage ratios $\tilde{V}_3/\tilde{V}_1$ and $\tilde{V}_2/\tilde{V}_1$ are used as starting values for optimizing the performance of the whole system – source volume, lens and spectrometer – by ray-tracing calculations. Based on these calculations good performance was predicted and verified experimentally for $\tilde{V}_3/\tilde{V}_1 = 30$ and $\tilde{V}_2/\tilde{V}_1 = 3.1$ [Elm92].

† Of course, the actual values of M_{lin}, M_α, and $\tilde{\Phi}_\ell/\tilde{\Phi}_r$ have to fulfil the Helmholtz–Lagrange relation.

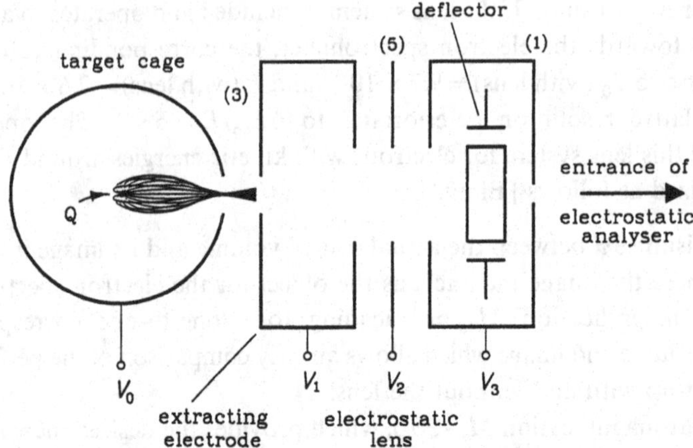

Figure 4.36 Schematic drawing of a threshold-energy electron spectrometer using a three-aperture lens at its entrance (the spectrometer itself is not shown). Ionization takes place at Q in the middle of the target cage. Field penetration is produced by the extracting electrode, which causes 2 meV electrons created at Q to follow the trajectories as indicated. The potentials applied to the target cage and the electrodes of the lens are given by V_0–V_3. The numbers in parentheses above the electrodes indicate the diameters of the apertures in mm. From [KZR87].

4.4.2.3 *Lens systems for 0 eV electrons*

High resolution *and* high acceptance are essential prerequisites for electron spectrometry in the threshold region. Beyond this, the difficulties in handling low-energy electrons (0 eV electrons) due to residual electric and magnetic fields increase the high demands that such investigations place on the experimenter. One solution to these problems for experiments which use a lens system is shown in Fig. 4.36 (For a high-resolution electron spectrometer with $\Delta E_{sp} < 2.7$ meV at 2 eV pass energy, see [MBB94].) There are four important features which will be explained with reference to this figure:

(i) Ionization takes place within a cage surrounding the interaction region. In the middle of such a cage, potential differences from disturbing electric fields are reduced to a few tens of mV. In addition, if the cage is made from thin wires so that it is as transparent as possible, the photon beam can traverse it, and the photoelectrons emitted with non-zero kinetic energy can leave it without producing too many secondary electrons at the surfaces that they hit.

(ii) High acceptance of nearly 0 eV electrons is achieved by the penetrating field method. (For electron collection by field penetration see [CRe74]; for a different solution of the problem see [MSS84] and Section 5.7.) On the lens side of the cage there is a small hole, facing which is an extracting electrode at positive potential. Hence, the electron field penetrates the cage and, depending on the kinetic energy of the electrons emitted as compared to the strength of the penetrating field, low-energy electrons are extracted with high efficiency. This can be seen in Fig. 4.36 where the trajectories of 2 meV electrons in the cage moving towards the small hole are shown. These electrons are collected over the whole 4π solid angle, i.e., one has full transmission with $T_0 \approx 1$.

(iii) From Fig. 4.36 it can also be seen that the trajectories of the extracted electrons cross over near the exit hole of the target cage. This crossover 'point' then is imaged onto the entrance slit of the electron spectrometer. (The deflector plates in the last element of the lens help to steer the electrons into the analyser.) The lens is a system with three apertures which also accelerates the electrons towards the analyser. Hence, the difficulties of handling 0 eV electrons will vanish as soon as these electrons enter the lens and the electron spectrometer where they have a higher kinetic energy. In addition, the inserted lens provides another advantage. The photon beam passes the cage perpendicular to the plane of the drawing in Fig. 4.36. Even though the target gas emanates from a needle positioned perpendicular and close to the photon beam, the intersection volume between them will be line-shaped and the source volume is not restricted to a certain length Δz. This means that the crossover 'point' shown in the figure is also line-shaped, perpendicular to the plane of the drawing, and not confined in Δz. However, a small Δz and, therefore, a good performance of the whole system is achieved by using a real aperture (1 mm diameter) at the entrance slit of the spectrometer, thus confining the actual accepted source volume via the image properties of the lens.

(iv) The electron spectrometer then analyses the accelerated threshold electrons according to the selected acceleration voltage and adapted pass energy. At these settings strong discrimination is achieved against electrons of higher kinetic energy which are produced in the source volume and which cannot be prevented from entering the electron spectrometer. Since these electrons provide a source of low-energy electrons when they are scattered at solid surfaces, special care in the design and operation of this analyser is necessary to obtain good performance characteristics for 0 eV electrons. In particular, a highly transparent mesh has to be used for the outer cylinder of the analyser, special corrections for the fringing fields at the entrance and exit of the analyser have to be applied, all parts exposed to the electron beam have to be made from molybdenum, and continuous baking of the whole spectrometer at 100 °C during data acquisition is necessary.

In spite of all these difficulties, the threshold spectrometer described has demonstrated high performance capabilities, and one special application will be presented in Section 5.7.

4.5 Disturbances of the performance

The ideal performance of an electron spectrometer can be disturbed considerably by several effects which will be discussed below. These include influences from electron scattering, residual magnetic fields, unwanted electric fields, and thermal Doppler broadening. Disturbances due to mechanical imperfections and/or jitter in the spectrometer voltage are not considered. Since postcollision interaction can severely affect the position and shape of photo- and Auger lines, a discussion of these particular aspects is also included.

4.5.1 Scattering losses

Electron spectrometry requires a good vacuum in order to allow a free passage of the electrons on their way from the source volume through the analyser to the detector, i.e., scattering losses should be negligible. In addition, for a good performance of the channeltron or channelplates, these detectors should work at pressures below 10^{-4} and 10^{-5} mb, respectively. However, a high count rate of recorded electrons depends on the target density n_v in the source volume which increases the backing pressure in the analyser and in the vacuum chamber. Therefore, in electron spectrometry with gaseous species there are usually two pressure regions, one in the source volume with high pressure and the other one outside this region with low pressure. (Working with synchrotron radiation and gaseous targets, there is a third region with an even lower pressure ($\approx 10^{-8}$ mb) between the experiment chamber and the monochromator in order to protect the monochromator's optical elements (at pressures of approximately 10^{-9} mb) from deleterious gasloads and deposits. Usually, this pressure region is provided by differential pumping between the vacuum chamber and the exit of the monochromator using a capillary which surrounds the photon beam (for an example of the use of a capillary array to transmit vacuum-ultraviolet radiation and to partition the pressure regions see [LMR79]).) Three ways of fulfilling the pressure constraint are: First, the interaction region is enclosed in a source cell which retains the target gas except for some flow through the small holes needed for the undisturbed passage of the photon beam and the escape of the electrons towards the electron spectrometer (for a device allowing a target pressure of about 5×10^{-4} mb by maintaining in the main vacuum chamber about 5×10^{-6} mb see [KCW81]). Second, the regions of different pressure can be further separated by placing a lens system between them. Good differential pumping is then achieved with suitable diaphragms. Third, special gas-inlet structures, e.g., needles or collimated hole-structures, are used which provide a collimation of the gas beam (see Section 10.7). By these methods a pressure factor of about 100 can be maintained between the regions. An experimental determination of this pressure factor is possible if the counting rate obtained with a sophisticated gas-inlet system is compared for equal pressures in the vacuum chamber with that for a uniform target pressure.

When the electrons pass from the source volume through the analyser to the detector, they are subject to *elastic* and *inelastic* scattering processes. Three effects may result from electron scattering:

(i) In general, each scattering event will cause such a perturbance in the direction and/or the energy of the electron that the regular trajectory is changed. As a consequence, the scattered electron normally will not reach the detector, i.e., scattering will lead to a decrease of the recorded counting rate I_{reg}. This is discussed in detail below.

(ii) Some of the electrons which are scattered out of their original trajectory may still reach the detector, but at spectrometer voltages which differ from the regular value U^0_{sp} (they may be at lower or higher values). Such processes manifest themselves in the spectrum of ejected electrons as additional background. Obviously, it is extremely difficult to separate this background from true continuous energy distributions.

(iii) Inelastic scattering of the electrons is connected with an excitation of the collision partner and, hence, the original energy E^0_{kin} is reduced by the excitation energy. Such a process is of special importance if it happens in the high-pressure regime of the source volume, because then the inelastically scattered electron can still follow a principal trajectory. As a result, additional structures (*inelastic energy loss lines*) appear in the spectrum of ejected electrons; these occur on the low-energy side of the intense electron lines. An example is shown in Fig. 4.37. The main interests of this study were the photolines of helium, i.e., the normal 1s photoline at 24.59 eV binding energy and the discrete satellite lines for 1s ionization and simultaneous excitation

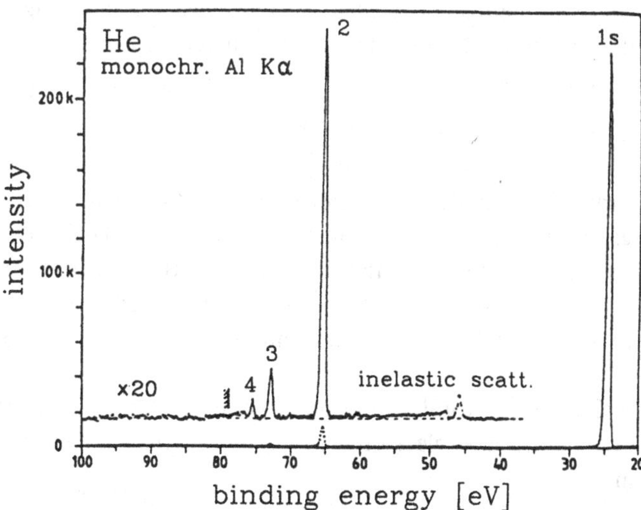

Figure 4.37 Spectrum of electrons ejected from helium after photoionization with monochromatized Al Kα radiation. The main 1s photoline and (magnified by a factor of 20) discrete ($n = 2, 3, 4$) and continuous satellites (above the threshold indicated at 79 eV) are shown as well as structures resulting from the inelastic scattering of 1s photoelectrons in the source volume. Reprinted from *J. Electron Spectrosc. Relat. Phenom.* **47**, Svensson *et al.*, 327 (1988) with kind permission from Elsevier Science – NL, Sara Burgerhartstraat 25, 1005 KV Amsterdam, The Netherlands.

to $n \geq 2$ (characterized by n in the figure). However, lines from inelastic scattered 1s photoelectrons are seen in addition. The first of these lines can be easily identified with the help of optical data for the target atom [Moo71]. In the example shown in Fig. 4.37 $1s^2 \rightarrow 1sn\ell$ excitations are involved with the lowest one being the $1s^2 \rightarrow 1s2p\ {}^1P_1^o$ excitation with an excitation energy of 21.2 eV. This process appears at an apparent binding energy of 45.8 eV which is the sum of the binding energy for photoionization and the lowest excitation energy, and the structure can be clearly identified.

It should be noted that a normal line and its accompanying lines due to inelastic energy losses have different dependences on the target density. In the limit of low scattering losses one has

$$I_{reg}(\text{normal line}) \approx N_{ph} n_v \sigma\, \Delta z\, T\varepsilon, \tag{4.49a}$$

but

$$I_{reg}(\text{inel. loss line}) = (N_{ph} n_v \sigma\, \Delta z)\sigma_{inel.} n_v\, T\varepsilon, \tag{4.49b}$$

i.e.,

$$\frac{I_{reg}(\text{inel. loss line})}{I_{reg}(\text{normal line})} \sim n_v. \tag{4.49c}$$

Hence, these different dependences can be used to identify or suppress inelastic scattering effects.

For quantitative electron spectrometry the most important disturbance caused by scattering processes is the loss of intensity in the registered counting rate I_{reg} as compared to the undisturbed counting rate I_0. The effect can be described by a scattering factor f_s defined by

$$I_{reg} = f_s I_0, \tag{4.50}$$

which describes how much the original electron intensity is reduced by scattering processes. In order to evaluate the scattering factor, the total cross section for electron scattering, $\sigma_{e\text{-scatt.}}$, is of interest because both elastic and inelastic scattering are responsible for the intensity loss. For the rare gases such cross sections are shown in Fig. 4.38. Two properties can be noted: the huge scattering cross section of xenon, and the Ramsauer minimum [Ram21] at lower kinetic energies.

With given values of the total scattering cross section $\sigma_{e\text{-scatt.}}$ and the density distribution $n_v(x)$ along the electron's path, the scattering factor f_s can be calculated from

$$f_s(E_{kin}^0) = \exp\left\{ - \int_{\text{electron path}} \sigma_{e\text{-scatt.}}[E_{kin}(x)] n_v(x)\, dx \right\}. \tag{4.51a}$$

In this equation the change in the kinetic energy of the electrons along their trajectory x has to be taken into account. For example, in the deflection field of an electrostatic spectrometer the electrons lose kinetic energy as they move

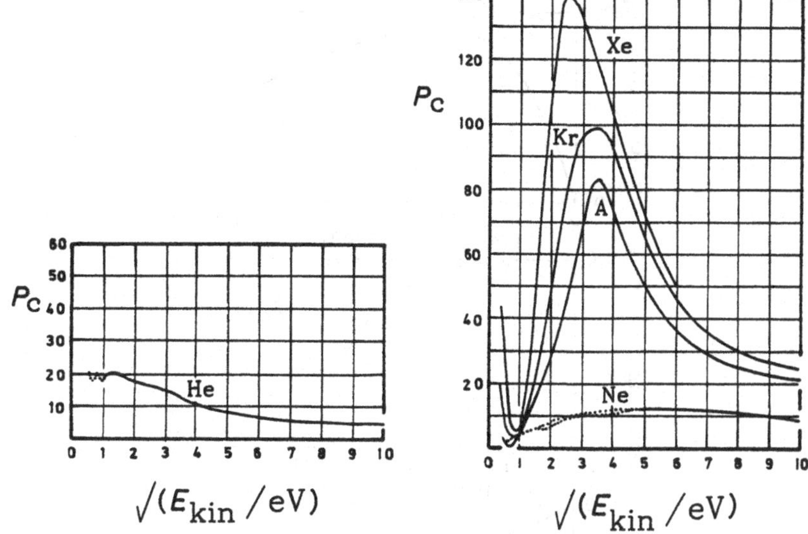

Figure 4.38 Total cross sections for the scattering of electrons on rare gases. This cross section is given by $\sigma_{\text{e-scatt.}}\,(\text{cm}^2) = 0.283 \times 10^{-16}\,P_c$. The dotted curves for neon are different sets of experimental data; the extremely small scattering probability for low-speed electrons found in argon, krypton and xenon is known as the *Ramsauer* effect [Ram21]. From [Bro33].

against the electric field up to the turning point, but after the turning point they regain energy. At the turning point the original kinetic energy E^0_{kin} reduces to

$$E_{\text{kin}}|_{\text{turning point}} = \tfrac{1}{2}m_0 v_\perp^2, \tag{4.52a}$$

where $v_\perp$ is the transverse velocity component (perpendicular to the direction of the analyser field). Since this velocity component is given by $v_0 \cos\vartheta_0$, where ϑ_0 is the entrance angle, one obtains

$$E_{\text{kin}}|_{\text{turning point}} = E^0_{\text{kin}} \cos^2\vartheta_0. \tag{4.52b}$$

This equation becomes important if, for example, one considers scattering of 1 eV electrons on xenon in a deflection analyser with $\vartheta_0 = 45°$. From Fig. 4.38 one would conclude that scattering effects are small because the scattering cross section is small (Ramsauer minimum), but the reduction of kinetic energies towards the turning point increases $\sigma_{\text{e-scatt.}}$ considerably.

To demonstrate the scattering losses, the scattering factor f_s will be calculated for 50 eV electrons scattered on argon assuming a path length $\ell = 50$ cm for the electron trajectory and, for simplicity, an energy-independent average cross section $\bar{\sigma}_{\text{e-scatt.}} = 10^{-15}\,\text{cm}^2$ and a constant average target density n_v or equivalently a constant pressure p. For a given temperature T the pressure p is related to the target density n_v by

$$n_v/\text{cm}^{-3} = \frac{0.965 \times 10^{19}\,p/\text{Torr}}{T/K}, \tag{4.53}$$

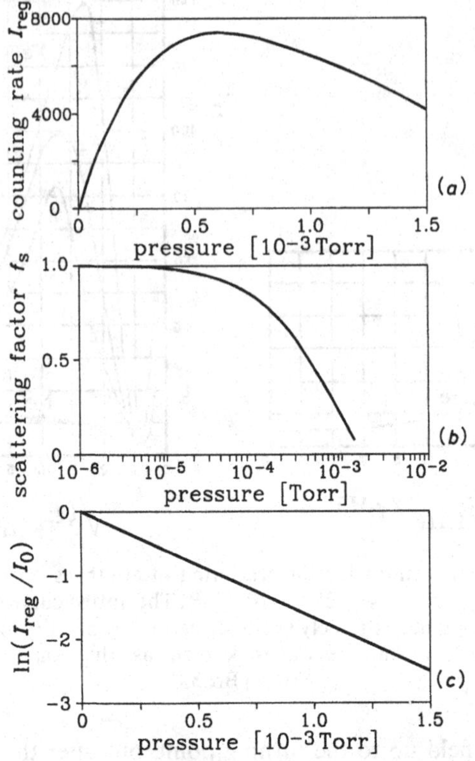

Figure 4.39 Different ways to demonstrate scattering losses in electron spectrometry. The data refer to 50 eV electrons scattered on argon at varying pressures with $\sigma_{\text{e-scatt.}} = 10^{-15}\,\text{cm}^2$ and along a path of 50 cm length. (*a*) Dependence of the registered count rate I_{reg} on the pressure p shown on a linear scale (the undisturbed count rate I_0 would follow from $I_0 = \text{const}\ p$ with const $= 33 \times 10^6$ counts/(s Torr) in the example shown). (*b*) Scattering factor f_s as a function of pressure (logarithmic pressure scale). (*c*) Plot of $\ln(I_{reg}/I_0) = \ln f_s$ as a function of the pressure. For details see main text.

with 1 Torr = 1.333 mb = 0.1333 kPa. Equ. (4.51a) then leads to

$$f_s(E_{kin}) = \exp\{-\bar{\sigma}_{\text{e-scatt.}} n_v \ell\}, \tag{4.51b}$$

and the resulting scattering effects are shown in Fig. 4.39 as a function of the pressure p. Fig. 4.39(*a*) shows that scattering processes limit the registered counting rate. Any further increase in pressure leads to a reduction of the counting rate. This behaviour is reflected also in the drop of the scattering factor f_s at higher pressure (Fig. 4.39(*b*); note the different scales for the pressure). Hence, it follows that the registered counting rate is proportional to the pressure only in a low-pressure region where scattering losses can be neglected and f_s is close to unity. In this example the proportional relationship holds for pressures up to approximately 10^{-5} Torr. The special plot in Fig. 4.39(*c*) is useful for an experimental determination of scattering losses: if p is substituted by n_v, the slope

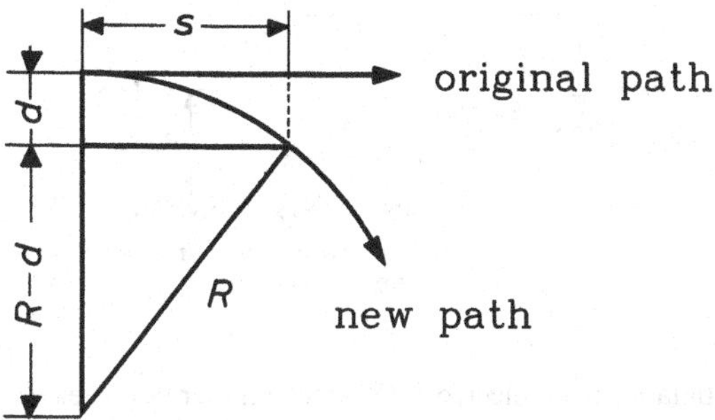

Figure 4.40 Deflection of an electron in a magnetic field which is perpendicular to its original path and leads in the plane of drawing to a new path with characteristic radius R. The distance s along the original path is related to the distance d as indicated.

m of the straight line is

$$m = \{-\bar{\sigma}_{\text{e-scatt.}}\ell\} \tag{4.54}$$

which can then be used in equ. (4.51b) to calculate f_s.

4.5.2 Magnetic fields

The electron trajectories in a spectrometer are affected by magnetic fields. This is demonstrated with the help of Fig. 4.40: the formerly straight path of an electron with velocity v is deflected, under the action of the magnetic field B perpendicular to the trajectory, through a distance d given by

$$d = R - \sqrt{(R^2 - s^2)}, \tag{4.55a}$$

where R is the radius of the electron's path. R can be calculated by setting the Lorentzian force equal to the centrifugal force:

$$e_0 vB = m_0 v^2/R. \tag{4.55b}$$

For a weak magnetic field the resulting small deflection d can be approximated by

$$d \approx s^2/2R, \tag{4.55c}$$

i.e.,

$$d/\text{cm} = \frac{0.148\ s^2/\text{cm}^2\ H/\text{Oe}}{\sqrt{(E_{\text{kin}}/\text{eV})}}. \tag{4.55d}$$

(Conventionally, the magnetic field H is measured in oersted, but in a vacuum the B field measured in gauss has the same value.) For example, consider an electron of 1 eV energy that travels 30 cm in a residual magnetic field of 1 mOe. This results in a deflection d of 1.3 mm which must be compared with critical slit openings, such as the width and/or length of the dispersion slit of the analyser. Depending on the direction of the magnetic field and, hence, on the deflection, such a disturbance can lead to a shift and/or loss of intensity for the observed electron line. Since the residual magnetic field will also have local inhomogeneities which affect the bundle of electron trajectories differently, a broadening of the

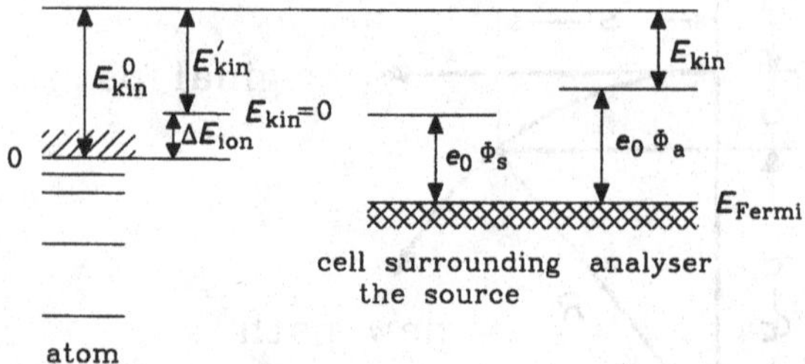

Figure 4.41 Energy balance in an electrostatic spectrometer. Some excited states of an atom converging towards the ionization threshold which defines the reference point for the kinetic energy E^0_{kin} of ejected electrons are shown on the left. The electrons pass three different potential regions when they leave the source volume and travel towards and through the electron energy analyser. In these regions the reference point of their original kinetic energy E^0_{kin} changes as indicated. ΔE_{ion} is due to the positive charge of photoions in the source volume; Φ_S and Φ_a are the work functions of the cell surrounding the source volume and of the analyser plate at its entrance, respectively, both referring to the common Fermi level. From [Meh68b].

observed line will generally accompany the loss of intensity. As a consequence, a careful shielding of the large earth field (between 500 and 600 mOe) down to values below 1 mOe is necessary for electron spectrometry at low energies. This can be achieved by the use of several layers of μ-metal and/or Helmholtz coils around the source volume and the electron spectrometer (see Section 10.6). In addition, a careful selection of non-magnetic materials is necessary for all parts of the equipment which are inside the shielding, in particular for all components of the spectrometer itself and the electric cables.

4.5.3 Electric fields

While the controlled electric field is used in the deflection analyser to produce the desired electron trajectories, other electric fields present will disturb these trajectories. As with residual magnetic fields, these can produce a shift and/or broadening of the observed line with possibly a loss of transmission. The most important disturbance of electron spectrometry by residual electric fields arises from differences in the static potentials experienced by the electrons on their way from the source to the field of the analyser. This effect will be discussed with the help of Fig. 4.41. Photoionization of an electron with ionization energy E_I creates a photoelectron with kinetic energy E^0_{kin}. It is this energy which is of interest in electron spectrometry, but in an actual experiment it is E_{kin} that is measured which differs from E^0_{kin} due to the following effects. First, the photoions created have thermal velocities. This means that the ions stay in the source region for a long time and form in a dynamical equilibrium a positive potential which has to be overcome by the escaping photoelectrons. As a result, the photoelectrons leave the source region with $E'_{kin} = E^0_{kin} - \Delta E_{ion}$. Second, the value $E'_{kin} = 0$ fixes the level of zero kinetic energy for the electrons which are in the cell surrounding the source volume. The cell material might differ from that of the analyser, and in this case one gets a non-vanishing contact potential which is the difference between the work functions Φ_s and Φ_a for the source cell and the analyser, respectively. Combining these effects with the help of Fig. 4.41, the following energy balance can be obtained:

$$E_{kin} + e_0\Phi_a = E^0_{kin} - \Delta E_{ion} + e_0\Phi_s, \tag{4.56a}$$

i.e.,

$$fU_{sp} = E_{kin} = E^0_{kin} + c, \tag{4.56b}$$

with

$$c = e_0(\Phi_s - \Phi_a) - \Delta E_{ion}. \tag{4.56c}$$

Here U_{sp} represents the actual spectrometer voltage at which the electron line with *nominal* energy E^0_{kin}, but *actual* energy E_{kin}, has its maximum. For photoionization the correction, ΔE_{ion}, is of the order of 10 meV and plays a role only in high-resolution work (see the discussion in [BKL93]). The quantity c takes into account the contact potential, and this disturbance can be kept small by covering all surfaces around the source volume and at the entrance of the analyser with the same material, i.e., graphite. However, for precise determinations of binding energies and/or for studies with metal vapours which pollute the surfaces and change the contact potential, careful energy calibrations are required (one example is described in Section 5.5.2).

4.5.4 Thermal Doppler broadening

The thermal motion of the atoms in the source volume leads to a broadening of the observed line, because the thermal velocity $\mathbf{v}_{th}$ of the electron-emitting atom is added to the electron velocity $\mathbf{v}_0$. For an estimation of the resulting disturbance by this kinematical effect it is sufficient to select for the thermal velocity the two directions at which the Doppler effect becomes extreme, i.e.,

$$v(\text{electron}) = v_0 \pm v_{th}. \tag{4.57a}$$

These velocity components v_{th} in a given direction follow from the Maxwell distribution

$$f(v_{th})\,\mathrm{d}v_{th} = \sqrt{\left(\frac{m}{2\pi kT}\right)}\,\mathrm{e}^{-mv_{th}^2/2kT}\,\mathrm{d}v_{th}, \tag{4.58}$$

where m is the mass of the atom, k the Boltzmann constant, and T the temperature. In a further approximation one takes as representative of all the possible thermal velocities v_{th} the fwhm values Δv_{th} of the Maxwell distribution

$$v_{th} \approx \Delta v_{th} = 2\sqrt{[2kT\ln(2)/m]}. \tag{4.57b}$$

Within these approximations, the broadening effect of an electron line expressed as the fwhm value ΔE_{th} on the energy scale follows from (note $v_{th} \ll v_0$)

$$\Delta E_{th} = \frac{m_0}{2}v^2 - \frac{m_0}{2}v_0^2 = m_0 v_0\,\Delta v_{th} \tag{4.59a}$$

as

$$\Delta E_{\text{th}} = 4\sqrt{\left(\frac{E^0_{\text{kin}} kT \ln 2}{1836M}\right)} \tag{4.59b}$$

where M is the atomic mass number. As an example, the broadening effect on K–LL Auger electron emission in neon will be calculated. With $E^0_{\text{kin}} = 800$ eV, $M = 20$ and $kT = 1/40$ eV one gets $\Delta E_{\text{th}} = 78$ meV. This value is not negligibly small compared to the natural line width $\Gamma = 270$ meV. Hence, their combined effect has to be considered. Under the conditions [Vol68]

$$8E^0_{\text{kin}}kT/(1836M) \ll \Gamma^2 \quad \text{and} \quad \Gamma \ll E^0_{\text{kin}} \tag{4.60a}$$

one obtains

$$\mathit{\Gamma}_{\text{eff}} \approx \Gamma + \frac{8E^0_{\text{kin}}kT}{1836M\Gamma}, \tag{4.60b}$$

and for the current example one has $\mathit{\Gamma}_{\text{eff}} = 286$ meV.

4.5.5 Postcollision interaction (PCI)

PCI differs from the disturbances of electron spectrometry treated above, because it is an *inherent* property of photon-induced two-step double ionization, and not an external disturbance. Nevertheless, PCI is included here because the effects on lineshape, line broadening and line position belong to the present discussion (for more details on PCI see Section 5.5).

When the photon energy comes close to the threshold value for the ejection of an inner-shell electron, the subsequent Auger decay may be influenced by the presence of the slowly receding photoelectron. The resulting interaction between the escaping electrons is called *postcollision interaction.* (For review articles on PCI see [Nie78, AKS79, Amu80, Abe81, Sch82, Cra87, Sch87, KSh89].) PCI belongs to the class of 'simple' electron correlation effects which can be described approximately as a change in the mutual shielding of the two outgoing electrons: close to threshold, the slow photoelectron can shield the doubly charged ion such that the faster Auger electron gains energy and the slower photoelectron loses the same amount of energy. (The PCI energy shift is also sometimes called the Berry shift [Ber62, BBe66].) This energy exchange has two consequences which can be inferred from Fig. 4.42 in which a PCI-affected Auger line is shown. First, the resulting PCI energy distribution for the emitted electrons is no longer described by a Lorentzian lineshape. Instead, it becomes asymmetric and broadened, and its maximum is shifted in energy. Second, the energy exchange may be so large that the slow photoelectron can be captured in a bound orbital of the remaining ion (*shake down effect*, this energy region is indicated in the figure by the shaded area). Both manifestations of PCI have fundamental consequences:

(i) The change of the lineshape can impose serious problems for the appropriate

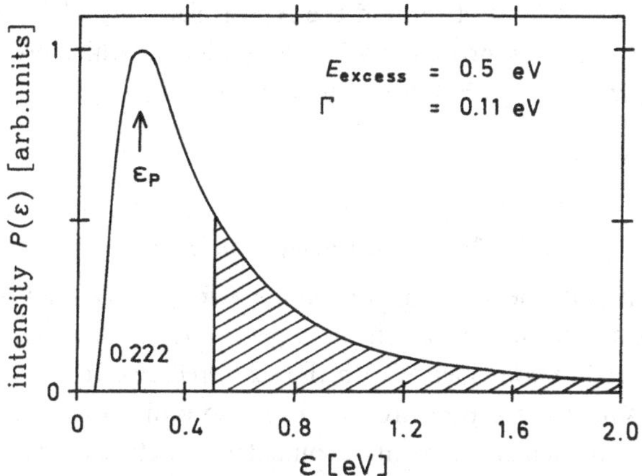

Figure 4.42 Energy distribution $P(\varepsilon)$ caused by PCI in inner-shell photoionization. The data refer to the N_5–$O_{2,3}O_{2,3}$ $^1S_0$ Auger line in xenon following photoionization with an excess energy of 0.5 eV. ε is the energy gained by the Auger electron due to PCI, the value $\varepsilon = 0$ corresponds to the nominal energy of these Auger electrons, $E^0_{kin} = 29.97$ eV. The energy distribution has been calculated in the sudden PCI model [Nie77] using an inherent level width Γ of the photoionized state of 0.11 eV. It can be seen that all Auger electrons gain energy. The maximum of $P(\varepsilon)$ lies at $\varepsilon_p = 0.222$ eV ($\varepsilon_p = \Delta E_{PCI}$ in the main text). For $\varepsilon > 0.5$ eV the energy gain becomes larger than the available excess energy, and the photoelectron is shaken down (shaded area). From [Sch82].

separation of neighbouring Auger or corresponding photoelectron lines and for the correct determination of inner-shell binding energies.

(ii) The PCI-induced shake down mechanism establishes a natural link between inner-shell photoionization with subsequent Auger decay on one side and outer-shell ionization processes accompanied by simultaneous excitation on the other (see Section 5.1.2.2).

In the present context it is important to note that PCI effects in inner-shell photoionization vanish if the photoelectron becomes faster than the Auger electron. Due to this simple criterion, inner-shell photoionization with subsequent Auger decay represents a unique case, because proper selection of the photon energy allows a controlled tuning of PCI effects from negligible influence at high photon energies to maximum disturbances close to threshold. If this statement is expressed quantitatively, one obtains for the most probable energy shift ΔE_{PCI} induced by PCI (see references given in [Sch82]; Γ is the inner-shell level width, all quantities refer to atomic units)

$$\Delta E_{PCI} = \begin{cases} 0.43\Gamma^{2/3} & \text{for } E_{phe} \approx 0, \\ 0 & \text{for } E_{phe} > E_A. \end{cases} \tag{4.61}$$

(ΔE_{PCI} is the energy gain for the Auger electron, but the energy loss for the photoelectron.) It should be pointed out that this situation differs considerably from the case of electron impact ionization with subsequent Auger decay. Following electron impact, there are three free particles present in the final state (the projectile, the ejected electron and the Auger electron) which can be subject to PCI and, due to the high probability for soft collisions which create an electron with very low kinetic energy, PCI does not vanish for high impact energies [San86, VSS88, SVö89].

4.6 Electron–electron coincidences

Electron–electron coincidence measurements are frequently used to study single ionization processes caused by electron impact, and in most cases the ejected electron is measured in coincidence with the scattered electron in order to fix the kinematics of a single ionization process. For photoionization the kinematics is simple† and, therefore, a measurement of coincidences between two electrons yields direct information on double photoionization. Two limiting cases are distinguished: *direct* double photoionization (both electrons are emitted simultaneously) and photon-induced *two-step* double ionization (the ejection of a photoelectron is followed by Auger electron emission). Due to the low cross section for direct double photoionization and the many decay branches in most cases of two-step ionization, the coincidence signal is rather low, and such studies are limited. However, only by means of electron–electron coincidences can the full emission pattern of a double ionization process be viewed, with this pattern showing features characteristic of the underlying mechanism of double ionization as well as the role of electron correlations. In particular, direct double photoionization in helium presents the ideal test case to study the non-trivial and still unsolved many-body problem where three charged particles are subject to their mutual long-range Coulomb interactions. Hence, such coincidence experiments provide the basis for testing newly developed theories for energy- and angle-dependent two-electron emission. In the following, fundamental aspects of electron–electron coincidence experiments used as a tool for the study of photon-induced double ionization will be presented. They involve the parametrization of the many-fold differential cross section for two-electron emission, the general aspects of true and accidental coincidences, and the optimization of the experimental set-up for electron–electron coincidence measurements using synchrotron radiation. An example of a specific two-step double ionization electron–electron coincidence experiment will be described in Section 5.6.

4.6.1 Parametrization of the differential cross section for two-electron emission

As with single photoionization, the observables for two-electron emission depend on the properties of: the incident light (energy, intensity and polarization), the

† In the dipole approximation and for linearly polarized incident light, the natural reference axis is the electric field vector, the energy transfer is equal to the photon energy, and the momentum transfer is negligible.

target (differential cross section) and the two electron spectrometers (positions in space, acceptance solid angles, transmissions, resolutions, and efficiency of the detectors). However, since of all possibilities for detected electrons only *coincident* pairs are of interest, the characteristic properties of the coincidence circuit also need to be considered. In order to obtain a transparent description of these dependences, one first needs a parametrization of the differential cross section for two-electron emission in which dynamical and geometrical effects are separated.

Neglecting a spin-analysis of the registered electrons, the energy- and angle-dependent differential cross section describing the ejection of two electrons with wave vectors $\boldsymbol{\kappa}_a$ and $\boldsymbol{\kappa}_b$ can be written as

$$\left.\frac{d^4\sigma}{d\Omega_a\, d\Omega_b\, dE_a\, dE_b}(\boldsymbol{\kappa}_a, \boldsymbol{\kappa}_b)\right|_{\mathbf{S}} \delta(E_{ph} - E_a - E_b - E_I^{++}). \tag{4.62a}$$

This is a four-fold differential cross section, because it depends on four continuous variables: two energies and two directions in space. (This is often also called a six-fold differential cross section because the two directions in space are fixed by four angles.) The expression given contains the following dependences:

(i) Dependence on the polarization of the incident light; this is indicated by the Stokes parameters $\{S_1, S_2, S_3\} = \mathbf{S}$. As has been shown in Fig. 1.15, the direction of the photon beam and its polarization properties determine a convenient reference frame (tilted collision frame) against which other geometrical quantities are measured. This reference frame is used also for the case of two-electron emission. (For linearly polarized incident light the direction of the electric field vector is frequently selected as reference axis.)

(ii) Dependence on the angles for electron emission into the solid angles $d\Omega_a$ and $d\Omega_b$ centred around the directions $\hat{\boldsymbol{\kappa}}_a$ and $\hat{\boldsymbol{\kappa}}_b$ of electron emission where these directions are described in the coordinate frame selected above.

(iii) Dependence on the energies E_a and E_b of the two electrons, within the energy intervals dE_a and dE_b. Energy conservation is ensured by the δ-function. For direct double photoionization (dd) both electrons share the available excess energy E_{exc} continuously

$$E_{exc} = E^0_{ph} - E_I^{++}, \tag{4.63a}$$

where

$$E_a + E_b = E_{exc}, \tag{4.63b}$$

E^0_{ph} is the photon energy, and E_I^{++} is the energy necessary for double ionization. The δ-function restricts the energies to $E_b = E_{exc} - E_a$, so equ. (4.62a) is replaced by the triple-differential cross section (TDCS)

$$\left.\frac{d^3\sigma}{d\Omega_a\, d\Omega_b\, dE}(\boldsymbol{\kappa}_a, \boldsymbol{\kappa}_b)\right|_{\mathbf{S}}^{dd}. \tag{4.62b}$$

Here the differential dE with $dE = |dE_a| = |dE_b|$ indicates that the requirement of energy conservation always has to be fulfilled (the matrix element is taken 'on-the-energy-shell').

In the case of photon-induced two-step double ionization the formulation differs in two points. First, the existence of the intermediate hole-state leads to the appearance of a Lorentzian function $L_\Gamma(E_\ell, E_\mathrm{I}^+)$, compare equ. (2.19c) with $E_\mathrm{I}^+ = E_{n\ell_j}$, $E_\ell = E$, and $\Gamma = \Gamma_{n\ell j}$. Second, the δ-function splits into two factors

$$\delta(E_\mathrm{ph} - E_\mathrm{a} - E_\mathrm{b} - E_\mathrm{I}^{++}) = \delta(E_\mathrm{ph} - E_\mathrm{a} - E_\ell)\,\delta(E_\ell - E_\mathrm{b} - E_\mathrm{I}^{++}), \quad (4.64)$$

i.e. the energies of the photoelectron (subscript a) and of the Auger electron (subscript b) are fixed. Therefore (4.62b) can be replaced by

$$\frac{\mathrm{d}^3\sigma}{\mathrm{d}\Omega_\mathrm{a}\,\mathrm{d}\Omega_\mathrm{b}\,\mathrm{d}E}(\boldsymbol{\kappa}_\mathrm{a}, \boldsymbol{\kappa}_\mathrm{b})\Bigg|_\mathrm{S}^{\text{2-step}} = L_\Gamma(E_\ell, E_\mathrm{I}^+)\,\frac{\mathrm{d}^2\sigma}{\mathrm{d}\Omega_\mathrm{a}\,\mathrm{d}\Omega_\mathrm{b}}(\boldsymbol{\kappa}_\mathrm{a}, \boldsymbol{\kappa}_\mathrm{b})\Bigg|_\mathrm{S}^{\text{2-step}}. \quad (4.62c)$$

If the atoms are assumed to be randomly oriented in space (e.g., gaseous target in a field-free region) and polarization of the ion after double ionization is not observed, a parametrized form of the differential cross section can be obtained [KFe92]. (See also for direct double photoionization [YPN85, HSW91]; for the related case of continuous double Auger decay [ALK92]; and for two-step double ionization [Kab92, KSc95].) For the case of linearly polarized incident light, the basic formula will be reproduced here and discussed for two examples: direct double photoionization in helium and photon-induced two-step double ionzation in magnesium.

4.6.1.1 *Basic formula*

On the grounds of very general symmetry arguments the triple differential cross section for two-electron emission following photon impact can be represented as

$$\frac{\mathrm{d}^3\sigma}{\mathrm{d}\Omega_\mathrm{a}\,\mathrm{d}\Omega_\mathrm{b}\,\mathrm{d}E} = \sum_{k_1k_2kq} A(k_1, k_2, k)\,B_{kq}^{k_1k_2}(\hat{\boldsymbol{\kappa}}_\mathrm{a}, \hat{\boldsymbol{\kappa}}_\mathrm{b})\,\rho_{kq}(E1). \quad (4.65)$$

This expression disentangles the properties of the light polarization (coefficients $\rho_{kq}(E1)$), the geometry of two-electron emission (coefficients $B_{kq}^{k_1k_2}(\hat{\boldsymbol{\kappa}}_\mathrm{a}, \hat{\boldsymbol{\kappa}}_\mathrm{b})$) and the dynamical parameters of the double photoionization process (coefficients $A(k_1, k_2, k)$). The $\rho_{kq}(E1)$ are the *statistical tensors* of the incident light which describe its polarization properties in the electric dipole approximation represented by $E1$. For linearly polarized light in which the electric field vector defines the z-axis of the coordinate frame, one has only two non-vanishing components given by (see equ. (8.99b))†

$$\rho_{00}(E1) = 1/\sqrt{3} \quad \text{and} \quad \rho_{20}(E1) = -2/\sqrt{6}. \quad (4.66)$$

The B-coefficients are *bipolar spherical harmonics* defined by

$$B_{kq}^{k_1k_2}(\hat{\boldsymbol{\kappa}}_\mathrm{a}, \hat{\boldsymbol{\kappa}}_\mathrm{b}) = \sum_{q_1q_2} (k_1q_1k_2q_2\,|\,kq)\,Y_{k_1q_1}(\hat{\boldsymbol{\kappa}}_\mathrm{a})\,Y_{k_2q_2}(\hat{\boldsymbol{\kappa}}_\mathrm{b}), \quad (4.67)$$

† For right- and left-circularly polarized incident light, the emission patterns differ in general [BKl92, BKH93, KSc95]. This phenomenon is called *circular dichroism* in the angular distribution.

where $(k_1q_1k_2q_2\,|\,kq)$ is a Clebsch–Gordan coefficient, and the $Y_{kq}(\hat{\mathbf{\kappa}})$ are spherical harmonics. Since the A-coefficients are the dynamical parameters of two-electron emission, they contain the matrix elements with relative phases and depend on the energies E_a and E_b of the ejected electrons.

Making use of the restrictions brought in by the statistical tensors of the incident light, equ. (4.65) can be rewritten as

$$\frac{d^3\sigma}{d\Omega_a\, d\Omega_b\, dE} = A_{000} + \sum_{k>0} A_{kk0} P_k(\cos\vartheta_{ab}) + \sum_{k_1,k_2} A_{k_1k_22} B_{20}^{k_1k_2}(\hat{\mathbf{\kappa}}_a, \hat{\mathbf{\kappa}}_b). \quad (4.68)$$

Note in this transformation

$$B_{00}^{kk}(\hat{\mathbf{\kappa}}_a, \hat{\mathbf{\kappa}}_b) = (-1)^k \sqrt{(2k+1)}\, P_k(\cos\vartheta_{ab})/4\pi \quad (4.69)$$

where ϑ_{ab} is the relative angle between the directions $\hat{\mathbf{\kappa}}_a$ and $\hat{\mathbf{\kappa}}_b$, and furthermore, some numerical factors are incorporated into the dynamical parameters $A_{k_1k_2k}$ so that they differ from $A(k_1, k_2, k)$, but

$$A_{k_1k_2k} \sim A(k_1, k_2, k). \quad (4.70)$$

In the parametrization of equ. (4.68) the terms associated with the Legendre polynomials $P_k(\cos\vartheta_{ab})$ represent that part of the angular correlation which is independent of the light beam, while the terms associated with the bipolar harmonics are due to the multipole expansion of the interactions of the electrons with the electric field vector. The link between geometrical angular functions and dynamical parameters is made by the summation indices k_1, k_2 and k. These quantities are related to the orbital angular momenta ℓ_i of the two individual emitted electrons, and they are subject to the following conditions:

(i) for direct double photoionization

$$\left.\begin{aligned} k_1 &= 0, 1, 2, \ldots, 2\ell_a(\max), \\ k_2 &= 0, 1, 2, \ldots, 2\ell_b(\max), \\ k &= k_1 \text{ for } k_1 = k_2, \\ k_1 + k_2 &= \text{even}; \end{aligned}\right\} \quad (4.71a)$$

(ii) for two-step double ionization the condition of $k_1 + k_2 =$ even breaks down to the condition that each k_i value must be even, and one gets

$$\left.\begin{aligned} k_1 &= 0, 2, \ldots, 2\ell_a(\max), \\ k_2 &= 0, 2, \ldots, 2\ell_b(\max), \\ k &= k_1 \text{ for } k_1 = k_2, \\ k_1 &= \text{even}, k_2 = \text{even}. \end{aligned}\right\} \quad (4.71b)$$

Depending on the individual orbital angular momenta ℓ_a and ℓ_b involved, the summation can go up to infinity. Such a situation occurs for double photoionization in helium where the $^1P^o$ state of the continuum pair wavefunction can be obtained by an unlimited coupling of individual orbital momenta ($\varepsilon s\varepsilon p$, $\varepsilon p\varepsilon d$, $\varepsilon d\varepsilon f, \ldots$). However, in the case of photon-induced two-step double ionization the formulation in two steps not only fixes the kinetic energies E_a and E_b of the ejected electrons as given in equ. (4.64), but also restricts their orbital angular momenta ℓ_a and ℓ_b to a few values, and the summations remain finite (see the magnesium example, below).

4.6.1.2 *Direct double photoionization in helium*

As a first application of the parametrization derived, the case of direct double photoionization in helium by completely linearly polarized light will be discussed. Here the electron-pair wavefunction for the two continuum electrons has to represent a $^1P^o$ state, because *LS*-coupling holds well, and according to equs. (2.5) and (2.9b) this implies that photoionization from the ground state of helium leaves the singlet spin state unaffected, but changes the total orbital angular momentum L and the parity, thus giving the final state $^1P^o$. Using individual electron orbitals characterized by $\varepsilon_a\ell_a$ and $\varepsilon_b\ell_b$, the final state can be built in different ways. Within the single-electron picture and including relaxation one can understand the two channels $\varepsilon_a s\varepsilon_b p\ ^1P^o$ and $\varepsilon_a p\varepsilon_b s\ ^1P^o$. However, due to electron correlations, higher orbital angular momenta also contribute [Fan74], giving (in principle) an infinite number of possibilities for the $(\ell_1 + \ell_2 = 1)$-combinations, i.e., infinite summation indices k_1, k_2 and k in equ. (4.68). This situation is demonstrated in Table 4.1 which includes the channels $\varepsilon_a p\varepsilon_b d\ ^1P^o$, $\varepsilon_a d\varepsilon_b p\ ^1P^o$, and $\varepsilon_a d\varepsilon_b f\ ^1P^o$. From this table certain systematics can be inferred for the *A*-coefficients. Starting with the lowest combinations of εs and εp electrons, the coefficients A_{000}, A_{022} and A_{202} with $A_{022} = A_{202}$ appear. Taking the next group of εp and εd electrons, these coefficients are again necessary (which implies that their actual values will be modified due to this new group of electrons) but in addition, new coefficients appear. This systematic process continues and leads to the correct numerical value of each *A*-coefficient only being found if all $(\ell_1 + \ell_2 = 1)$-combinations are included. This feature means that the parametrization of two-electron emission in helium based on the correct equ. (4.68) is not very tractable, and two other approaches will be discussed next: one in which the idea of parametrization is dropped and a direct numerical calculation is performed, and one in which a simpler, but then only approximate, parametrization is sought.

The result of a numerical calculation for direct double photoionization in helium taking into account electron correlations in the initial and final states is shown in Fig. 4.43 and compared there with experimental data. Good agreement between theoretical and experimental data can be noted. (This statement holds for the shape of this angular correlation pattern, because the intensities have been adapted by an overall scaling factor. For an experimental determination of absolute values see [SSc95]. For additional experimental and theoretical work on helium see [MBr93, MBr94, SKS94, DAM95, KOs95, LMA95, MPB95, PSh95] with further references therein.) The explanation of the observed features characteristic for the differential cross section of this direct double photoionization process will be

Table 4.1. *The lowest* A*-coefficients needed for the description of double photoionization in helium according to equ. (4.68).*

ℓ_a	k_1	ℓ_b	k_2	Responsible for A_{kk0}	Responsible for $A_{k_1k_22}$
0	0	1	0, 1, 2	$\mathbf{A_{000}}$	$\mathbf{A_{022}}$
1	0, 1, 2	0	0	A_{000}	$\mathbf{A_{202}}$
1	0, 1, 2	2	0, 1, 2, 3, 4	A_{000} $\mathbf{A_{110}}$ $\mathbf{A_{220}}$	A_{022} $\mathbf{A_{112}}$ $\mathbf{A_{132}}$ A_{202} $\mathbf{A_{222}}$ $\mathbf{A_{242}}$
2	0, 1, 2, 3, 4	1	0, 1, 2	A_{000} A_{110} A_{220}	A_{022} A_{112} A_{202} A_{222} $\mathbf{A_{312}}$ $\mathbf{A_{422}}$
2	0, 1, 2, 3, 4	3	0, 1, 2, 3, 4, 5, 6	A_{000} A_{110} A_{220} $\mathbf{A_{330}}$ $\mathbf{A_{440}}$	A_{022} $\mathbf{A_{042}}$ A_{112} $\mathbf{A_{132}}$ A_{202} A_{222} A_{242} A_{312} $\mathbf{A_{322}}$ $\mathbf{A_{352}}$ $\mathbf{A_{422}}$ $\mathbf{A_{462}}$

The two-electron wavefunction in the continuum is represented by $\varepsilon_a\ell_a\varepsilon_b\ell_b$ partial waves which couple to ${}^1P^o$. Starting on top with the lowest ℓ values possible, one derives according to equ. (4.71a) the allowed k_1 and k_2 coefficients. These and the condition for non-vanishing Clebsch–Gordan coefficients in equ. (4.67) then lead to the possible A_{kk0}- and $A_{k_1k_22}$-coefficients. A-coefficients which appear (from top to bottom) for the first time are written in bold. For higher ℓ values in the allowed combinations $\varepsilon_a\ell_a\varepsilon_b\ell_b\ {}^1P^o$, further A-coefficients will appear.

postponed at the moment. It follows rather naturally from an approximate parametrization which differs from that in equ. (4.68).

Starting in a manner similar to the treatment of single photoionization described in Section 2.1, double photoionization in helium caused by linearly polarized light will be treated first with uncorrelated wavefunctions. A calculation of the differential cross section for double photoionization then requires the evaluation

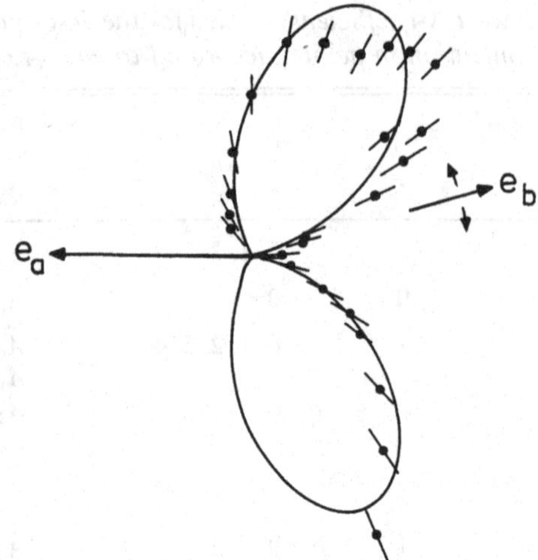

Figure 4.43 Energy- and angle-resolved triple-differential cross section for direct double photoionization in helium at 99 eV photon energy. The diagram shows the polar plot of relative intensity values for one electron (e_a) kept at a fixed position while the angle of the coincident electron (e_b) is varied. The data refer to electron emission in a plane perpendicular to the photon beam direction for partially linearly polarized light (Stokes parameter $S_1 = 0.554$) and for equal energy sharing of the excess energy, i.e., $E_a = E_b = 10$ eV. Experimental data are given by points with error bars, theoretical data by the solid curve. From [SKE93] using theoretical data from [MBr93]; see also [HLA94].

of the matrix element M_{fi} given by

$$M_{fi} = \langle \boldsymbol{\kappa}_a \boldsymbol{\kappa}_b{}^{(-)} | z_1 + z_2 | 1s^2\, S^e \rangle, \tag{4.72a}$$

where $\boldsymbol{\kappa}_a$ and $\boldsymbol{\kappa}_b$ stand for the wavenumber vectors of the two ejected electrons, and the superscript $(-)$ indicates the correct asymptotic boundary conditions for two-electron emission. The electron spin is omitted in this expression because the electrons remain in the singlet state (see equ. (2.5)), i.e., the spatial wavefunctions are symmetric with respect to electron exchange. (For a two-electron system the antisymmetric determinantal wavefunction can always be described as the product of a symmetrical spatial and an unsymmetrical spin wavefunction (this result holds for spin singlets; the reverse statement is valid for spin triplets).) Therefore, one has

$$|1s^2\, S^e\rangle = \varphi_{1s0}(\mathbf{r}_1)\varphi_{1s0}(\mathbf{r}_2), \tag{4.73a}$$

$$|\boldsymbol{\kappa}_a \boldsymbol{\kappa}_b\rangle = \frac{1}{\sqrt{2}}(\varphi_{\boldsymbol{\kappa}_a}(\mathbf{r}_1)\varphi_{\boldsymbol{\kappa}_b}(\mathbf{r}_2) + \varphi_{\boldsymbol{\kappa}_a}(\mathbf{r}_2)\varphi_{\boldsymbol{\kappa}_b}(\mathbf{r}_1)), \tag{4.73b}$$

which give

$$M_{fi} = \sqrt{2}\langle \boldsymbol{\kappa}_a | z | 1s0\rangle\langle \boldsymbol{\kappa}_b | 1s0\rangle + \sqrt{2}\langle \boldsymbol{\kappa}_b | z | 1s0\rangle\langle \boldsymbol{\kappa}_a | 1s0\rangle. \tag{4.72b}$$

This result reflects the fact that the photon operator, as a one-particle operator, interacts with the 'active' 1s-electron only, ejecting it into a wave characterized asymptotically by $\boldsymbol{\kappa}_a$ or, alternatively, $\boldsymbol{\kappa}_b$, while the 'passive' 1s-electron leads to an overlap matrix element with the corresponding function of the other continuum electron. If all single-particle wavefunctions are calculated using the same model Hamiltonian (the *frozen* atomic structure approximation) these overlap matrix elements vanish due to the orthogonality of the wavefunctions. Hence, the differential cross section for double photoionization in helium is zero. If, however, the single-particle functions in the initial and final states are calculated from different model Hamiltonians, the overlap has a finite value, provided the orbital angular momenta involved are the same (the *relaxed* atomic structure calculation). Due to the change of shielding of the nuclear charge during photoionization, the assumption of relaxation is justified in the present case. Using, for simplicity, plane wavefunctions for the emitted electrons and hydrogenic wavefunctions for the initial state, one then has to calculate integrals of the following forms (the discussion follows the description in [MBr93] which includes extensions to other wavefunctions):

$$\langle \boldsymbol{\kappa} | 1s0 \rangle = \frac{1}{\sqrt{[(2\pi)^3]}} \frac{1}{\sqrt{(4\pi)}} 2\, Z_{\text{eff}}^{3/2} \int \mathrm{e}^{-\mathrm{i}\boldsymbol{\kappa}\cdot\mathbf{r}}\, \mathrm{e}^{-Z_{\text{eff}} r}\, \mathrm{d}\mathbf{r} \tag{4.74a}$$

and

$$\hat{\boldsymbol{\varepsilon}} \cdot \langle \boldsymbol{\kappa} | \mathbf{r} | 1s0 \rangle = \frac{1}{\sqrt{[(2\pi)^3]}} \frac{1}{\sqrt{(4\pi)}} 2\, Z_{\text{eff}}^{3/2}\, \hat{\boldsymbol{\varepsilon}} \cdot \int \mathrm{e}^{-\mathrm{i}\boldsymbol{\kappa}\cdot\mathbf{r}}\, \mathbf{r}\, \mathrm{e}^{-Z_{\text{eff}} r}\, \mathrm{d}\mathbf{r}, \tag{4.75a}$$

where the dipole operator z has been replaced by $z = \hat{\boldsymbol{\varepsilon}} \cdot \mathbf{r}$ with the unit vector $\hat{\boldsymbol{\varepsilon}}$ giving the direction of the electric field vector of the incident linearly polarized radiation. The first integral, as the overlap integral between the two single-particle functions, directly proves the s-character contained in the plane wave wavefunction, and the second integral, through the action of the dipole operator on the 1s orbital, measures the p-character. Since these integrals are simply Fourier transforms, one derives

$$\int \mathrm{e}^{-\mathrm{i}\boldsymbol{\kappa}\cdot\mathbf{r}}\, \mathrm{e}^{-Z_{\text{eff}} r}\, \mathrm{d}\mathbf{r} = \frac{8\pi Z_{\text{eff}}}{(\kappa^2 + Z_{\text{eff}}^2)^2} \tag{4.74b}$$

and

$$\int \mathrm{e}^{-\mathrm{i}\boldsymbol{\kappa}\cdot\mathbf{r}}\, \mathbf{r}\, \mathrm{e}^{-Z_{\text{eff}} r}\, \mathrm{d}r = \frac{-32\mathrm{i}\pi Z_{\text{eff}}}{(\kappa^2 + Z_{\text{eff}}^2)^3}\, \boldsymbol{\kappa}. \tag{4.75b}$$

(The expansion of the plane wave into partial waves yields $Y_{\ell m}^{*}(\hat{\boldsymbol{\kappa}})\, Y_{\ell m}(\hat{\mathbf{r}})$ components. If these are multiplied by $Y_{00}(\hat{\mathbf{r}})\sqrt{(4\pi)}$, the orthogonality condition for spherical harmonics then leads to the result that only the $(\ell = 0)$-component remains. Hence, there is no dependence on $\hat{\boldsymbol{\kappa}}$ in equ. (4.74b). Similarly, in equ. (4.75b) only the $(\ell = 1)$-component is proved. Note, in addition, the typical overlap property of these integrals: if a Coulomb wave with $Z = 2$ were used for the continuum electron the result of the integration would be proportional to $Z - Z_{\text{eff}}$ and would vanish for $Z_{\text{eff}} = 2$. In other words, the dependence on $Z - Z_{\text{eff}}$ reflects the fact that the final and initial wavefunctions belong to different sets of wavefunctions.) With this information the dipole matrix element can be evaluated, giving

$$M_{fi}(\text{length}) = \frac{-i32\sqrt{2}\,Z_{eff}^5}{\pi^2(\kappa_a^2 + Z_{eff}^2)^2(\kappa_b^2 + Z_{eff}^2)^2}\left(\frac{\hat{\boldsymbol{\varepsilon}}\cdot\boldsymbol{\kappa}_a}{\kappa_a^2 + Z_{eff}^2} + \frac{\hat{\boldsymbol{\varepsilon}}\cdot\boldsymbol{\kappa}_b}{\kappa_b^2 + Z_{eff}^2}\right). \tag{4.76a}$$

A similar calculation using the velocity form of the dipole operator leads to

$$M_{fi}(\text{velocity}) = \frac{-i\sqrt{28}\,Z_{eff}^5}{\pi^2(\kappa_a^2 + Z_{eff}^2)^2(\kappa_b^2 + Z_{eff}^2)^2}(\hat{\boldsymbol{\varepsilon}}\cdot\boldsymbol{\kappa}_a + \hat{\boldsymbol{\varepsilon}}\cdot\boldsymbol{\kappa}_b). \tag{4.76b}$$

Two observations about the matrix element M_{fi} can be made from these results. First, length and velocity forms exhibit different behaviours. For example, if κ_a is large compared to κ_b, the velocity form result gives a dominant contribution when this electron interacts with the polarization vector while the opposite behaviour is found for the length form result. (For a general discussion of length and velocity forms see Section 8.1.3. In particular it should be noted that according to equ. (8.19b) the forms also differ by a factor which has been omitted here.) Second, for equal sharing of the available excess energy between the ejected electrons, both forms lead to a common factor given by

$$M_{fi}(E_a = E_b = E_{exc}/2) \sim (\hat{\boldsymbol{\varepsilon}}\cdot\boldsymbol{\kappa}_a + \hat{\boldsymbol{\varepsilon}}\cdot\boldsymbol{\kappa}_b) = (\cos\vartheta_a + \cos\vartheta_b), \tag{4.77}$$

which yields in the differential cross section the *angular* factor

$$W(\hat{\boldsymbol{\kappa}}_a, \hat{\boldsymbol{\kappa}}_b) = (\cos\vartheta_a + \cos\vartheta_b)^2. \tag{4.78}$$

The content of this angular factor is in agreement with two selection rules derived rigorously for vanishing intensity in helium double photoionization [MBr93]:

(1) Both electrons cannot be emitted perpendicular to the electric field vector of the linearly polarized light ($\vartheta_a = \vartheta_b = 90°$):

$$W(\boldsymbol{\kappa}_a \perp \hat{\boldsymbol{\varepsilon}} \text{ and } \boldsymbol{\kappa}_b \perp \hat{\boldsymbol{\varepsilon}}) = 0. \tag{4.79}$$

(2) For equal energies the electrons cannot be emitted in opposite directions ($\vartheta_b = 180° - \vartheta_a$):

$$W(\boldsymbol{\kappa}_a = -\boldsymbol{\kappa}_b) = 0. \tag{4.80}$$

(This special symmetry property of the singlet wavefunction is also called the 'unfavoured' character of the two-electron wavefunction [KSc76, Sta82b, GRa82, GRa83, HSW91].)

The effect of these selection rules can be seen clearly in Fig. 4.43 where no intensity is observed if the electrons are emitted in opposite directions. (In the case shown the incident light is only partially linearly polarized, and the observed zero results from both selection rules.)

The matrix element M_{fi} derived so far for the differential cross section of double photoionization in helium is based on uncorrelated wavefunctions in the initial and final states. For simplicity the initial state will be left uncorrelated, but electron correlations in the final state will now be included. The significance of final state correlations can be inferred from Fig. 4.43: without these correlations an intensity maximum would be predicted when both electrons are emitted in the same direction, which contradicts experimental observation. Employing a correlated three-body wavefunction, known as the *3C wavefunction*, because it treats all three mutual two-body Coulomb interactions (electron–nucleus, other electron–nucleus, electron– electron) on equal footing,[†] it can be shown [MBr93] that the major parts of the electron–electron interactions (the correlated motion of the escaping electrons) arise from the normalization factor of this wavefunction. Its absolute squared value, called the *correlation* or *Sommerfeld* [Som19; Vol. II, p. 127] factor $C(\boldsymbol{\kappa}_a, \boldsymbol{\kappa}_b)$, is given by [BBr86]

$$C(\boldsymbol{\kappa}_a, \boldsymbol{\kappa}_b) = \frac{\pi/\kappa_{ab}}{(e^{\pi/\kappa_{ab}} - 1)} \tag{4.81a}$$

with

$$\kappa_{ab} = \tfrac{1}{2}|\boldsymbol{\kappa}_a - \boldsymbol{\kappa}_b| = \tfrac{1}{2}\sqrt{(\kappa_a^2 + \kappa_b^2 - 2\kappa_a\kappa_b \cos\vartheta_{ab})}. \tag{4.81b}$$

This factor weighs the simultaneous two-electron emissions according to their relative angles ϑ_{ab} and leads to a third selection rule which requires vanishing intensity for two-electron emission for $\vartheta_{ab} = 0$ [MBr93]. This can be understood naively: due to the Coulomb repulsion between the electrons no intensity is expected if both electrons emerge with equal energy into the same direction, but maximum intensity is possible for observation at $\vartheta_{ab} = 180°$, i.e.,

$$C(\boldsymbol{\kappa}_a = \boldsymbol{\kappa}_b) = 0, \qquad C(\boldsymbol{\kappa}_a = -\boldsymbol{\kappa}_b) = \max. \tag{4.82}$$

This correlation factor is responsible for the low intensity observed in Fig. 4.43 at angles ϑ_{ab} close to 0° which suggests vanishing intensity at $\vartheta_{ab} = 0°$ (where no measurement is possible due to the finite size of the spectrometers).

Often the correlation factor $C(\boldsymbol{\kappa}_a, \boldsymbol{\kappa}_b)$ is approximated for equal energy sharing by a Gaussian function peaked at $\vartheta_{ab} = 180°$ (for the derivation see [BBK91]):

$$C(\boldsymbol{\kappa}_a, \boldsymbol{\kappa}_b)|_{E_a = E_b} \approx G(180° - \vartheta_{ab}) = \exp[-4\ln 2\,(180° - \vartheta_{ab})^2/\vartheta_{fwhm}^2] \tag{4.83a}$$

with

$$\vartheta_{fwhm} = \Theta_0 E_{exc}^{1/4}, \quad \Theta_0 = \text{characteristic parameter}. \tag{4.83b}$$

(Starting from a different treatment of double photoionization in helium, based on properties of the wavefunctions in the threshold region, and special coordinates (hyperspherical coordinates) for the description of the correlated motion of the electrons, different predictions for this Θ_0 parameter have been obtained (see [HSW91, KOs92] with references therein).)

Collecting together the information contained in equs. (4.78) and (4.81), the triple-differential cross section for double photoionization can be represented as

$$\left.\frac{d^3\sigma}{d\Omega_a\, d\Omega_b\, dE}\right|_{\text{lin.pol.}} (E_a = E_b) = \text{const}\,(\cos\vartheta_a + \cos\vartheta_b)^2\, C(\boldsymbol{\kappa}_a, \boldsymbol{\kappa}_b). \tag{4.84a}$$

(It should be emphasized that the electron correlations in the final state as expressed in the correlation factor $C(\boldsymbol{\kappa}_a, \boldsymbol{\kappa}_b)$ provide an additional source of high

[†] The wavefunction also fulfils the correct asymptotic boundary condition if all three particles are at infinitely large distances from each other [Red73, Mer77, GMi80, BBK89].

orbital angular momenta as contained and needed in the parametrization of equ. (4.68). This can be seen if the correlation factor and the angular factor are both expanded in terms of spherical harmonics for the directions $\hat{\kappa}_a$ and $\hat{\kappa}_b$. The multiplication of these two expansions then leads for each direction to sums over products of spherical harmonics. Using

$$Y_{\ell_1 m_1}(\hat{\kappa})\, Y_{\ell_2 m_2}(\hat{\kappa}) = \sum_{k,\rho} \sqrt{\left[\frac{(2\ell_1+1)(2\ell_2+1)}{4\pi(2k+1)}\right]}(\ell_1 0 \ell_2 0|k0)(\ell_1 m_1 \ell_2 m_2|k\rho)\, Y_{k\rho}(\hat{\kappa}), \tag{4.85}$$

each such product can be replaced by *one* spherical harmonic only, but at the expense of terms with higher k value which imply combinations of higher orbital angular momenta. Unfortunately, the slow convergence of equ. (4.68) makes this expansion intractable (see the related case of slow convergence of the ground-state wavefunction of helium in the CI presentation as compared to the Hylleraas ansatz, Section 7.5).)

The parametrization in equ. (4.84a) agrees with one derived for the near-threshold region where one gets in the standard formulation [HSW91]

$$\left.\frac{d^3\sigma}{d\Omega_a\, d\Omega_b\, dE}\right|_{\text{lin.pol.}} (E_a = E_b) = 2\,|a(^1P^o)|^2\, E_{exc}^{n-2}\,(\cos\vartheta_a + \cos\vartheta_b)^2\, G(180^\circ - \vartheta_{ab}), \tag{4.84b}$$

where $a(^1P^o)$ is an unknown complex constant, n is the Wannier exponent ($n = 1.056$; see Section 5.7.1) and $G(180^\circ - \vartheta_{ab})$ is given by equ. (4.83).

According to the parametrization in equs. (4.84) the shape of the triple-differential cross section for equal energy sharing is given by the product of an angular factor and a correlation factor. All three quantities are shown in Fig. 4.44 as polar plots of the coincident intensities for some examples. The cases selected refer to double photoionization in helium, induced by linearly polarized light, with equal sharing of the excess energy of 1 eV between the electrons. Both electrons are observed in a plane perpendicular to the photon beam direction, one electron (e_a) is kept at a fixed position (with $\vartheta_a = 180^\circ$, $\vartheta_a = 150^\circ$ and $\vartheta_a = 90^\circ$), the angle ϑ_b of the other electron is varied. The following features characteristic for direct double photoionization in helium can be noted:

(i) The angular factors plotted in the top row show the characteristic zero intensity required by the selection rules in equs. (4.79) and (4.80). Going from $\vartheta_a = 180^\circ$ to $\vartheta_a = 90^\circ$, their shapes change as follows: the patterns shrink† and develop an additional small lobe along the positive x-axis which is largest for $\vartheta_a = 90^\circ$. All angular factors possess axial symmetry around the x-direction of the oscillating electric field vector of the incident light.

(ii) The correlation factor is plotted in the middle row of the figure. It shows the

† The maximum for the pattern with angle-setting $\vartheta_a = 180^\circ$ is four times the maximum value for the pattern with $\vartheta_a = 90^\circ$.

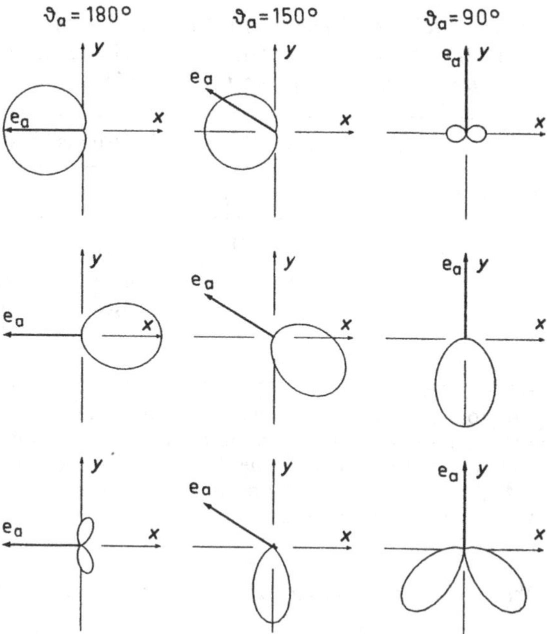

Figure 4.44 Illustration of energy- and angle-resolved two-electron emission following direct double photoionization in helium at 80 eV photon energy using linearly polarized light (electric field vector along the x-axis). Both electrons are detected in a plane perpendicular to the photon beam direction, the direction of one electron (e_a) is fixed at $\vartheta_a = 180°$, $\vartheta_a = 150°$ and $\vartheta_a = 90°$ (first, second and third columns in the figure), while the angle of the other electron is varied. Equal sharing of the excess energy has been selected, i.e., $E_a = E_b = 0.5$ eV. The polar plots in the top row show the angular factor of equ. (4.78), those in the middle row the correlation factor of equ. (4.83a) with $\Theta_0 = 91° \text{ eV}^{-1/4}$, and those in the bottom row the product of both which gives according to equ. (4.84b) the triple-differential cross section for the selected parameters (within each row relative intensities are to scale). For experimental data in this energy region see [DAM95, LMA 95].

characteristic zero intensity for $\vartheta_{ab} = 0°$ (selection rule in equ. (4.82)), and its maximum always points to $\vartheta_{ab} = 180°$. All correlation factors have the same size and shape, and axial symmetry around the direction of electron e_a.

(iii) The triple-differential cross section, the product of the angular and correlation factors, is plotted in the bottom row. Each pattern shows a dominance for emitting the second electron into the half-plane opposite to the one containing the first electron. This feature is characteristic of direct double photoionization in helium and differs from photon-induced two-step double ionization (see below). The pattern in which electron e_a is observed along the direction of the electric field vector ($\vartheta_a = 0°$ or $\vartheta_a = 180°$) retains axial symmetry around this axis. For $0 < \vartheta_a < 90°$ or $90° < \vartheta_a < 180°$ this symmetry is destroyed, and for $\vartheta_a = 90°$ there is reflection symmetry with respect to the yz-plane.

Table 4.2. *A-coefficients needed for the description of* $2p_{3/2}$ *photoionization and subsequent* L_3–M_1M_1 *Auger decay in magnesium, according to equ.* (4.68).

ℓ_a	k_1	ℓ_b	k_2	Responsible for A_{kk0}	Responsible for $A_{k_1k_22}$
0	0	1	0, 2	$\mathbf{A_{000}}$	$\mathbf{A_{022}}$
2	0, 2, 4	1	0, 2	A_{000} $\mathbf{A_{220}}$	A_{022} $\mathbf{A_{202}}$ $\mathbf{A_{222}}$ $\mathbf{A_{422}}$

Since photoionization can have only εs and εd partial waves, and the attached Auger decay has only εp, the number of k values is restricted (equ. (4.71b)), and only the A-coefficients listed exist. As in Table 4.1, A-coefficients which appear (from top to bottom) for the first time are written in bold.

4.6.1.3 *Photon-induced two-step double ionization in magnesium*

Photoionization in magnesium leading to a $2p^{-1}\,{}^2P_J$ final ionic state and the subsequent $L_{2,3}$–M_1M_1 Auger decay provides an illustrative example for the parametrization of two-step double ionization. This is due mainly to the simple structure of the system and the applicability of *LSJ*-coupling. (This example has not yet been studied experimentally, as has the similar case of $4d^{-1}\,{}^2D_{5/2}$ photoionization in xenon with subsequent N_5–$O_{2,3}O_{2,3}\,{}^1S_0$ Auger decay (see Section 5.6).) Here photoionization of a 2p electron is described by two dipole matrix elements D_s and D_d which belong to the two photoionization channels with εs and εd partial waves of the photoelectron, and the Auger decay is described by one matrix element only which belongs to an εp partial wave of the Auger electron (see Section 5.2). Assuming for simplicity completely linearly polarized incident light with the quantization axis along the direction of the electric field vector, equ. (4.68) can be applied. Because only a few orbital angular momenta are involved, the number of A-coefficients is limited, and these are listed in Table 4.2. One then obtains for the differential cross section of the coincident observation of a $2p_{3/2}$ photoelectron† and an L_3-M_1M_1 Auger electron the following expression

$$\left.\frac{d^2\sigma}{d\Omega_a\, d\Omega_b}\right|^{\text{2-step}}_{\text{lin.pol.}} \mathbf{Mg} = A_{000} + A_{220}P_2(\cos\vartheta_{ab}) + A_{022}\frac{1}{\sqrt{(4\pi)}}\,Y_{20}(\hat{\mathbf{\kappa}}_b)$$
$$+ A_{202}\frac{1}{\sqrt{(4\pi)}}\,Y_{20}(\hat{\mathbf{\kappa}}_a) + A_{222}\sum_q (2q2-q\,|\,20)\,Y_{2q}(\hat{\mathbf{\kappa}}_a)Y_{2-q}(\hat{\mathbf{\kappa}}_b)$$
$$+ A_{422}\sum_q (4q2-q\,|\,20)\,Y_{4q}(\hat{\mathbf{\kappa}}_a)Y_{2-q}(\hat{\mathbf{\kappa}}_b), \tag{4.86a}$$

† To talk of a $2p_{3/2}$ photoelectron suggests an underlying single-particle description, but this is not necessarily the case. Here and in the following it is understood to be a shorthand notation for photoionization leading to the $2p^{-1}\,{}^2P_{3/2}$ final ionic state.

where $\hat{\kappa}_a = (\vartheta_a, \varphi_a)$ and $\hat{\kappa}_b = (\vartheta_b, \varphi_b)$ define the directions of the emitted electrons. Even though this parametrization provides the desired separation of 'geometry' and 'dynamics', it contains no detailed information about the internal structure of the dynamical parameters $A_{k_1k_2k}$. For this, the two-step process of photoionization and subsequent Auger decay must be worked out, including explicitly the necessary matrix elements. Using the powerful method of statistical and efficiency tensors sketched in Section 8.4, the inherent structure of the $A_{k_1k_2k}$-coefficients is revealed (equ. (8.133)), and one gets

$$\left.\frac{d^2\sigma}{d\Omega_a\, d\Omega_b}\right|^{\text{2-step}}_{\text{lin.pol.}} \mathbf{Mg} = A_{000}\Big\{1 + \tfrac{1}{2}\beta_{ph}(2p_{3/2})P_2(\cos\vartheta_{ab})$$
$$+ \beta_{ph}(2p_{3/2})P_2(\cos\vartheta_a) + \beta_A(L_3\text{–}M_1M_1)P_2(\cos\vartheta_b)$$
$$+ \left[\frac{\pi}{5}\sqrt{\left(\frac{2}{7}\right)}(18r - 7\beta_{ph}(2p_{3/2})\right]$$
$$\times\left[\sqrt{\left(\frac{2}{7}\right)}\,Y_{22}(\hat{\kappa}_a)\,Y_{2-2}(\hat{\kappa}_b) + \frac{1}{\sqrt{(14)}}\,Y_{21}(\hat{\kappa}_a)Y_{2-1}(\hat{\kappa}_b)\right.$$
$$-\sqrt{\left(\frac{2}{7}\right)}\,Y_{20}(\hat{\kappa}_a)Y_{20}(\hat{\kappa}_b) + \frac{1}{\sqrt{(14)}}\,Y_{2-1}(\hat{\kappa}_a)Y_{21}(\hat{\kappa}_b)$$
$$\left.+\sqrt{\left(\frac{2}{7}\right)}\,Y_{2-2}(\hat{\kappa}_a)\,Y_{22}(\hat{\kappa}_b)\right] + \frac{144\pi}{5\sqrt{(70)}}$$
$$\times r\left[\sqrt{\left(\frac{5}{42}\right)}Y_{42}(\hat{\kappa}_a)Y_{2-2}(\hat{\kappa}_b) - \sqrt{\left(\frac{5}{21}\right)}\,Y_{41}(\hat{\kappa}_a)Y_{2-1}(\hat{\kappa}_b)\right.$$
$$+\sqrt{\left(\frac{2}{7}\right)}Y_{40}(\hat{\kappa}_a)Y_{20}(\hat{\kappa}_b) - \sqrt{\left(\frac{5}{21}\right)}Y_{4-1}(\hat{\kappa}_a)Y_{21}(\hat{\kappa}_b)$$
$$\left.+\sqrt{\left(\frac{5}{42}\right)}\,Y_{4-2}(\hat{\kappa}_a)Y_{22}(\hat{\kappa}_b)\right]\Big\}, \tag{4.86b}$$

where A_{000} describes the overall intensity normalized against $(4\pi)^2$, i.e.,

$$A_{000} = \frac{\sigma_{ph}(2p_{3/2})}{4\pi}\,\frac{\omega_A(L_3\text{–}M_1M_1)}{4\pi}. \tag{4.87}$$

The dynamics are contained in the photoionization cross section $\sigma_{ph}(2p_{3/2})$, in the Auger yield $\omega_A(L_3\text{–}M_1M_1)$ and in the angular distribution parameters $\beta_{ph}(2p_{3/2})$ and $\beta_A(L_3\text{–}M_1M_1)$ of the non-coincident photo- and Auger electrons (see Section 5.2). For convenience, a ratio r is used in equ. (4.86b) which, however, is not a new parameter since in the present case it depends on $\beta_A(L_3\text{–}M_1M_1)$ (see equ. (8.133g)):

$$r = [1 - \beta_A(L_3\text{–}M_1M_1)]/1.80. \tag{4.88}$$

In order to discuss equ. (4.86b), the specific case of photoionization at 80 eV photon energy is chosen, because for this case the dipole matrix elements and their relative phases are known from experimental data (see Table 5.1) and all other necessary dynamical parameters are known. Restricting the discussion to the relative shapes of the angular correlation patterns, the coefficient A_{000} in equ. (4.86b) is of no relevance, and only the two dynamical angular distribution parameters of non-coincident photo- and Auger electrons are needed, which at a photon energy of 80 eV have the values $\beta_{ph}(2p_{3/2})=0.74$ and $\beta_A(L_3\text{–}M_1M_1)=0.16$. In order to derive the angular correlation pattern for an explicit case, it will be assumed that the emitted electrons are observed in a plane perpendicular to the direction of the photon beam ($\varphi_a = \varphi_b = 0$),† keeping the direction of the photoelectron fixed at $\vartheta_a = 150°$, but varying the direction of observation for the coincident Auger electron (ϑ_b). With these preconditions one obtains from equ. (4.86b)

$$\left.\frac{d^2\sigma}{d\Omega_a\, d\Omega_b}\right|^{\text{2-step}}_{\text{lin.pol.}} (\vartheta_a = 150°, \varphi_a = 0°; \vartheta_b, \varphi_b = 0°)\mathbf{Mg} \sim (1.959 + 0.370 P_2(\cos\vartheta_{ab}) - 0.446\cos^2\vartheta_b - 1.604\cos\vartheta_b \sin\vartheta_b), \quad (4.89a)$$

and, using† $\vartheta_{ab} = 150° - \vartheta_b$ and renormalizing this expression, the equivalent form

$$\left.\frac{d^2\sigma}{d\Omega_a\, d\Omega_b}\right|^{\text{2-step}}_{\text{lin.pol.}} (\vartheta_a = 150°, \varphi_a = 0°; \vartheta_b, \varphi_b = 0°)\mathbf{Mg} \sim (1 - 0.088\cos^2\vartheta_b - 1.090\cos\vartheta_b \sin\vartheta_b). \quad (4.89b)$$

Before a polar plot of this angular pattern of coincident electrons is shown, a slightly different parametrization for photon-induced two-step double ionization will be introduced, because it provides a deeper insight into the constituents from the first and second steps of the process which finally determine the angular pattern of coincident electrons. Recalling that for non-coincident observation of Auger electrons it is the alignment of the photoionized state which is responsible for the angular distribution (see Section 3.5), it is natural also to introduce a quantity corresponding to the alignment if the Auger electrons are measured in coincidence with their preceding photoelectrons. In this more general case one needs the full alignment tensor $\mathcal{A}_{2q}(J, \hat{\mathbf{\kappa}}_a)$ with all its q-components, and the tensor depends on the direction $\hat{\mathbf{\kappa}}_a$ of the emitted photoelectron. Reproducing equ. (8.129a) one obtains (see [Kab92])

$$\left.\frac{d^2\sigma}{d\Omega_a\, d\Omega_b}\right|^{\text{2-step}}_{\text{lin.pol.}} \mathbf{Mg} = \tilde{\rho}_{00}\left[1 + \alpha_2 \sum_q \tilde{\mathcal{A}}_{2q}(J = 3/2; \hat{\mathbf{\kappa}}_a)\sqrt{\left(\frac{4\pi}{5}\right)}\, Y_{2q}(\hat{\mathbf{\kappa}}_b)\right]. \quad (4.90)$$

In this expression $\tilde{\rho}_{00}$ describes the angular distribution of non-coincident

† Originally, the polar and azimuthal angles ϑ_i and φ_i were restricted to be within the ranges $0° \le \vartheta_i \le 180°$ and $0 \le \varphi_i \le 360°$, respectively. In the present geometry, however, φ_i can have only two values, $\varphi_i = 0°$ or $\varphi_i = 180°$, related to electron detection in the first and fourth quadrant or the second and third quadrant of the observation plane, respectively. Inspection of the angle-dependent terms in equ. (4.86b) then shows that the case of $\varphi_i = 180°$ with $0° \le \vartheta_i \le 180°$ can be described alternatively by $\varphi_i = 0°$ and $-180° \le \vartheta_i \le 0°$ or $180° \le \vartheta_1 \le 360°$. The last representation is used here.

photoelectrons given by (see equ. (8.129b))

$$\tilde{\rho} = \frac{\sigma_{\text{ph}}\omega_{\text{A}}}{4\pi 4\pi}[1 + \beta_{\text{ph}}P_2(\cos\vartheta_{\text{a}})], \tag{4.91}$$

α_2 is the Auger decay parameter and has the value -1 in the present example (see equ. (8.115b)), and the alignment tensor, defined by

$$\tilde{\mathscr{A}}_{2q}(J; \vartheta_{\text{a}}, \varphi_{\text{a}}) = \tilde{\rho}_{2q}(J; \vartheta_{\text{a}}, \varphi_{\text{a}})/\tilde{\rho}_{00}, \tag{4.92}$$

is known if the *statistical tensors* $\tilde{\rho}_{2q}(J; \vartheta_{\text{a}}, \varphi_{\text{a}})$ are known. These last quantities have been worked out (equs. (8.130) and (8.131)) and, hence, numerical values can be derived for the chosen example of $2p_{3/2}$ photoionization followed by the L_3–M_1M_1 Auger transition in magnesium. One gets

$$\tilde{\mathscr{A}}_{20}(\vartheta_{\text{a}}) = \frac{-1.000 + 8.238\cos^2\vartheta_{\text{a}} - 10.00\cos^4\vartheta_{\text{a}}}{1 + 1.762\cos^2\vartheta_{\text{a}}}, \tag{4.92a}$$

$$\tilde{\mathscr{A}}_{21}(\vartheta_{\text{a}}) = -\tilde{\mathscr{A}}_{2-1}(\vartheta_{\text{a}}) = \frac{\sin\vartheta_{\text{a}}\cos\vartheta_{\text{a}}(-3.004 + 8.165\cos^2\vartheta_{\text{a}})}{1 + 1.762\cos^2\vartheta_{\text{a}}}, \tag{4.92b}$$

$$\tilde{\mathscr{A}}_{22}(\vartheta_{\text{a}}) = \tilde{\mathscr{A}}_{2-2}(\vartheta_{\text{a}}) = -\frac{4.082\sin^2\vartheta_{\text{a}}\cos^2\vartheta_{\text{a}}}{1 + 1.762\cos^2\vartheta_{\text{a}}}. \tag{4.92c}$$

($\tilde{\mathscr{A}}_{2q}(\vartheta)$ is an abbreviation of $\tilde{\mathscr{A}}_{2q}(\vartheta, \varphi = 0)$.) These functions are plotted in Fig. 4.45 together with the other terms in the square bracket of equ. (4.90): the uppermost part of the figure is the isotropic intensity of the number '1'; below this on the left-hand side are the components of the ϑ_{a}-dependent alignment tensor, and on the right-hand side the ϑ_{b}-dependent spherical harmonics. For each selected direction ϑ_{a} of the photoelectron, the pattern of the coincident Auger electron can then be read from the figure. For example, at $\vartheta_{\text{a}} = 0°$ (or 90°, 180°, 270°) the alignment parameter $\tilde{\mathscr{A}}_{20}(\vartheta_{\text{a}} = 0°)$ is the only non-vanishing component of the full alignment tensor. Its value $\tilde{\mathscr{A}}_{20}(\vartheta_{\text{a}} = 0°) = -1$ is a pure geometrical quantity, independent of the photoionization matrix elements. The product $\tilde{\mathscr{A}}_{20}(\vartheta_{\text{a}} = 0°)$ with $\alpha_2 = -1$ then provides the weight for the ϑ_{b}-dependent spherical harmonics (with the prefactor $\sqrt{(4\pi/5)}$), and this together with the number '1' finally leads to the angular distribution of Auger electrons measured in coincidence with the preceding photoelectron. The result is shown in Fig. 4.46(*a*). The numerical expression for these patterns is

$$W(\vartheta_{\text{a}} = 0°, 90°, 180° \text{ or } 270°, \varphi_{\text{a}} = 0°; \vartheta_{\text{b}}, \varphi_{\text{b}} = 0°) \sim [1 + P_2(\cos\vartheta_{\text{b}})], \tag{4.93}$$

i.e., one gets a pure geometrical function (see [VBe92]). Hence, no information on the photo- or non-radiative decay process is obtained, and this rather cumbersome coincident experiment can only be used to check the correct performance of the apparatus.

In order to derive information on *dynamical* quantities from such a coincidence experiment, one has to select angles ϑ_{a} at which other components of the alignment tensor also contribute, because then the different dependences on the dipole matrix elements, including their relative phases, are involved. As can be seen from

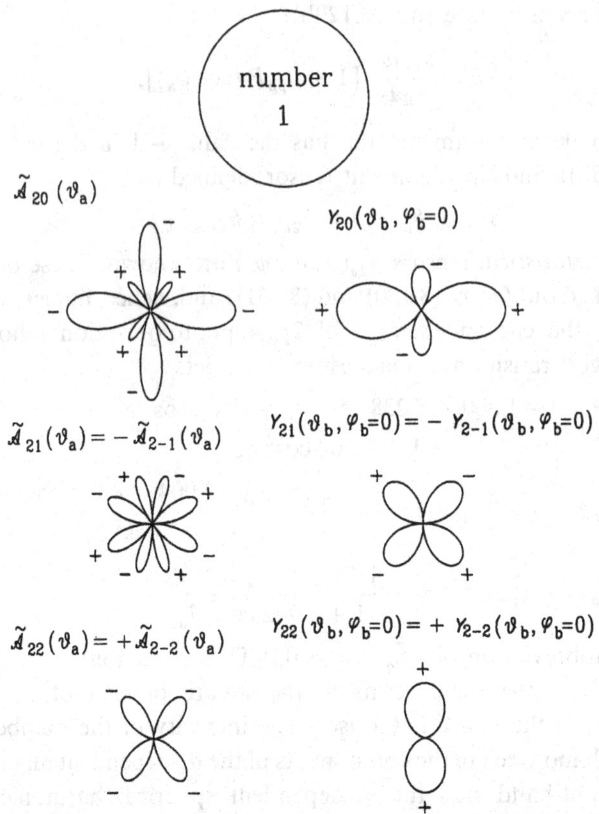

Figure 4.45 Illustration of the content of equ. (4.90) which describes the angular distribution of Auger electrons (e_b) in coincidence with the preceding photoelectron (e_a). The data refer to $2p_{3/2}$ ionization of magnesium by linearly polarized photons of 80 eV and subsequent L_3–M_1M_1 Auger decay, with emission of both electrons in a plane perpendicular to the photon beam direction. The alignment tensor $\tilde{\mathscr{A}}_{2q}(\vartheta_a, \varphi_a = 0)$ is abbreviated to $\tilde{\mathscr{A}}_{2q}(\vartheta_a)$. Positive and negative values of this tensor and of the spherical harmonics $Y_{2q}(\vartheta_b, \varphi_b = 0)$ are indicated by (+) and (−) on the corresponding lobes. For further details see main text. Reprinted from *Nucl. Instr. Meth. B* **87**, Schmidt, 241 (1994) with kind permission from Elsevier Science - NL, Sara Burgerhartstraat 25, 1055 KV Amsterdam, The Netherlands.

Fig. 4.45, such a case can be found, e.g., for $\vartheta_a = 150°$. Here all components of the alignment tensor are needed, and they give positive weight to the ϑ_b-dependent spherical harmonics:

$$\left.\begin{aligned} \alpha_2\tilde{\mathscr{A}}_{20}(\vartheta_a = 150°) &= 0.192, \\ \alpha_2\tilde{\mathscr{A}}_{21}(\vartheta_a = 150°) &= 0.582 = -\alpha_2\tilde{\mathscr{A}}_{2-1}(\vartheta_a), \\ \alpha_2\tilde{\mathscr{A}}_{22}(\vartheta_a = 150°) &= 0.330 = \alpha_2\tilde{\mathscr{A}}_{2-2}(\vartheta_a). \end{aligned}\right\} \tag{4.94}$$

(Note $\alpha_2 = -1$ and the sign compensation in $\tilde{\mathscr{A}}_{2-1}(\vartheta_a)$ with $Y_{2-1}(\vartheta_b)$.) The resulting angular distribution of coincident Auger electrons is then determined, according to the square bracket in equ. (4.90), by the properly weighted contribu-

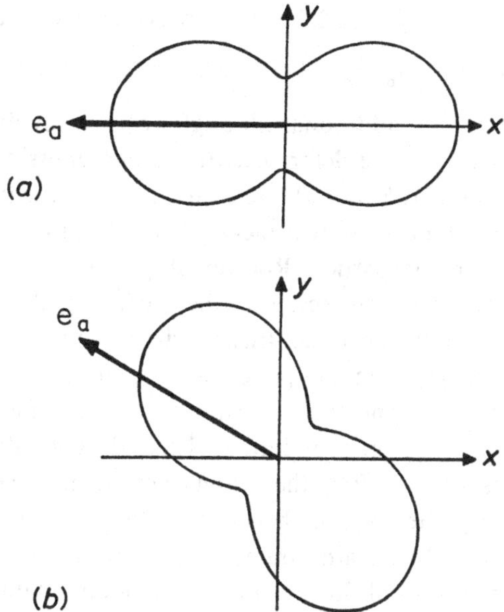

Figure 4.46 Energy- and angle-resolved patterns for two-electron emission in the two-step process of $2p_{3/2}$ photoionization of magnesium with subsequent L_3–M_1M_1 Auger decay induced by 80 eV photons with linear polarization (electric field vector along the x-axis). Both electrons are detected in a plane perpendicular to the photon beam direction: the direction of the photoelectron (e_a) is fixed at (*a*) $\vartheta_a = 180°$ and (*b*) $\vartheta_a = 150°$, while the angle of the Auger electron is varied. Reprinted from *Nucl. Instr. Meth. B* **87**, Schmidt, 241 (1994) with kind permission of Elsevier Science – NL, Sara Burgerhartstraat 25, 1055 KV Amsterdam, The Netherlands.

tions of the spherical harmonics together with the isotropic intensity of the number '1'. Due to the plus and minus parts of these spherical harmonics it can be seen with the help of Fig. 4.45 that a greater intensity will occur in the second and fourth quadrants than in the first and third quadrants. Hence, the observed pattern no longer possesses axial symmetry about the electric field vector as in the former example. Instead, as can be seen in Fig. 4.46(*b*), the pattern is tilted. Quantitatively, the coincident angular distribution is

$$\begin{aligned} W(\vartheta_a = 150°, \varphi_a = 0°; \vartheta_b, \varphi_b = 0°) &= 1 + 0.192 P_2(\cos\vartheta_b) \\ &+ 0.582\sqrt{(4\pi/5)}\, Y_{21}(\vartheta_b, \varphi_b = 0) - 0.582\sqrt{(4\pi/5)}\, Y_{2-1}(\vartheta_b, \varphi_b = 0) \\ &+ 0.330\sqrt{(4\pi/5)}\, Y_{22}(\vartheta_b, \varphi_b = 0) + 0.330\sqrt{(4\pi/5)}\, Y_{2-2}(\vartheta_b, \varphi_b = 0) \\ &= 1.308 - 0.115\cos^2\vartheta_b - 1.426\sin\vartheta_b\cos\vartheta_b, \end{aligned} \tag{4.95a}$$

i.e.,

$$W(\vartheta_a = 150°, \varphi_a = 0°; \vartheta_b, \varphi_b = 0°) \sim (1 - 0.088\cos^2\vartheta_b - 1.090\cos\vartheta_b\sin\vartheta_b), \tag{4.95b}$$

where the last expression is identical to the result obtained in equ. (4.89b). It gives the angular correlation pattern shown in Fig. 4.46(*b*). (In the present example there is only one tilt-angle, $\chi_b = 132.7°$, and expression (4.95b) can be recast as

$$W(\vartheta_a = 150°, \varphi_a = 0°; \vartheta_b, \varphi_b = 0°) \sim 0.956 + 0.547 \cos 2(\vartheta_b - \chi_b). \quad (4.95c)$$

In a more general situation, additional tilt-angles can be involved, and then the overall pattern will also become deformed.) If the two emission patterns shown in Fig. 4.46 for $\vartheta_a = 180°$ and $\vartheta_a = 150°$ are compared with the two corresponding emission patterns for helium (first two pictures in the bottom row of Fig. 4.44), one can note the following properties.† Rotational symmetry appears in both cases if one electron is emitted in the direction of the oscillating electric field vector of the incident light. However, and in contrast to direct double photoionization of helium, there is for magnesium no preference for electron emission into the opposite half-plane. Instead, and the case with rotational symmetry is a striking example, the probabilities for emission into the left and right half-planes are equal. There are two reasons for this. First, the existence of the intermediate state makes the alignment tensor $\tilde{\mathscr{A}}_{2q}$ the only link between the two steps of electron emission, and this tensor is subject to certain symmetry properties (see Fig. 4.45). Second, the correlation factor responsible for the mutual Coulomb interaction between the two escaping electrons loses its influence in two-step double ionization, because the electrons differ in energy and are emitted one after the other. Of course, such distinct features will only occur in the well-defined limiting cases of photon-induced two-electron emission, which are direct double photoionization and two-step ionization. Processes between these limits also exist in many cases.

4.6.2 True and accidental coincidences

In order to study two-electron emission in double photoionization, one has to isolate the two specified electrons from all other ionization processes by measuring them in coincidence. (For an overview of coincidence measurements in photoionization studies, see [ESc96].) 'In coincidence' in this context means that both events occur simultaneously, or nearly so, i.e., the time delay between the emission of both electrons has to be small compared to the resolving time Δt of the coincidence counting device. The standard experimental set-up for coincidence measurements is shown in Fig. 4.47. Two coincident electrons emitted towards and analysed in two separate analysers give two signals. By means of electronic or cable delay it is always possible to retard the arrival of one signal with respect to the other and then to distinguish a START and a STOP signal. These signals are fed into the corresponding entrances of the time-measuring device, usually a time-to-digital converter (TDC), where the time difference between the STOP and START signal is measured. (It is also possible to use a time-to-amplitude converter (TAC) with a multichannel analyser (MCA) to obtain the desired time spectrum.)

† It should be added that in the case of two emitted electrons with unequal energies, which electron is selected to be at the fixed direction plays a role in the shape of the emission pattern (see Fig. 5.33).

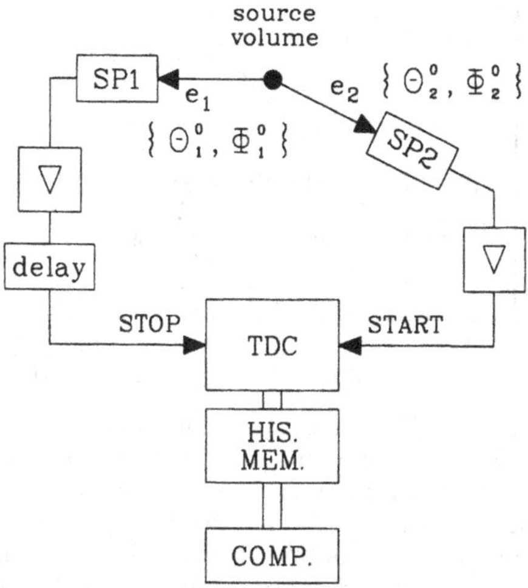

Figure 4.47 Typical electronic circuit for the measurement of electron–electron coincidences with two spectrometers (SP1, SP2) placed at the positions $\{\Theta_1^0, \Phi_1^0\}$ and $\{\Theta_2^0, \Phi_2^0\}$, respectively. The pre- and main amplifiers are together represented by a triangle. The delay retards the signal from SP1, thus providing a STOP of the time-to-digital converter (TDC) if this time measuring device has been initiated by a START signal from a time-correlated event registered in SP2. The output of the TDC, i.e., the number of time-correlated events as function of the correlation time is stored in a histogramming memory (HIS. MEM.) which then is read out by a computer (COMP.).

This time difference, called the *correlation time*, is stored in a histogramming memory, and after the necessary processing time (dead time T_{dead}) the TDC is reset and waits for the next START signal. The histogramming memory collects all individual coincidences by sorting them according to their correlation times. In this way, a time spectrum is obtained which can be transferred to a computer.

Even if each electron pair were emitted at exactly the same moment, the associated START–STOP time difference would vary from pair to pair. This is because of the different travelling times of the coincident electrons from their places of origin to the electron detector as well as the different processing times of the electron detector, including the construction of the electronic pulse. Hence, the coincident electrons will lead to a certain time spectrum as shown in the section of the spectrum in Fig. 4.48 which is labelled 'true' coincidences. The name *true* coincidences implies that these are the coincidences in which one is interested.

Obviously, the *coincidence resolving time* Δt, shown in Fig. 4.48, has to be large enough to accept all these true coincidences. However, this finite value of Δt then leads to the recording of not only the desired true coincidences, but also of *accidental* coincidences (also called *random* or *false* coincidences). As indicated by the name, these accidental coincidences accidentally follow one another within the time Δt. Hence, they are due to any two electrons which match the conditions set by the experimenter for the selected double ionization process; they might originate from two different double-ionization processes, two single-ionization processes, or

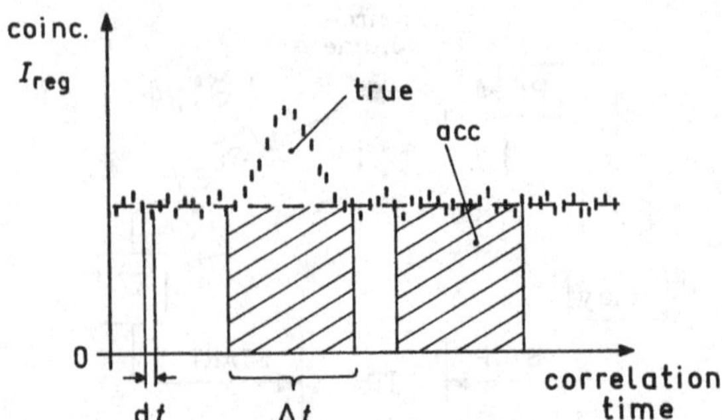

Figure 4.48 Typical spectrum of electron–electron coincidences recorded with a TDC. The data refer to a situation in which the photon beam has no time structure. True coincidences are collected in the peak while accidental coincidences give a flat and smooth background. Δt indicates the coincidence resolving time and dt the time resolution of the time-measuring device. The two shaded areas represent accidental coincidences, measured on the left-hand side together with the desired true coincidences, but on the right-hand side separately (and simultaneously) in the full time spectrum.

from one double- and one single-ionization processes, or from scattered electrons. In the lowest approximation, true and accidental coincidences can be considered to be independent of each other. (In this approximation, one can expect that the accidental coincidences will depend on the counting rates I_1 and I_2 of the two coincidence channels, and on the coincidence resolving time (see equ. (4.100c)).) One then gets the following relation between recorded true and accidental counting rates:

$$I_{tot} = I_{true} + I_{acc}, \tag{4.96}$$

where I_{tot} and I_{acc} can be measured (see below), i.e., the desired rate of true coincidences I_{true} simply follows by a subtraction of the measured quantities. (It should be added that this simple approach does not hold generally. Coincidences between more than two partners need special treatment. Even for two-fold coincidences a completely different situation can arise if there is a high probability for more than one ionization process during the ionization interval such as in the single-bunch mode of operation of an electron storage ring; see [ESc96] and references therein.) Due to their random nature, the disturbing accidental coincidences can be calculated with Poisson statistics which describe the photoionization process on a random time scale. The calculation involves three steps (see [RBK67]):

(i) formulation of the probability for accepted START signals;
(ii) formulation of the probability for a STOP signal occurring in the time interval between t and t + dt; and
(iii) a multiplication of these two probabilities which leads to the desired probability of START and STOP of uncorrelated events.

In the calculation for the probability of START signals accepted by the coincidence device, one has to take into account that due to the dead time T_{dead} of the coincidence unit not every signal at the START entrance will initiate a start of the coincidence unit. If I_1 represents the mean counting rate of START pulses, the probability that within T_{dead} no pulse enters the START entrance follows from Poisson statistics as

$$P(\text{no START signal within } T_{\text{dead}}) = \mathrm{e}^{-I_1 T_{\text{dead}}}. \tag{4.97a}$$

Hence, the probability of START signals within T_{dead} is given by

$$P(\text{START signals within } T_{\text{dead}}) = 1 - \mathrm{e}^{-I_1 T_{\text{dead}}}, \tag{4.97b}$$

and this probability is the same as the probability of lost START signals during T_{dead}. Therefore, the probability of accepted START signals is

$$\begin{aligned} P(\text{accepted START signals}) &= 1 - P(\text{lost START signals}) \\ &= P_{\text{START}} = \exp(-I_1 T_{\text{dead}}), \end{aligned} \tag{4.97c}$$

and for the rate of accepted START signals, denoted as I_{START}, one gets

$$I_{\text{START}} = I_1 P_{\text{START}} = I_1\,\mathrm{e}^{-I_1 T_{\text{dead}}}. \tag{4.98}$$

To determine the probability of having a STOP signal after a certain time t one needs first to find the probability that no STOP occurs during the time interval between 0 and t, which is given by (I_2 denotes the mean counting rate of STOP signals)

$$P(\text{no STOP within } t) = \mathrm{e}^{-I_2 t}. \tag{4.99a}$$

This probability then has to be multiplied by the probability that one event will happen to occur within the short *resolution* time interval dt of the measuring device (dt is different from the resolving time Δt introduced above). According to the fundamental probability of the Poisson distribution, this probability dP for observing one event in an infinitesimally small time interval dt obeys the relation

$$\mathrm{d}P = I_2\,\mathrm{d}t. \tag{4.99b}$$

Hence, the desired STOP probability becomes

$$P_{\text{STOP}}(t)\,\mathrm{d}t = \begin{cases} I_2\,\mathrm{e}^{-I_2 t}\,\mathrm{d}t & \text{for } t \text{ smaller than the operation range of the time measuring device} \\ 0 & \text{otherwise.} \end{cases} \tag{4.99c}$$

From the relations quoted, the rate of accidental coincidences, $I_{\text{acc.}}$, recorded within the finite resolving time interval Δt (see Fig. 4.48) can be calculated from

$$I_{\text{acc.}} = I_1 P_{\text{START}} \int_{\Delta t} P_{\text{STOP}}(t)\,\mathrm{d}t \tag{4.100a}$$

and gives for $I_2\,\Delta t \ll 1$

$$I_{\text{acc.}} = I_1 I_2\,\mathrm{e}^{-I_1 T_{\text{dead}}}\,\mathrm{e}^{-I_2 t_0}\,\Delta t, \tag{4.100b}$$

where t_0 is the conversion time for processing of the signal. For $I_1 T_{\text{dead}} \ll 1$ and $(I_2 t_0) \ll 1$, one then derives the standard formula

$$I_{\text{acc.}} = I_1 I_2\,\Delta t. \tag{4.100c}$$

All the quantities in equ. (4.100c) can be measured separately in a coincidence experiment, and then the rate of accidental coincidences can be calculated. However, it is also possible to extract the accidental coincidences from the time spectrum accumulated with the time-measuring device. This is possible, because the time dependence necessary for the derivation of equ. (4.100c) disappears in the final result. As a consequence, the accidental coincidences give a constant 'background' in the time spectrum. As indicated in Fig. 4.48 by the two shaded regions, the accidental coincidences can then be obtained easily from a suitable selected range of the measured time spectrum. This becomes important if working with synchrotron radiation, because there the photon flux decreases with time which implies a decrease of the source strength. As will be demonstrated below, true and accidental coincidences depend differently on the source strength and, therefore, equ. (4.96) can be applied only if accidental and total coincidences are measured simultaneously.

In analogy with the treatment of electron detection after single ionization (Sections 2.4 and 10.4), expressions for the counting rates I can be derived. Characterizing the two selected branches of detected electrons by the indices 1 and 2, one gets

$$I_1(E_{\text{pass}1}) = N_{\text{ph}} n_{\text{v}}\, \Delta z_1 \sigma_1 T_1 \varepsilon_1 \langle F(E_{\text{pass}1}, E_{\text{kin}1})\rangle, \tag{4.101a}$$

$$I_2(E_{\text{pass}2}) = N_{\text{ph}} n_{\text{v}}\, \Delta z_2 \sigma_2 T_2 \varepsilon_2 \langle F(E_{\text{pass}2}, E_{\text{kin}2})\rangle, \tag{4.101b}$$

$$I_{\text{true}}(E_{\text{pass}1}, E_{\text{pass}2}) = N_{\text{ph}} n_{\text{v}}(\Delta z_1 \cap \Delta z_2)\sigma_{12} T_1 \varepsilon_1 T_2 \varepsilon_2 \langle F(E_{\text{pass}1}, E_{\text{kin}1}, E_{\text{pass}2}, E_{\text{kin}2})\rangle, \tag{4.102}$$

$$I_{\text{acc.}}(E_{\text{pass}1}, E_{\text{pass}2}) = I_1(E_{\text{pass}1})\, I_2(E_{\text{pass}2})\, \Delta t. \tag{4.103}$$

Most of the notation used in these expressions is explained in connection with equ. (2.37) for the case of single ionization, but three further comments are in order which are specific for coincidence experiments:

(i) The cross sections σ_i are a measure of the strengths in the two branches of electrons. For example, σ_1 might stand for a partial photoionization cross section leading to photoelectrons, and σ_2 might stand for the subsequent non-radiative decay leading to Auger electrons, i.e., $\sigma_2 = \sigma_1 \omega_{\text{A}}$, where ω_{A} is the Auger yield. Since the electrons are coincident, one also has $\sigma_{12} = \sigma_1 \omega_{\text{A}}$. However, the non-coincident counting rates usually contain large contributions from other (and disturbing) processes which have to be included in σ_1 and σ_2, but not in σ_{12}. As a result, such disturbances will increase the number of accidental coincidences, but leave the true coincidences unaffected.

(ii) The symbol $\Delta z_1 \cap \Delta z_2$ describes the sectional length between Δz_1 and Δz_2 accepted by each of the two spectrometers. The best-adapted situation follows for $\Delta z_1 = \Delta z_2 = \Delta z_1 \cap \Delta z_2 =: \Delta z$, which demonstrates the important requirement that the two spectrometers must view the same source volume. Otherwise the coincident signal decreases, and the accidental signal becomes relatively high. Assuming that this condition is always fulfilled, one can introduce the source strength S as

$$S = \mathbb{N}\sigma = N_{ph} n_v \,\Delta z\, \sigma. \tag{4.104}$$

(iii) The $\langle\ \rangle$ framing the spectrometer function $F(\ldots)$ indicate that the effects of all energy-dependent distribution functions have to be taken into account, for example, the energy distribution of the photons, the level width of an intermediate state and the spectrometer function. In particular, the function $\langle F(E_{pass1}, E_{kin1}, E_{pass2}, E_{kin2})\rangle$ in the coincident signal is a crucial quantity, because one also has to consider the relation between the kinetic energies required by energy conservation. It leads to the condition that for an optimum signal the pass energy should be adjusted properly (index '0'; for details see Section 10.4.3).

Usually, the pass energies E_{pass1} and E_{pass2} are kept fixed in an actual coincidence experiment, and the accumulated counts follow from equ. (4.102), where the numerical values of $\langle F(\ldots)\rangle$ depend on the energy distribution functions involved. In the case of non-coincident electron detection, these dependences are overcome by scanning the spectrometer voltage across the corresponding electron line, thus obtaining peak areas A. Following the same procedure, one gets here (see Section 10.4.3)

$$A_1 := \int I_1(E_{pass1})\, dE_{pass1} = N_{ph} n_v \,\Delta z_1 \sigma_1 T_1 \varepsilon_1 \,\Delta E_{sp1}, \tag{4.105a}$$

$$A_2 := \int I_2(E_{pass2})\, dE_{pass2} = N_{ph} n_v \,\Delta z_2 \sigma_2 T_2 \varepsilon_2 \,\Delta E_{sp2}. \tag{4.105b}$$

Using then equ. (2.42) one can obtain an estimate for the single counting rates obtained at the properly selected pass energy E^0_{pass}:

$$I_1(E^0_{pass1}) \approx N_{ph} n_v \,\Delta z_1 \sigma_1 T_1 \varepsilon_1 \,\Delta E_{sp1}/\Delta E_{exp1}, \tag{4.106a}$$

$$I_2(E^0_{pass2}) \approx N_{ph} n_v \,\Delta z_2 \sigma_2 T_2 \varepsilon_2 \,\Delta E_{sp2}/\Delta E_{exp2}. \tag{4.106b}$$

In principle it is also possible to get rid of the complicated $\langle F(E_{pass1}, E_{kin1}, E_{pass2}, E_{kin2})\rangle$ function in the coincidence case by defining an area A_{true} of coincident events: for each selected E_{pass1} value the coincident events have to be measured for all E_{pass2} values, and the individual results must be added. According to equ. (10.64) one gets

$$A_{true} = N_{ph} n_v \,\Delta z_1 \cap \Delta z_2 \,\sigma_{12}\, T_1 \varepsilon_1\, T_2 \varepsilon_2 \,\Delta E_{sp1}\, \Delta E_{sp2}. \tag{4.107}$$

However, this approach is rather time consuming, and in many cases only the coincidence signals at the nominal spectrometer settings are of interest. The corresponding rate of true coincidences is then given by

$$I_{true}(E^0_{pass1}, E^0_{pass2}) = N_{ph} n_v \,\Delta z_1 \cap \Delta z_2 \,\sigma_{12}\, T_1 \varepsilon_1\, T_2 \varepsilon_2\, f(\text{e.c.}), \tag{4.108}$$

where the numerical factor f(e.c.) contains both the quality of energy conservation realized at a given setting for the pass energies and the influence of all finite resolutions of the relevant energy distribution functions (see Section 10.4.3).

4.6.3 Optimization

In general the counting rate in electron–electron coincidence experiments is extremely low, and rather long times T_{coll} are needed to collect sufficient data. This becomes particularly important for coincidence experiments using synchrotron radiation where only limited beam time is available and it might be difficult to ensure the required stability of the electron storage ring over a long time period. Hence, a careful optimization of the experimental set-up is required.

An important quantity is the ratio r of true to accidental coincidences:

$$r = I_{true}/I_{acc.}, \tag{4.109a}$$

which is approximately given by

$$r \approx \frac{\sigma_{12}}{\sigma_1 \sigma_2} \frac{1}{\mathbb{N}} \frac{1}{\Delta t}. \tag{4.109b}$$

From this relation it follows that a good, i.e., large, ratio of true to accidental coincidences requires a small coincidence resolving time Δt and a small source strength ($\mathbb{N}\sigma$). However, for small values of $\mathbb{N}$ all counting rates, I_1, I_2, and I_{true}, are small, and therefore the ratio r is not well suited as a criterion of the quality or feasibility of coincidence experiments. Indeed, a more appropriate figure of merit follows if the relative error α of true coincidences, defined by

$$\alpha = \Delta N_{true}/N_{true}, \tag{4.110a}$$

is considered [Wap66, GvW83, VSa83], where N is the number of counts accumulated at the counting rate I within the collection time $T_{coll.}$,

$$N = I\, T_{coll.}. \tag{4.111}$$

N_{true} is calculated, according to equ. (4.96), as the difference between the total number of recorded coincidences, N_{tot}, and the number of accidental coincidences, $N_{acc.}$, measured simultaneously. Therefore, one gets for the relative error α

$$\alpha = \frac{1}{N_{true}} \sqrt{(N_{tot.} + (\Delta N_{acc.})^2)}. \tag{4.110b}$$

The relative error $\Delta N_{acc.}$ for the measurement of accidental coincidences is retained in this expression, because its value depends on the efforts undertaken for measuring $N_{acc.}$. For example, from Fig. 4.48 it can be inferred that it is possible to use for the accidental coincidences a larger time interval, $\Delta t_{acc.}$, than the Δt relevant for the true coincidences. This then helps to reduce the statistical error of $\Delta N_{acc.}$, and one gets two limits for $\Delta N_{acc.}$:

$$\Delta N_{acc.} = \sqrt{(\delta N_{acc.})} \quad \text{with } \delta = \begin{cases} 1 & \text{for } \Delta t_{acc.} = \Delta t, \\ 0 & \text{for } \Delta t_{acc.} \gg \Delta t. \end{cases} \tag{4.112}$$

(If the accidental coincidences are calculated using equ. (4.103), $\Delta N_{acc.}$ becomes negligible, because I_1 and I_2 are usually large numbers. However, due to the decreasing photon flux at an electron storage ring this equation would have to be applied at every instant of time. Hence, it is preferable to measure the accidental coincidences and desired total coincidences simultaneously.) With equ. (4.112) and replacing N_{tot} in equ. (4.110b) by $N_{true} + N_{acc.}$, one derives

$$\alpha^2 = \frac{1}{N_{true}}\left(1 + \frac{1+\delta}{r}\right). \tag{4.113}$$

Substituting for N_{true} using equs. (4.111) and (4.102), and resolving the expression for the data collection time $T_{coll.}$, one finally gets

$$T_{coll.} = \frac{1}{\alpha^2}\frac{1}{T_1\varepsilon_1 T_2\varepsilon_2}\frac{1}{\langle F(E_{pass1}, E_{kin1}, E_{pass2}, E_{kin2})\rangle}\frac{1}{\mathbb{N}\sigma_{12}}\left(1 + \frac{1+\delta}{r}\right), \tag{4.114a}$$

or, incorporating the approximate equ. (4.109b),

$$T_{coll.} = \frac{1}{\alpha^2}\frac{1}{T_1\varepsilon_1 T_2\varepsilon_2}\frac{1}{\langle F(E_{pass1}, E_{kin1}, E_{pass2}, E_{kin2})\rangle}\frac{1}{\sigma_{12}}\left[\frac{1}{\mathbb{N}} + (1+\delta)\frac{\sigma_1\sigma_2}{\sigma_{12}}\Delta t\right]. \tag{4.114b}$$

Equs. (4.114) then describe the time required for an actual coincidence experiment to reach an accuracy α of true coincidences. This time proves to be a better figure of merit than the ratio r. Noting that $\mathbb{N}$ depends on the ratio r, one can take the difference between $T_{coll.}(r)$ and $T_{coll.}(r = 1)$, using equ. (4.114b), and normalize this quantity against $T_{coll.}(r = 1)$, using equ. (4.114a). This yields

$$T_{coll.}|_{norm.} = \frac{T_{coll.}(r) - T_{coll.}(r=1)}{T_{coll.}(r=1)} = \frac{1}{2+\delta}\left[\frac{\mathbb{N}(r=1)}{\mathbb{N}(r)} - 1\right]. \tag{4.115}$$

This normalized collection time is plotted in Fig. 4.49 as a function of the relative source strength $\mathbb{N}(r)/\mathbb{N}(r = 1)$ and for $\delta = 0$. It can be seen that for fixed accuracy α the normalized collection time approaches its asymptotic value at a relative

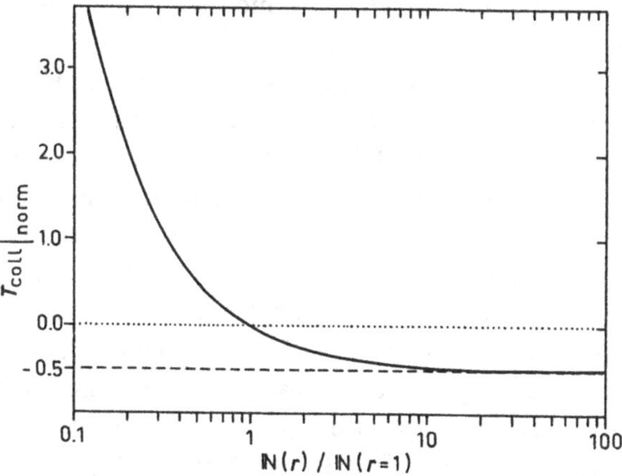

Figure 4.49 The normalized collection time $T_{coll}|_{norm.}$ of equ. (4.115) as a function of the relative source strength $\mathbb{N}(r)/\mathbb{N}(r = 1)$, and for $\delta = 0$. The parameter r is the ratio of true-to-accidental coincidences, and $r^{-1} = \mathbb{N}(r)/\mathbb{N}(r = 1)$. The parameter δ describes the error attached to the measured number of accidental coincidences (see equ. (4.112)). The collection time decreases for increasing source strength; for practical applications the asymptotic limit is reached well enough for $\mathbb{N}(r)/\mathbb{N}(r = 1) = 10$. From [Krä94]; see [VSa83].

source strength of approximately 10. Since equ. (4.109b) means $r = [\mathbb{N}(r)/(\mathbb{N}(r=1)]^{-1}$, this implies a true-to-accidental ratio r of approximately 0.1. Therefore, it is advantageous to work with $r < 1$, and not at $r > 1$, as might have been expected from the ratio r alone.

After this detailed treatment of true and accidental coincidences, all the statements derived so far for a good performance for coincidence experiments can be summarized:

(i) The coincidence resolving time Δt should be as small as possible, but large enough to collect the coincident events.
(ii) The product of the transmission and the detection efficiency, in particular the solid angle accepted by each of the electron spectrometers, should be as large as possible.
(iii) The source volumes seen by the spectrometers should be the same.
(iv) An appropriate data collection time $T_{\text{coll.}}$ results if the ratio r of true-to-accidental coincidences is smaller than 1.
(v) The pass energies of the electron analysers and their spectrometer functions have to be selected properly, taking care of energy conservation, to collect a large portion of the true coincidences. (In direct photoionization it can be worthwhile to use large ΔE_{sp1} and ΔE_{sp2} values in order to accept a larger portion of the double-ionization continuum.)
(vi) Because true and accidental coincidences depend differently on the source strength, for changing strength they should be measured simultaneously.

Of these requirements, two topics will be discussed further. One concerns ways to get a small coincidence resolving time Δt, and the other addresses how large acceptance angles of the spectrometers can modify the shape of observed angular distribution patterns.

As demonstrated in connection with Fig. 4.48, the lower limit for the coincidence resolving time Δt follows from the time spread of coincident events, i.e., from different travelling times t_1 and t_2 of the coincident electrons through the respective spectrometers and from a time spread in the corresponding electronic pulses. (Note that it is the *jitter* in an electronic timing signal which contributes to Δt, and not different processing times in the electronics or travelling times in the cables (these times produce a *delay*, i.e., a shift of the coincidence peak).) Of all these possibilities, the largest contribution comes from the different geometrical paths of the electrons in the spectrometers, $\Delta t_{\text{geometry}}$. (Variations in the transit time of the charge cloud inside the electron detector also lead to a time spread $\Delta t_{\text{detector}}$. One has $\Delta t_{\text{detector}}$(channeltrons) $\approx$ 2–3 ns [GKB72, HBW80]; $\Delta t_{\text{detector}}$(channelplates) $\approx$ 0.1 ns [GKB75, HBW80]; see also [VSa83].) The time spread $\Delta t_{\text{geometry}}$ can be calculated for a given electron spectrometer, and as an example the formula for a CMA with an acceptance angle Ω_0 for a point source and for electrons with given energy E^0_{kin} will be derived. It is then the angular spread $\pm\Delta\vartheta$ which causes the time spread $\Delta t_{\text{geometry}}$. The calculation is simplified considerably by the fact that the z-component of the electron's momentum p_z along the symmetry axis of the analyser must be conserved, because this component is perpendicular to the direction of the electric field in the spectrometer. Therefore, one gets the time t_{geometry} needed for an electron to travel from the source to the focus from (z_{sp} is the source–detector distance along the z-axis, see Fig. 4.4)

$$p_z = m_0 v_z = m_0 v_0 \cos \vartheta_0 = \text{const} \tag{4.116a}$$

to be

$$t_{\text{geometry}} = \frac{z_{\text{sp}}}{v_0 \cos \vartheta_0}. \tag{4.116b}$$

From this relation one obtains

$$\Delta t_{\text{geometry}} = \frac{z_{\text{sp}}}{v_0} \frac{\sin \vartheta_0}{\cos^2 \vartheta_0} \Delta\hat{\vartheta} \tag{4.116c}$$

with

$$v_0 = \sqrt{(2E^0_{\text{kin}}/m_0)}. \tag{4.116d}$$

For $z_{\text{sp}} = 40$ cm, $\vartheta_0 = 42.3°$ and $\Delta\vartheta = \pm 3°$ this gives $\Delta t_{\text{geometry}} = \pm 31$ ns, ± 9.7 ns, ± 3 ns for 2 eV, 20 eV and 200 eV electrons, respectively, which is a rather large time spread for electrons of low kinetic energy.

Several possibilities have been worked out for reducing the time spread $\Delta t_{\text{geometry}}$:

(i) a geometrical compensation in which a large detector (channelplate) is aligned parallel to the surface of equal flight time (see Fig. 4.50);
(ii) a geometrical compensation in which two analysers are combined such that longer/shorter electron paths in the first analyser become the shorter/longer paths in the second analyser (see Fig. 4.51);
(iii) an electronic compensation in which the position sensitivity of a channelplate detector with discrete multianodes is used to sort different flight times and to compensate them by properly adjusted time delays (see Fig. 4.52);
(iv) an acceleration of the electrons before they enter the analyser (see Fig. 4.35);
(v) proper compensation by lens-coupled spectrometers [BZh84].

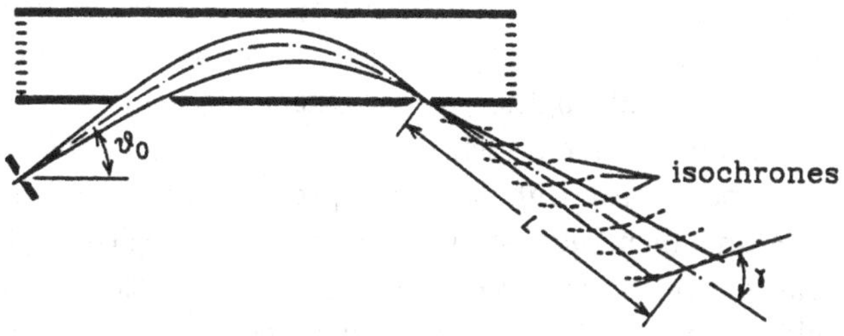

Figure 4.50 Calculated surfaces of equal flight times (isochrones, broken curves) intersecting the electron paths behind the exit slit of an electrostatic parallel plate analyser ($\vartheta_0 = 30°$). An individual isochrone may be labelled by its distance L from the exit slit together with its angle of inclination γ with respect to the principal trajectory. From [VSa83] where numerical data for γ at preselected distances L are also given.

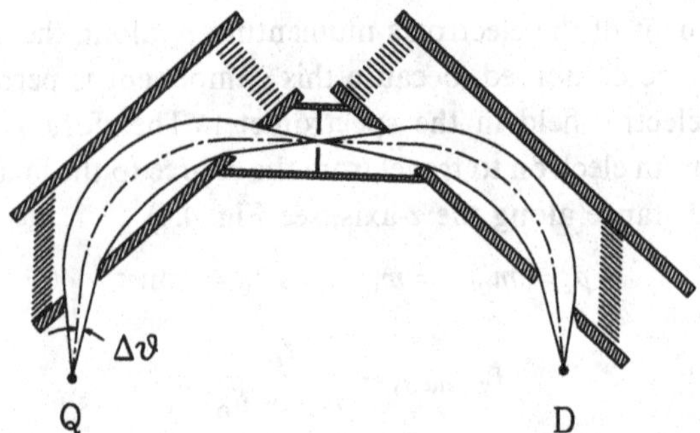

Figure 4.51 Geometrical time compensation (in $\Delta\vartheta$) using two sector CMAs. The source volume and the focal point are indicated by Q and D, respectively. From [Sch92b].

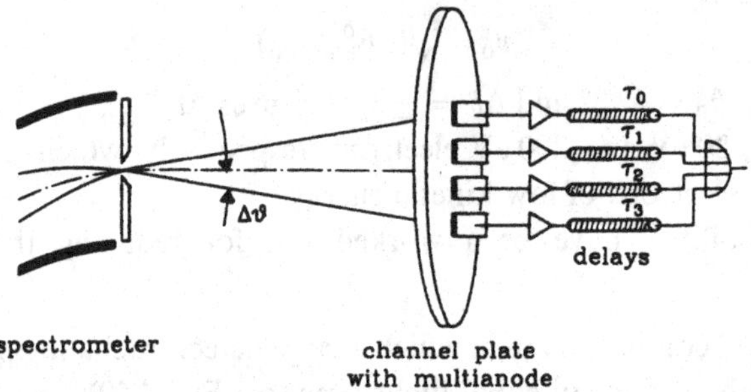

Figure 4.52 Block diagram for electronic compensation of the geometric time spread induced by electron trajectories with $\Delta\vartheta$ in the spectrometer. For the special configuration shown the delays must be chosen such that the delay times τ_i, with $\tau_3 < \tau_2 < \tau_1 < \tau_0$, compensate for these flight time differences. From [VSa83].

Finally, the large acceptance angles of the electron spectrometers, which are of great importance for a good signal in electron–electron coincidence experiments, will be considered with regard to their effect on the shape of the coincident angular pattern. Any finite acceptance angle of an electron spectrometer also leads to the recording of electrons at angles which differ from that of the principal ray (see Fig. 4.4). Therefore, any observed angular distribution pattern contains weighted contributions from different angles, and in general its form will change when compared with the form for the principal ray only. Usually, angular distribution measurements are performed with electron spectrometers which have a small acceptance angle, and then such solid-angle effects are negligible. However, in coincidence measurements larger acceptance angles are needed, and the influence upon the coincident angular distribution of two increased angular ranges has to be worked out quantitatively, if a correct interpretation of the experimental data is required. The importance of this effect is demonstrated with the help of Fig. 4.53. This comes from an experiment described in detail in Section 5.6 in which photoelectrons and subsequent Auger electrons are measured in coincidence in a

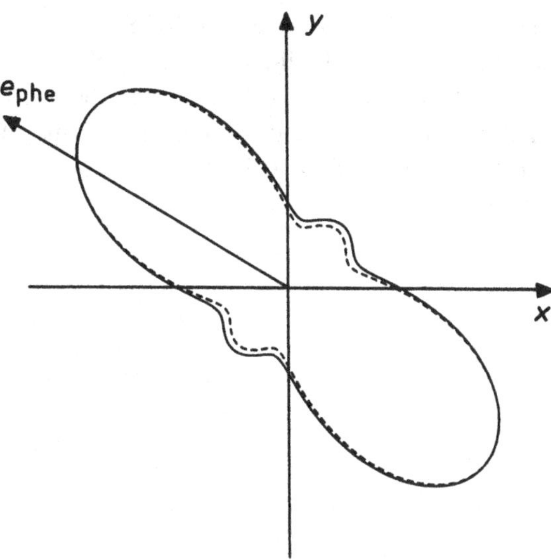

Figure 4.53 Illustration of the influence of finite acceptance angles on the angular correlation pattern between coincident photoelectrons and Auger electrons (called the *solid-angle* effect). Polar plots of coincident intensities are shown. Solid curve = inclusion of finite acceptance angles; dashed curve = infinitesimally small acceptance angles, adapted to the same value for the maximum intensity. The data refer to $4d_{5/2}$ photoionization at 94.5 eV in xenon with subsequent N_5–$O_{2,3}O_{2,3}$ $^1S_0$ Auger decay, i.e., to the example discussed in Section 5.6. The photoelectron and the Auger electron are observed in a plane perpendicular to the photon beam direction, the emission direction of the photoelectron is kept fixed (e_{phe}), but the direction of the coincident Auger electron is varied. From [KSc93].

plane perpendicular to the photon beam. The direction of the photoelectrons is kept fixed as indicated by e_{phe} in the figure, and the angle-dependent intensity of coincident Auger electrons is shown as a polar plot. The solid curve includes solid-angle effects, but they are neglected in the dashed curve, i.e., the dashed curve corresponds to a theoretical pattern referring to point-like detectors. Both curves have been normalized to the same maximum value, because in the present context it is the shape which is of interest, and not the absolute value of intensities. (Another possibility for normalization is to divide the solid-angle affected intensity by the solid angle (see equ. (10.77)).) It can be seen that the solid-angle effect produces a considerable disturbance of the angular distribution pattern, in particular in the small lobes. There the finite acceptance angles smear out the small minima and lead to an increase of the observed intensity which is much more pronounced than the effects on the smoother and larger lobes.

The quantitative treatment of solid-angle effects is rather cumbersome. A concise formulation can be derived only if the theoretical angular correlation is known, for example, by the general parametrization of equ. (4.68) with known A-coefficients, and if a point source can be assumed. (The condition for simulating a 'point' source is not so much an infinitely small dimension, but an equal angular acceptance of the analyser for electrons emerging from points within the finite source volume.) If these conditions are fulfilled, one can apply general transformation properties of the spherical harmonics, calculate the solid-angle effect for the given theoretical angular distribution, and then compare this result with the experimental data (for details see Section 10.5).

5
Recent examples

In the preceding chapters the material necessary for studying photoionization processes in atoms using synchrotron radiation and electron spectrometry was presented. The discussion will now be completed with some examples of current research activities. These include:

(1) photon-induced electron emission around the 4d ionization threshold in xenon from which a complete mapping of these spectra can be obtained and many features characteristic of inner- and outer-shell photoprocesses are well visualized;
(2) a complete experiment for 2p photoionization in magnesium which also provides a detailed illustration of the role that many-electron effects have on main photolines;
(3) an investigation of discrete satellite lines in the outer-shell photoelectron spectrum of argon which demonstrates for a simple case the origin of satellite processes in electron correlations, and also the importance that instrumental resolution has on the determination of satellite structures;
(4) a complete experiment for $5p_{3/2}$ photoionization in xenon which includes a measurement of the photoelectron's spin polarization;
(5) a quantitative study of postcollision interaction (PCI) between $4d_{5/2}$ photoelectrons and N_5–$O_{2,3}O_{2,3}$ $^1S_0$ Auger electrons in xenon which also serves as an example of energy calibration in accurate experiments;
(6) the determination of coincidences between $4d_{5/2}$ photoelectrons and N_5–$O_{2,3}O_{2,3}$ $^1S_0$ Auger electrons in xenon which allows a spatial view of the angular correlation pattern for this two-electron emission process;
(7) a near-threshold study of state-dependent double photoionization in the 3p shell of argon in which the cross section approaches zero and two electrons of extremely low kinetic energy have to be measured in coincidence.

Though the examples selected do not cover all possible aspects of electron spectrometry of atoms using synchrotron radiation, they serve to provide a sound basis for understanding the exciting research activities that are presently being undertaken (for recent reviews see [Sch92a, SZi92, BSh96]).

5.1 Electron emission around the 4d ionization threshold in xenon

Depending on the photon energy chosen, the spectrum of ejected electrons contains electrons originating from different processes. As explained in Section 1.2, in well-defined cases these electrons can be attributed to photoionization (main and satellite processes), autoionization decay of resonances (doubly excited states, outer- and inner-shell electron excitation) and Auger decay (main and satellite processes), and some of these have already been considered in connection with the discussion of Fig. 3.2. In this section photon-induced electron emission around the 4d ionization threshold in xenon will be treated in detail because so many experimental data exist for this that a complete mapping of the spectra can be presented [CWS93]. How this large amount of data is accumulated experimentally, represented visually and explained generally will be described. With this information a detailed analysis can be given demonstrating specific aspects around this inner-shell ionization threshold. The cases selected are resonance Auger transitions in comparison to normal Auger transitions and postcollision interaction (PCI) as a link between inner- and outer-shell ionization processes.

5.1.1 Experimental details

In order to get a complete mapping of photon-induced spectra of ejected electrons one has to record the electron intensity as a function of both the photon energy and the electron's kinetic energy. This can be achieved by measuring at one photon energy the corresponding spectrum of ejected electrons, then increasing the photon energy by a small amount, and repeating the measurement of the electron spectrum, and so on (for such a study see [BGW87]). Obviously, this is a rather time consuming procedure, and it is essential to improve the efficiency of data accumulation by equipping the electron spectrometer with a position-sensitive detector. In the experiment described, the derivation of the position-dependent information is based on the concept of independent multiple anodes (see Fig. 4.23). In this novel design the disadvantages associated with discrete external electronic circuits have been overcome by building a monolithic, self-scanning detection/amplification device. On the top face of an integrated circuit and adjacent to the exit of the second channelplate there is a linear array of electron-sensing electrodes which collect the charge pulses from the channelplate. Each electrode is periodically sampled by connecting it to a separate on-chip amplifier whose output controls an eight-bit counter. Hence, each electron pulse from the channelplates increments the relevant counter so that the spectrum is accumulated in these counters on the detector chip. The data are read out sequentially on an eight-bit data bus and fed to the computer (for details see [HBC92]). In this way, a dispersed electron image, 2 cm in length, can be detected at once, and this corresponds in this case to an energy range of approximately 1 eV, which provides a remarkable gain in the efficiency of the data accumulation.

The sequential recording of electron spectra for small changes in the photon energy finally leads to the desired complete data set for electron intensities $I = I(h\nu, E_{kin})$. The data can be mapped in three dimensions, or in two dimensions if a grey scale is used for the intensities (black and white for extreme values of

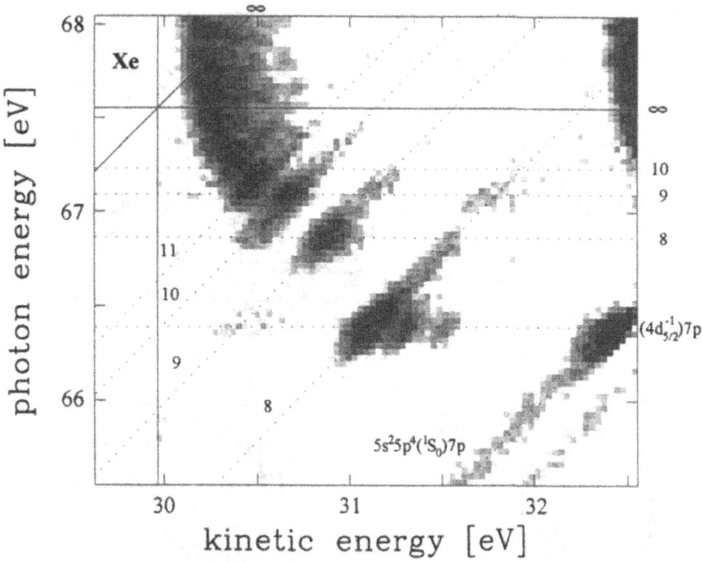

Figure 5.1 Two-dimensional grey-scale plot for the intensity $I(h\nu, E_{kin})$ of electrons ejected by photon impact on xenon in the vicinity of the $4d_{5/2}$ ionization threshold and the N_5–$O_{2,3}O_{2,3}$ $^1S_0$ Auger line. Increasing electron intensity is shown using the grey scale from white to black. Since the relevant processes are related to $4d_{5/2} \to np$ excitations and $4d_{5/2}$ ionizations, individual excitation energies are marked on the right-hand photon energy scale. Also, the energy positions of photosatellites $5p^4(^1S_0)np$ which are affected by the decay of these resonances are indicated by the dotted diagonal lines and the principal quantum number n. The vertical line at 29.97 eV kinetic energy gives the nominal kinetic energy of the N_5–$O_{2,3}O_{2,3}$ $^1S_0$ Auger line; due to PCI this line appears at slightly higher energy as can be seen in the left upper part of the figure. For a detailed explanation of the whole figure see main text. From [CWS93].

maximum and zero intensity). An example for electron emission in the vicinity of the xenon N_5–$O_{2,3}O_{2,3}$ $^1S_0$ Auger line and the $4d_{5/2}$ threshold is shown in Fig. 5.1. Several interesting features can be noted. First, the intensities appear in two limiting forms, as smoothly changing mountain chains and as suddenly emerging isolated peaks. Second, these limiting forms are connected in certain ways; they merge and/or separate if their directions in the ($h\nu$–E_{kin})-plane and their intensities are considered. Third, three directions of preference exist for these patterns: parallel to the axes of photon and kinetic energy, respectively, and parallel to the diagonal of these axes. Fourth, from such a complete mapping of the data it is easy to extract by certain projections the conventional plots of electron intensities, usually obtained in different modes of operation (see Section 4.2.4): keeping $h\nu$ constant, one gets the energy distribution (EDC), keeping E_{kin} constant, one obtains curves of the constant-final state mode (CFS), and along lines parallel to the diagonal one has curves of the constant-ionic state mode (CIS).

Within the energy ranges covered in Fig. 5.1 the dominant photoprocesses are 4d excitation/ionization with subsequent autoionization/Auger decay, and the

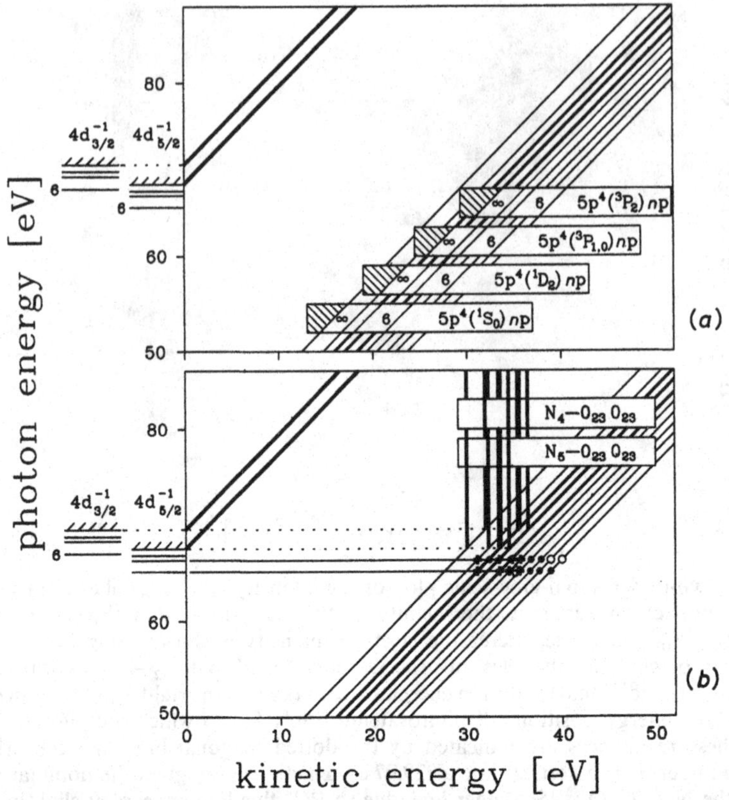

Figure 5.2 Schematic two-dimensional plot of the electron intensity as a function of the kinetic energy and photon energy in the vicinity of xenon $N_{4,5}$–$O_{2,3}O_{2,3}$ Auger lines and the $4d_{5/2,3/2}$ ionization thresholds. (*a*) $4d_j$ photolines and $5p^4(^{2S+1}L_J)np$ satellites (for $n = 6, 7, \ldots$ converging to the respective double-ionization thresholds). It should be noted that many other satellites exist in the energy ranges shown, in particular $5s5p^5np$ and $5s^{-2}5p^6np$ satellites, but also satellites with $n\ell = ns$ and $n\ell = nd$ instead of np only. These cases have been omitted for simplicity. (*b*) Photolines and satellites as shown in (*a*) plus Auger/autoionization decay of these inner-shell excited states. Again, the discussion is simplified by omitting Auger transitions like $N_{4,5}$–$O_1O_{2,3}$ and $N_{4,5}$–O_1O_1, and corresponding Auger satellites. The special region between 29.5 and 32.5 eV kinetic energy and 65.5 and 68 eV photon energy is used to explain Fig. 5.1. The electron intensities are shown as black lines, except for the satellite decay, where open circles, crosses and full circles refer to shake-down, normal spectator, and shake-up spectator transitions, respectively. For a detailed explanation see main text.

production of outer-shell satellites belonging to $5s^25p^4k'\ell'$ electron configurations.† These processes will be discussed with the help of Fig. 5.2(*a*) for the primary atom–photon interaction and Fig. 5.2(*b*) for those processes including non-radiative decay. Starting with Fig. 5.2(*a*), one can see on the left-hand side the energies necessary to excite/ionize a $4d_{5/2}$ or $4d_{3/2}$ electron in xenon. Postponing

† $5s^25p^4k'\ell'$ is an abbreviation of $1s^22s^22p^63s^23p^63d^{10}4s^24p^64d^{10}5s^25p^4k'\ell'$; another way of writing this electron configuration is $5p^{-2}k'\ell'$. Similar short versions are used for other electron configurations.

a discussion of autoionization decay of the excited states, photoionization leads to electron emission, and the corresponding $4d_{5/2}$ and $4d_{3/2}$ main photolines appear at threshold with zero kinetic energy and finite intensity. An increase of the photon energy then produces the straight line (mountain chain) positioned along the diagonal in the $(h\nu - E_{kin})$-plane (see equ. (1.29a)). A similar behaviour would be found for the outer-shell 5p and 5s main photolines which, however, are outside of the given energy range, and for the outer-shell satellites where, for simplicity, only the $5s^2 5p^4(^{2S+1}L_J)np$ satellites are included in the figure. Insets help to identify for these satellites the Rydberg levels $n = 6, 7, 8$, and the onset of the double-ionization continuum. Neglecting possible disturbance by autoionization decay, each satellite line follows its own straight line which represents a smooth mountain chain positioned along the diagonal in the $(h\nu - E_{kin})$-plane. Towards their respective double-ionization thresholds, both inherent and experimental resolution prevent the appearance of individual mountain chains. Above the double-ionization continuum the two electrons share the available excess energy (see equ. (4.63b)), and this results in a smooth intensity distribution aligned parallel to the E_{kin}-axis (not shown in the figure). Due to such processes, the whole left-hand region in the double-ionization continuum becomes covered with a smooth 'background'.

Hitherto the discussion of Fig. 5.2 has neglected the possibility of non-radiative decay following 4d shell excitation/ionization. These processes are explained with the help of Fig. 5.2(*b*) which also reproduces the photoelectron emission discussed above, because both photo- and autoionization/Auger electrons will finally yield the observed pattern of electron emission. (In this context it should be noted that in general such direct photoionization and non-radiative decay processes will interfere (see below).) As can be inferred from Fig. 5.2(*b*), two distinct features arise from non-radiative decay of 4d excitation/ionization. First, $4d_j \to n\ell$ resonance excitation, indicated on the photon energy scale on the left-hand side, populates certain outer-shell satellites, the so-called *resonance Auger* transitions (see below), via autoionization decay. An example of special interest in the present context is given by

$$4d^9 5s^2 5p^6(^2D_{5/2})np \to 4d^{10} 5s^2 5p^4(^{2S+1}L_J)np + e, \tag{5.1a}$$

$$4d^9 5s^2 5p^6(^2D_{3/2})np \to 4d^{10} 5s^2 5p^4(^{2S+1}L_J)np + e. \tag{5.1b}$$

Since these processes arise only at specific photon and satellite energies, they appear in the $h\nu - E_{kin}$ diagram as isolated points (see equ. (1.29c)). These points are marked by open circles, full circles, and crosses in order to distinguish *shake down*, *normal spectator*, and *shake-up spectator* transitions, respectively (see below). (Shake-off transitions are also possible; they give rise to a continuous energy distribution which is indicated by lines parallel to the E_{kin}-axis and positioned at $h\nu_{resonance}$.) Second, as soon as the photon energy is equal to (dotted lines in Fig. 5.2(b)) or greater than the threshold for $4d_j$ ionization, the $4d_j$ hole produced will be filled by a normal Auger transition. (For the influence of PCI see below.)

Restricting the discussion again to $5s^25p^4$ electron configurations of the final state, one has two groups of Auger transitions classified as

$$N_5\text{–}O_{2,3}O_{2,3}\text{:}\quad 4d_{5/2}^{-1} \rightarrow 5s^25p^4(^{2S+1}L_J) + e_A, \tag{5.2a}$$

$$N_4\text{–}O_{2,3}O_{2,3}\text{:}\quad 4d_{3/2}^{-1} \rightarrow 5s^25p^4(^{2S+1}L_J) + e_A. \tag{5.2b}$$

According to equ. (1.29b), the energy of these Auger transitions will depend only on the energy differences between the inner-shell one-hole state and the outer-shell two-hole state. Hence, these Auger lines will start at threshold with finite values for their intensity and kinetic energy, i.e., they appear as straight lines (mountain chains) parallel to the photon energy axis (see Fig. 5.2(*b*) with inserts for the selected Auger transitions; for the influence of PCI see below). Due to the presence of high $4d_j \rightarrow np$ Rydberg excitations just below the $4d_j$ ionization thresholds, the mountain chain of Auger transitions is initiated by the autoionization decay of these resonances.

After this schematic discussion of possible processes around the 4d ionization shell in xenon, Fig. 5.2(*b*) can be compared with the experimental results shown in Fig. 5.1. The main structures can be related to $4d_j$ photoexcitation/ionization, and autoionization/Auger decay to $5s^25p^4k\ell$ electron configurations can be identified in the energy regions of overlap.

5.1.2 Specific aspects

From the pieces of information contained in Fig. 5.1, two are selected for further discussion: resonance Auger transitions and a qualitative demonstration of PCI above and below the threshold of inner-shell ionization.

5.1.2.1 *Resonance Auger transitions*

Many features typical of atomic inner-shell photoexcitation with its autoionization decay have been explored for the case of $4d \rightarrow np$ excitation in xenon. Though the $4d_{5/2} \rightarrow 6p$ resonance at 65.11 eV photon energy is just outside the energy range shown in Fig. 5.1, it will be considered here because it can be treated as an isolated resonance, and is the best-studied case; the $4d_{5/2} \rightarrow 7p$ resonance seen in Fig. 5.1 at 66.37 eV photon energy shows rather similar features. The notation $4d_{5/2} \rightarrow 6p$ is an abbreviation for

$$h\nu_{res} + \text{Xe} \rightarrow \text{Xe}^*(4d^95s^25p^6(^2D_{5/2})6p\ J = 1), \tag{5.3}$$

because in the single-electron model with $(\ell + s \rightarrow j)$-classified orbitals a $4d_{5/2}$ electron is excited to the 6p level. According to the dipole selection rules of equ. (2.9a), excitation to 6f is also allowed, but the shape of the atomic potential leads to a negligible probability for this channel, i.e., only $4d \rightarrow np$ excitations occur. All $4d \rightarrow np$ excitations with their autoionization 'decay' are called *resonances*, because photoabsorption takes place only at certain energies $h\nu_{res}$. The description of these phenomena given suggests a two-step process, however; the quotation marks used for the word decay indicate that it ought to be treated as a one-step process. (Compare the corresponding discussion of photon-induced Auger decay in Section 8.3; in the present case one calls these resonances, described by a one-step formula, *Auger resonant Raman* processes ([BCC80], see Fig. 5.3).) A selected decay channel, for example, $5s^2 5p^4(^{2S+1}L_J)6p$, can be reached via two pathways, one direct and the other going over the resonance:

$$h\nu_{res} + \mathrm{Xe} \left\{ \begin{array}{c} \xrightarrow{\qquad\qquad\qquad} \\ \rightarrow \mathrm{Xe}^*(4d_{5/2}^{-1} \rightarrow 6p) \rightarrow \end{array} \right\} \mathrm{Xe}^+(5s^2 5p^4(^{2S+1}L_J)6p) + e. \quad (5.4)$$

Since these pathways are indistinguishable, the corresponding amplitudes interfere, and the effect of the resonance on the observed single-ionization channel is between enhancement and reduction. In the present examples, however, the direct ionization channel is weak, and the presence of a resonance usually leads to an enhancement (see Fig. 5.1). The discussion of xenon 4d → 6p resonance excitation and decay is very special because interference effects due to equ. (5.4) are negligible. Often such interference effects are strong, and they give rise to peculiar lineshapes, frequently called *Fano* line profiles ([Fan61], for details see references in [Sch92a]).

The inner-shell hole $4d_j$ produced by photoexcitation is filled by a non-radiative transition (radiative decay is negligible). Therefore, the resonance has a certain mean lifetime τ which corresponds to a natural width Γ of approximately 114 meV for the xenon $4d_{5/2} \rightarrow 6p$ resonance. Autoionization produces many final ionic states, and the important branches are given by

$$(a)\ \mathrm{Xe}(4d_{5/2}^{-1}6p\ J = 1) \rightarrow \mathrm{Xe}^+(4d^{10}5s^2 5p^5\ ^2P) + e_1 \quad (5.5a)$$

$$(b)\ \rightarrow \mathrm{Xe}^+(4d^{10}5s^1 5p^6\ ^2S) + e_1 \quad (5.5b)$$

$$(c)\ \rightarrow \mathrm{Xe}^+(4d^{10}5s^2 5p^4 6p\ ^{2S+1}L) + e_1 \quad (5.5c)$$

$$(d)\ \rightarrow \mathrm{Xe}^+(4d^{10}5s5p^5 6p\ ^{2S+1}L) + e_1 \quad (5.5d)$$

$$(e)\ \rightarrow \mathrm{Xe}^+(4d^{10}5p^6 6p\ ^2P) + e_1 \quad (5.5e)$$

$$(f)\ \rightarrow \mathrm{Xe}^+(4d^{10}5s^2 5p^4 7p\ ^{2S+1}L) + e_1 \quad (5.5f)$$

$$(g)\ \rightarrow \mathrm{Xe}^{2+}(4d^{10}5s^2 5p^4\ ^{2S+1}L) + e_1 + e_2. \quad (5.5g)$$

There are more complicated branches which are due to final-ionic-state configuration interaction. These have been omitted for simplicity (for details see [Sch92a]). Some of these final states proceed further via a second-step autoionization or Auger decay; for example,

$$(h)\ \mathrm{Xe}^+(4d^{10}5s5p^5 6p) \rightarrow \mathrm{Xe}^{2+}(4d^{10}5s^2 5p^4\ ^{2S+1}L) + e_2. \quad (5.5h)$$

In addition, even triple-ionization is possible during the resonance decay leading to

$$(i)\ \mathrm{Xe}(4d_{5/2}^{-1}6p\ J = 1) \rightarrow \mathrm{Xe}^{3+}(4d^{10}5s^2 5p^3\ ^{2S+1}L) + e_1 + e_2 + e_3. \quad (5.5i)$$

In (*a*) and (*b*) the 6p Rydberg electron takes part in the autoionization process. Therefore, these processes are *participator* or *involved* transitions. In (*c*), (*d*), and (*e*), the excited electron remains in its orbital when an ($n = 5$)-electron fills the 4d vacancy and another ($n = 5$)-electron is ejected. Due to the similarity to a normal Auger process, these transitions are called *resonance Auger* or *autoionization spectator* transitions and, roughly speaking, these resonance Auger spectra are expected to be the same as the normal Auger spectra recorded above the 4d ionization threshold, except for an energy shift and a fine-structure splitting caused by the presence of the spectator electron.

In addition to the spectator transitions discussed so far, relaxation during the autoionization decay can promote the spectator electron from its original orbital 6p into different orbitals. In the lowest approximation, relaxation is due to a change in the nuclear shielding if one of the electrons is emitted. The resulting rearrangement of the electrons can then be calculated within the *shake* model. As the name implies, this model treats the shake effect on electron orbitals caused by a sudden change of nuclear shielding. Shake theory considers the overlap matrix element of passive electrons and assumes that the wavefunctions in the bra-vector, $\langle\Psi|$, and the ket-vector, $|\Psi\rangle$, belong to different shieldings [SDe66, Abe67, CNT68, Abe69, CNe73]. In the present example the essential quantity of this overlap matrix element is the one-electron overlap $\langle n\ell_{\text{after}} | 6\text{p}_{\text{before}}\rangle$. Its numerical value squared gives the probability for a 'transition' $6\text{p}_{\text{before}} \rightarrow n\ell_{\text{after}}$. Due to the orthogonality in the angular functions (spherical harmonics), one gets the one-particle *monopole* 'selection' rules of shake processes:

$$\Delta n = \text{arbitrary}, \ \Delta\ell = 0, \ \Delta m_\ell = 0. \tag{5.6}$$

In the present example a change of 6p to np with different principal quantum numbers n is possible. Hence, *shake-modified* spectator transitions or *resonance shake* transitions can occur, and (*f*) and (*g*) are examples of *shake-up* and *shake-off* resonance double Auger processes, respectively. (If there is a lower unoccupied orbital, *shake-down* is also possible).

Summarizing the individual decay branches of the $4\text{d}_{5/2} \rightarrow 6\text{p}$ resonance, one finds that all final ionic states can also be reached by outer-shell photoionization, in (*a*) and (*b*) by main processes, and in (*c*)–(*i*) by discrete and continuous satellite processes. The effect of the resonance decay will then be a modification of these otherwise undisturbed direct outer-shell photoionization processes which turns out to be an enhancement in the present case. Therefore, these outer-shell satellites are called *resonantly enhanced satellites.* In this context it is important to note that outer-shell photoionization also populates other satellites, attached, for example, to electron configurations $5\text{s}^2 5\text{p}^4 n\text{s}$ and $5\text{s}^2 5\text{p}^4 n\text{d}$. However, the parity of these satellites is even, while the decay branches (*c*)–(*f*) lead to odd parity. Therefore, both groups of final ionic states can be treated independently of each other (if configuration interaction in the continuum is neglected).

The electron spectrum of the decaying $4\text{d}_{5/2} \rightarrow 6\text{p}$ resonance can be obtained, keeping the photon energy fixed at 65.11 eV and scanning through the grey-scale plot, which is equivalent to the direct observation of an electron spectrum in

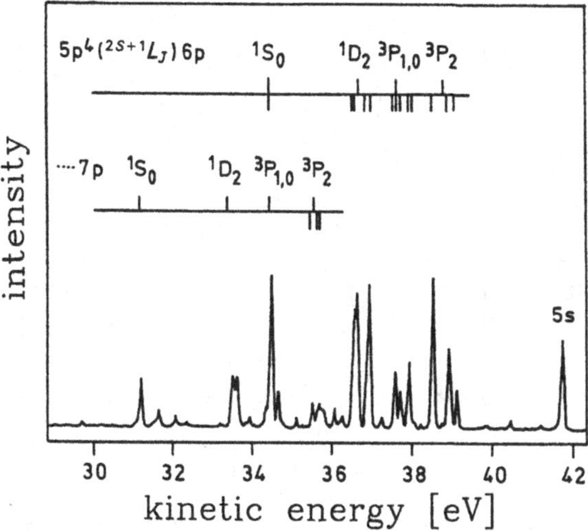

Figure 5.3 Resonance Auger spectrum of xenon for $4d_{5/2} \rightarrow 6p$ excitation and autoionization decay to $5p^4(^{2S+1}L_J)np$, and the 5s photoline. The spectrum was obtained with a bandpass of the monochromatized undulator radiation of only 19 meV and a spectrometer resolution of 50 meV; both are small compared to the natural width of 114 meV of the decaying resonance state. As a consequence, and in contrast to the discussion given in Section 3.3 for normal Auger lines, the observed width of the resonance Auger lines becomes narrower than the natural level width (this phenomenon is called the *Auger resonant Raman effect* [BCC80]; see [KNA93] for references to this topic), thus allowing a clear identification of individual fine-structure resolved lines. The energy positions of these resonance Auger lines are given at the top of the figure, for the normal spectator transitions, $np = 6p$, and for the shake-up transitions, $np = 7p$. The upper ticks refer to the energy spacing of the $5p^4\ {}^{2S+1}L_J$ case only; the lower ticks show for the available optical data the splitting due to the additional np electron. From [KNA93].

the energy distribution mode. In Fig. 5.3 such an electron spectrum is shown for the region of resonance Auger decay and the normal 5s photoline. (In principle, the 5s photoline can also be affected by the involved autoionization decay of the resonance, cf. equ. (5.5b), and normal off-resonance photosatellites contribute to the spectrum. However, both aspects are neglected in the present case.) The remarkable intensity of $5s^25p^4(^{2S+1}L_J)np$ satellites manifests in the many peaks with kinetic energies up to approximately 39 eV. The bar diagrams mark the energy positions of these satellites with $n = 6$ and $n = 7$ and help to identify individual lines as belonging to these states. The upper ticks come from the energy splitting of N_5–$O_{2,3}O_{2,3}$ Auger lines, shifted because of the binding energy of the additional electron. The lower ticks give for $n = 6$ the energy positions of the corresponding fine-structure resolved components $5s^25p^4(^{2S+1}L_J)6p\ {}^{2S'+1}L'_{J'}$, which are due to the coupling of the spectator electron with the double-hole core. A one-to-one correspondence between observed peaks and energy positions of individual fine-structure components can be noted. Surprisingly, there are also a remarkable number of $6p \rightarrow 7p$ shake-up transitions. This interesting property can be understood, and is reproduced fairly well, within the shake model which takes

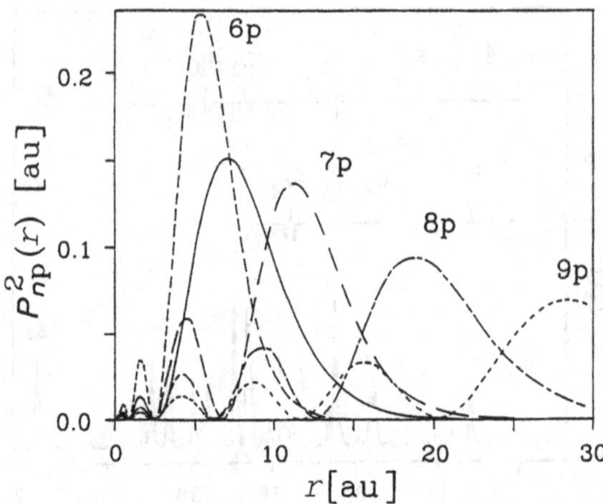

Figure 5.4 Radial charge densities $P^2_{np}(r)$ of np electrons as a function of the radial distance r (all quantities are in atomic units). The solid line gives $P^2_{6p}(r)$ with the 6p orbital computed in the field of the $4d^9 5s^2 5p^6$ electron configuration, the broken lines give $P^2_{np}(r)$ with the 6p, 7p, 8p, and 9p orbitals calculated in the field of the $4d^{10} 5s^2 5p^4$ electron configuration. The corresponding shake probabilities $|\langle np(\text{dashed})|6p(\text{solid})\rangle|^2$ are given by $|\langle 6p'|6p\rangle|^2 = 0.801$, $|\langle 7p'|6p\rangle|^2 = 0.196$, $|\langle 8p'|6p\rangle|^2 = 0.001$, $|\langle 9p'|6p\rangle|^2 = 0.0005$. From [AAk95] and [SAA95].

into account the different wavefunctions in the one-electron overlap matrix element $\langle np|6p\rangle$. The radial charge densities of these orbitals calculated in the potentials of the initial and final electron configurations $4d^9 5s^2 5p^6 6p$ and $4d^{10} 5s^2 5p^4 np$, respectively, are shown in Fig. 5.4. It can be seen that in comparison to the initial 6p charge density (solid curve) the change of shielding experienced by the 6p Rydberg orbital during the autoionization decay pulls all relaxed orbital functions (charge densities, dashed curves) towards the inner atomic region. (For cases other than xenon $4d_{5/2} \to 6p$ excitation this effect can be very strong. In analogy to the 'collapse' of wavefunctions in the aufbau principle of atoms this phenomenon is also a *collapse* of the $n\ell$ orbital (see Section 5.3.1).) As a consequence, the overlap $\langle 6p(\text{dashed})|6p(\text{solid})\rangle$ is no longer unity, and the overlap $\langle 7p(\text{dashed})|6p(\text{solid})\rangle$ no longer vanishes. The numerical values then confirm that there is still a high probability for the 6p Rydberg electron to stay in its orbital, but also that there is a significant amount of $6p \to 7p$ shake-up.

So far the decay paths of a selected resonance have been discussed: $4d_{5/2} \to 6p$ excitation in xenon with decay to $5s^2 5p^4 (^{2S+1}L_J) np$ satellites. A slightly different way of looking at resonance phenomena follows if the whole series of resonances is considered as being embedded in certain continua. In the present case these are the $4d \to 6p$ resonances with the continua of photoionization satellites, mainly $5s^2 5p^4 (^{2S+1}L_J) np$, and the question arises of how these resonances might affect the satellite processes. Such a consideration corresponds to the analysis of electron spectra parallel to the diagonal in the $(h\nu - E_{kin})$-plane (see Figs. 5.1 and 5.2;

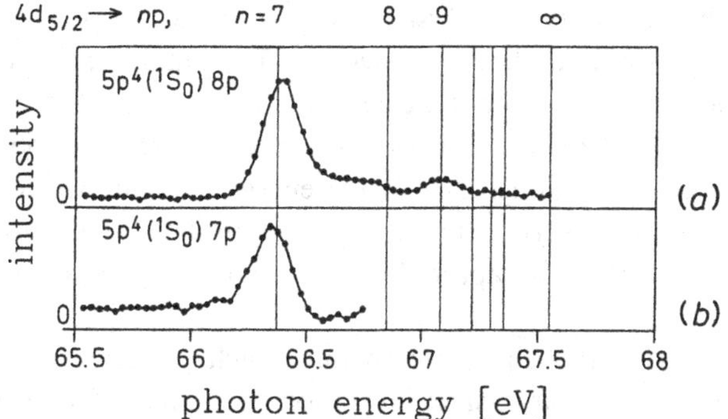

Figure 5.5 Relative intensities of xenon photosatellites as functions of photon energy, in the region of 4d $\rightarrow$ *n*p excitations: (*a*) $4d^{10}5s^25p^4(^1S_0)7p\ ^2P^o$ satellite; (*b*) $4d^{10}5s^25p^4(^1S_0)8p\ ^2P^o$ satellite. The energy positions of $4d_{5/2} \rightarrow np$ excitations are indicated by the vertical lines, numbered $n = 7, 8, 9, \ldots$. For a detailed explanation of the effect that these resonance excitations have on the satellite intensities see main text; for a related study see [BSK89b]. From [CWS93].

constant-ionic-state spectroscopy). One then obtains, as a function of the photon energy, partial cross sections of these satellites, and at photon energies fitting a resonance excitation, one can expect modifications in these otherwise smooth cross sections. Two examples are shown in Fig. 5.5 (for the corresponding grey-scale plot see Fig. 5.1). Within the energy range shown, and starting at the lower photon energy, it can be seen that the cross section $\sigma(5s^25p^4(^1S_0)7p)$ has some finite value and is considerably enhanced at the position of the $4d_{5/2} \rightarrow 7p$ resonance (in the former discussion this is due to a true spectator decay of this resonance). In contrast, the cross section $\sigma(5s^25p^4(^1S_0)8p)$ starts with a much lower value, is also most strongly enhanced at the $4d_{5/2} \rightarrow 7p$ resonance excitation (spectator shake-down decay) with a wing towards higher photon energies, and shows a small effect at the $4d_{5/2} \rightarrow 8p$ resonance (normal spectator decay), but a noticeable enhancement at the $4d_{5/2} \rightarrow 9p$ resonance (spectator shake-up decay). Hence, rather distinct influences of resonance excitations on selected partial photoionization cross sections become visible in such plots.

5.1.2.2 *PCI as a link between inner- and outer-shell ionizations*

Using the foregoing discussion, it is possible to take a closer look at the grey-scale plot in Fig. 5.1, concentrating on the $4d_{5/2}$ threshold region with decay to the $5s^25p^4(^1S_0)kp$ channel only (k stands for processes above, $k = \varepsilon$, and below, $k = n$, the ionization threshold, respectively). As will be demonstrated in the following, this plot depicts well how inner-shell ionization with subsequent Auger decay evolves to outer-shell ionization accompanied by excitation, if inner-shell resonance excitations and PCI are taken into account. The characteristic features of PCI have been listed in Section 4.5.5, and only the three points needed for this discussion will be recapitulated. First, PCI phenomena start when the photoelectron becomes slower than the Auger electron. Their influence increases towards the inner-shell ionization threshold. Second, the main effect of PCI is an energy gain for the Auger electron which is compensated by the corresponding energy loss of the photoelectron. Third, due to the energy gain the shape of an Auger line is shifted and becomes asymmetric, and part of this line can belong to processes in which the photoelectron is captured (shaken-down) in a Rydberg orbital (see Fig. 4.42).

In order to describe the link between inner-shell photoionization with subsequent Auger decay and outer-shell ionization accompanied by excitation, one has to tune the photon energy across the inner-shell threshold, and to observe how the mountain chain of Auger electrons (usually along a straight line parallel to the photon-energy axis) evolves into a specific, possibly resonance-affected pattern of certain outer-shell satellites (usually parallel to the diagonal in the $(h\nu - E_{kin})$-plane). At high photon energies the photoelectron is faster than the Auger electron, and nothing special occurs. This range is still outside the one shown in Figs. 5.1 and 5.2, but one would get a mountain chain parallel to the photon-energy axis, positioned at the nominal Auger energy of $E_A^0 = 29.97$ eV, and for $h\nu = $ const one would get a cross-cut with a Lorentzian profile convoluted with the instrumental resolution function. However, on lowering the photon energy to values at which the photoelectron moves slower than the Auger electron, PCI effects start and become increasingly important as the ionization threshold is approached. The energy gain of the Auger electrons bends the mountain chain away from its original position and makes it broader and asymmetrical. These effects can be seen clearly in Fig. 5.1: at the ionization threshold the centre of the peak is at approximately $E_A = 30.24$ eV and not at $E_A^0 = 29.97$ eV, and there is a larger grey-tone range on its high-energy side than on its low-energy side. As the ionization threshold is approached, the probability for shake-down of the escaping photoelectron becomes more and more important, and in this way the $4d_{5/2}$ ionization process above threshold starts to feed outer-shell satellites, in the present case $5s^2 5p^4 (^1S_0) n\ell$ satellites with high n values. In Fig. 5.1 this effect can be seen in the small grey islands positioned at approximately 67.5 eV photon energy and 31 eV kinetic energy. For photon energies just below the inner-shell ionization threshold it is still justifiable to think within the PCI model [Nie77]. The excitation process with PCI taken into account can be described by the movement of the slow 'ejected' photoelectron which travels towards the classical turning point of the potential curve of the ion. During this motion energy exchange with the faster Auger electron is also possible. However, the 'ejected' photoelectron can move out only up to its classical turning point which implies the boundary condition that only standing wave solutions lead to a population of truly bounded Rydberg levels. (This condition means in a time-dependent description of PCI that the interaction is possible only at specific times.) Hence, only discrete values of the energy exchange occur, and the Auger peak becomes more and more structured. For high Rydberg states their inherent level widths and their small energy separations will smear out these structures, but the structures will become more and more visible if the region of isolated resonances is reached. This can be clearly identified in Fig. 5.1 in the smooth intensity propagation in the transition region at about 67.4 eV photon energy, and in the lower isolated resonances $4d_{5/2} \rightarrow np$ for $n = 9, 8, 7$. In particular, two important features can be seen. First, the formerly smooth mountain chain of Auger transitions merges to give the region where dramatic intensity jumps are caused by the isolated resonances. Second, the intensity is spread out into many channels which have been classified before as belonging to true and modified spectator transitions. In this way the strong inner-shell ionization process with its Auger decay flares up before it disappears completely because of too low photon energies. As has been demonstrated for two particular cases in Fig. 5.5 these features of the dying Auger process find their expression in typical resonance structures in the partial cross sections of outer-shell photoionization accompanied by excitation. For still lower photon energies these cross sections become smooth energy-dependent quantities which can, however, show new interesting energy variations as they approach their own ionization thresholds.

5.2 2p photoionization in atomic magnesium

An experiment is said to be, within a certain theoretical framework, *perfect* or *complete* if from different experimental observations complementary information can be obtained which finally allows the determination of all matrix elements involved and, therefore, all possible observables. This will be illustrated for 2p photoionization in atomic magnesium (ground state electron configuration $1s^2 2s^2 2p^6 3s^2$).

Within the dipole approximation the selection rules given in equs. (2.9) and (2.10) allow the following channels:

$$\left.\begin{array}{l} 2p^{-1}\,{}^2P_{3/2}\,\varepsilon s_{1/2}\; J = 1\; \pi = (-1), \\ 2p^{-1}\,{}^2P_{3/2}\,\varepsilon d_{3/2}\; J = 1\; \pi = (-1), \\ 2p^{-1}\,{}^2P_{3/2}\,\varepsilon d_{5/2}\; J = 1\; \pi = (-1), \end{array}\right\} \tag{5.7a}$$

and

$$\left.\begin{array}{l} 2p^{-1}\,{}^2P_{1/2}\,\varepsilon s_{1/2}\; J = 1\; \pi = (-1), \\ 2p^{-1}\,{}^2P_{1/2}\,\varepsilon d_{3/2}\; J = 1\; \pi = (-1), \end{array}\right\} \tag{5.8a}$$

and attached to each of these one has a certain photoionization matrix element D_γ (see equ. (8.35)):

$$\left.\begin{array}{lll} 2p^{-1}\,{}^2P_{3/2}\,\varepsilon s_{1/2}\; J = 1\; \pi = (-1) & \text{with} & D_-, \\ 2p^{-1}\,{}^2P_{3/2}\,\varepsilon d_{3/2}\; J = 1\; \pi = (-1) & \text{with} & D_0, \\ 2p^{-1}\,{}^2P_{3/2}\,\varepsilon d_{5/2}\; J = 1\; \pi = (-1) & \text{with} & D_+, \end{array}\right\} \tag{5.7b}$$

and

$$\left.\begin{array}{lll} 2p^{-1}\,{}^2P_{1/2}\,\varepsilon s_{1/2}\; J = 1\; \pi = (-1) & \text{with} & \tilde{D}_0, \\ 2p^{-1}\,{}^2P_{1/2}\,\varepsilon d_{3/2}\; J = 1\; \pi = (-1) & \text{with} & \tilde{D}_+. \end{array}\right\} \tag{5.8b}$$

A convenient shorthand notation describes these channels in a single-particle nomenclature

$$\left.\begin{array}{lll} 2p_{3/2}\,\varepsilon s_{1/2} & \text{channel with} & D_-, \\ 2p_{3/2}\,\varepsilon d_{3/2} & \text{channel with} & D_0, \\ 2p_{3/2}\,\varepsilon d_{5/2} & \text{channel with} & D_+, \end{array}\right\} \tag{5.7c}$$

and

$$\left.\begin{array}{lll} 2p_{1/2}\,\varepsilon s_{1/2} & \text{channel with} & \tilde{D}_0, \\ 2p_{1/2}\,\varepsilon d_{3/2} & \text{channel with} & \tilde{D}_+. \end{array}\right\} \tag{5.8c}$$

Even though these channels have been formulated within the single-particle model and within the *jjJ*-coupling scheme, the attached dipole matrix elements are very general (see Section 8.2.2). Hence, these matrix elements D_γ can be understood to include all possible kinds of many-electron interactions as well as intermediate coupling, and in this most general way the three dipole matrix elements will be three complex quantities, or alternatively six real parameters. Since, according to the principles of quantum mechanics, an absolute phase is not relevant for the observables, only five real parameters will determine the photoionization process; the squares of three absolute values of the matrix elements, and two relative phases. Therefore, five complementary measurements are sufficient for the determination of these relevant quantities, a convenient choice being the partial cross section $\sigma(2p_{3/2})$, the angular distribution parameter $\beta(2p_{3/2})$, and the three components of the spin polarization vector of the photoelectron. In certain cases, however, only two matrix elements contribute, such as in the 2p photoionization in magnesium where spin–orbit effects are small and Russell–Saunders coupling applies. The two photoionization channels of relevance are then given by

$$2p \rightarrow \varepsilon s \quad \text{with} \quad D_s, \tag{5.9a}$$

$$2p \rightarrow \varepsilon d \quad \text{with} \quad D_d, \tag{5.9b}$$

and both are involved in the fine-structure resolved transitions

$$1s^2 2s^2 2p^6 3s^2\ {}^1S^e \rightarrow 1s^2 2s^2 2p^5 3s^2\ {}^2P^o_{3/2} + e, \tag{5.10a}$$

$$1s^2 2s^2 2p^6 3s^2\ {}^1S^e \rightarrow 1s^2 2s^2 2p^5 3s^2\ {}^2P^o_{1/2} + e'. \tag{5.10b}$$

The example of 2p photoionization in magnesium has been selected for a more detailed discussion because a complete experiment can be performed without having to measure the photoelectron's spin, which is difficult, and because a rather transparent analysis can be given for the matrix elements by comparing theoretical data of different approximations with the experimental results.

5.2.1 Observables for 2p photoionization

The special role of 2p photoionization in magnesium originates in the simplicity of the system which, however, is still complicated enough to exhibit typical electron–electron interactions. The main advantages are the following:

(i) The system can be well described within the *LSJ*-coupling scheme, and for photon energies not too far away from the 2p ionization energy (57.63 eV) the dipole approximation applies. Therefore, the two complex dipole matrix elements D_s and D_d introduced above completely describe the photoionization process. This result also holds if the fine-structure components of the ionic state after 2p photoionization, $^2P_{1/2}$ and $^2P_{3/2}$, are taken into account, a fact that will be important below. It is sometimes convenient to relate these dipole matrix elements to the radial dipole integrals $R_{\varepsilon\ell,2p}$ introduced in equ. (2.11) for the corresponding case of 1s photoionization. Using the length form for the dipole matrix elements and measuring phases in radians and all other quantities in atomic units, one obtains

$$\left.\begin{aligned} |D_s| &= \sqrt{2}\ R_{\varepsilon s,2p} = \sqrt{2}\int_0^\infty P_{\varepsilon s}(r)\, r P_{2p}(r)\,dr \\ &\text{and} \\ |D_d| &= 2\, R_{\varepsilon d,2p} = 2\int_0^\infty P_{\varepsilon d}(r)\, r P_{2p}(r)\,dr, \end{aligned}\right\} \qquad (5.11a)$$

and the relative phase Δ is given by

$$\Delta = (\sigma_{\varepsilon s} - \sigma_{\varepsilon d}) + (\delta_{\varepsilon s} - \delta_{\varepsilon d}), \qquad (5.11b)$$

where the phases $\sigma_{\varepsilon\ell}$ and $\delta_{\varepsilon\ell}$ result from the influence of the Coulomb and short-range potentials, respectively.

(ii) The ground state of magnesium is $1s^2 2s^2 2p^6 3s^2\ {}^1S_0$, and the closed-shell structure with ${}^1S_0^e$ symmetry considerably simplifies the theoretical treatment.

(iii) The two fine-structure components of the ionic state after 2p photoionization, ${}^2P^o_{1/2}$ and ${}^2P^o_{3/2}$, are 0.278 eV apart which is large compared to the natural level width of $\Gamma \approx 1$ meV of these states. This means that a well-defined intermediate state $1s^2 2s^2 2p^5 3s^2\ {}^2P_J$ exists from which Auger decay can proceed, and this has the consequence that this Auger decay can be treated as a two-step process with negligible disturbance from PCI.

(iv) The two possible Auger transitions are L_2–M_1M_1 and L_3–M_1M_1. In both cases the electron configuration after Auger decay is $1s^2 2s^2 2p^6$ which again simplifies the theoretical treatment by coupling to a ${}^1S_0$ state. In particular, for each Auger transition only one non-radiative matrix element exists. This matrix element determines the Auger yield ω_A and the α_2-parameter. In the present case the Auger yield has been measured to be $\omega_A = 1.00 \pm 0.02$, and the α_2-parameter becomes a simple numerical value given by

$$\alpha_2(L_2\text{–}M_1M_1) = 0 \qquad \text{and} \qquad \alpha_2(L_3\text{–}M_1M_1) = -1. \qquad (5.12)$$

(v) Due to the two-step mechanism, the angular distribution parameter β_A factorizes into the alignment parameter $\mathcal{A}_{20}$ of the photoionized state and the Auger decay parameter α_2 (see equ. (3.29)) which gives the simple result

$$\beta(L_2\text{–}M_1M_1) = 0 \qquad (5.13a)$$

(the alignment parameter $\mathcal{A}_{20}(2p^5 3s^2\ {}^2P^o_{1/2})$ is also zero, cf equ. (3.30c)),

$$\beta(L_3\text{–}M_1M_1) = +2\mathcal{A}_{20}(2p^5 3s^2\ {}^2P^o_{3/2}). \tag{5.13b}$$

Hence, the emission of L_2–M_1M_1 Auger electrons is isotropic, while that of L_3–M_1M_1 Auger electrons has an angular distribution coefficient proportional to the alignment parameter. Therefore, a measurement of $\beta(L_3\text{–}M_1M_1)$ is equivalent to a determination of the alignment of the photoionized state, and this quantity can be included in the list of observables in 2p photoionization of magnesium.

All possible observables related to 2p photoionization in magnesium can be expressed in terms of the dipole matrix elements $|D_s|$ and $|D_d|$ and their relative phase Δ. This will now be demonstrated by listing the corresponding expressions for non-coincident observation of photo- and Auger electrons (for the coincident case see Section 4.6.1.3). If the two fine-structure components ${}^2P_{3/2}$ and ${}^2P_{1/2}$ of the photoionized state are resolved experimentally, the corresponding quantity will be characterized by $2p_{1/2}$ (or L_2) and $2p_{3/2}$ (or L_3), otherwise only by 2p (or $L_{2,3}$). (It should be stressed that the identification of the ${}^2P^o_J$ final ionic state with the ejection of a $2p_j$-electron with $j = J$ is a simplification valid in the single-particle limit. In the present case it is used only as a convenient notation. In general, electron correlations will make the final ionic state, still characterized by ${}^2P^o_J$, much more complicated; see the example of 2p excitations in magnesium [WHK94].) Using atomic units one has (E_{ph} is the photon energy, and α the fine structure constant):

(i) for the partial photoionization cross section:

$$\sigma_{2p_{3/2}} = \frac{2}{3}\frac{4\pi^2}{3}\,\alpha\,E_{ph}\,(|D_s|^2 + |D_d|^2), \tag{5.14a}$$

$$\sigma_{2p_{1/2}} = \frac{1}{3}\frac{4\pi^2}{3}\,\alpha\,E_{ph}\,(|D_s|^2 + |D_d|^2), \tag{5.14b}$$

$$\sigma_{2p} = \sigma_{2p_{3/2}} + \sigma_{2p_{1/2}}; \tag{5.14c}$$

(ii) for the angular distribution parameter of photoelectrons:

$$\beta_{2p_{3/2}} = \frac{|D_d|^2 - \sqrt{8}\,|D_s||D_d|\cos\Delta}{|D_s|^2 + |D_d|^2}, \tag{5.15a}$$

$$\beta_{2p_{1/2}} = \frac{|D_d|^2 - \sqrt{8}\,|D_s||D_d|\cos\Delta}{|D_s|^2 + |D_d|^2}, \tag{5.15b}$$

$$\beta_{2p} = \tfrac{2}{3}\beta_{2p_{3/2}} + \tfrac{1}{3}\beta_{2p_{1/2}} = \beta_{2p_{3/2}} = \beta_{2p_{1/2}}; \tag{5.15c}$$

(iii) for the alignment parameter $\mathcal{A}_{20}({}^2P^o_J)$ of the intermediate photoionized state (quantization axis along the direction of the incident light):

$$\mathcal{A}_{20}(2p^5 3s^2\ {}^2P^o_{3/2}) = \frac{1}{2}\frac{|D_s|^2 + 0.1|D_d|^2}{|D_s|^2 + |D_d|^2}, \tag{5.16a}$$

$$\mathscr{A}_{20}(2p^5 3s^2\ {}^2P^o_{1/2}) = 0, \tag{5.16b}$$

$$\mathscr{A}_{20}(2p^5 3s^2\ {}^2P^o) = \tfrac{2}{3}\,\mathscr{A}_{20}(2p^5 3s^2\ {}^2P^o_{3/2}); \tag{5.16c}$$

(iv) for the three spin-polarization parameters, ξ, η, and ζ (for the definition of these parameters and their relation to the components of the spin-polarization vector see Section 5.4):

$$\xi_{2p_{3/2}} = -\frac{1}{2}\,\frac{-|D_s|^2 + |D_d|^2 + (1/\sqrt{2})|D_s||D_d|\cos\Delta}{|D_s|^2 + |D_d|^2}, \tag{5.17a}$$

$$\xi_{2p_{1/2}} = \frac{-|D_s|^2 + |D_d|^2 + (1/\sqrt{2})|D_s||D_d|\cos\Delta}{|D_s|^2 + |D_d|^2}, \tag{5.17b}$$

$$\xi_{2p} = \tfrac{2}{3}\,\xi_{2p_{3/2}} + \tfrac{1}{3}\,\xi_{2p_{1/2}} = 0, \tag{5.17c}$$

$$\eta_{2p_{3/2}} = \frac{3}{2\sqrt{2}}\,\frac{|D_s||D_d|\sin\Delta}{|D_s|^2 + |D_d|^2}, \tag{5.18a}$$

$$\eta_{2p_{1/2}} = -\frac{3}{\sqrt{2}}\,\frac{|D_s||D_d|\sin\Delta}{|D_s|^2 + |D_d|^2}, \tag{5.18b}$$

$$\eta_{2p} = \tfrac{2}{3}\,\eta_{2p_{3/2}} + \tfrac{1}{3}\,\eta_{2p_{1/2}} = 0, \tag{5.18c}$$

$$\zeta_{2p_{3/2}} = -\frac{1}{2}\,\frac{|D_s|^2 + 0.5|D_d|^2 + \sqrt{2}\,|D_s||D_d|\cos\Delta}{|D_s|^2 + |D_d|^2}, \tag{5.19a}$$

$$\zeta_{2p_{1/2}} = \frac{|D_s|^2 + 0.5|D_d|^2 + \sqrt{2}\,|D_s||D_d|\cos\Delta}{|D_s|^2 + |D_d|^2}, \tag{5.19b}$$

$$\zeta_{2p} = \tfrac{2}{3}\,\zeta_{2p_{3/2}} + \tfrac{1}{3}\,\zeta_{2p_{1/2}} = 0. \tag{5.19c}$$

These equations show that several distinct experiments can be selected to determine the desired values $|D_s|$, $|D_d|$, and Δ. The standard procedure has been to measure the partial cross section, the angular distribution parameter of photoelectrons and one suitable spin-polarization component of the ejected photoelectrons. The low efficiency of Mott detectors used for the determination of spin-polarization makes such a measurement tedious and difficult. In addition, to observe a non-vanishing spin-polarization parameter, the experimental resolution must be good enough to resolve the two fine-structure components in the photoelectron spectrum, because otherwise the observed spin-polarization parameters vanish. (Compare equs. (5.17c), (5.18c), and (5.19c). This result reflects the fact that the spin parameters are due to spin–orbit effects.) Often such a good resolution is difficult to achieve for photoelectrons because the bandpass of the monochromatized light contributes to their actual linewidth. However, Auger electrons are not burdened by this problem, and even for insufficient resolution of the non-isotropic L_3–M_1M_1 and the isotropic L_2–M_1M_1 Auger lines, the weighted angular distribution parameter still refers to the alignment parameter (see equ. (5.13)). Therefore, it will be advantageous to select in the present case the following observables: the partial photoionization cross section σ_{2p}, the angular distribution parameter β_{2p} of photoelectrons, and the angular distribution parameter $\beta(L_3\text{–}M_1M_1)$ of the non-isotropic $L_3\text{–}M_1M_1$ Auger electrons. A general comment has to be added for such a selection about the relative phase Δ which appears only in β_{2p}, as $\cos \Delta$. For an unambiguous determination of Δ one needs both the cosine and the sine in order to decide the quadrant in which the Δ value lies. However, it is possible to solve this problem within a given experimental data set if the energy dependence of Δ is known and the fact that the main part of the phase difference comes from the analytically formulated Coulomb phases is taken into account. Since the experiment described below is performed at only one photon energy; this approach cannot be applied, and theory is used as a guide to the sign of Δ.

5.2.2 Experimental details and results

In order to measure the selected observables σ_{2p}, β_{2p}, and $\beta(L_3\text{–}M_1M_1)$ for atomic magnesium, electron spectrometry of magnesium vapour has to be performed. Since it is more complicated to work with metal vapours than with gases, some details about the evaporation of magnesium will be given (for a general discussion of metal atomic beam sources see [RSo95]). The oven required is shown in Fig. 5.6. Its chamber ('O' in the figure) is heated resistively by the quadro-filar wire ('H') surrounding it, so that no magnetic field from the heating process disturbs the path of the ejected electrons. In order to reduce the heat transfer to the environment, the oven chamber is fixed by a few rubin balls to a double layer of heat shielding (also separated by rubin balls; these are not shown in the figure), and the whole system is covered by a water-cooled jacket ('C'). The metal vapour leaves the oven chamber as a divergent beam. A collimated atomic beam is produced geometrically by means of the walls in the oven chamber† and two diaphragms. One diaphragm ('D1') is made from molybdenum and is mounted on the oven chamber which heats it, the other ('D2') is made from copper and is mounted on the cooling jacket. In the middle ('S') of the source region seen by the electron spectrometer the vapour beam crosses the photon beam (perpendicular to the plane of the drawing). The vapour is then stopped by a plate ('B') which is maintained at liquid-nitrogen temperature in order to reduce the rebounding of the magnesium atoms. Electrons created in the source volume can enter either of two analysers, one fixed in space ('mon.'), the other rotatable around the photon beam, with the two extreme positions of the rotatable analyser shown ('pos.1' and 'pos.2'). It can be seen that instead of the 90° range necessary to exhaust the full intensity variation in the angle-dependent intensity, only a limited range of 65° is accessible which, however, suffices to extract β-parameters with high accuracy. The fixed analyser is used to measure, for the successive angle-settings of the

† Note in Fig. 5.6 a special cylinder inserted into the oven chamber on top of the magnesium rod.

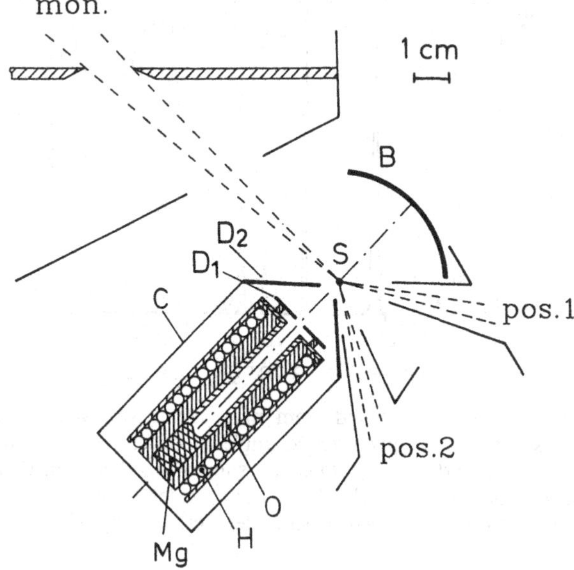

Figure 5.6 Schematic drawing of the oven used for the evaporation of magnesium (Mg) and the essential apparatus for angle-resolved electron spectrometry around the oven. The source volume which also provides a centre of reference is marked by the symbol S. The oven itself is shown in the third quadrant; for the symbols characterizing the oven see main text. The direct vapour beam is stopped by the baffle B (placed in the first quadrant). The dashed curves starting from S indicate the angular ranges in which the emitted electrons are accepted by the monitor analyser (second quadrant, 'mon.', fixed in space), and by the two extreme positions of a rotatable sector CMA (fourth quadrant, 'pos. 1', 'pos. 2', respectively). Reprinted from *J. Electron Spectrosc. Relat. Phenom.* **67**, Kämmerling *et al.*, 363 (1994) with kind permission of Elsevier Science – NL, Sara Burgerhartstraat 25, 1055 KV Amsterdam, The Netherlands.

rotatable analyser, a reference signal which monitors the actual target pressure† and the incident light flux. Typical parameters for the performance of this vapour source during the magnesium experiment are 30 W heating power, 750 K oven temperature and a target density of more than 10^9 atoms/cm^3. Due to the closed-shell structure of magnesium and because of the low pressure gradient in the production of the divergent beam, the magnesium vapour contains only ground-state atoms and no molecules or clusters. Depending on the element and the experimental set-up used for producing the vapour, such favourable conditions will not, in general, be met.

In Fig. 5.7 the spectrum of ejected electrons obtained at 80 eV photon energy is shown. One can see the 3s and 2p main photolines (2s ionization is not possible here because the 2s binding energy (96.5 eV) is larger than the photon energy

† This information is extremely important for quantitative measurements with metal vapours, because even for constant oven temperature the target density will not necessarily be constant due to the gradual closing of the diaphragms ('D1','D2') due to vapour deposition.

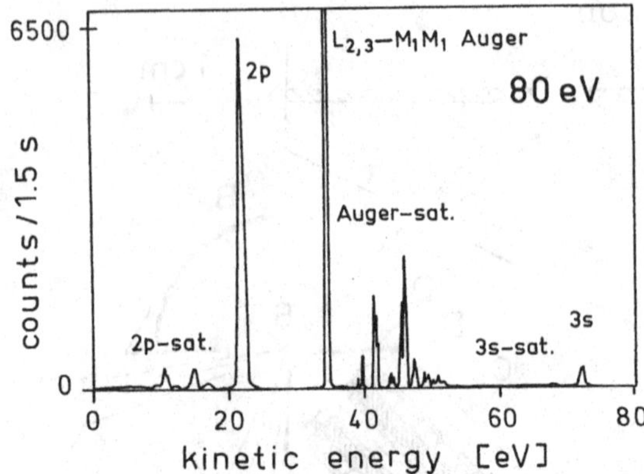

Figure 5.7 Spectrum of electrons ejected from magnesium atoms after interaction with 80 eV photons (measured at the quasi-magic angle in order to allow the extraction of relative intensities). The 3s and 2p photolines are shown together with their satellites and the radiationless transitions following the 2p main and satellite processes, i.e., $L_{2,3}$–M_1M_1 normal Auger transitions and Auger satellites, respectively. From [HKK88].

used), 2p photosatellites with a high intensity (they belong to electron configurations $2p^53sn\ell$ and $2p^43s^2n\ell$), normal Auger lines $L_{2,3}$–M_1M_1 following 2p ionization, and Auger satellites following the decay of 2p photosatellites. Due to the simple structure of magnesium, both kinds of Auger transitions are the mirror images of the 2p main and satellite photolines. However, these Auger lines are much narrower than their photolines because they are not affected by the bandpass of the monochromatized light. The Auger lines have larger peak heights than the corresponding photolines for two reasons. First, only the peak areas contain information on intensities, and a sharper line must be higher than a broader line in order to have comparable intensity (see equ. (2.42)). Second, the spectrum shown in Fig. 5.7 has not been corrected for the energy dispersion of the electron spectrometer.

After interpreting all the features observed in the spectrum of ejected electrons, one can concentrate on the photolines separately. From the dispersion-corrected areas, and taking into account a smooth decrease of the analyser transmission and detection efficiency towards lower kinetic energies (see Fig. 4.30), one obtains at 80 eV photon energy the following ratios of partial cross sections:

$$\sigma_{3s}/\sigma_{2p} = 0.018 \pm 0.002, \tag{5.20a}$$

$$\sigma_{3s}(\text{sat})/\sigma_{2p} \approx 0.003 \pm 0.001, \tag{5.20b}$$

$$\sigma_{2p}(\text{discr. sat})/\sigma_{2p} = 0.285 \pm 0.006. \tag{5.20c}$$

The contribution of continuous 2p satellites where two electrons share the available excess energy is difficult to assess, and an approximate value only can be given:

$$\sigma_{2p}(\text{cont. sat})/\sigma_{2p} \approx 0.04 \pm 0.02. \tag{5.20d}$$

The large ratio $\sigma_{2p}(\text{discr. sat})/\sigma_{2p}$ quantifies the remarkable intensity of 2p satellites and is a direct indicator for strong electron correlation effects in 2p photoionization.

For a complete investigation of 2p photoionization, the absolute cross section σ_{2p} is also required. For metal vapours a direct determination of this cross section fails because of the difficulties in determining the vapour density in the source region, and only the total absorption cross section σ_{tot} of solid magnesium can be measured. However, agreement between the total cross sections for solid and gaseous samples has been found in some studies provided the photon energies are larger than about 50 eV and not near thresholds or resonances [HLT82]. Assuming that this result also holds for photoionization in magnesium at 80 eV, one can use the value

$$\sigma_{tot} = 6.0 \pm 0.5 \text{ Mb} \tag{5.21}$$

for solid magnesium [HLT82], where the large uncertainty is due to the atom-solid problem. With this information the relative intensities of photolines from the electron spectrum discussed above can be placed on an absolute scale by using the relation (at 80 eV)

$$\begin{aligned} \sigma_{tot} &= \sigma_{3s} + \sigma_{3s}(\text{sat}) + \sigma_{2p} + \sigma_{2p}(\text{discr. sat}) + \sigma_{2p}(\text{cont. sat}) \\ &= \sigma_{2p}\left[\frac{\sigma_{3s}}{\sigma_{2p}} + \frac{\sigma_{3s}(\text{sat})}{\sigma_{2p}} + 1 + \frac{\sigma_{2p}(\text{discr. sat})}{\sigma_{2p}} + \frac{\sigma_{2p}(\text{cont. sat})}{\sigma_{2p}}\right]. \end{aligned} \tag{5.22}$$

This then gives

$$\sigma_{3s} = 0.080 \pm 0.011 \text{ Mb}, \tag{5.23a}$$

$$\sigma_{3s}(\text{sat}) = 0.014 \pm 0.005 \text{ Mb}, \tag{5.23b}$$

$$\sigma_{2p} = 4.46 \pm 0.40 \text{ Mb}, \tag{5.23c}$$

$$\sigma_{2p}(\text{sat}) = 1.45 \pm 0.16 \text{ Mb}. \tag{5.23d}$$

Within experimental error the experimental values of $[\sigma_{3s} + \sigma_{3s}(\text{sat})] = 0.094 \pm 0.012$ Mb and $[\sigma_{2p} + \sigma_{2p}(\text{sat})] = 5.91 \pm 0.43$ Mb agree with the results from a calculation using the many-body perturbation theory, MBPT, which gives $\sigma_{3s}(\text{MBPT}) = 0.093$ Mb and $\sigma_{2p}(\text{MBPT}) = 6.25$ Mb [Alt89] and $[\sigma_{2p} + \sigma_{2p}(\text{sat})]$ agrees with the prediction from a calculation using the relativistic random-phase approximation, RRPA, which gives $\sigma_{2p}(\text{RRPA}) = 6.37$ Mb. This result turns out to be rather general, i.e., such calculations lead to partial cross sections $\sigma(n\ell)$ which are roughly equal to the sum of the cross sections of the main lines and their satellites. This will be taken up below when numerical values for the dipole matrix elements involved are discussed.

In Figs. 5.8 and 5.9 the angular distribution parameters β_{2p}, $\beta(L_3\text{–}M_1M_1)$ and $\beta(L_2\text{–}M_1M_1)$ are shown as a function of the photon energy. Within experimental error, $\beta(L_2\text{–}M_1M_1)$ is zero as predicted by theory, and this underlines the quality of the experimental data. The angular distribution of photoelectrons cannot be measured closer to the 2p ionization threshold than approximately 2.4 eV, because

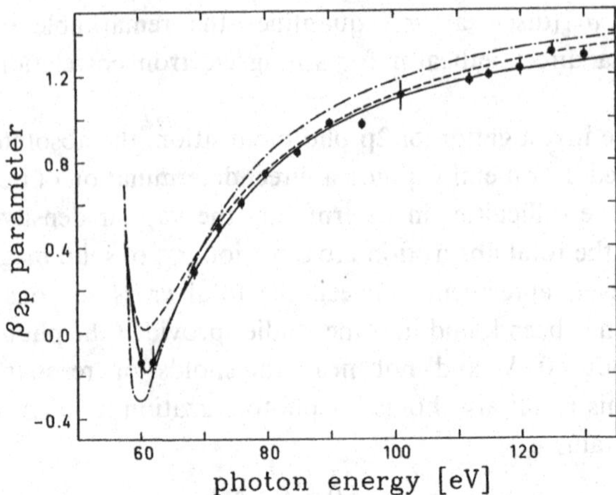

Figure 5.8 Angular distribution parameter β_{2p} for 2p photoionization in atomic magnesium as a function of photon energy. Experimental data: points with error bars. Theoretical data (adapted to the experimental threshold for 2p ionization; 2s → np resonances between 94 and 97 eV photon energy omitted): full curve, RRPA results [DMa83]; within uncertainties of the drawing the same result is obtained in the MBPT approach [Alt89]; broken curve, HF(^{1}P) results [Völ88]; chain curve, Herman–Skillman results [DSa73]. From [KHL92].

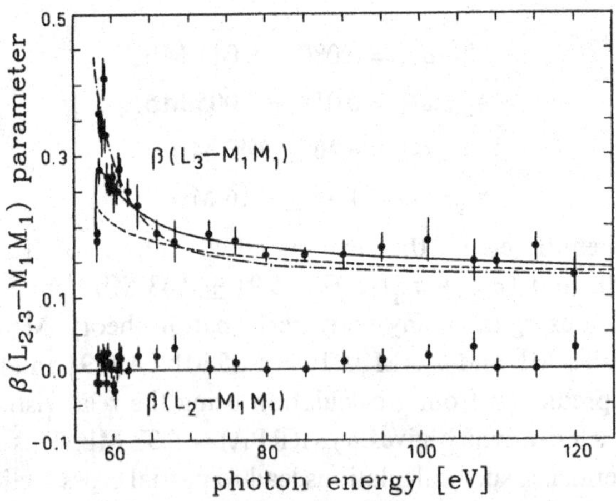

Figure 5.9 Angular distribution parameters for L_2–M_1M_1 and L_3–M_1M_1 photon-induced Auger lines in atomic magnesum (these two Auger lines are well resolved in the experiment, and individual β_A-parameters can be given; in Fig. 5.7 these individual Auger lines are not visible because of the compressed energy scale). Experimental data: points with error bars. Theoretical data for $\beta(L_2–M_1M_1)$: zero (full straight line); for $\beta(L_3–M_1M_1)$, adapted to the experimental threshold for 2p ionization: full curve, RRPA results [DMa83]; within the uncertainties of the drawing the same result is obtained in the MBPT approach [Alt89]; broken curve, HF(^{1}P) results [Völ88]; chain curve, Herman–Skillman results [DSa73]. From [KHL92].

vapour deposits cause changing contact potentials which severely affect low-energy electrons (see Fig. 4.41). This is not a serious problem for the Auger electrons, because they have higher kinetic energies, $E_{kin}(L_2-M_1M_1) = 35.15$ eV and $E_{kin}(L_3-M_1M_1) = 34.87$ eV. However, as the photoionization threshold is approached one can note rather pronounced structures in the $\beta(L_3-M_1M_1)$ parameter at specific photon energies. These are attributed to disturbances by doubly excited states with electron configurations $2p^53s3p^2$ which influence the 2p photoionization and its subsequent Auger decay. Keeping in mind the problems in the threshold region, one can compare the experimental β_{2p} and $\beta(L_3-M_1M_1)$ data with theoretical predictions based on several approximations (full, broken and chain curves in Figs. 5.8 and 5.9). Over the whole range of photon energies accessible in this experiment, one finds excellent agreement between the experimental data and the predictions from the MBPT and RRPA approaches (full curves; for the other curves see below). (The special values at 80 eV photon energy needed in Section 4.6.1.3 are $\beta_{2p} = 0.74 \pm 0.02$ and $\beta(L_3-M_1M_1) = 0.16 \pm 0.01$.)

5.2.3 Comparison of experimental matrix elements with theoretical results from mean-field theories

The statement that there is good agreement between theoretical and experimental data is conventionally the conclusion drawn from the results presented. However, a deeper insight can be obtained when the dipole matrix elements involved are determined experimentally, and compared with results from different theoretical approximations of the influences of electron–electron interactions. In order to make such a comparison, the experimental matrix elements will be derived first for the case of 2p photoionization in magnesium at 80 eV photon energy. From equs. (5.14), (5.15), and (5.16) one can calculate $|D_s|$, $|D_d|$ and Δ from the measured observables. The results are listed in Table 5.1, together with data from several theoretical calculations.

The calculation labelled HS in Table 5.1 is based on the model of uncorrelated motion of the electrons (the *independent-particle* or *single-particle picture*). HS stands for Herman–Skillman and indicates that the radial parts $P_{\varepsilon s}(r)$ and $P_{\varepsilon d}(r)$ of the photoelectron wavefunction have been calculated in the atomic potential (Herman–Skillman potential; [HSk63]), modified at large distances for the $1/r$ dependence of the ion [Lat55]. For the corresponding single-particle Schrödinger equation see equ. (7.68) with $V_{HF}(r_i)$ replaced by the modified Herman–Skillman potential. In diagrammatical language, the single-particle photoionization process is represented in Fig. 5.10(a): the wavy line describes the incident photon, and it is annihilated when interacting with an atomic electron leading to a 2p-hole (arrow pointing downwards) and an $\varepsilon\ell$-electron (arrow pointing upwards).

Results from a HF(^{1}P) approach are also shown in Table 5.1. Here HF stands for Hartree–Fock, and the label ^{1}P indicates that the $P_{\varepsilon s}(r)$ and $P_{\varepsilon d}(r)$ functions are calculated in a *state-dependent* Hartree–Fock potential which takes into

Table 5.1. *Dipole matrix elements, absolute values $|D_s|$ and $|D_d|$, radial integrals R_s and R_d, and relative phase Δ for 2p photoionization in magnesium at 80 eV photon energy. From [HKK88, KHL92].*

	HS [DSa73]	HF(^{1}P) [Völ88]	RRPA [DMa83]	MBPT [Alt89]	Exp.
$|D_s|$/au	0.214	0.218	0.254	0.247	0.194(17)
R_s/au	0.151	0.154	0.180	0.175	0.137(12)
$|D_d|$/au	0.907	0.846	0.861	0.854	0.726(35)
R_d/au	0.454	0.423	0.431	0.427	0.363(18)
Δ/rad	4.92	4.82	5.03	5.02	4.99(16)[(a)]

[(a)] Because $\cos(2\pi \pm \Delta) = \cos \Delta$ there are two solutions, $\Delta_{exp} = 1.29(4)$ rad and $\Delta_{exp} = 4.99(16)$ rad; this ambiguity has been resolved by calculating for both phases the spin polarization parameters $\eta_{2p_{3/2}}$ and $\eta_{2p_{1/2}}$, which contain $\sin \Delta$, and comparing the signs of these results with theoretical data [DMa83]. The phase difference Δ contains two contributions, the Coulomb phase difference, $\sigma(\varepsilon s) - \sigma(\varepsilon d)$, and the phase difference caused by the short-range potential, $\delta(\varepsilon s) - \delta(\varepsilon d)$; $\sigma(\varepsilon s) - \sigma(\varepsilon d)$ can be calculated analytically, giving here 1.03 rad. For details see text.

account the $^1P^o$ symmetry of the complete final state (ion plus photoelectron). This method is expected to give better results than the HS approach, because *intra*channel electron–electron interactions are incorporated automatically [DVi66, ACC71]. Such intrachannel interactions are depicted diagrammatically in their lowest order in Fig. 5.10(*e*). The initial process of photon interaction creates a 2p hole and an electron in a k-orbital (cut the loop in the middle and compare the lower half of this diagram with Fig. 5.10(*a*)). During its escape, the 'outgoing' k-electron 'is scattered' via Coulomb interaction, indicated in the figure by a dashed line, by another electron from the 2p shell such that this k-electron falls back into a 2p-orbital (this closes the loop), while the other 2p-electron is ejected, thus creating a 2p-hole (arrow pointing downwards) and an $\varepsilon\ell$-electron (arrow pointing upwards). If such diagrams are summed to infinite order, they describe the dynamical effect of the dipole interaction between the dipole moment of the photoelectron–ion system and the remaining electrons of the photoion. Due to the spherical symmetry of the εs partial wave, there is no change in the corresponding $|D_s|$-amplitude, but a modification of $|D_d|$ is expected as a result of such intrachannel interactions. This behaviour is reflected in the corresponding matrix elements given in Table 5.1: comparing the HF(^{1}P) results with those from the HS calculation, one sees that $|D_s|$ remains practically the same, but $|D_d|$ is reduced.

Next the results from the relativistic random-phase approximation (RRPA) and the many-body perturbation theory (MBPT), also shown in Table 5.1, will be discussed. Because both calculations include basically the same electron–electron interactions, rather good agreement exists, and it is sufficient to concentrate only on the RRPA model.

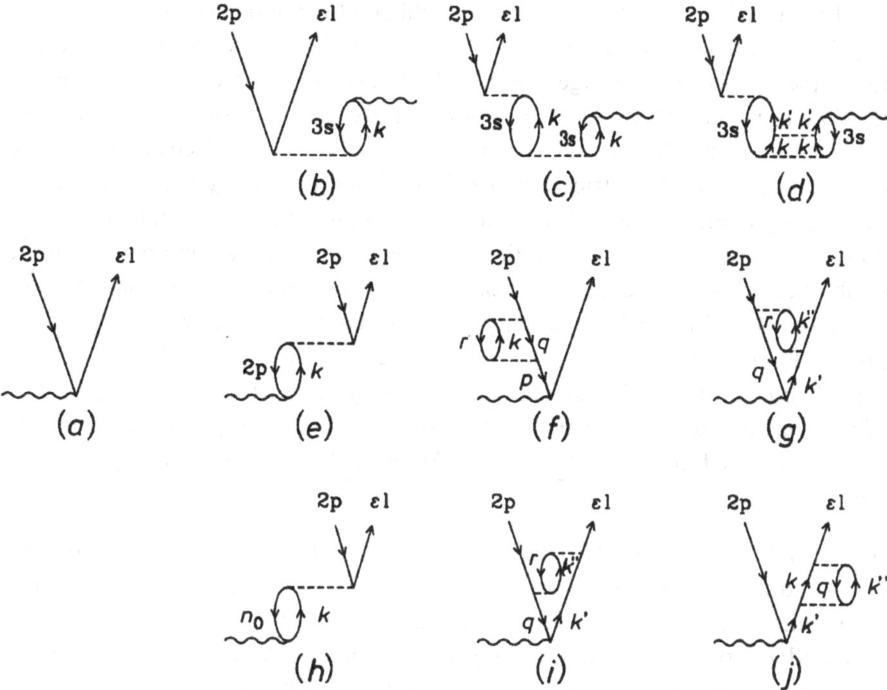

Figure 5.10 Some typical low-order diagrams which are relevant for 2p photoionization in magnesium; exchange counterparts ought to be included. The wavy lines represent the dipole–photon interaction, the broken lines the Coulomb interaction (the number of broken lines gives the order of the corresponding diagram). The diagrams are read from bottom to top with respect to the time axis. In this way, electron–hole pairs can be plotted conveniently: electrons propagate forwards (arrows up), holes backwards (arrows down) in time; if a particle is created at the point P of interaction, its bra-vector $\langle k|$ in the matrix element is shown as a line originating from P; if the particle is destroyed at P, its ket-vector $|k\rangle$ is shown as a line going to P; if a hole is created at P, its bra-vector $\langle i|$ is represented as a line coming from P; if the hole is destroyed at P, its ket-vector $|i\rangle$ is represented as a line going to P. The symbols 2p and $\varepsilon\ell$ characterize the $2p \to \varepsilon\ell$ photoionization process under consideration. The other symbols stand for other discrete/continuous orbitals out of a given set of single-particle orbitals. For a detailed description of the individual diagrams see main text and [Wen81, Wen84].

The relativistic or non-relativistic random-phase approximation (RRPA or RPA)† is a generalized self-consistent field procedure which may be derived making the Dirac/Hartree–Fock equations time-dependent. Therefore, the approach is often called *time-dependent* Dirac/Hartree–Fock. The name *random phase* comes from the original application of this method to very large systems where it was argued that terms due to interactions between many alternative pairs of excited particles, so-called *two-particle–two-hole* interactions ((2p–2h); see below) tend to

† It is often called RPA with exchange (RPAE) in order to make it explicit that exchange effects are included. This, however, will be understood always to be the case here.

cancel because their coefficients have quite different ('random') phases (for details see [CFa76]). A time-dependent Dirac/Hartree–Fock procedure means that not only is the effect of the average *static* field on each single-particle state incorporated, like in normal Dirac/Hartree–Fock, but also the necessary rearrangements of these states introduced by the movement of the atomic electrons under the action of the *external* electric field. This additional oscillatory field of the other electrons represents an *induced* field which screens the external field, and it contains an average from the correlated motion of the electrons. This selection from all the possibilities for correlated motion is made via expansions in one-particle–one-hole excitations (1p–1h) which are usually depicted diagrammatically. (Note that the Hartree–Fock approach is stable to 1p–1h excitations (the *Brillouin* theorem [Bri32]), but not to 1p–1h excitations induced by an external field.)

For a detailed description of the RPA method in atomic photoionization the reader is referred to [ACh75, Wen84, Amu90], but the following points are mentioned here:

(i) The solutions of the RPA equations have to be found by a self-consistent procedure. Hence, the approach is an *ab initio* method without adjusted parameters and is the analogue of the static case of Hartree–Fock calculations.

(ii) Based on the diagrammatical representation of the RPA equations, well-defined diagrams of 1p–1h excitations are selected and taken into account up to infinite order. Depending on the Coulomb interaction between the electrons, *before* or *after* the photon interaction, these diagrams then describe the average effect of electron–electron interactions in the *initial* or *final* state (often identified with initial-state configuration interaction (ISCI) and final-state configuration interaction (FSCI)).

(*Many-body perturbation theory* (MBPT, see [Kel64, Kel85] and further references in [Sch92a]) treats such correlation effects in a perturbation expansion up to a certain order, i.e., many of these diagrams appear there too. Provided that this expansion converges rapidly, it has the advantage that processes beyond the conventional RPA approach as well as the extension to open-shell systems can be straightforwardly incorporated in MBPT.)

(iii) Due to the self-consistent procedure, interference effects are incorporated. This concerns not only interference between RPA diagrams, but also between one-electron ionization channels.

(iv) Calculations of the dipole matrix elements in the length or velocity form (see Section 8.1.3) give identical results for the observables of photoionization provided that the basis functions are selected properly (Dirac–Fock functions for RRPA, Hartree–Fock functions for RPA; exchange effects are always included).

Following this discussion of the RPA method the results for 2p photoionization in magnesium can be interpreted. From the foregoing discussion one expects that the RRPA method should give better results than the HF (^{1}P) approach, because it accounts for many essential electron–electron interactions (and for relativistic effects, but these are unimportant in the present example). The typical lowest-order RPA diagrams are shown in Fig. 5.10(b), (e) and (h). It can be seen that the intrachannel interactions of Fig. 5.10(e) are again included, as well as certain kinds of *inter*channel interactions in the continuum (Fig. 5.10(h)), and electron–electron interactions in the initial state (Fig. 5.10(b)). In addition, in the RPA approach the εs and εd partial waves are not treated separately in Fig. 5.10(a), and therefore the interference between these continuum channels is taken into account. The meaning of Fig. 5.10(b) will be discussed in the next paragraph, but that of Fig. 5.10(h) will be explained here. In Fig. 5.10(h) the photon interaction occurs with an electron n_0 of the basis set which is not a 2p-electron. In spite of this the process is a 2p photoionization provided that the Coulomb interaction finally scatters the (n_0–k)-hole–electron pair to the (2p–$\varepsilon\ell$) pair under consideration. It is natural to consider for these n_0-electrons the closest neighbours of 2p ionization, i.e., electrons from the 3s and 2s shells. This gives two diagrams of the type shown in Fig. 5.10(h) where $n_0 \rightarrow k$ stands for 2s $\rightarrow k$p and 3s $\rightarrow k$p, respectively. The first describes 2s photoexcitation/ionization with subsequent Auger decay, because the Coulomb interaction destroys the 2s-hole–k-electron excitation and produces instead a 2p-hole–$\varepsilon\ell$-electron excitation. The second can be interpreted in the same way, even though for energy reasons at first glance it is difficult to imagine an Auger decay filling an *outer*-shell hole. However, for such diagrams energy conservation is not necessarily required at the point of an intermediate Coulomb interaction. If energy conservation is violated, as is the case for Fig. 5.10(h) with n_0 = 3s, the excitations are called *virtual* (in this case a *virtual Auger 'transition'*) and the diagram describes a particular effect of electron–electron interactions in the final state. If energy conservation is fulfilled, as is the case for Fig. 5.10(h) with n_0 = 2s and a sufficiently high photon energy, a real Auger transition takes place which can be observed experimentally.

In 2p photoionization in magnesium the partial cross section σ_{2p} is large compared to σ_{3s} and σ_{2s}. Therefore, the effect of the 3s and 2s interchannel interactions on the 2p photoionization channel will be small. (In contrast, the effect of 2p photoionization on the 3s and 2s channels can be expected to be significant.) More important here is the continuum mixing of Fig. 5.10(a) with εs and εd partial waves. Since $|D_d|$ is larger than $|D_s|$, it is essentially the $|D_d|$-amplitude which modifies the $|D_s|$-amplitude. This can be verified from Table 5.1 by comparing the RRPA results with those of the HF(^{1}P) calculation: $|D_s|$ is increased while $|D_d|$ remains nearly the same.

Summarizing the foregoing discussion for the dipole matrix elements of 2p photoionization in magnesium, the following conclusions can be drawn. The relative phase Δ is not very sensitive to the different theoretical models. It agrees sufficiently well with the experimental value. In contrast, the magnitudes of the dipole matrix elements show significant differences depending on the actual treatment of electron–electron interactions. When compared to the experimental value, even the most sophisticated theoretical models (RRPA and MBPT) give values which are too high. This reflects that in both calculations the treatment of electron correlation with a mean-field approach is not sufficient to describe the experimental data correctly.

5.2.4 Inclusion of ISCI and FISCI correlations

The time-dependent mean field in the RPA treatment of photoionization is common to all electrons. Therefore, a coherent motion of the electrons is possible, and this is self-evidently a strong many-particle effect. (It can give rise to *giant* resonances, also called *shape* resonances which have been the subject of considerable interest over the last decades (see [CEK87, Sch92a] and references therein).) However, electron correlations, which, in their strict definition, impose a certain structure on the spatial part of the many-electron wavefunction (angular correlation, radial correlation or both; see Fig. 1.2) are not taken into account in conventional RPA. In order to include such electron correlations in a diagrammatical language, alternative *pairs* of *excited* particles must scatter one another (2p–2h interactions). The difference between 1p–1h and 2p–2h interactions in the initial state can be explained with the help of Fig. 5.10(*c*) and (*d*). Figure 5.10(*c*) is one of the second-order RPA diagrams for 2p photoionization in magnesium. Starting from the bottom, one can see that the Coulomb interaction causes *virtual double excitations* of the 3s-electrons to k orbitals (no energy conservation is required, see above). The photon interaction then occurs with one of these k orbitals which is deexcited to the 3s orbital, closing the right loop. The final state of 2p ionization is then reached via another Coulomb interaction in which the other k orbital is deexcited to 3s, closing the left loop, and a 2p–$\varepsilon\ell$ hole–electron pair is created. Due to the mean-field character of RPA the virtual double excitations in the initial step describe displacements of charge densities out of the manifold of alternative two-electron excitations. In order to include the *correlated* motion of such virtual excitations, one needs a scattering (Coulomb interaction) between at least two excited orbitals. One such 2p–2h diagram relevant for 2p photoionization in magnesium is shown in Fig. 5.10(*d*). There the correlated motion can be seen clearly between those parts of the loops where the Coulomb interaction intervenes twice and modifies k^2 to k'^2. This correlated motion of two electrons in space induced by 2p–2h excitations is described conveniently by the initial state configuration interaction (ISCI).

As well as such electron correlations in the initial state, general *relaxation* processes in the final state of the system are also beyond conventional RRPA. (Relaxation effects are included in the *generalized* random-phase approximation (GRPA); see [Amu80].) The most important relaxation diagrams are also shown in Fig. 5.10 in their lowest order. Figure 5.10(*f*) describes the *self-energy* correction term which is responsible for the correct binding energy of the photoionized electron. For 2p photoionization in magnesium monopole relaxation ($q = p = 2p$)

is sufficient, and then the net effect of this diagram taken in all orders can be approximated if the theoretical binding energy is replaced by the experimental value. This approximation implies a shift of the theoretically predicted observables as a whole. (Care has to be taken if the subsequent decay process is important, as, for example, in 4p photoionization in xenon [WOh76].) Figures 5.10(g) and (i) represent the effect of *relaxation* and *polarization* on the electron–hole interaction. For $q = 2p$ the process describes monopole relaxation, and both diagrams with their higher-order contributions can be approximated by using relaxed orbitals in the final-state wavefunction, where for the calculation of these relaxed orbitals configuration interactions in the final-ionic state (FISCI) have to be taken into account. (In this context compare the discussion of shake theory (equ. (5.6)) which also relies on relaxed orbitals.) Finally, Fig. 5.10(j) is called a *polarization* or *photoelectron's self-energy* diagram, because the outgoing photoelectron excites the core and interacts with this excitation, i.e., it represents a self-induced polarization potential seen by the photoelectron. Because this diagram directly affects the wavefunction of the escaping photoelectron, no approximation seems to exist which avoids a direct calculation of the diagram. Therefore, when 2p photoionization in magnesium is discussed with regard only to ISCI and FISCI, the influence of the photoelectron's self-energy diagram is not included.

After this necessary excursion into the broad field of electron–electron interactions as described in diagrammatic language and by certain approximations, the effects of electron correlations beyond the RRPA approach can be summarized by using the following rules (neglecting the photoelectron's self-energy diagram): ISCI and FISCI must be taken explicitly into account, and the theoretical ionization threshold has to be adapted to the experimental value.

The correlated wavefunction which incorporates ISCI follows in analogy to equ. (1.25a) as an expansion into independent-particle wavefunctions for the ground state and contributions from virtual two-electron excitations:

$$\begin{aligned}\Psi_{\mathrm{corr,i}} = {} & A_0|\dots 2\mathrm{p}^6 3\mathrm{s}^2\ {}^1\mathrm{S}_0\rangle + b_1|\dots 2\mathrm{p}^6 3\mathrm{p}^2\ {}^1\mathrm{S}_0\rangle + \cdots \\ & + c_1|\dots 2\mathrm{p}^4 3\mathrm{s}^2 k\ell k'\ell'\ {}^1\mathrm{S}_0\rangle + \cdots .\end{aligned} \tag{5.24}$$

The mixing coefficients can be determined by a *multiconfigurational* Dirac or Hartree–Fock procedure (MCDF, MCHF). In the present case, however, numerical values are not of interest, only the fact that $|A_0|$ is smaller than unity due to the presence of virtual excitations in the normalized correlated wavefunction.

Similarly, FISCI can be incorporated into the wavefunction of the final ionic state:

$$\begin{aligned}\Psi_{\mathrm{corr}}(2\mathrm{p}^{-1}\ \text{ion};\ {}^2\mathrm{P}) = {} & \tilde{\alpha}_0\, |1\mathrm{s}^2 2\mathrm{s}^2 2\mathrm{p}^5 3\mathrm{s}^2\ {}^2\mathrm{P}\rangle_{\mathrm{relaxed}} + \tilde{\eta}_i\, |1\mathrm{s}^2 2\mathrm{s}^2 2\mathrm{p}^5 3\mathrm{s} k\ell\ {}^2\mathrm{P}\rangle_{\mathrm{relaxed}} + \cdots \\ & + \tilde{\zeta}_i\, |1\mathrm{s}^2 2\mathrm{s}^2 2\mathrm{p}^5 k'\ell' k''\ell''\ {}^2\mathrm{P}\rangle_{\mathrm{relaxed}} + \cdots .\end{aligned} \tag{5.25}$$

The coefficient $\tilde{\alpha}_0$ belongs to the process under consideration; if FISCI effects are not dramatic it will have the largest value. As well as the characteristic 2p–2h excitations attached to the coefficients $\tilde{\zeta}_i$, terms of 1p–1h excitations have also been included (coefficients $\tilde{\eta}_i$), because they describe the very interesting process of 2p ionization accompanied by outer-shell excitation, i.e., satellite processes, which are known to appear with appreciable intensity (see equ. (5.23d)). However, these $\tilde{\eta}_i$-coefficients cannot be determined by a MCDF/MCHF approach performed for the ion, because the admixed electron configurations belong to 1p–1h excitations, and the DF/HF procedure is stable to such excitations (Brillouin theorem, see above). In spite of this difficulty, a detailed calculation of such 1p–1h excitations can be avoided. The final ionic state considered so far belongs to a fully relaxed ion, and its relaxed single-orbital functions differ from the unrelaxed functions of the initial state. It might, therefore, be more appropriate to expand the wavefunction of equ. (5.25) in terms of unrelaxed single-orbital functions (frozen atomic orbitals). This procedure leads to an expression similar to that of equ. (5.25), but with different mixing coefficients:

$$\begin{aligned}\Psi_{\mathrm{corr}}(2\mathrm{p}^{-1}\ \mathrm{ion};\ {}^2\mathrm{P}) = {}& \alpha_0|1\mathrm{s}^2 2\mathrm{s}^2 2\mathrm{p}^5 3\mathrm{s}^2\ {}^2\mathrm{P}\rangle_{\mathbf{unrelaxed}} \\ & + \eta_i|1\mathrm{s}^2 2\mathrm{s}^2 2\mathrm{p}^5 3\mathrm{s}k\ell\ {}^2\mathrm{P}\rangle_{\mathbf{unrelaxed}} \\ & + \cdots + \zeta_i|1\mathrm{s}^2 2\mathrm{s}^2 2\mathrm{p}^5 k'\ell' k''\ell''\ {}^2\mathrm{P}\rangle_{\mathbf{unrelaxed}} + \cdots. \end{aligned} \tag{5.26}$$

The mixing coefficients are still unknown, but as will be seen below, in the present context only the α_0-coefficient, where $|\alpha_0| < 1$, is needed. Hence one can add to the wavefunction in equ. (5.26) the photoelectron's contribution, in order to calculate the matrix element between the initial state (equ. (5.24)) and the selected 2p photoionization channel. Working out the relevant dipole matrix elements, one then has the advantage that the overlap factors are unity if there the electron configurations are the same on both sides, and zero for all other cases. Hence only two ionization channels remain, and they are given by

$$D_{\mathrm{s}}^{\mathrm{corr}} = A_0\, \alpha_0\, \langle 1\mathrm{s}^2 2\mathrm{s}^2 2\mathrm{p}^5 \varepsilon\mathrm{s}|Op|1\mathrm{s}^2 2\mathrm{s}^2 2\mathrm{p}^6\rangle, \tag{5.27a}$$

$$D_{\mathrm{d}}^{\mathrm{corr}} = A_0\, \alpha_0 \langle 1\mathrm{s}^2 2\mathrm{s}^2 2\mathrm{p}^5 \varepsilon\mathrm{d}|Op|1\mathrm{s}^2 2\mathrm{s}^2 2\mathrm{p}^6\rangle, \tag{5.27b}$$

where the electron configurations on both sides of the transition matrix elements belong to the unrelaxed basis functions. If these dipole matrix elements are replaced by the corresponding RRPA values, the typical RRPA mean-field electron–electron interactions are brought in, giving finally

$$D_{\mathrm{s}}^{\mathrm{corr}} = A_0\, \alpha_0\, D_{\mathrm{s}}^{\mathrm{RRPA}}, \tag{5.28a}$$

$$D_{\mathrm{d}}^{\mathrm{corr}} = A_0\, \alpha_0\, D_{\mathrm{d}}^{\mathrm{RRPA}}. \tag{5.28b}$$

This is a very important result. It states that both dipole amplitudes from the RRPA calculation are modified by a common factor that reflects the influences of electron correlations in the initial and final ionic states which are beyond mean-field electron–electron interactions. The $|A_0\alpha_0|^2$-value is called the *spectroscopic factor* (or the quasi-particle strength or the pole strength or the renormalization factor) and describes the weight given to the improved 2p photoionization cross section as compared to a calculation which does not include these specific electron correlations. The remaining intensity is transferred to satellite processes accompanying 2p photoionization. In this way the satellites 'borrow' their intensity from the process assumed to be uncorrelated, which explains the name *the intensity borrowing model* given to this approach (see, for example, [CDS80]). Formulating the result quantitatively, one obtains instead of equ. (5.14)

$$\sigma_{2p} = \frac{4\pi^2}{3}\alpha\,\omega\,(|D_s^{corr}|^2 + |D_d^{corr}|^2)$$

$$= (A_0\alpha_0)^2\frac{4\pi^2}{3}\alpha\,\omega\,(|D_s^{RRPA}|^2 + |D_d^{RRPA}|^2) \tag{5.29a}$$

and

$$\sigma_{2p_{sat}} = [1-(A_0\alpha_0)^2]\frac{4\pi}{3}\alpha\,\omega\,(|D_s^{RRPA}|^2 + |D_d^{RRPA}|^2) \tag{5.29b}$$

with

$$\sigma_{2p} + \sigma_{2p_{sat}} = \frac{4\pi^2}{3}\alpha\,\omega\,(|D_s^{RRPA}|^2 + |D_d^{RRPA}|^2). \tag{5.29c}$$

Using the intensity borrowing model, natural explanations now arise for the observations given above for the comparison of experimental and theoretical photoionization data of magnesium:

(i) According to equ. (5.29c), the RRPA and MBPT calculations performed for 2p photoionization provide the sum of partial cross sections of main and attached satellite lines, and not the 2p partial cross section alone.

(ii) Using the ratio $\sigma_{2p_{sat}}/\sigma_{2p}$ = Int(2p)/Int(all 2p sat.), one can extract the factor $(A_0\alpha_0)^2$ from the experimental data which gives 0.75, and then, using equ. (5.28), $|D_s^{corr}| = 0.220$ and $|D_d^{corr}| = 0.746$. These values are rather close to the data from direct experimental determination given in Table 5.1. Therefore, the origin of the discrepancy between experimental and theoretical values for the RRPA and MBPT calculations for the magnitude of the involved dipole matrix elements is explained.

(iii) The excellent agreement between the experimental data and the theoretical predictions from RRPA/MBPT for the angular distribution parameters β_{2p} and $\beta(L_3\text{–}M_1M_1)$ of photo- and Auger electrons (Figs. (5.8) and (5.9); adapted to the same binding energy value) can be explained: the spectroscopic factor, which is responsible for electron correlations beyond mean-field calculations, cancels in the theoretical expressions of these observables because the spectroscopic factor appears in the expression for the β-parameter in both the numerator *and* the denominator and, therefore, drops out.

5.3 3s satellite spectrum of argon

The existence of satellite lines has already been mentioned several times, for example, in the explanation of the photoelectron spectra of rare gases (see Fig. 2.4), and in the discussion of 2p photoionization in magnesium. In this section the satellite spectrum related to outer-shell photoionization in argon will be treated in detail for two reasons. First, it provides a case in which the intense satellite lines can be understood nearly quantitatively on the basis of configuration interaction in the final ionic state only (FISCI). Second, it is a good example of the progress achieved in electron spectrometry with synchrotron radiation from the beginning of such studies until now.

5.3.1 General background

The ground state of the argon atom is written as $1s^2 2s^2 2p^6 3s^2 3p^6\ {}^1S_0^e$. An early photoelectron spectrum obtained with monochromatized synchrotron radiation at 77.2 eV photon energy and with a bandpass of 1 eV is shown in Fig. 5.11. The dominant features are the 3p and 3s photolines at binding energies of 15.82 eV and 29.24 eV, respectively (note the change in the energy and intensity scales). In addition, satellite lines appear at higher binding energies. These satellites arise from ionization accompanied by simultaneous excitation. In the energy range

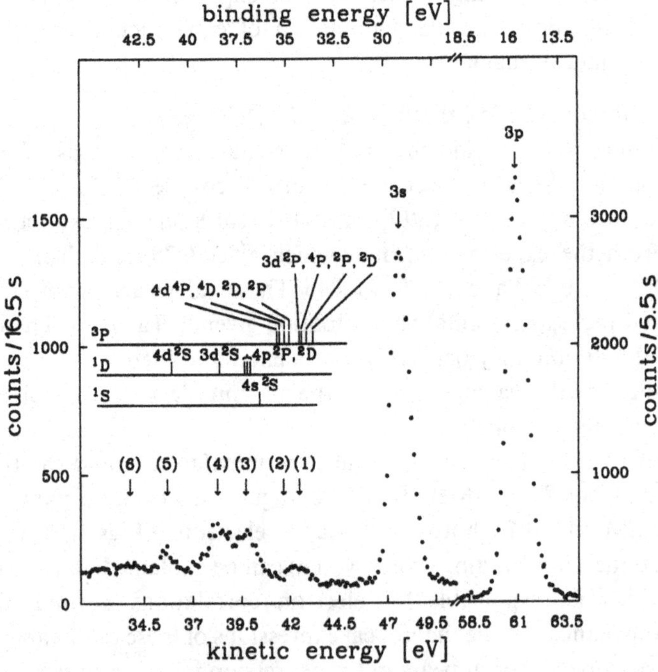

Figure 5.11 Outer-shell photoionization in argon at 77.2 eV photon energy (bandpass 1.0 eV). The 3p and 3s main photolines (note the interruption of the energy scale and the different intensity scales) with their correlation satellites numbered (1)–(6) are shown. On the left-hand side of the figure the vertical bars give the positions of Ar^{+*} states taken from optical data [Moo71] for the electron configuration $3s^2 3p^4 n\ell$ (the terms ^3P, ^1D and ^1S refer to the core configuration $3s^2 3p^4$ alone). With this information an attempt can be made to identify some of the observed satellites, for example, lines (4) and (5) belong to $3s^2 3p^4(^1D)3d\ ^2S^e$ and $3s^2 3p^4(^1D)4d\ ^2S^e$, and line (3) to $3s^2 3p^4(^3P)4p\ ^2P^o$ and/or $3s^2 3p^4(^3P)4p\ ^2D^o$, and/or $3s^2 3p^4(^1S)4s\ ^2S^e$. From [AWK78].

considered, they have the electron configuration $1s^22s^22p^63s^23p^4n\ell$, abbreviated to$3s^23p^4n\ell$. Help important for the identification of these satellites can come from tabulated data for the binding energies of $3s^23p^4(^{2S'+1}L')n\ell\ ^{2S+1}L_J$ fine-structure resolved states which are derived from the analysis of optical spectra [Moo71]. Some energy positions predicted from such optical data are marked as the bar diagram in Fig. 5.11. There it can be noted that the limited experimental resolution prevents an unambiguous correspondence between the bar diagram and the observed structures numbered (1)–(6). Because of this difficulty, theoretical attempts were undertaken to find guidance in the interpretation of such satellites.

As demonstrated in the previous section, satellites are due to electron correlation effects, and, in principle, all types which are classified in a configurational picture as *initial state configuration interaction* (ISCI), *final ionic state configuration interactions* (FISCI) and *final state configuration interactions* (FSCI which includes interchannel interactions in the continuum) have to be taken into account. In certain cases, however, one type of correlation is more important than the others, and in the present case of 3s and 3p photoionization in argon this is FISCI. This property allows a rather transparent analysis of the implications which these correlations have on the corresponding satellites.

The FISCI effects in the 3s and 3p photoelectron spectrum of argon are dominant because the 3d orbital, even though it is not yet bounded, belongs to the same ($n = 3$) shell as the 3s and 3p orbitals. Before describing the FISCI approach, the special role of this 3d orbital needs to be explained. It goes back to irregularities in the aufbau principle for atoms which are due to the special shape of the effective atomic potential [Fer28, GMa41, Kar81].† In the elements potassium and calcium which follow argon in the periodic table, the atomic potential is not strong enough to accommodate the 3d orbital and, therefore, they have the electron configurations $1s^22s^22p^63s^23p^64s$ and $1s^22s^22p^63s^23p^64s^2$, respectively, in which the 3d orbital is bypassed. However, after completion of the 4s shell, the increased nuclear charge allows the atomic potential to bound the bypassed 3d orbital, and this later filling of the 3d shell gives the *first series* of the *transition group* elements (starting with scandium, $1s^22s^22p^63s^23p^63d4s^2$). This shows that the relatively small change of one unit in the atomic number can cause a rather dramatic change in the wavefunction of a bypassed orbital: a *diffuse* orbital located far outside of the atomic region becomes a *bound* orbital located at the position of the n-shell electrons. This phenomenon is called the 'collapse' of the wavefunction. Because a photoionization process also changes the atomic potential, it is understandable that bypassed orbitals are potential candidates for electron correlation effects during photoionization, i.e., for the production of satellites. In the case of argon, one can expect FISCI between $3s3p^6$ and $3s^23p^43d$ electron configurations to play a dominant role in 3s photoionization in argon

† Note that the potential in equ. (7.72) includes a centrifugal term which for large values of the orbital angular momentum ℓ keeps the electrons away from the nucleus.

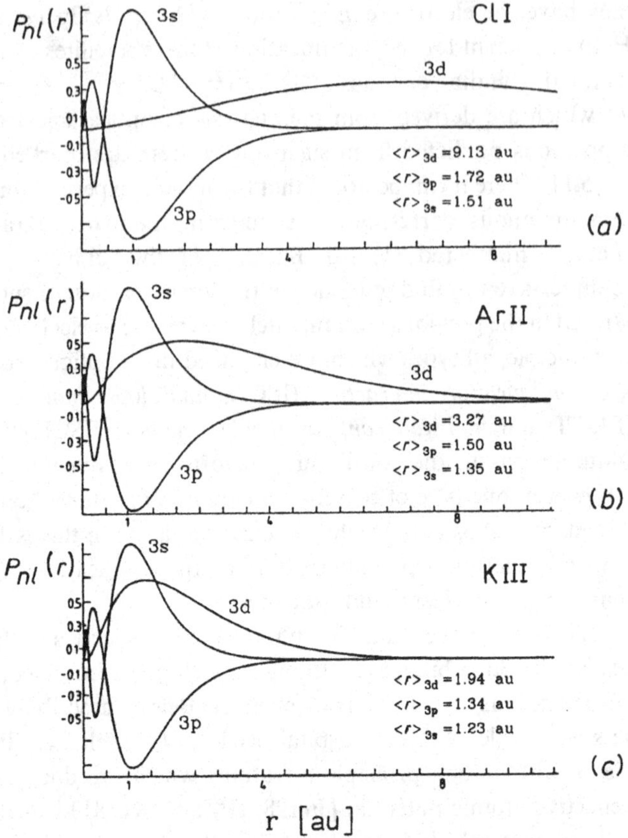

Figure 5.12 Radial wavefunctions $P_{n\ell}(r)$ with $n = 3$ for neutral chlorine (Cl I), singly-ionized argon (ArII), and doubly-ionized potassium (KIII) which demonstrate the collapse of the 3d orbital (all these systems have the electron configuration $1s^22s^22p^63s^23p^5$ in the ground state while the results apply to the ... $3s^23p^4(^1D)3d\ ^2S$ states). Note that the horizontal scales are slightly different, but the distance from the origin to $(\langle r\rangle_{3s} + \langle r\rangle_{3p})/2$ is the same in all three cases. It is the average position $\langle r\rangle_{3d}$ of the $P_{3d}(r)$ function (and its shape) that is of interest. For Cl I this is located far outside the maximum of $\langle r\rangle_{3s}$ and $\langle r_{3p}\rangle$, i.e., the 3d orbital is clearly uncollapsed. The opposite is true for KIII where the 3d orbital has clearly collapsed. The 3d orbital of ArII falls between these two cases. From [SHa83].

because the bypassed 3d orbital might collapse when forming singly-ionized argon. From Fig. 5.12 it can be seen that such a collapse at least partly occurs: all ($n = 3$) orbitals for the isoelectronic sequence of neutral chlorine (Fig. 5.12(*a*)), singly-ionized argon (Fig. 5.12(*b*)) and doubly-ionized potassium (Fig. 5.12(*c*)) are shown. While in all cases the locations of 3s and 3p orbitals are the same, the 3d orbital is clearly located outside the atomic region for chlorine, has clearly collapsed in doubly-ionized potassium, and is somewhere between these extremes for singly-ionized argon. This intermediate position for argon has the consequence that not only 3d orbitals, but also higher members of the *k*d series including the continuum must be considered in the FISCI approach.

Due to the symmetry requirements of the interacting configurational wavefunctions of equ. (1.24), 3s photoionization in argon can mix in the FISCI model only with $3s^2 3p^4 kd\ ^2S^e$ (and $3s^2 3p^4 ks\ ^2S^e$), but not with $3s^2 3p^5\ ^2P^o$.† Similarly, 3p photoionization in argon will mix with $3s^2 3p^4 kp\ ^2P^o$. Therefore, two *manifolds*, $^2S^e$ and $^2P^o$, can be distinguished and treated separately within FISCI. For simplicity, only the $^2S^e$ manifold will be considered further (see (Sha83]).

The interaction matrix element relevant for FISCI in the $^2S^e$ manifold is given by

$$H_{kd3s} = \langle 1s^2 2s^2 2p^6 3s^2 3p^4 kd\ ^2S^e | \sum_{i<j} \frac{1}{r_{ij}} | 1s^2 2s^2 2p^6 3s 3p^6\ ^2S^e \rangle, \tag{5.30a}$$

where k runs over all discrete quantum numbers (n) and extends into the continuum (ε). If calculated with single-particle wavefunctions, H_{kd3s} becomes a numerical factor and a Slater integral (see equs. (7.57)–(7.60)),

$$H_{kd3s} = -\sqrt{(\tfrac{2}{3})} R^1(3p3p, 3skd). \tag{5.30b}$$

In diagrammatic language, this matrix element can be interpreted as a non-radiative decay where the 3s-hole is filled through deexcitation of a 3p-electron, with another 3p-electron excited to kd. (Compare Fig. 5.10(h) with $n_0 = 3s$, $k = 3p$, and the (2p–$\varepsilon\ell$)–(hole–electron) pair replaced by the (3p–kd)–(hole–electron) pair.) Since the primary hole is filled with an electron of the same principal quantum number, such a non-radiative decay is called a *super Coster–Kronig* transition, the special case for $k = 3$ is called a *giant* super Coster–Kronig transition because of its exceptionally large interaction strength (see below). Since in the present example energy conservation is not fulfilled for an actual transition, the matrix element in equ. (5.30) just describes a virtual 'transition' or, in other words, a pure effect of electron correlation. As can be seen in equ. (5.30b), the numerical value of this virtual Coster–Kronig matrix element is determined by the spatial overlap of the electrons involved, and at this point the important role of the virtual giant super Coster–Kronig matrix element can be realized: all electrons have the same principal quantum number $n = k = 3$, and as discussed with Fig. 5.12, the nearly collapsed 3d orbital overlaps strongly with the 3s and the 3p orbitals, leading to a large value for the matrix element upon integration over space. Similar arguments hold for the other matrix elements with $k > 3$, which have smaller, but still significant values. The generally large magnitude of these interaction matrix elements explains why FISCI is the dominant correlation mechanism in the outer-shell photoionization in argon.

Based on early theoretical calculations, which included FISCI, the most intense satellite structures of Fig. 5.11 were attributed to the following final ionic states (see [AWK78] and references therein):

$$\text{line (4): } 3s^2 3p^4(^1D)3d\ ^2S^e, \tag{5.31a}$$

$$\text{line (5): } 3s^2 3p^4(^1D)4d\ ^2S^e \text{ and/or } 3s^2 3p^4(^1S)5s\ ^2S^e, \tag{5.31b}$$

$$\text{line (6): } 3s^2 3p^4(^1D)5d\ ^2S^e. \tag{5.31c}$$

† Note that here $3p^4 kd\ ^2S^e$ is an abbreviation for $3p^4(^1D^e)kd\ ^2S^e$.

Similar arguments give

$$\text{line (3): } 3s^23p^4(^1S)4s\ ^2S^e \text{ and/or } 3s^23p^4(^1D)4p\ ^2P^o \text{ and/or } 3s^23p^4(^1D)4p\ ^2D^o. \tag{5.31d}$$

For later work see [DLa82, SHa83, AMW85, HHa87, SLS92, SLP94].

5.3.2 $^2S^e$ manifold treated by FISCI

The FISCI method will be considered in more detail for the $^2S^e$ manifold of 3s photoionization in argon, even though finer details of the satellite spectrum can be understood only if further electron correlations and spin–orbit effects are included (see below). In the FISCI approach, uncorrelated functions $\tilde{\Phi}_i$ with single-particle orbitals provide the starting point. They are abbreviated to their characteristic electron orbital, i.e.,

$$\tilde{\Phi}_{3s} = \tilde{\Phi}(1s^2 \ldots 3s3p^6\ ^2S^e), \tag{5.32a}$$

$$\tilde{\Phi}_{kd} = \tilde{\Phi}(1s^2 \ldots 3s^23p^4kd\ ^2S^e). \tag{5.32b}$$

According to equ. (1.24), the correlated wavefunctions $\tilde{\Psi}_I$ then follow as expansions in the basis functions $\tilde{\Phi}_i$:

$$\left.\begin{aligned} \tilde{\Psi}_{3s} &= A_{3s,3s}\tilde{\Phi}_{3s} + b_{3s,3d}\tilde{\Phi}_{3d} + b_{3s,4d}\tilde{\Phi}_{4d} + \cdots, \\ \tilde{\Psi}_{3d} &= a_{3d,3s}\tilde{\Phi}_{3s} + B_{3d,3d}\tilde{\Phi}_{3d} + b_{3d,4d}\tilde{\Phi}_{4d} + \cdots, \\ \tilde{\Psi}_{kd} &= a_{kd,3s}\tilde{\Phi}_{3s} + b_{kd,3d}\tilde{\Phi}_{3d} + b_{kd,4d}\tilde{\Phi}_{4d} + \cdots \end{aligned}\right\} \tag{5.33a}$$

abbreviated as

$$\tilde{\Psi}_I = a_{I,3s}\tilde{\Phi}_{3s} + b_{I,3d}\tilde{\Phi}_{3d} + b_{I,4d}\tilde{\Phi}_{4d} + \cdots. \tag{5.33b}$$

The dots indicate contributions from other electron configurations, including the continuum. (In order to describe the continuum, a certain number of energy values are selected: in the present example, approximately 40 energies from the ionization limit up to about 10 au with the energies more closely spaced near the limit than at higher energies are chosen [SHa83].) The capital-letter expansion coefficients denote large contributions of the corresponding electron configuration which, therefore, are also used to name the correlated wavefunction $\tilde{\Psi}_I$. It should, however, be kept in mind that $\tilde{\Psi}_{3s}$ does not represent a pure 3s-hole-state as described by $\tilde{\Phi}_{3s}$ (see below). From the short outline in Section 7.4.2, the energies E_I of the correlated states and the mixing coefficients can be found by solving equ. (7.91). Adapting the nomenclature to the present example, one gets

$$\left.\begin{aligned} \varepsilon_{3s}\ a_{I,3s} + H_{3s,3d}\, b_{I,3d} + H_{3s,4d}\, b_{I,4d} + \cdots &= E_I\, a_{I,3s} \\ H_{3d,3s}\, a_{I,3s} + \varepsilon_{3d}\ b_{I,3d} + 0 \quad + 0 &= E_I\, b_{I,3d} \\ H_{4d,3s}\, a_{I,3s} + 0 \quad + \varepsilon_{4D}\ b_{I,4d} + 0 &= E_I\, b_{I,4d} \\ \vdots \qquad & \end{aligned}\right\} \tag{5.34}$$

Note that in the frozen core approximation the orthonomality of the basis functions leads for $k \neq k'$ to vanishing matrix elements $\langle\tilde{\Phi}_{kd}| H | \tilde{\Phi}_{k'd}\rangle$ [Wei69]. The single-particle eigenvalues are given by

$$\varepsilon_{3s} = \langle\tilde{\Phi}_{3s}|H|\tilde{\Phi}_{3s}\rangle \qquad \text{and} \qquad \varepsilon_{kd} = \langle\tilde{\Phi}_{kd}|H|\tilde{\Phi}_{kd}\rangle, \tag{5.35}$$

and the interaction matrix elements H_{3skd} are defined in equ. (5.30). With appropriate single-particle basis functions one can evaluate the single-particle properties and get numerical values for the matrix H_{ij} (atomic units, diagonal values are given relative to the ion core; from [SHa81]):

$$H_{ij} = \begin{pmatrix} -0.3201 & -0.1902 & -0.1117 & -0.0736 & \dots \\ -0.1902 & -0.2902 & 0 & 0 & 0 \\ -0.1117 & 0 & -0.1561 & 0 & 0 \\ -0.0736 & 0 & 0 & -0.0959 & 0 \\ \dots & & & & \end{pmatrix} \tag{5.36}$$

The values given refer to a *frozen* core calculation, i.e., the core $3s^2 3p^4\,{}^1D^e$ orbitals taken from a Hartree–Fock calculation were kept fixed for the determination of the kd orbitals. From both equ. (5.34) and equ. (5.36) it can be seen that the matrix H_{ij} is a bordered matrix with non-vanishing elements only in the first row, in the first column, and in the diagonal elements. This is due to the interaction of the one 3s-hole-state with the series of $3p^{-2}kd$ excitations. (The 3s-hole-state can be considered as a 'perturber' of the Rydberg series of kd excitations.) The solutions of such an eigenvalue problem with a bordered matrix can be easily found and interpreted in a graphical way (see also [Wei69]). The first row of equ. (5.34) means that

$$(E_{\text{I}} - \varepsilon_{3s})a_{\text{I},3s} = \sum_k H_{3s,kd} b_{\text{I},kd} \tag{5.37a}$$

and all other rows can be included in the expression

$$b_{\text{I},kd} = \frac{H_{3s,kd}}{E_{\text{I}} - \varepsilon_{kd}} a_{\text{I},3s}. \tag{5.37b}$$

The last expression relates the $b_{\text{I},kd}$-coefficients to the $a_{\text{I},3s}$-coefficients. Substitution into equ. (5.37a) gives a formal solution for the eigenvalues:

$$E_{\text{I}} = \varepsilon_{3s} + \sum_k \frac{H^2_{3s,kd}}{E_{\text{I}} - \varepsilon_{kd}}. \tag{5.38}$$

The eigenvalues E_{I} can be found graphically by plotting separately the functions $f_1(E)$ and $f_2(E)$ defined as

$$f_1(E) = E - \varepsilon_{3s}, \tag{5.39a}$$

$$f_2(E) = \sum_k \frac{H^2_{3s,kd}}{E - \varepsilon_{kd}} \tag{5.39b}$$

and seeking their intersections. This is demonstrated with the help of Fig. 5.13.

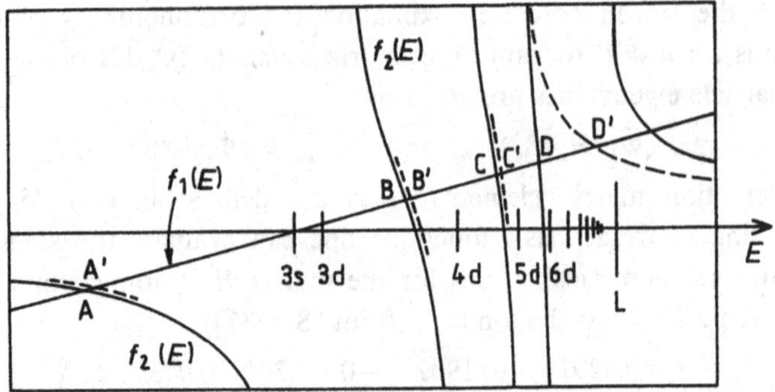

Figure 5.13 Graphical solution of the eigenvalue problem for the $^2S^e$ manifold of 3s photoionization and accompanying $3s^23p^4(^1D)nd\ ^2S^e$ satellites in argon. The single-particle eigenvalues of the basis functions from a frozen Hartree–Fock calculation are shown on the energy axis E using the symbols 3s and nd; L indicates the position of the ionization limit $3s^23p^4(^1D)$. As described in the main text, the eigenvalues E_I from the configuration interaction approach follow as intersections between the straight-line function $f_1(E)$ and the tangent-like functions $f_2(E)$, and the eigenvector component $a_{I,3s}$ is given by the slope of the tangent-like curve at the specific intersection. The results of two calculations are shown. The dashed curve is from a limited configuration iteration calculation using the four basis functions $3s3p^6$, $3s^23p^43d$, $3s^23p^44d$, and $3s^23p^45d$, all with $^2S^e$ symmetry; the corresponding eigenvalues are given by A′, B′, C′, and D′. The full curve is from a limited configuration interaction calculation using all bound nd states; here the four lowest eigenvalues (A, B, C, D) and the highest one are shown. For numerical values see Table 5.2. From [SHa83].

Here $f_1(E)$ gives a straight line crossing the energy axis E at the position of the single-particle energy ε_{3s}, while $f_2(E)$ yields a set of tangent-like curves which go to $\pm\infty$ at the positions of the single-particle energies ε_{kd}. The resulting lowest eigenvalues E_I are called A, B, C, and D in the figure, and they belong to $3s3p^6\ ^2S^e$, $3s^23p^43d\ ^2S^e$, $3s^23p^44d\ ^2S^e$, and $3s^23p^45d\ ^2S^e$, respectively (for the dashed functions $f_2(E)$ and their intersections A′, B′, C′, D′ see the figure caption).

It is also possible to obtain graphically the special $a_{I,3s}$-coefficients of the eigenvectors (the other components then follow from equ. (5.37b)), because the normalization of each correlated wavefunction $\tilde{\Psi}_I$ imposes the following condition on its eigenvector components:

$$(a_{I,3s})^2 + \sum_k (b_{I,kd})^2 = 1. \tag{5.40a}$$

Using equ. (5.37b), this expression is equivalent to

$$(a_{I,3s})^2 = \left(1 + \sum_k \frac{(H_{3s,kd})^2}{(E_I - \varepsilon_{kd})^2}\right)^{-1} \tag{5.40b}$$

or

$$(a_{I,3s})^2 = (1 - f_2'(E)|_{E_I})^{-1}, \tag{5.40c}$$

i.e., these mixing coefficients are related to the slope of the tangent-like curves $f_2(E)$ at the positions of the corresponding eigenvalues E_I. A flat slope yields a

Table 5.2. *Results of three configuration interaction calculations for the $3s3p^6\,^2S^e \leftrightarrow 3s^23p^4kd\,^2S^e$ interaction in singly ionized argon.*

State	Energy			Spectroscopic factor		
	lim. CI ($n \leq 5$)	lim. CI (all n)	compl. CI$^{(*)}$	lim. CI ($n \leq 5$)	lim. CI (all n)	comp.CI$^{(*)}$
$3s3p^6\,^2S^e$	30.91	30.63	29.32	0.5579	0.5658	0.6291
$3s^23p^43d\,^2S^e$	39.56	39.43	39.15	0.0855	0.0856	0.0847
$3s^23p^44d\,^2S^e$	42.02	41.88	41.68	0.0391	0.0433	0.0436
$3s^23p^45d\,^2S^e$	44.56	43.03	42.90	0.3175	0.0243	0.0242
$3s^23p^46d\,^2S^e$		43.67	43.58		0.0147	0.0144
Contribution of all higher bound states					0.2663	0.0372
Contribution of continuum						0.1668
Sum (normaliz.)				1.000	1.000	1.000

lim. CI($n \leq 5$) is limited to nd = 3d, 4d, 5d; lim. CI(all n) is limited to all discrete excitations; compl. CI stands for a complete configuration interaction approach in which the continuum is also included. Energy values are ionization energies in eV; for the spectroscopic factors see equ. (5.40c). The asterisk $^{(*)}$ indicates that these values are from a two-core approach where separate Hartree–Fock calculations were performed for the states $3s^23p^4(^1D)kd\,^2S^e$ with the core frozen at the orbitals determined for $3s^23p^4\,^1D$, and for $3s3p^6\,^1S^e$, in order to account better for relaxation effects. In comparison to Fig. 5.13 this leads to a change of the single-particle energy ε_{3s} from ε_{3s}(unrelaxed) = -0.3201 au to ε_{3s}(relaxed) = -0.3584 au, and these values, in turn, modify the eigenvalues as listed in the table. The data are from [SHa81, SHa83].

large value, a steep slope a small value. The squared $a_{\mathrm{I},3s}$-coefficients describe the probability of the 3s basis state being represented in the correlated state $\tilde{\Psi}_\mathrm{I}$. Hence, these eigenvector coefficients squared are *spectroscopic factors* (see the discussion after equ. (5.28)). In the present case they are defined not only for the 3s main photoline, $(a_{3s,3s})^2$, but also for the attached individual satellites, $(a_{3s,kd})^2$. Some of these spectroscopic factors (and of the corresponding eigenvalues) are given in Table 5.2 for three calculations, one limited to discrete excitations with $n \leq 5$ only, one for the inclusion of all discrete excitations, and one for a complete calculation including the continuum. Two special points can be noted. First, the spectroscopic factor $(a_{3s,3s})^2$ amounts in the complete calculation to 63%, i.e., naming the $\tilde{\Psi}_{3s}$ state the '3s state' is still justified, but 37% of the total oscillator strength available in the single-particle model for 3s photoionization is redistributed into discrete and continuous satellites. Second, even in the limited calculations the lowest eigenvalues and spectroscopic factors for the $3s^23p^4nd\,^2S^e$ states[†] are close to those

[†] The primary reason for the large energy difference for the $3s3p^6\,^2S^e$ state in the complete and limited calculations is not so much the effect of the continuum, but the use of different core orbitals and, hence, single-particle eigenvalues ε_{3s} (see Table 5.2).

of the complete calculation, but there is always a considerable discrepancy for the values for the state with the highest eigenvalue. This means that this state from the truncated configuration interaction approach cannot be associated with a physical state and should, therefore, not be considered.

5.3.3 *Experimental results from well-resolved spectra*

With known wavefunctions $\tilde{\Psi}_I$ for the final ionic states of the $^2S^e$ manifold one can calculate the photoionization matrix elements for all processes leading to a selected $\tilde{\Psi}_I$. Using for simplicity an uncorrelated wavefunction of the initial state

$$\tilde{\Phi}_i = \tilde{\Phi}(1s^2 \ldots 3s^2 3p^6\ {}^1S), \tag{5.41}$$

one gets for linearly polarized light

$$\langle \tilde{\Psi}_I;\ \text{photoelectron}\ {}^1P | \sum_j z_j | \tilde{\Phi}_i \rangle = a_{I,3s} \langle \varepsilon p^{(I)} | z | 3s \rangle, \tag{5.42}$$

i.e., the matrix element factorizes into the mixing coefficient $a_{I,3s}$ whose value squared is the spectroscopic factor and a one-particle dipole matrix element which depends on the selected state I through the specific ionization energy E_I. Four predictions were made based on this result, which are of particular relevance for a comparison experimental data:

(i) The partial cross sections for the 3s photoline and the accompanying $3p^4kd$ $^2S^e$ satellites are given by

$$\sigma_I \sim (a_{I,3s})^2\, |\langle \varepsilon p^{(I)} | z | 3s \rangle|^2, \tag{5.43}$$

i.e., due to the normalization property of the spectroscopic factor (see Table 5.2) the satellites 'borrow' their intensity from the main line.

(ii) The partial cross sections of the satellite lines follow that of the main line, possibly with shifted energy because the ionization energies differ. These satellites are called 'shadow' satellites [Amu85].

(iii) At high photon energies, differences in the ionization energies of the main and satellite lines disappear, and the matrix elements $\langle \varepsilon p^{(I)} | z | 3s \rangle$ can be approximated by the same numerical value. This gives

$$\left. \frac{\sigma_I}{\sum \sigma_I} \right|_{\substack{\text{high}\\ \text{energy}}} = (a_{I,3s})^2, \tag{5.44a}$$

i.e., spectroscopic factors can be determined. (In general, eigenvectors depend on the basis set used and have no physical significance as such. Therefore, the interpretation of measured cross section ratios in terms of calculated $(a_{I,3s})^2$ values depends on the selected basis ([Han82]; see also Section 1.1.2); for the important influence of ISCI on spectroscopic factors see [AKh85, Khe85].) A quantity simpler to obtain from the experimental observables is

$$\left. \frac{\sigma_I}{\sigma_{3s}} \right|_{\substack{\text{high}\\ \text{energy}}} = \frac{(a_{I,3s})^2}{(a_{3s,3s})^2}. \tag{5.44b}$$

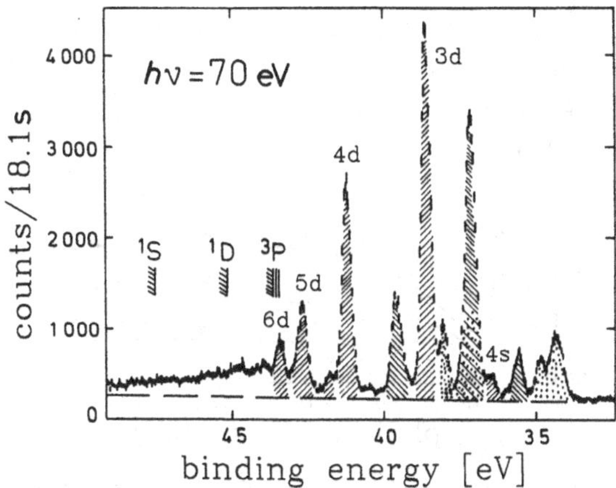

Figure 5.14 Valence-shell satellite spectrum of argon measured at 70 eV photon energy (bandpass 0.2 eV), and for a spectrometer setting close to the quasi-magic angle. Satellites of different groups (see main text) are characterized by different shading. For the prominent $3s^2 3p^4(^1D)nd\ ^2S^e$ satellites their nd labels are given. The estimated background is shown as a broken line, the onset of the double ionization continua $3s^2 3p^4\ ^{2S+1}L$ is indicated by the ^{2S+1}L symbols. From [Sch86]; see also [BBT87, BLK88, KKS87, SHG87, BBT88, SEM88].

(iv) Because the dipole matrix elements contain the same angular functions, the angular distribution of the satellite lines is equal to that of the main line (possibly shifted on the energy scale because of different ionization energies). Hence, all $3p^4kd\ ^2S^e$ satellites are expected to have an angular distribution factor $\beta = 2$.

In order to check these predictions for the argon satellites, improved experimental conditions with respect to resolution and counting statistics were necessary. One example of such spectra is shown in Fig. 5.14. Guided by theory, and with the help of optical data, three groups of satellite lines can be identified:

Group (a) satellites (shaded ////). These contain final ionic states from the $^2S^e$ manifold, and they are due to FISCI associated with 3s ionization. One can clearly see the $3s^2 3p^4(^1D)nd\ ^2S^e$ satellites with $n = 3, 4, 5, 6$ and the small $3s^2 3p^4(^1S)4s\ ^2S^e$ satellite.

Group (b) satellites (shaded \\\\). These contain final ionic states from the $^2P^o$ manifold which result from FISCI associated with 3p ionization. One can clearly see the $3p^4(^3P)4p\ ^2P^o$, $3p^4(^1D)4p\ ^2P^o$, and $3p^4(^1S)4p\ ^2P^o$ satellites.

Group (c) satellites (dotted). These arise from electron correlation effects which are beyond the FISCI mechanism. The large spatial overlap between orbitals with the same principal quantum number $n = 3$ still favours electron correlation effects between $3s3p^6$ and $3s^2 3p^4 3d$, but ISCI and FSCI are responsible for the presence of $3s^2 3p^4 3d\ ^{2S+1}L^e$ final ionic states with symmetries different from $^2S^e$. Most of these states can be identified in the electron spectrum.

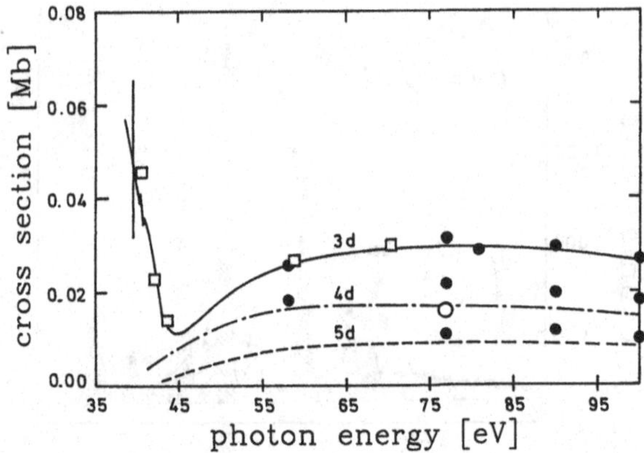

Figure 5.15 Partial photoionization cross sections of argon $3s^23p^4(^1D)nd\ ^2S^e$ satellites with $n = 3, 4, 5$. Experimental data: solid circles, relative intensities from [KKS87] placed on an absolute scale by using the calculated value of $\sigma(3s)$, see also data from [BBT87, BBT88]; open squares, data from [BLK88]; open circle, data point from [KKS87] for $3s^23p^44d\ ^2S^e$, but reduced by 25% due to the better-resolved results from fluorescence spectroscopy [SCL88]. Theoretical values: solid, dash-dotted and dashed lines for 3d, 4d, and 5d satellites, respectively (at 39.5 eV the $3s^23p^4(^1D)4d(^1S)5p$ resonance is also shown); for details see [WKe87, WKe89]. From [WKe87, WKe89], see also [KWC92, SLS92, SLP94].

Based on spectra similar to the one shown in Fig. 5.14, relative intensities of the dominant nd satellites can be obtained and placed on an absolute scale with the help of known cross sections for the 3s reference line. The result is shown in Fig. 5.15 and compared with theoretical data. It can be seen that there is good agreement between the experimental and theoretical values.† In particular, the minimum in the cross section for the $3s^23p^43d\ ^2S^e$ satellite mimics the minimum in the cross section $\sigma(3s)$, which is characteristic for a shadow satellite. It is reasonable to suppose that the cross sections of the other $3p^4nd\ ^2S^e$ satellites with higher binding energies would also show a minimum if their oscillator strengths were extended into the region of bound Rydberg states. In addition, the angular distribution parameter of the $3p^43d\ ^2S^e$ satellite was found to be $\beta = 2$ [AMW85, KWC92]. Summarizing these experimental results, one then can conclude that FISCI is the dominant mechanism for the production of $3p^4nd\ ^2S^e$ satellites in argon, even in the presence of other electron correlations. (For photon energies closer to the individual ionization thresholds, rather complicated and resonance-affected changes in the relative intensities occur (see discussion of Fig. 5.5). In these cases the whole complex of the slightly excited 18-electron system Ar* has to be considered taking into account all electron–electron interactions.)

Of course, the dominance of FISCI in the argon satellite spectrum at intermediate and high photon energies only provides a rough orientation. In general, electron

† In addition to FISCI the theoretical calculation also includes ISCI and FSCI which have a significant effect on the absolute values, but only a slight influence on the intensities relative to the 3s main line. These relative intensities agree to within a few per cent with those of a pure FISCI calculation (see Table 5.2) if the energy dependence of the dipole matrix elements is also included.

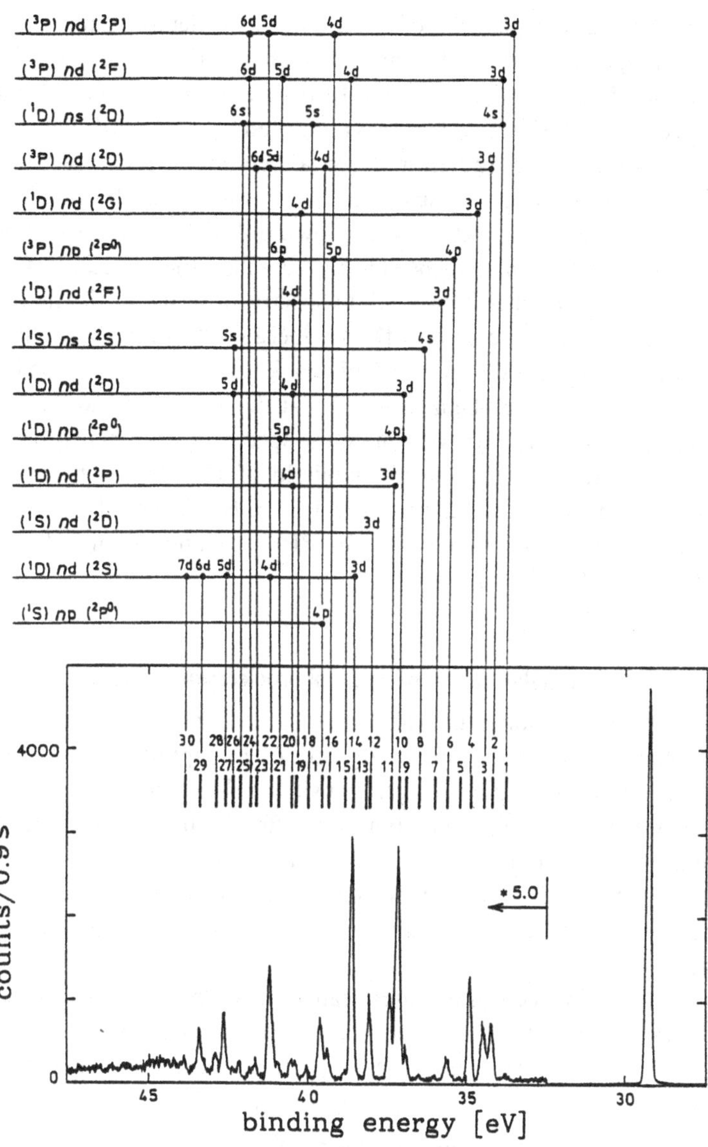

Figure 5.16 High-resolved valence-shell satellite spectrum of argon covering the 3s photoline at 29.24 eV binding energy and the 3s/3p correlation satellites. The experimental spectrum measured with monochromatized synchrotron radiation from an undulator at 58 eV photon energy (bandpass ≈ 0.100 eV), and under the quasi-magic angle is shown at the bottom. The scheme for the identification of individual final ionic states according to optical data [Moo71] is given at the top; the $3s^23p^4(^{2S'+1}L')n\ell(^{2S+1}L)$ ionic states (the $3s^23p^4$ core has to be added) with individual $n\ell$ values marked explicitly at the corresponding energy positions are shown. From [Kos92], see also [BBT88, SEM88 and KWC92]; some small differences in these spectra and in the interpretation of lines are not discussed here.

correlation will yield to a plethora of satellites. This is demonstrated in Fig. 5.16 which shows a spectrum obtained with monochromatized undulator radiation (bandpass approximately 100 meV). Due to the superior brilliance of undulator radiation (see equ. (1.35)) the bandpass of monochromatized light is much improved as compared to radiation from a bending magnet (with the new generation of electron storage rings even better values can be obtained). The comparison of this spectrum with the one shown in Fig. 5.11 clearly demonstrates the experimental progress that has been achieved within the last 15 years, allowing a detailed assignment of such rich satellite spectra in terms of the final angular momentum coupling of the core and the Rydberg electron. A further discussion of these satellites is beyond the scope of this book. In the context of the foregoing discussion it should, however, be mentioned that the $3p^4nd\ ^2S^e$ and the $3p^4np\ ^2P^o$ satellites arising from FISCI still dominate the spectrum, and their properties are the same as discussed above. (The $^2S^e$ manifold of nd satellites belongs in Fig. 5.16 to the numbers 14, 22, 27, 29 for $n = 3, 4, 5, 6$, respectively; the lowest 4p satellites of the $^2P^o$ manifold are no. 6 for $3p^4(^3P)4p\ ^2P^o$, no. 10 for $3p^4(^1D)4p\ ^2P^o$, and no. 17 for $3p^4(^1S)4p\ ^2P^o$.)

5.4 Spin-polarization of $5p_{3/2}$ photoelectrons in xenon

The complete experiment for 2p photoionization in magnesium described previously depends on the validity of the non-relativistic *LSJ*-coupling scheme and on the existence of a simple subsequent Auger transition. However, such conditions are rarely met, since in heavier elements spin–orbit effects cannot be neglected, and for outer-shell photoionization no subsequent decay is possible. In order to perform a complete experiment for such cases,† measurement of the spin-polarization of the photoelectrons is necessary. As an example, 5p photoionization in xenon will be discussed.

The ejection of a 5p-electron in xenon leads to two fine-structure resolved ionic states, $5p^5\ ^2P_{3/2}$ and $5p^5\ ^2P_{1/2}$, which will be identified for simplicity with the ejection of a $5p_{3/2}$ and a $5p_{1/2}$ electron, respectively. One then has the following channels for the dipole matrix elements D_y (see equs. (5.7), (5.8), and (8.36)):

$$5p_{3/2}\varepsilon s_{1/2} \text{ with } D_-, \tag{5.45a}$$

$$5p_{3/2}\varepsilon d_{3/2} \text{ with } D_0, \tag{5.45b}$$

$$5p_{3/2}\varepsilon d_{5/2} \text{ with } D_+ \tag{5.45c}$$

and

$$5p_{1/2}\varepsilon s_{1/2} \text{ with } \tilde{D}_0, \tag{5.46a}$$

$$5p_{1/2}\varepsilon d_{3/2} \text{ with } \tilde{D}_+. \tag{5.46b}$$

$5p_{3/2}$ photoionization in xenon is governed by three complex dipole amplitudes,

† Closed-shell atoms are considered here, because in open-shell systems the photoionization process is generally determined by more than three matrix elements (five parameters).

instead of the two amplitudes discussed in the magnesium example. Consequently, five complementary measurements are needed for the determination of the three absolute values and the two relative phases of the matrix elements involved. These pieces of information are provided by the partial cross section for photoionization, $\sigma(5p_{3/2})$, the angular distribution parameter $\beta(5p_{3/2})$, and the three components of the spin-polarization vector **P** of the photoelectron. This five-to-five relation underlines the importance of spin-polarization measurements which will be discussed here for the case of $5p_{3/2}$ photoionization in xenon. The main topics will be the parametrization of observables for a spin measurement, including the general principle of spin detection by Mott scattering, the experimental set-up for spin-dependent measurements with synchrotron radiation, and finally the presentation of the results in terms of dipole matrix elements and relative phases.

5.4.1 Parametrization for spin observation

In analogy with the parametrization of the angle-dependent partial cross section for photoionization discussed in connection with equ. (1.50), the parametrization for the additional observation of the electron spin has to be worked out. It depends on

(i) the Stokes parameters of the incident light (included in the quantity **S**);
(ii) the polar and azimuthal angles $\{\Theta, \tilde{\Phi}\}$ of the emitted photoelectron (in the tilted collision frame introduced in Fig. 1.15);
(iii) the Stokes parameters of the polarization-sensitive detector (included in the quantity **Q**), they describe for a preselected detector system (x', y' and z'; see Fig. 1.5) the sensitivity of the electron detector to the spin of the electrons (see Section 9.2.1).

With these introductory definitions, the general parametrization can be represented as [HJC81]:

$$\begin{aligned} I(\mathbf{S}, \Theta, \tilde{\Phi}, \mathbf{Q}) = \frac{\sigma}{4\pi}\, [& I^{(0)}(\mathbf{S}, \Theta, \tilde{\Phi}) \\ & + (-S_3\xi + \tilde{S}_1 \sin 2\tilde{\Phi}\, \eta) \sin\Theta\, Q_{x'} \\ & + (1 + \tilde{S}_1 \cos 2\tilde{\Phi})\eta \sin\Theta \cos\Theta\, Q_{y'} \\ & + (-S_3\zeta \cos\Theta) Q_{z'}]. \end{aligned} \tag{5.47a}$$

(Note the optical definition of light polarization **S** used here (equ. (9.30)).) $I^{(0)}(\mathbf{S}, \Theta, \tilde{\Phi})$ is the angular distribution function for a measurement without spin observation ($Q_{x'} = Q_{y'} = Q_{z'}$; see equ. (1.50)) given by

$$I^{(0)}(\mathbf{S}, \Theta, \tilde{\Phi}) = 1 - \frac{\beta}{2}[P_2(\cos\Theta) - \tfrac{3}{2}\tilde{S}_1 \cos 2\tilde{\Phi} \sin^2\Theta]. \tag{5.47b}$$

The spin-dependent terms of the photoionization process are those associated with the spin-dependent Stokes parameters of the detector. There it can be seen that three new dynamical parameters appear, ξ, η, ζ. They join the list of observables like the cross section σ and the angular distribution parameter β, and like these quantities they depend on the dipole matrix elements involved (for explicit expressions in the LSJ-coupling limit see equs. (5.17)–(5.19)).

From the formulas listed, the components $P_{x'}$, $P_{y'}$ and $P_{z'}$ of the photoelectron's spin-polarization vector **P** defined in the x', y', z' detector frame can be calculated (for the definition of **P** see Fig. 1.5 and equ. (9.15)). If *tr* stands for transverse, *long* for longitudinal, and if the subscripts $\|$ and $\perp$ indicate the component within or perpendicular to the 'scattering' plane, respectively, one gets†

$$P_{x'}(\mathrm{S}, \Theta, \tilde{\Phi}) = P_{\mathrm{tr}\|}(\mathrm{S}, \Theta, \tilde{\Phi}) = \frac{I(\mathrm{S}, \Theta, \tilde{\Phi}, Q_{x'} = +1) - I(\mathrm{S}, \Theta, \tilde{\Phi}, Q_{x'} = -1)}{I(\mathrm{S}, \Theta, \tilde{\Phi}, Q_{x'} = +1) + I(\mathrm{S}, \Theta, \tilde{\Phi}, Q_{x'} = -1)}$$
$$= (-S_3\xi + \tilde{S}_1 \sin 2\tilde{\Phi}\, \eta) \sin\Theta / I^{(0)}(\mathrm{S}, \Theta, \tilde{\Phi}), \tag{5.48a}$$

$$P_{y'}(\mathrm{S}, \Theta, \tilde{\Phi}) = P_{\mathrm{tr}\perp}(\mathrm{S}, \Theta, \tilde{\Phi}) = \frac{I(\mathrm{S}, \Theta, \tilde{\Phi}, Q_{y'} = +1) - I(\mathrm{S}, \Theta, \tilde{\Phi}, Q_{y'} = -1)}{I(\mathrm{S}, \Theta, \tilde{\Phi}, Q_{y'} = +1) + I(\mathrm{S}, \Theta, \tilde{\Phi}, Q_{y'} = -1)}$$
$$= (1 + \tilde{S}_1 \cos 2\tilde{\Phi})\eta \sin\Theta \cos\Theta / I^{(0)}(\mathrm{S}, \Theta, \tilde{\Phi}), \tag{5.48b}$$

$$P_{z'}(\mathrm{S}, \Theta, \tilde{\Phi}) = P_{\mathrm{long}}(\mathrm{S}, \Theta, \tilde{\Phi}) = \frac{I(\mathrm{S}, \Theta, \tilde{\Phi}, Q_{z'} = +1) - I(\mathrm{S}, \Theta, \tilde{\Phi}, Q_{z'} = -1)}{I(\mathrm{S}, \Theta, \tilde{\Phi}, Q_{z'} = +1) + I(\mathrm{S}, \Theta, \tilde{\Phi}, Q_{z'} = -1)}$$
$$= (-S_3\zeta \cos\Theta) / I^{(0)}(\mathrm{S}, \Theta, \tilde{\Phi}). \tag{5.48c}$$

From these relations it can be seen that only circularly polarized light (non-vanishing S_3) allows the determination of all spin-polarization parameters, ξ, η, ζ; linearly polarized ($\tilde{S}_1 = 1$) or unpolarized light ($\tilde{S}_1 = S_3 = 0$) yields only η. Therefore, a complete experiment for $5p_{3/2}$ photoionization in xenon, where σ, β, ξ, η, ζ are needed, relies on the availability of circularly polarized synchrotron radiation.

5.4.2 Spin detection by Mott scattering

There are in principle several interaction processes suitable for measuring the spin projection of an electron but only few of them such as Mott scattering [Mot29] have turned out to be practical (for an extensive discussion on polarized electrons see [Kes85]). Mott scattering is an elastic scattering process of an electron in the Coulomb potential of an atom in which the spin–orbit interaction plays an important role. This process will be explained in detail with the help of Fig. 5.17. In Fig. 5.17(*a*) the path of the electron under the influence of the ionic potential inside the scattering atom is shown in the laboratory frame. The electron not only moves under the influence of the Coulomb field, it also 'feels' a magnetic field

† In these relations the interest lies in the spin-polarization vector of the photoelectron itself, i.e., the 'detector' response **Q** is assumed to be always perfect, $Q_i = \pm 1$. The same procedure applies if the response of an actual detector with $|Q_i| < 1$ to polarized electrons is calculated (see below where, for the case of Mott scattering, Q_i has to be identified with the Sherman function S_S).

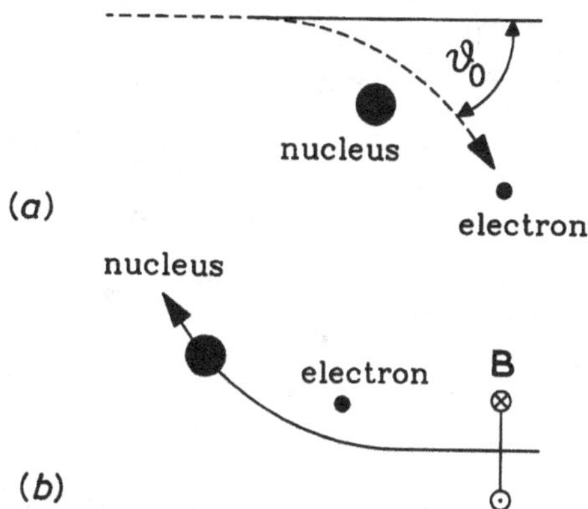

Figure 5.17 Electron scattering in the electric field of a nucleus: (*a*) picture in the laboratory frame where the nucleus is at rest; (*b*) frame of the electron in which the current of the moving nucleus produces a magnetic field **B** as indicated.

induced by its movement in the electrical field. This additional interaction can be understood best by considering the scattering process in the rest frame of the electron shown in Fig. 5.17(*b*). There the motion of the ion produces a magnetic field **B** which interacts with the magnetic moment $\boldsymbol{\mu}$ of the electron. (Note that the magnetic moment $\boldsymbol{\mu}$ and the spin **s** of the electron are related to each other by

$$\boldsymbol{\mu} = -2\mu_B \mathbf{s}, \tag{5.49}$$

where μ_B is the Bohr magneton and **s** is measured in units of $\hbar$.) Hence, an additional energy term is obtained given by

$$\Delta E|_{\text{rest frame}} = -\boldsymbol{\mu}\cdot\mathbf{B}. \tag{5.50a}$$

The transformation back to the laboratory frame brings in the so-called Thomas factor, and one derives [Tho26, ERe85]

$$\Delta E|_{\text{lab frame}} = -\tfrac{1}{2}\boldsymbol{\mu}\cdot\mathbf{B}. \tag{5.50b}$$

If worked out quantitatively one gets for the **B** field

$$\mathbf{B} = -\frac{1}{c^2}\mathbf{v}\times\mathbf{E}. \tag{5.51}$$

Replacement of **E** by the gradient of the atomic potential $V(r)$ gives a numerical factor and the vector **r**, and substitution of $\mathbf{r}\times\mathbf{v}$ by the orbital angular momentum $\boldsymbol{\ell}$ of the moving electron yields finally

$$\Delta E|_{\text{lab frame}} = \frac{1}{2m_0^2c^2}\frac{1}{r}\frac{\mathrm{d}V(r)}{\mathrm{d}r}\mathbf{s}\cdot\boldsymbol{\ell}. \tag{5.50c}$$

This expression says that the interaction energy is due to the spin–orbit interaction which, therefore, is responsible for the spin-dependent effect in Mott scattering.

The energy of the spin–orbit interaction adds to the potential of the Coulomb

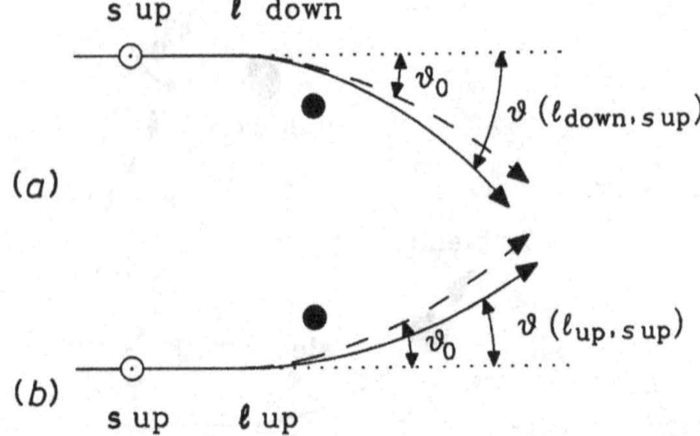

Figure 5.18 The spin–orbit effects during electron scattering in the field of a nucleus (see also [Hof58]). Two situations shown are: in (*a*) electron scattering to the right and in (*b*) electron scattering to the left. The nucleus is indicated by the full circle, the electron trajectories by the solid lines. These solid lines define the scattering plane which coincides with the drawing plane of the figure. With respect to the scattering plane the incident electron is assumed to have in both cases its spin pointing upwards, but depending on the direction of scattering, its orbital angular momentum ℓ can have the values 'up' and 'down' as indicated. The dashed curves describe the trajectory of the electron neglecting spin–orbit effects; they yield the same scattering angle ϑ_0 independent of whether scattering occurs to the right or to the left. The solid curves describe the electrons' trajectories if spin–orbit interaction is taken into account; they yield a left–right asymmetry in the scattering angle as indicated in the figure.

interaction, leading to

$$V_{\text{total}} = -\frac{1}{4\pi\varepsilon_0}\frac{e_0^2 Z}{r} + \frac{1}{2m_0^2c^2}\frac{1}{r}\frac{\mathrm{d}V(r)}{\mathrm{d}r}\,\mathbf{s}\cdot\boldsymbol{\ell}. \tag{5.52}$$

From this equation two conclusions relevant for spin-dependent measurements by Mott scattering can be drawn. First, due to the dot-product in the spin–orbit interaction term only the transverse component of the electron spin can be measured (the orbital angular momentum $\boldsymbol{\ell}$ is always perpendicular to the scattering plane). Second, depending on the spin direction (up or down) and on the orbital angular momentum $\boldsymbol{\ell}$ of the incident electron, the spin–orbit interaction term weakens or increases the strength of the Coulomb interaction. This leads to a left–right asymmetry in the scattering angles, which is demonstrated in Fig. 5.18 for an electron with spin up (s_{up}). If the scattering occurs to the right, $\boldsymbol{\ell}$ points down (ℓ_{down}), and for s_{up} and ℓ_{down} the interaction potential V_{total} becomes stronger, i.e., the scattering angle $\vartheta_{right}(\ell_{down}, s_{up})$ will increase in comparison to the scattering angle ϑ_0 without spin–orbit interaction. On the other hand, if the scattering occurs to the left, $\boldsymbol{\ell}$ points up (ℓ_{up}), for s_{up} and ℓ_{up} the interaction potential becomes weaker, and the scattering angle $\vartheta_{left}(\ell_{up}, s_{up})$ will decrease. Similarly, $\vartheta_{right}(\ell_{down}, s_{down})$ becomes smaller and $\vartheta_{left}(\ell_{up}, s_{down})$ larger than ϑ_0.

In the next step the left–right asymmetry in the scattering angles has to be transferred into a left–right asymmetry in the scattering cross section. Hereby one has to bear in mind that a change in the scattering angle means a shift of the

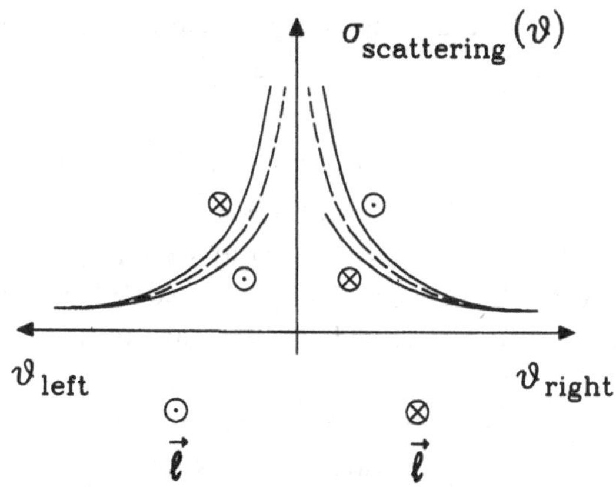

Figure 5.19 Schematic illustration of the spin–orbit dependence of the scattering cross section $\sigma_{scattering}(\vartheta)$. The cross section values for electron scattering to the right and left, respectively, are shown (see Fig. 5.18). Since these directions are related to the orbital angular momentum ℓ being 'down' or 'up' with respect to the scattering plane, the corresponding ℓ values are also given. The dashed curves refer to the cross section neglecting spin–orbit effects (Rutherford scattering); they are the same on both sides. The curves with higher/lower values than the dashed ones represent the cross section which includes spin–orbit effects. Attached to these curves is the up/down value of the spin projection of the electron (given with respect to the scattering plane). Depending on the actual spin projection, the different values of the scattering cross section for equal scattering angles ϑ, but different directions, can be seen. This results in a left–right asymmetry of observed scattering intensities.

original intensity, given at ϑ_0, to the new scattering angle. Hence, one derives four scattering cross sections, according to the combinations $\sigma(\text{left}, \ell_{up}, s_{up})$, $\sigma(\text{left}, \ell_{up}, s_{down})$, $\sigma(\text{right}, \ell_{down}, s_{up})$, $\sigma(\text{right}, \ell_{down}, s_{down})$. These are shown schematically in Fig. 5.19. With the help of this figure, the intensities I for the scattering of N incident electrons with transverse spins up or down can be calculated, giving[†]

$$I_{up}(\vartheta_{left}) = \text{const } N_{up}(1 - |S_S|), \tag{5.53a}$$

$$I_{down}(\vartheta_{left}) = \text{const } N_{down}(1 + |S_S|), \tag{5.53b}$$

$$I_{up}(\vartheta_{right}) = \text{const } N_{up}(1 + |S_S|), \tag{5.53c}$$

$$I_{down}(\vartheta_{right}) = \text{const } N_{down}(1 - |S_S|), \tag{5.53d}$$

where S_S describes the effect of the change in the scattering cross section. This quantity is called the *Sherman function* [She56], and its value $S_S = -0.4099 \pm 0.0044$ [GKe91] is well established for 120 keV electrons scattered on a gold foil and detected with scattering angles of $\pm 120°$ (for the influence of the finite thickness of the foil see [GKe91]; for earlier work see [Kli66]).

With the formulas given it is now possible to derive the required relation for the determination of the transverse spin component of an incident electron beam. If the beam contains N_{up} electrons with spin up and N_{down} electrons with spin

† These relations refer to a negative value of the Sherman function obtained in a pure Coulomb field.

down, its total intensity N_0 and transverse polarization $P_\perp$ follow from

$$N_0 = N_{up} + N_{down} \tag{5.54a}$$

and

$$P_\perp = \frac{N_{up} - N_{down}}{N_{up} + N_{down}}. \tag{5.54b}$$

In order to measure $P_\perp$ by spin-dependent Mott scattering, the incident electron has to be accelerated to 120 keV, scattered on a gold foil, and the backscattered intensities at $\pm 120°$, called $I(\vartheta_{left})$ and $I(\vartheta_{right})$, have to be recorded. This gives the signals

$$\begin{aligned} I(\vartheta_{left}) &= I_{up}(\vartheta_{left}) + I_{down}(\vartheta_{left}) \\ &= \text{const } N_0(1 - |S_S| P_\perp) \end{aligned} \tag{5.55a}$$

and similarly

$$\begin{aligned} I(\vartheta_{right}) &= I_{up}(\vartheta_{right}) + I_{down}(\vartheta_{right}) \\ &= \text{const } N_0(1 + |S_S| P_\perp), \end{aligned} \tag{5.55b}$$

and from the combination of these one obtains

$$P_\perp = -\frac{1}{|S_S|} \frac{I(\vartheta_{left}) - I(\vartheta_{right})}{I(\vartheta_{left}) + I(\vartheta_{right})} = \frac{1}{S_S} \frac{I(\vartheta_{left}) - I(\vartheta_{right})}{I(\vartheta_{left}) + I(\vartheta_{right})}, \tag{5.56}$$

i.e., the unknown transverse polarization $P_\perp$ follows from the observed left–right asymmetry of counting rates and the known value of the Sherman function.

5.4.3 *Experimental set-up for spin-dependent measurements*

Having established the parametrization for the angle- and spin-dependent partial cross section, the method of measuring transverse spin components by Mott scattering, and having access to monochromatized synchrotron radiation with right or left circular polarization as desired, the appropriate experimental set-up for spin-dependent measurements of photoelectrons can now be developed. The two basic components are an electron spectrometer through which a desired photoprocess may be selected (energy analysis according to equ. (1.29)) and the Mott detector for the detection of transverse spin polarizations. Since equs. (5.48) simplify greatly for $\tilde{\Phi} = 45°$, an experimental set-up as shown in Fig. 5.20 is convenient. The rotatable electron spectrometer accepts electrons in a small solid angle $\Delta\Omega$ with average angles $\tilde{\Phi}_0$ and Θ_0, where $\tilde{\Phi}_0 = 45°$, but Θ_0 can be selected by rotating the spectrometer. Two 90° deflectors are needed to allow the rotation of the electron energy analyser, while keeping the Mott detector at a fixed position in space. The first 90° deflection condensor directs the angle- and energy-analysed electrons back to the rotation axis of the electron spectrometer. When the electrons have traversed a zoom lens, they are directed by a second 90° deflector towards the Mott detector. On the way to the gold foil the electrons are accelerated, and after backscattering they are recorded by two pairs of detectors with respect to their transverse spin components (see below). When the electrons travel through

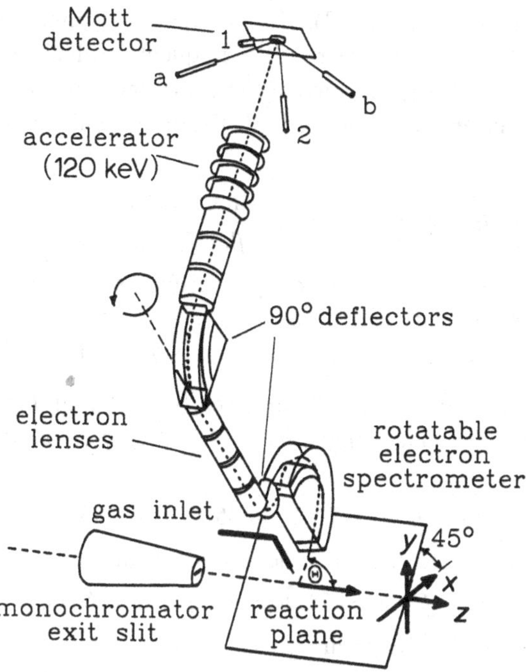

Figure 5.20 Schematic drawing of an apparatus for measuring the spin-polarization of photoelectrons. The *xyz* coordinate system refers to the tilted frame introduced in Fig. 1.15. For a detailed explanation see main text. From [HSS84].

all these optical elements, their spin components remain unchanged in space, because for non-relativistic velocities the electrostatic fields do not affect the magnetic moments and, hence, the spins. Therefore, one can record with one pair of electron detectors (numbers 1 and 2 in Fig. 5.20) the spin component $P_z(\mathbf{S}; \Theta_0, \tilde{\Phi}_0 = 45°)$, and with the other pair (a and b in Fig. 5.20) the component $P_{y'}(\mathbf{S}, \Theta_0, \tilde{\Phi}_0 = 45°)$. P_z is a measure of the spin polarization along the photon beam direction and is given by

$$P_z(\mathbf{S}, \Theta, \tilde{\Phi}) = P_{z'}(\mathbf{S}, \Theta, \tilde{\Phi}) \cos\Theta - P_{x'}(\mathbf{S}, \Theta, \tilde{\Phi}) \sin\Theta, \tag{5.57a}$$

which, using equs. (5.48a) and (5.48c), yields

$$P_z(\mathbf{S}, \Theta_0, \tilde{\Phi}_0) = \frac{-\eta \sin^2\Theta_0}{1-(\beta/2)P_2(\cos\Theta_0)} \tilde{S}_1 S_\mathrm{S} + \frac{A - \alpha P_2(\cos\Theta_0)}{1-(\beta/2)P_2(\cos\Theta_0)} (-S_3) S_\mathrm{S}, \tag{5.57b}$$

if the spin polarization parameters ξ and ζ are replaced by

$$A = \tfrac{1}{3}\zeta - \tfrac{2}{3}\xi \qquad \text{and} \qquad \alpha = -\tfrac{2}{3}(\xi + \zeta). \tag{5.58}$$

The parameter A, related to P_z, is the only spin-polarization component which does not vanish if all electrons are spin-analysed regardless of their direction of emission (the *Fano effect* [Fan69]). If P_z is measured angle-resolved, one also obtains the component α and the result is frequently called the *angle-dependent* Fano effect. $P_{y'}(\mathbf{S}, \Theta, \tilde{\Phi})$ is a measure of the transverse spin polarization perpendicular to the reaction plane defined by the photoionization process, and is given by

$$P_{y'}(\mathbf{S}, \Theta_0, \tilde{\Phi}_0 = 45^\circ) = \frac{\eta \sin \Theta_0 \cos \Theta_0}{1 - (\beta/2) P_2(\cos \Theta_0)} (-S_3) S_S. \qquad (5.59)$$

Of the many possibilities for performing experiments with the set-up described, three special measurements can be selected to determine the spin-polarization dynamical parameters A, α and η of a selected photoionization process [HSS86]:

(i) A measurement with circularly or linearly polarized light, and with the spectrometer at the magic angle $\Theta_0 = 54.7°$. Using circularly polarized light, one measures $P_{y'}$ and gets

$$P_{y'}(S_3, \Theta_0 = 54.7°, \tilde{\Phi}_0 = 45°) = 0.4716\, \eta\, (-S_3) S_S, \qquad (5.60a)$$

with linearly polarized light, one determines P_z and obtains

$$P_z(\tilde{S}_1, \Theta_0 = 54.7°, \tilde{\Phi}_0 = 45°) = -0.6667\, \eta\, \tilde{S}_1 S_S. \qquad (5.60b)$$

Either method yields the parameter η.

(ii) Two measurements with right- and left-circularly polarized light, and with the spectrometer at the magic angle Θ_0. Here one gets

$$P_z(\tilde{S}_1, S_3 = \text{rcp}, \Theta_0 = 54.7°, \tilde{\Phi}_0 = 45°) = (-0.6661\, \eta\, \tilde{S}_1 + A\, |S_3|) S_S, \quad (5.61a)$$

and

$$P_z(\tilde{S}_1, S_3 = \text{lcp}, \Theta_0 = 54.7°, \tilde{\Phi}_0 = 45°) = (-0.6661\, \eta\, \tilde{S}_1 - A\, |S_3|) S_S, \quad (5.61b)$$

The contribution from $\tilde{S}_1$ can be eliminated by subtraction, and the dynamical parameter A follows.

(iii) A series of measurements like those in (ii), but at other angle settings Θ_0 of the electron spectrometer. These measurements yield $P_z(S_1, S_3 = \text{rcp}, \Theta_0, \tilde{\Phi}_0 = 45°)$ and $P_z(S_1, S_3 = \text{lcp}, \Theta_0, \tilde{\Phi}_0 = 45°)$ as given in equ. (5.57b). Each of these quantities or their difference is then analysed with respect to the remaining unknown dynamical parameter α (and possibly β).

5.4.4 Results for $5p_{3/2}$ photoionization

The results of a spin-polarization measurement of xenon photoelectrons with $5p^5\ ^2P_{3/2}$ and $5p^5\ ^2P_{1/2}$ final ionic states are shown in Fig. 5.21 together with the results of theoretical predictions. Firstly, there is good agreement between the experimental data (points with error bars) and the theoretical results (solid and dashed curves, obtained in the relativistic and non-relativistic random-phase approximations, respectively). This implies that relativistic effects are small and electron–electron interactions are well accounted for. (In this context note that the fine-structure splitting in the final ionic states has also to be considered in

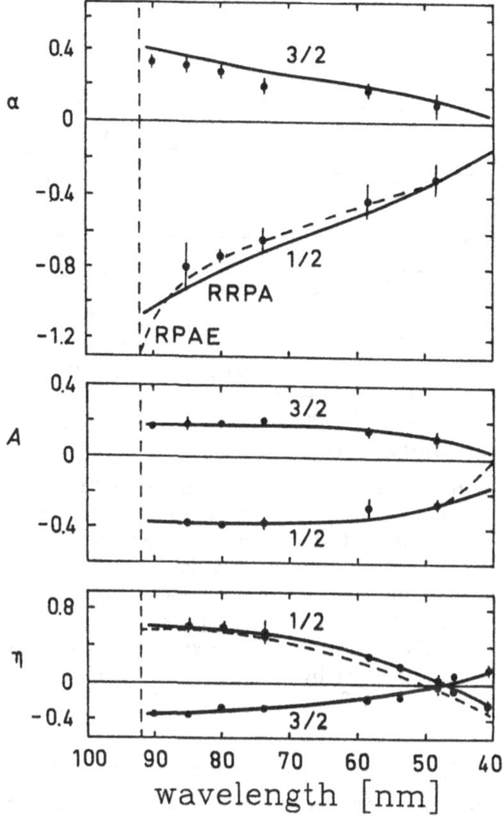

Figure 5.21 Spin-polarization parameters α, A, and η as functions of the photon wavelength in the continuous range for photoelectrons leaving the xenon ion in the $5p^5\ ^2P_{1/2}$ and $^2P_{3/2}$ states, respectively. The J value of the final state is indicated on the curves; the vertical dashed line shows the $J = 1/2$ ionization threshold. Experimental data: full circles [HSS86]. Theoretical data: full curves, relativistic random-phase calculation [HJC81]; dashed curves: non-relativistic random-phase calculation [Che79]. From [HSS86]; note ξ[HSS86] $= 0.5\,\eta$.

RPA (see below) and that the spin-polarization parameters are relative quantities like the β-parameter (matrix elements in the numerator and denominator).) Further, there are large differences in the spin-polarization parameters belonging to the $^2P_{3/2}$ and $^2P_{1/2}$ final ionic states, in magnitude as well as in sign. This behaviour can be traced back to the important fact that spin-polarization phenomena can be observed only if non-vanishing spin–orbit couplings exist, either in the initial atomic state and/or in the final state. This means, in particular, that no spin polarization can be observed if the two fine-structure components in the initial state are unresolved and spin–orbit effects in the continuum are negligible (see equs. (5.17)–(5.19)). This can be verified directly from Fig. 5.21: if the spin polarization components of the $^2P_{1/2}$ and $^2P_{3/2}$ final ionic states are added taking into account the corresponding statistical weights of 2:4, one gets nearly

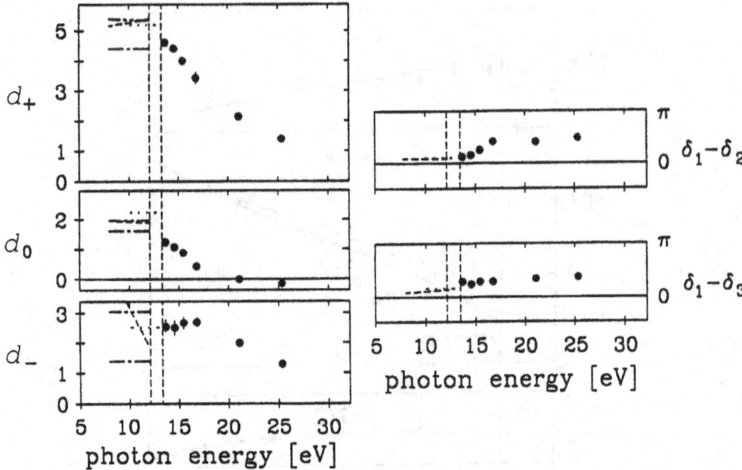

Figure 5.22 Reduced matrix elements d_γ, in atomic units, and relative phases $(\delta_\gamma - \delta_\nu)$, in radians, as functions of photon energy for 5p photoionization in xenon leading to the $5p^5\,{}^2P_{3/2}$ final ionic state. The dashed vertical lines give the ${}^2P_{3/2}$ and ${}^2P_{1/2}$ ionization thresholds at 12.13 and 13.44 eV, respectively. The data in the continuous range above the ${}^2P_{1/2}$ threshold are shown as full circles with error bars. The curves below the ${}^2P_{3/2}$ and ${}^2P_{1/2}$ ionization thresholds are expected to approach the values in the continuum continuously (for details see the discussion and references in the original publication). From [HSS86]; the d_i given here are larger by $\sqrt{3}$ in order to adapt them to the cross section defined in equ. (8.39a).

zero. The deviation is due to the non-vanishing spin–orbit interaction in the final state, for example, spin–orbit effects of the escaping photoelectron.

The experimental data in Fig. 5.21, combined with measured values for the partial cross section $\sigma({}^2P_{3/2})$ and the angular distribution parameter $\beta({}^2P_{3/2})$, finally provide the complete information on $5p^{-1}\,{}^2P_{3/2}$ photoionization in xenon required, and experimental values for the dipole matrix elements D_γ can be evaluated [Hei80, HSS86]. The results are shown in Fig. 5.22 using the notation (note the different signs from the factor $i^{-\ell_\gamma}$ in equ. (8.35))

$$D_+ = -d_+\,e^{i\sigma_d}\,e^{i\delta_{\varepsilon d5/2}} = -d_+\,e^{i\sigma_d}\,e^{i\delta_1}, \tag{5.62a}$$

$$D_0 = -d_0\,e^{i\sigma_d}\,e^{i\delta_{\varepsilon d3/2}} = -d_0\,e^{i\sigma_d}\,e^{i\delta_2}, \tag{5.62b}$$

$$D_- = d_-\,e^{i\sigma_s}\,e^{i\delta_{\varepsilon s1/2}} = d_-\,e^{i\sigma_s}\,e^{i\delta_3}, \tag{5.62c}$$

where d_γ is the magnitude, σ_γ the Coulomb phase and δ_γ the phase from the short-range potential (compare the related case of 2p photoionization in magnesium, equ. (5.11b)). From the figure the inherent dynamics of the photoionization process can be inferred. First, it can be seen that the data in the continuum smoothly match the values in the discrete region of the lower ionization threshold (curves below 12.13 eV photon energy). Such a smooth transition between the regions is a rather general phenomenon. In the present case it deserves special attention because extensive experimental and theoretical studies of the discrete region exist within the framework of the *multichannel quantum defect theory*

(MQDT, for details see references in [HSS86] and [Sch92a]). Second, for higher photon energies the reduced matrix elements d_γ and the relative phases show a pronounced energy dependence which is due to a strong spin–orbit interaction in the continuum. This interaction appears most clearly in the phase difference δ_1–δ_2 and in the amplitudes d_+ and d_0 which both refer to differences in the $\varepsilon d_{5/2}$ and $\varepsilon d_{3/2}$ ionization channels, because for neglected continuum spin–orbit interaction one expects $\delta_1-\delta_2 = 0$ and $d_+ = 3d_0$. Such a relation holds only approximately at the ionization threshold, but it fails completely at about 5 eV above threshold where d_0 even goes through zero while d_+ remains finite.

5.5 PCI between $4d_{5/2}$ photoelectrons and N_5–$O_{2,3}O_{2,3}$ $^1S_0$ Auger electrons in xenon

The basic aspects of PCI have been presented already in Sections 4.5.5 and 5.1.2.2. Here an experiment will be described the aim of which was the quantitative study of the energy position and shape of a selected Auger transition in order to allow a critical comparison with theoretical predictions. Special emphasis was placed on the question of whether, and possibly how, the finite velocity of the Auger electron affects PCI. In earlier PCI models it was assumed that the kinetic energy of the Auger electron is so large that it suddenly leaves the system ([Nie77], the *sudden* PCI model). As a result, PCI phenomena extend up to very high photon energies. In contrast, if the finite velocity of the Auger electron is included in the theoretical description, PCI phenomena are expected to vanish if the photoelectron becomes faster than the Auger electron, because then the photoelectron is not subject to a change of nuclear shielding ([Ogu83, RMe86], see also [ATA87, KSh89] and references therein). In a time-dependent picture, the time at which the Auger electron with finite velocity overtakes the photoelectron is delayed, and this description can be called the *retarded* PCI model. Due to the high accuracy needed in the experimental data to distinguish between these models, this study also serves as an example of the power of electron spectrometry with synchrotron radiation where absolute kinetic energies, correct lineshapes and a variable energy photon source are essential.

5.5.1 Experimental details

The system selected for the PCI study is the N_5–$O_{2,3}O_{2,3}$ $^1S_0$ Auger electron of xenon and its PCI 'inducer', the $4d_{5/2}$ photoelectron. In Fig. 5.23 the electron spectrum of xenon obtained after ionization with 108 eV photons is shown. The huge lines at 40.45 eV and 38.46 eV kinetic energy are the two photolines $4d_{5/2}$ and $4d_{3/2}$. Their decay leads to the $N_{4,5}$–OO Auger spectrum where the N_5–$O_{2,3}O_{2,3}$ and N_4–$O_{2,3}O_{2,3}$ groups are in the given energy range, and their positions are marked by the bar diagram in the figure. The N_5–$O_{2,3}O_{2,3}$ $^1S_0$ line at 29.97 eV has been selected for the PCI study because it is well separated from neighbouring lines. Furthermore, as can be seen from the photon flux curve of the

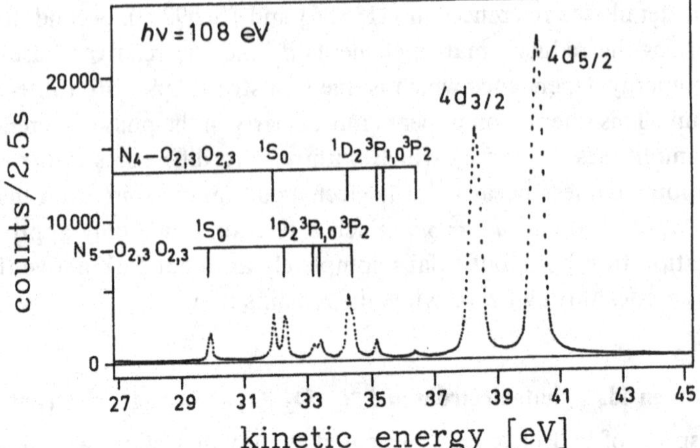

Figure 5.23 Part of the ejected electron spectrum in xenon following ionization with 108.0 eV photons (measured at the quasi-magic angle). The $4d_{5/2}$ and $4d_{3/2}$ photolines and individual Auger lines from $N_{4,5}$–$O_{2,3}O_{2,3}$ Auger transitions are indicated. From [Ehr91].

available monochromator (Fig. 1.12), PCI processes in xenon can be investigated for photon energies in the interesting region where the kinetic energies of the photoelectrons can be made larger or smaller than the Auger electron energy. Also, higher-order light from this monochromator is negligible in this energy region, a point of great importance because otherwise higher-order light would produce an Auger line with a different shape superimposed on the PCI lineshape of interest.

The experimental set-up was shown in Fig. 1.17. In particular, sector 2 was placed at the quasi-magic angle to avoid influences from the angular distribution of Auger and photoelectrons. Because PCI changes the energies of the electrons, it is convenient to distinguish between *actual* and *nominal* energies where the nominal values are not disturbed by PCI.† For the selected Auger transition one has $E^0_A = 29.97$ eV, for the corresponding photoelectrons $E^0_{ph} = h\nu - E_I$ with the ionization energy $E_I = 65.548$ eV; often it is advantageous to replace E^0_{ph} by the *excess* energy E_{exc} which is defined as $E_{exc} = h\nu - E_I$.

5.5.2 Special calibration measurements

In order to allow quantitative studies of the influence of PCI, detailed information on the performance of the electron spectrometer is required. This includes the spectrometer factor f, the energy shift caused by contact potentials (the c value introduced in equ. (4.56c)), the correct shape of the spectrometer function, and

† This use of actual and nominal differs from the use of these words in connection with equ. (4.56); the necessary adjustment is made in equ. (5.63) where E_{kin} replaces the former E^0_{kin}.

the spectrometer resolution. The determination of these quantities requires special calibration measurements which are described in the following.

According to equ. (4.56b) the PCI-modified actual energy E_{kin} of the electrons differs from the value which can be extracted from the spectrometer voltage at peak maximum, U_{sp}, multiplied by the spectrometer factor f, due to the presence of a contact potential c:†

$$E_{kin} = fU_{sp} - c. \tag{5.63}$$

Hence, in order to have control over all constituents of this relation, one has to use a calibrated source for the spectrometer voltage U_{sp} and to seek suitable methods of measuring f and c accurately. For the latter purpose, helium and xenon were brought simultaneously to the source region of the apparatus using two separate gas inlets, i.e., all measurements, not only the calibrations, were performed on a xenon/helium gas mixture with constant pressures in order always to have the same conditions. Two separate calibration measurements for f and c were performed. An example of each case is shown in Fig. 5.24. The spectrum in Fig. 5.24(*a*) was taken at a photon energy of 90.5 eV, and two different energy regions, separated by the dashed vertical line, were scanned for the ejected electrons. Interest is focused on the helium 1s photoline, $He^+(1s)$, positioned in the right-hand region, and in its $n = 2$ satellite line, $He^{+*}(n = 2)$, positioned in the left region, because these lines are separated by the well-known energy $\Delta E = 40.8135$ eV [Moo71]. Applying equ. (5.63) to the difference of these two photolines, the c value cancels, and one gets for known ΔE and accurately measured ΔU_{sp} an accurate value for the spectrometer factor f; in the given example, $f = 1.6827(4)$ eV/V. In order to determine the c value, the monochromator energy was then set in a second experiment to approximately 52 eV where there is about 7% second-order light. The helium 1s (and xenon 5s) photolines produced by first- and second-order light are shown in Fig. 5.24(*b*), again in separate regions (for an improvement of counting statistics, the spectrum from second-order light was registered four times and summed afterwards). Applying equ. (5.63) to the first- and second-order photolines, one gets the nominal kinetic energies

$$E^0_{kin}(\text{first-order}) = E_{ph} - E_I = fU_{sp}(\text{first-order}) - c, \tag{5.64a}$$

$$E^0_{kin}(\text{second-order}) = 2E_{ph} - E_I = fU_{sp}(\text{second-order}) - c. \tag{5.64b}$$

Hence, an accurate value for the photon energy E_{ph} follows, using the right-hand part of these equations and taking their difference:

$$E_{ph} = f\,\Delta U_{sp}. \tag{5.65}$$

(This method is well suited for the calibration of a monochromator.) With known photon energy and helium ionization energy $E_I = 24.5874$ eV [Moo71], the

† The use of actual and nominal differs from the use of these words in connection with equ. (4.56); the necessary adjustment is made in equ. (5.63) where E_{kin} replaces the former E^0_{kin}. Also, strictly speaking the contact potential is c/e_0, and not c.

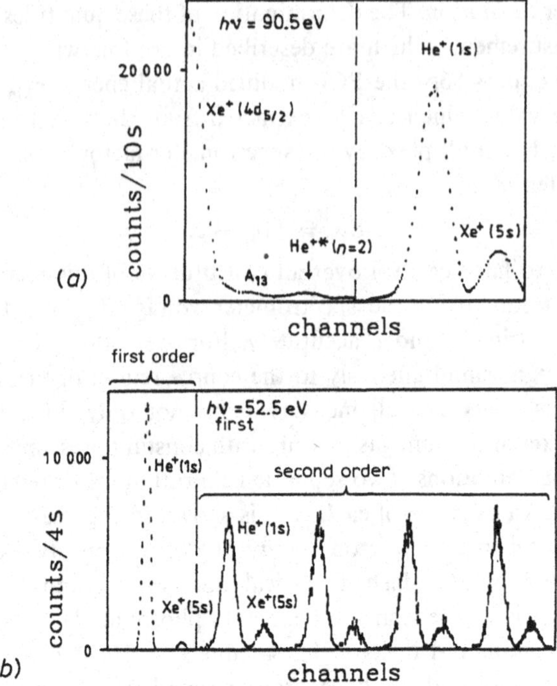

Figure 5.24 Photoelectron spectra used for calibration measurements with a target gas which is a mixture of helium and xenon. The abscissas are given in channels, i.e., in discrete steps of the spectrometer voltage which is ramped to cover certain energy regions of the ejected electrons. (*a*) The spectrum taken at a photon energy of 90.5 eV consists of two energy regions; that on the right-hand side contains the helium 1s and xenon 5s photolines, and that on the left part of the xenon $4d_{5/2}$ photoline, the so-called A_{13} Auger line in xenon, and the helium satellite for ionization and excitation to $n = 2$. From the position of the helium 1s and the $n = 2$ photolines the spectrometer factor f can be determined with high precision. (*b*) The spectrum taken at a photon energy of 52.5 eV also consists of two energy regions; that on the left contains the helium 1s and xenon 5s photolines, that on the right (magnified by a factor of 3 and repeated four times) again contains these two photolines, but obtained from the second-order light in the monochromatized synchrotron radiation (nominal photon energy 52.5 eV, see Fig. 1.12). From the positions of the helium 1s photolines measured in first and second order the exact photon energy can be evaluated (and with the spectrometer factor f the contact potential c/e_0 can be obtained). For details see main text.

left-hand side of equ. (5.64a) can then be used to calculate E^0_{kin} (first-order), and with accurate values for f and U_{sp}(first-order) the c value can be calculated. In the given example, $c = -37(14)$ meV. With known values of f and c, and the measured voltages U_{sp}, the absolute energy scale for arbitrary kinetic energies E_{kin} in equ. (5.63) can then be determined when performing specific measurements for the PCI effect (each individual study was accompanied by a set of such calibrations for f and c; within experimental error, all f and c values agreed well).

The shape observed for the Auger line results from the convolution of the

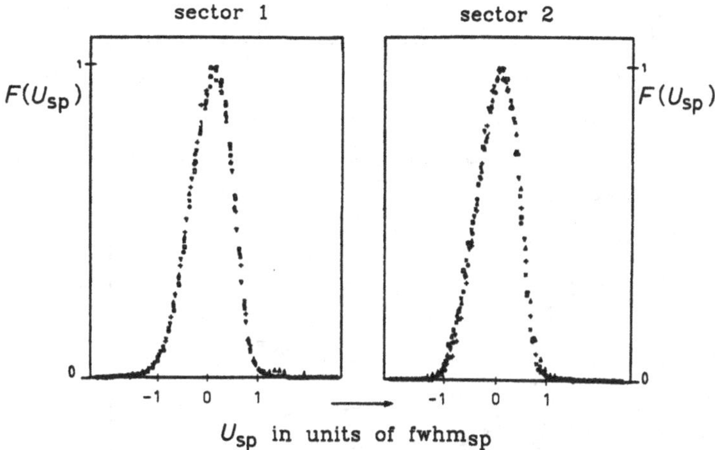

Figure 5.25 Shapes of spectrometer functions measured for two sectors of a double sector CMA. The individual symbols characterize data points from experiments performed at different kinetic energies between 5.64 eV (HeI light and $3\sigma_g$ ionization in the nitrogen molecule) and 28.68 eV (HeII light with 5p ionization in xenon). The individual spectra have been normalized to the same height and fwhm value in order to obtain a common lineshape. From [Mal82].

original lineshape, which is the possibly PCI distorted lineshape of interest, with the spectrometer function (see equ. (10.58b)). Therefore, one must know the correct shape of the spectrometer function before any statement can be made about the PCI lineshape. In the present experiment the shape of the spectrometer function was determined experimentally from outer-shell photoelectron spectra of rare gases and of the nitrogen molecule, using a HeI and HeII line source because their small line width is negligible compared to the spectrometer resolution (see Table 1.2). The individual energy scales of the photolines, measured between 5.6 eV and 28 eV kinetic energy, was then adapted to the same fwhm value in order to present a common lineshape. The results for both sectors of the analyser are shown in Fig. 5.25. It can be seen that all spectrometer functions have the same shape which is slightly asymmetrical with respect to a Gaussian.

Although the shape of the spectrometer function is well established, its width depends on the diameter of the photon beam (see Fig. 4.13). This diameter is difficult to determine for a divergent beam of HeI/II light, and the fwhm values from these measurements cannot be transferred to experiments with monochromatized synchrotron radiation. On the other hand, monochromatized synchrotron radiation has a low divergence, but there the relatively large photon bandpass ΔE_{ph} at the monochromator used prevents the extraction of the spectrometer resolution from an observed photoline ($\Delta E_{ph} \approx 0.2$ eV at 68 eV, $\Delta E_{ph} \approx 0.9$ eV at 125 eV photon energy). Therefore, an attempt was made to determine the correct width of the spectrometer function from an Auger line which is not burdened with the bandpass problem and does not show PCI effects. The Auger line selected

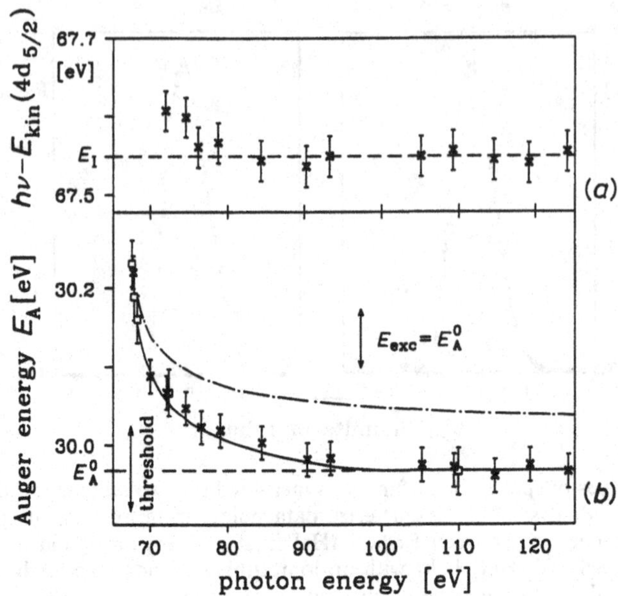

Figure 5.26 Absolute values for the kinetic energy of (*a*) $4d_{5/2}$ photoelectrons and (*b*) N_5–$O_{2,3}O_{2,3}$ $^1S_0$ Auger electrons in xenon as functions of photon energy. The symbols with error bars are experimental data. The horizontal broken lines indicate the values for the $4d_{5/2}$ ionization energy E_I, and the nominal energy E_A^0 of the Auger electron if PCI effects are absent. The solid and dash-dotted curves show the energy position of the maximum of the Auger line as predicted by the retarded and sudden PCI theories, respectively, taking into account the instrumental resolution. The characteristic value at which the excess energy E_{exc} becomes equal to the nominal Auger energy E_A^0 is also indicated in the figure. For lower photon energies, the photoelectron is slower than the Auger electron, and PCI effects are seen to become increasingly important as the $4d_{5/2}$ ionization threshold is approached; for higher photon energies the photoelectron is faster than the Auger electron, and PCI effects vanish in the retarded PCI model. From [Sch86], but adapted to $\Gamma = 120$ meV; see also [ASW87].

comes from the Auger transition of interest, measured at a photon energy of 125 eV, because there PCI disturbances are expected to be negligible (see below for a confirmation of this based on energy values). Hence, the measured Auger lineshape is compared with the convolution result between the spectrometer function (the experimental shape with a preselected ΔE_{sp} value) and a Lorentzian function (with a known Γ value for the natural width of the inner-shell hole-state, $\Gamma = 120$ meV[†]). The best fit to the experimental data then yields the desired ΔE_{sp} value. Before discussing in detail the results shown in Fig. 5.27, the quality of this fit as seen in the upper part of this figure should be noted: comparing the solid line (fit result) with the experimental data (points with error bars), very good agreement can be noted which implies a good value for ΔE_{sp}. In the present case,

[†] For this value see the measurement in [AOM95] which gives $\Gamma = 121(4)$ meV. The data shown in Figs. 5.26 and 5.27 have been adapted to this value.

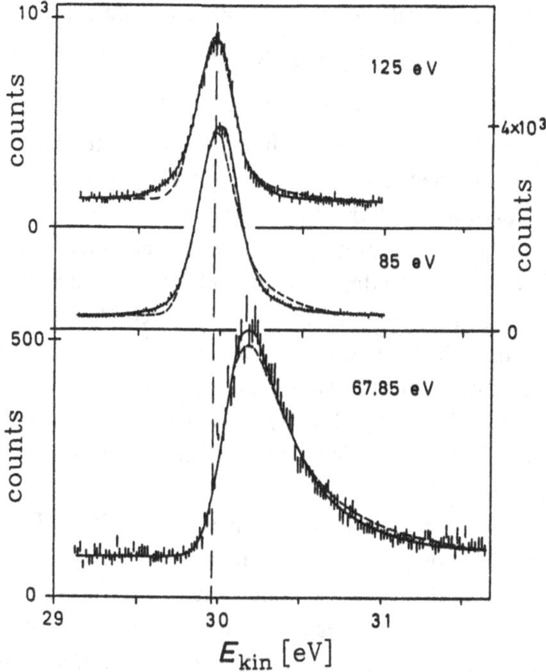

Figure 5.27 Shape and absolute energy position of the N_5–$O_{2,3}O_{2,3}$ $^1S_0$ Auger line of xenon at three different photon energies. At 125 eV photon energy PCI effects are absent, at 85 eV PCI effects start to become clearly visible, at 67.85 eV PCI effects are very pronounced. The solid curve is a fit based on predictions from the retarded theoretical PCI model [RMe86], taking into account the instrumental resolution. There is no free adaptable parameter except the height of the curves. The dashed curve is a fit based on predictions from the sudden PCI model [Nie77], again after convolution with the instrumental function, but keeping the height and the position as free parameters. From [BSc86, Sch86].

$\Delta E_{sp} = 204$ meV is obtained which corresponds to $(\Delta E/E)_{sp} = 0.68\%$. (The photon beam diameter was estimated to be approximately 2 mm; according to the ray-tracing results shown in Fig. 4.13 this corresponds to $(\Delta E/E)_{sp} = 0.64\%$.) Keeping fixed values for Γ and ΔE_{sp}, it is then possible to analyse the Auger lineshapes obtained at other, in particular lower, photon energies correctly.

5.5.3 Results

The position of the maximum of an electron line is the easiest quantity to extract from an experimental spectrum, and maxima of the xenon $4d_{5/2}$ photoline and N_5–$O_{2,3}O_{2,3}$ $^1S_0$ Auger line will be considered first for different photon energies. The results are shown in Fig. 5.26. For photon energies above 100 eV, i.e., for $E_{exc} > E_A^0$, the experimental values scatter within their error bars around a mean value. Such constant values are predicted in the retarded PCI model which includes the finite velocity of the Auger electron. Hence, this experimental result was a first confirmation of this theoretical model. The numerical data are

$$h\nu - E_{\text{kin}}(\text{phe})|_{E_{\text{exc}} > E^0_{\text{A}}} = E_{\text{I}}(4\text{d}_{5/2}) = 67.55 \pm 0.02\,\text{eV}, \tag{5.66a}$$

$$E_{\text{kin}}(A)|_{E_{\text{exc}} > E^0_{\text{A}}} = 29.97 \pm 0.02\,\text{eV}. \tag{5.66b}$$

Both values are in perfect agreement with values in the literature ([HPe82], see [Sch87] for a more extended discussion). Keeping these values fixed, and using $\Gamma = 120$ meV, the theoretical PCI energy distribution can be calculated for all excess energies,† and after convolution with the spectrometer function it can be compared with the experimental data. As a first characteristic property one obtains the maximum position which is plotted for the retarded PCI model in Fig. 5.26(*b*) as the solid curve. The values agree very well with the experimental data between $E_{\text{exc}} = E^0_{\text{A}}$ and the ionization threshold. In particular, at threshold one gets 265 ± 30 meV which is close to the 315 meV calculated from equ. (4.61). In contrast, the sudden PCI model yields the dashed-dotted curve which does not follow the experimental data.

In Fig. 5.26(*a*) and for $E_{\text{exc}} < E^0_{\text{A}}$ one can see that PCI effects also exist for the $4\text{d}_{5/2}$ photoline, because towards the ionization threshold $h\nu - E_{\text{kin}}(4\text{d}_{5/2})$ increases, i.e., $E_{\text{kin}}(4\text{d}_{5/2})$ decreases. A quantitative interpretation of these data, however, will be omitted for two reasons. First, the study requires quantitative electron spectrometry at low kinetic energies which is difficult due to the cutoff problem in the spectrometer transmission (see Fig. 4.15). Second, the cumulative effects of all possible Auger decays following $4\text{d}_{5/2}$ photoionization, not just a single chosen one, contribute to the photoline.

Finally, the lineshapes from the retarded and sudden PCI models, again convoluted with the spectrometer function, will be compared with the experimentally observed shape of the selected Auger transition. Three examples are shown in Fig. 5.27. At a photon energy of 125 eV the photoelectron is faster than the Auger electron and PCI effects are absent as has been discussed before. At a photon energy of 85 eV PCI effects are clearly seen, and at 67.85 eV which corresponds to $E_{\text{exc}} = 0.30$ eV, PCI is fully established, in particular the energy shift, the line broadening and the asymmetrical shape as discussed schematically in Fig. 4.42 can be seen clearly. The solid curve is the lineshape of the retarded model convoluted with the spectrometer function. It describes the experimental data very well. In this context it should be noted that the absolute energy scale and knowledge of all quantities relevant for the selected PCI model completely determine the position and shape of the observed Auger line. Hence, for the solid curves in Fig. 5.27 no free parameter, except the height, exists for adapting the convolution results to the experimental data. (This statement has to be weakened for the solid curve at 125 eV, because there information on ΔE_{sp} has been extracted, but it holds for all shapes of Auger lines taken at different excess energies.) For

† According to the PCI model in [RMe86], the PCI lineshape is fixed for given Γ, E_{exc} and E^0_{A} (and the average shell radii from which the photoelectron and Auger electron are ejected, but these quantities are not critical).

comparison, the dashed curves in Fig. 5.27 belong to the sudden PCI model, and their maxima have been shifted to the experimental peak positions in order to reduce the overall deviation, but even with this manipulation it can be seen that the sudden PCI model fails to reproduce the observed lineshapes. Therefore, the lineshape analysis also confirms the retarded PCI model, indicating the importance of including the finite velocity of the Auger electron.

5.6 Angular correlation between $4d_{5/2}$ photoelectrons and N_5–$O_{2,3}O_{2,3}$ $^1S_0$ Auger electrons in xenon

As an example of the measurement of electron–electron coincidences after photoionization using synchrotron radiation, the angular correlation between $4d_{5/2}$ photoelectrons and N_5–$O_{2,3}O_{2,3}$ $^1S_0$ Auger electrons in xenon at a photon energy of 94.5 eV is selected. This represents a special case of photon-induced two-electron emission for several reasons. The two-step model can be applied and allows a complete representation of the double-differential cross section of equ. (4.62c) in terms of 'geometry' and dynamical parameters. The dynamical parameters of the selected Auger transition are fixed, because only one partial wave contributes for the Auger electron. This has the consequence that the differential cross section depends only on the dynamical parameters of the photoionization process,† and a measurement of the angular correlation between these electrons gives additional information on the photoprocess. Together with data on the partial cross section $\sigma(4d_{5/2})$, the angular distribution parameter $\beta(4d_{5/2})$ and the alignment parameter $\mathscr{A}_{20}(4d_{5/2})$ of the inner-shell hole-state, a sufficient number of observables is then provided from which the dipole matrix elements D_γ and their relative phases $\Delta_{\gamma\nu}$ can be calculated. In other words, a complete experiment can be performed as discussed in Sections 5.2 and 5.4, and the experimental values for D_γ and $\Delta_{\gamma\nu}$ can be compared in the most direct way with theoretical predictions. Beyond that, with known values for D_γ and $\Delta_{\gamma\nu}$ the complete pattern for the energy-, angle- and even spin-resolved photon-induced process of two-electron emission can be established. Many of these aspects have been described in detail elsewhere (see [KSc91, JCh92, KSc93]), therefore, the subject to be treated here mainly concerns experimental aspects for electron–electron coincidences using synchrotron radiation. These topics not only underline the general remarks of Section 4.6, but they include many items which have been discussed in this book. In an overview of this new field of research, some predicted spatial views of angular patterns without spin-analysis for the ejected electrons will be presented.

† This statement holds except for the Auger yield, which governs the intensity, and for some simple and known numerical factors which result from the coupling of angular momenta connecting the initial and final ionic states of the Auger transition. Since all these quantities are energy-independent, they are not considered as real dynamical parameters.

5.6.1 Experimental details

The electron spectrum of xenon in this experiment is similar to the one shown in Fig. 5.23 except for the lower photon energy of 94.5 eV which places the $4d_{3/2}$ and $4d_{5/2}$ photolines on the left of the N_5–$O_{2,3}O_{2,3}$ $^1S_0$ Auger line. The lower photon energy was selected because this value was appropriate for an accompanying investigation of angle-dependent PCI (see [KKS93] and references therein). The experiment was performed on the undulator beam line at BESSY I with the toroidal grating monochromator TGM5 in order to have a high degree of linear polarization. The high degree of polarization is necessary because, as mentioned in connection with the general parametrization for the differential cross section of two-electron emission, equ. (4.62a), the coincident angular correlation depends in general on all three Stokes parameters of the incident light, i.e., on the linear and circular polarization. Though the linear polarization can be determined easily by angle-resolved non-coincident electron spectrometry (see Fig. 1.18), it is not yet possible to assess the circular polarization in this energy region with sufficient accuracy. Hence, it was essential to use monochromatized light from an undulator whose harmonic light is known to be linearly polarized up to 100%. With an appropriate setting of the undulator gap the second harmonics of the undulator radiation (see Fig. 1.8 for the principle) could be tuned to the desired photon energy of 94.5 eV, giving $\tilde{S}_1 = 0.957(5)$, $S_2 = 0$. This value seems to be sufficiently high that the influence of S_3 can be neglected. However, if the monochromatized light remains completely polarized, the large $\tilde{S}_1$ value would still lead to a circular component $S_3 = \pm 0.29$; if the remaining, not linearly polarized light is unpolarized, one would have $S_3 = 0$ (see equ. (1.37)). Hence the experiment is burdened with this uncertainty. (For details see [KSc93]; it would have been better to use an odd harmonics with $\tilde{S}_1 \geq 0.99$ but at that time this was not accessible in the required energy range.)

Coincidences between $4d_{5/2}$ photoelectrons and N_5–$O_{2,3}O_{2,3}$ $^1S_0$ Auger electrons have been measured with the experimental set-up shown in Fig. 1.17. The photoelectrons were recorded with the monitor analyzer placed at a fixed position ($\Theta_1^0 = 90°$, $\Phi_1^0 = 150°$), the Auger electrons were detected using both sectors of the rotatable double-sector CMA ($\Theta_2^0 = 90°$, Φ_2^0 variable). The relative transmission and detection efficiencies of these sectors were taken into account in order to use the signals from both sectors. For the coincidence experiment three modifications in the experimental set-up were necessary due to the points mentioned in Section 4.6.3 concerning optimization:

(i) The solid angles accepted by the spectrometers were enlarged by increasing the angular ranges of the entrance slits to $\Delta\vartheta_1 = \pm 4.5°$ and $\Delta\varphi_1 = \pm 20°$ for the monitor analyser, and to $\Delta\vartheta_2 = \pm 3.0°$ and $\Delta\varphi_2 = \pm 20°$ for the double-sector analyser (these angles refer to the cylindrical coordinate system of the analysers). The exit slits of the analysers where the electrons leave the electric field had to be increased correspondingly to ensure that all accepted

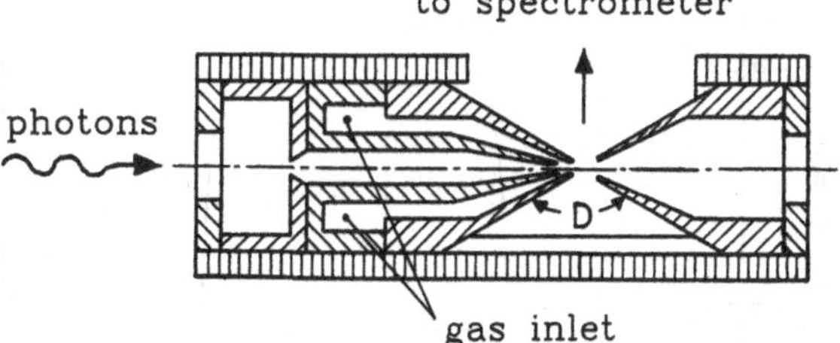

Figure 5.28 Schematic drawing of the limitation of the source volume by two conical diaphragms (D). The cones are aligned with their axes around the direction of the incident photon beam which enters the cone on the left-hand side and leaves the cone on the right-hand side. The cone on the left-hand side also serves as the inlet for the target gas as indicated in the figure. Electrons produced in the source volume between the cones can fly towards the electron spectrometer. From [KSc93].

electrons passed through. As usual, all large slits were covered by high-transmission gold meshes in order to avoid a distortion of the field and field-free regions.

(ii) The source volume was limited to $\Delta z = \pm 0.75$ mm by the system of diaphragms shown in Fig. 5.28, in order to simulate a point-like source volume. (Ray-tracing calculations show that full transmission is obtained for all points within a range of ± 2.5 mm relative to the symmetry axis of the spectrometer (see Fig. 4.6).) Then both spectrometers view the same source volume, and solid-angle corrections can be applied. The solid-angle correction was incorporated in the calculation of the angular correlation pattern (see equ. (10.73) and Fig. 4.53).

(iii) The monitor analyser was operated with a large-scale detector consisting of two channelplates in a Chevron mounting with a large position-insensitive anode (see Fig. 4.18). This had the advantage that the $4d_{5/2}$ photoline could be collected at once without changing the spectrometer voltage U_{sp1}. The responce function of this detector was measured, and the result is shown in Fig. 5.29. It can be seen that the response function, centred at the nominal pass energy for $4d_{5/2}$ photoelectrons with 27 eV kinetic energy, is flat within a range of approximately $\pm 5\%$. This range is large enough to accept the full photoline with equal efficiency, including all broadening effects due to photon bandpass, spectrometer resolution, and natural line width. To record the coincident Auger line it was then necessary to scan only the voltage U_{sp2} of the double-sector analyser over the Auger peak to get information on the coincident intensity (equ. (4.107)). In order to reduce the total time for data collection, the coincident Auger peak was recorded at some selected spectrometer voltages, typically at seven points, and its full lineshape was established by a least-squares fit of the experimental data to the convolution result of the spectrometer function and the theoretical distribution of the angle-dependent PCI function. An example is shown in Fig. 5.30. The desired intensity of

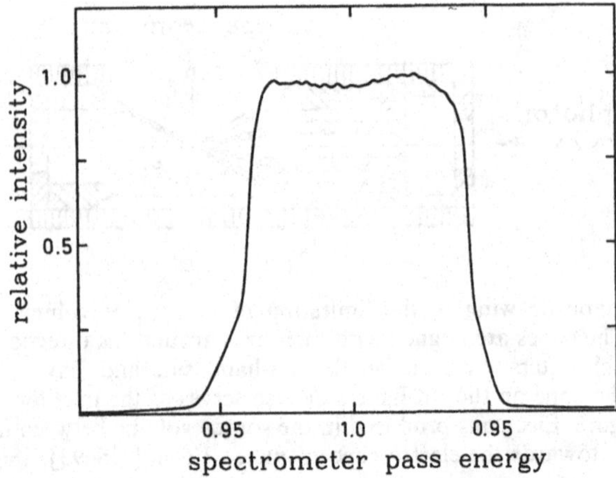

Figure 5.29 Response function of a photoelectron spectrometer equipped with a large-scale detector to ensure the recording of the whole xenon $4d_{5/2}$ photoline with constant efficiency. The accepted range of approximately $\pm 5\%$ corresponds to ± 1.35 eV at a photoelectron kinetic energy of 27 eV, and this value is large compared to the photon bandpass (0.4 eV at 94.5 eV), the contribution from the electron spectrometer (0.22 eV), and the natural linewidth ($\Gamma = 0.12$ eV). From [KKS93].

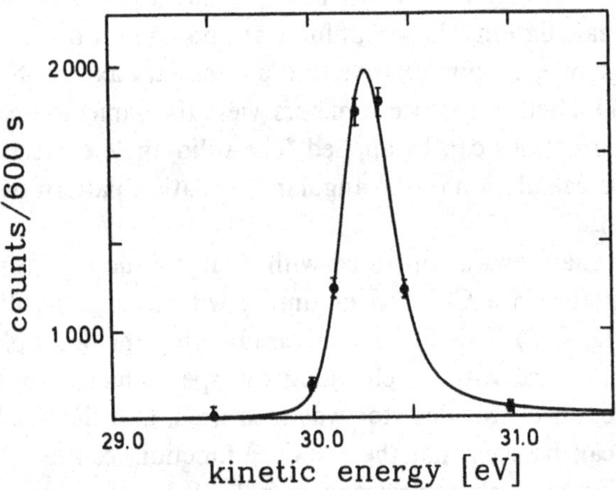

Figure 5.30 Illustration of the coincident xenon N_5–$O_{2,3}O_{2,3}\,{}^1S_0$ Auger line. Points with error bars are experimental data; the solid curve is the least-squares fit to the experimental data with the height and position of the fit function as free parameters; the fit function results from the convolution of the theoretical lineshape (angle-dependent PCI model with $\Gamma = 0.12$ eV) and the experimental spectrometer function (modified Gaussian function, see Fig. 5.25, with $\Delta E_{sp} = 0.19$ eV). From [KSc93].

true coincidences was then evaluated from the area of the adapted full lineshape.

After amplification, the signals from the photo- and Auger electrons were fed to the time-measuring device, a time-to-digital converter (see Fig. 4.47). To reduce the total dead time of the device, the Auger electrons which had a lower counting

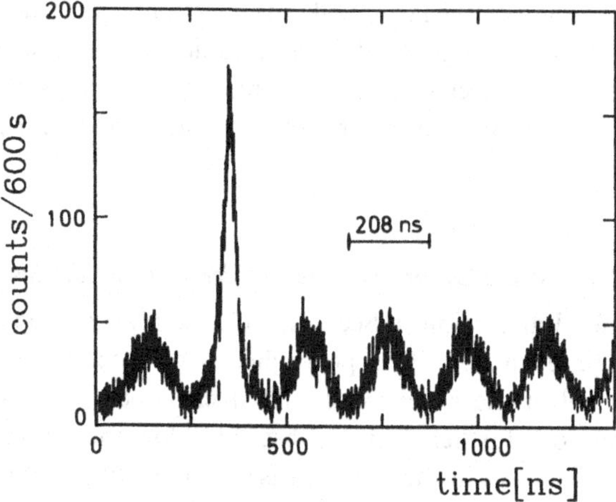

Figure 5.31 Time correlation spectrum between $4d_{5/2}$ photoelectrons and N_5–$O_{2,3}O_{2,3}$ $^1S_0$ Auger electrons in xenon, recorded with a time-to-digital converter. Note the repetition rate, 208 ns, of the circulating electron bunches in the storage ring. The large second peak contains true and accidental coincidences, and the periodic structure is due to accidental coincidences only. From [KSc93].

rate than the photoelectrons were used to give the START signal, and the corresponding photoelectron to provide the STOP signal. In Fig. 5.31 an example of the spectrum for time-correlated events is shown. The large second peak contains true and accidental coincidences and the periodic structure only accidental coincidences. The structure of accidental coincidences is not flat as shown in Fig. 4.48, because the storage ring BESSY I was operated in a special mode. In this 'multi'-bunch mode the ring is only partly filled with electron bunches, as it was found that this partial filling increases the lifetime of the electrons in the storage ring. Within the circulation time of one electron bunch, 208 ns, the storage ring provides approximately 65 light flashes separated by 2 ns, with no light afterwards. This structure is reflected in the periodic behaviour of accidental coincidences, where the nearly rectangular shape of the time distribution of light flashes leads to the triangular shape of the accidental coincidences. Depending on the overall delay between START and STOP, the true coincidences are superimposed on one of the wiggles, but always at the maximum. (For both kinds of coincidences, the maximum position corresponds to equal times for the creation of two electrons in the source volume.) As explained already for Fig. 4.48, the width of the true coincidence peak comes from the time spreads of coincident events introduced by the different travelling times of the electrons through the spectrometers and the pulses in the electronics. (The largest contribution results from the different paths of the electrons within the accepted angular range (see equ. (4.116c)).) A coincidence resolving time Δt equal to the circulation time of the electrons, $\Delta t = 208$ ns, is large enough safely to include all true coincidences. The time spectrum of Fig. 5.31 can then be separated into individual regions in units of

208 ns. From the region with the large peak, the contribution of true and accidental coincidences, and, from the region with the periodic structures, an average value for the accidental coincidences can then be obtained. Finally, subtraction of the latter number of coincidences from the former one gives the number of true coincidences.

5.6.2 Estimation of true and accidental coincidences

Following the general discussion in Section 4.6.2, the rates of true and accidental coincidences can be estimated. This possibility is of special interest because it allows better control of the experiment, and it can open ways to improve the experimental conditions further. Using subscript 1 for photoelectrons and subscript 2 for Auger electrons, in the present case the following parameters for such an estimation of coincidences are needed:

number N_{ph} of photons/s incident on the sample within a bandpass of $\sim$0.3 eV and a diameter of $\sim$0.5 mm

$$N_{ph} \sim 3 \times 10^{12}/\text{s}; \tag{5.67a}$$

target density n_v in the source region (approximately a factor 300 higher than the xenon density in the chamber)

$$n_v \sim 1 \times 10^{12}/\text{cm}^3; \tag{5.67b}$$

length Δz of the source volume along the photon beam direction (see Fig. 5.28)

$$\Delta z = 0.15 \text{ cm}; \tag{5.67c}$$

partial cross section σ_1 for $4d_{5/2}$ photoionization at 94.5 eV

$$\sigma_1 = 12.2 \times 10^{-18} \text{ cm}^2; \tag{5.67d}$$

cross section σ_2 for emission of N_5–$O_{2,3}O_{2,3}\ ^1S_0$ Auger electrons (product between σ_1 and the Auger yield $\omega_A = 0.053$)

$$\sigma_2 = 6.47 \times 10^{-19} \text{ cm}^2; \tag{5.67e}$$

cross section σ_{12} for coincident events between $4d_{5/2}$ photo- and N_5–$O_{2,3}O_{2,3}\ ^1S_0$ Auger electrons (equal to σ_2)

$$\sigma_{12} = 6.47 \times 10^{-19} \text{ cm}^2; \tag{5.67f}$$

transmission of the fixed analyser (product of spectrometer transmission from ray-tracing calculations, 5.57×10^{-3}, and the transmission through several meshes used to separate field and field-free regions, 0.6)

$$T_1 = 3.34 \times 10^{-3}; \tag{5.67g}$$

transmission of the rotatable analyser (product of spectrometer transmission from ray-tracing calculations, 3.80×10^{-3}, and the transmission through several meshes used to separate field and field-free regions, 0.6)

$$T_2 = 2.28 \times 10^{-3}; \tag{5.67h}$$

fwhm ΔE_{sp1} of the fixed analyser at 27 eV

$$\Delta E_{sp1} \approx 0.22\,\text{eV}; \tag{5.67i}$$

fwhm ΔE_{sp2} of the rotatable analyser at 30 eV

$$\Delta E_{sp2} \approx 0.19\,\text{eV}; \tag{5.67j}$$

detection efficiency of the channelplate detector of the fixed analyser

$$\varepsilon_1 \approx 0.45; \tag{5.67k}$$

detection efficiency of the channeltron detector of the rotatable analyser

$$\varepsilon_2 \approx 0.90; \tag{5.67l}$$

coincidence resolution time Δt

$$\Delta t = 208\,\text{ns} \tag{5.67m}$$

(this value could be reduced, setting in the time correlation spectrum of Fig. 5.31 a mask around the peak of true coincidences);

average data collection time T_{coll}

$$T_{coll} \approx 90\,\text{min} = 3.24 \times 10^5\,\text{s}. \tag{5.67n}$$

Since the full photoelectron line is detected with the large-scale channelplate detector, each photoelectron contributes to the single counting rate I_1 which, therefore, becomes independent of the instrumental resolution and is given by

$$I_1(\text{plate}) = N_{ph} n_v \,\Delta z\, \sigma_1\, T_1 \varepsilon_1. \tag{5.68}$$

According to the Auger yield ω_A, these photoelectrons are coincident with a subsequent Auger electron provided one scans over E_{pass2} of the coincident Auger electrons (see Fig. 10.6). This yields the area A_{true} of true coincidences (see equ. (10.65e))

$$A_{true} = \int I_{true}(\text{plate}, E_{pass2})\,\mathrm{d}E_{pass2}$$

$$= N_{ph} n_v \,\Delta z\, \sigma_{12}\, T_1\varepsilon_1\, T_2\varepsilon_2\, \Delta E_{sp2}, \tag{5.69a}$$

and from this expression one obtains with (see equ. (2.42))

$$A_{true} \approx I_{true}\, \Delta E_{exp} \tag{5.69b}$$

and the approximation $\Delta E_{exp} \approx \Delta E_{sp2}$ as estimate of the true coincidence rate

$$I_{true} \approx N_{ph} n_v \,\Delta z\, \sigma_{12}\, T_1\varepsilon_1\, T_2\varepsilon_2. \tag{5.70}$$

Finally, the single counting rate I_2 for the Auger electrons and the rate I_{acc} of accidental coincidences follow from equs. (4.106b) and (4.103), respectively. Using the numerical values from above one gets

$$I_{true} \approx 1\,\text{count/s} \tag{5.71a}$$

(for comparison, equs. (10.63a) and (10.63f) give $I(E^0_{pass1}, E^0_{pass2}) \approx 0.5$ count/s which is the same order of magnitude)

$$I_{acc} \approx 1\,\text{count/s}, \tag{5.71b}$$

and these estimates are close to the actual experimental values averaged over different angle settings.

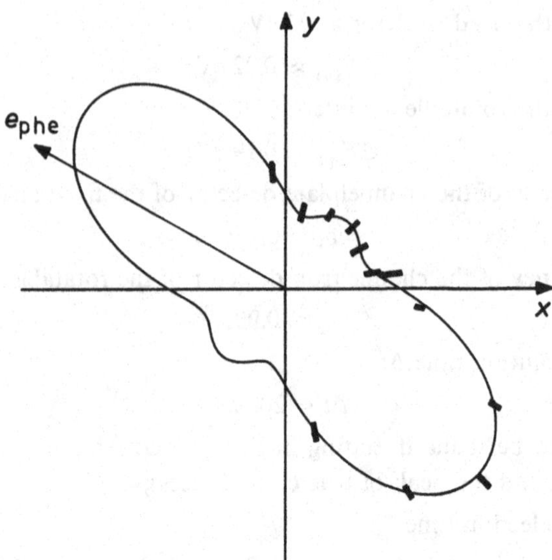

Figure 5.32 Angular correlation between $4d_{5/2}$ photo- and N_5–$O_{2,3}O_{2,3}$ $^1S_0$ Auger electrons in xenon. A polar plot of the coincident intensity is shown for the following experimental conditions: photon energy 94.5 eV, Stokes parameter $S_1 = 0.957$, $S_2 = 0$, S_3 unknown; both electrons are detected in a plane perpendicular to the incident photon beam, the photoelectron at fixed position (e_{phe}) with $\Theta_1^0 = 90°$, $\Phi_1^0 = 150°$, the Auger electron at $\Theta_2^0 = 90°$ and Φ_2^0 = variable (these angles refer to the tilted coordinate frame of Fig. 1.15). Points with error bars are experimental data; the solid line is a least-squares fit (see main text). From [KSc93].

5.6.3 Results

As an example of the electron–electron coincidence experiment between xenon $4d_{5/2}$ photo- and N_5–$O_{2,3}O_{2,3}$ $^1S_0$ Auger electrons, the measured angular correlation pattern (points with error bars) is shown in Fig. 5.32 together with the best fit (solid curve) based on the theoretical expression for the shape of this angular pattern. As was demonstrated in Section 4.6.1.3 in the context of the general parametrization for the case of two-step double photoionization in magnesium, this shape follows from (still unknown) numerical values of the *A*- and *B*-coefficients and attached angular functions. If these angular functions are worked out for the selected two-step double ionization process and for the geometry of the given experimental set-up, one obtains for the double-differential cross section

$$\frac{d^2\sigma}{d\Omega_1\, d\Omega_2}(\Theta_1^0 = 90°, \Phi_1^0 = 150°, \Theta_2^0 = 90°, \Phi_2^0 = \text{variable})$$

$$= A_0 + A_2 \cos 2\Phi_2^0 + B_2 \sin 2\Phi_2^0 + A_4 \cos 4\Phi_2^0 + B_4 \sin 4\Phi_2^0, \quad (5.72a)$$

and the observed number of true coincidences, normalized to constant photon flux and target density, is proportional to this quantity:

$$N_{true}(\Phi_2^0) \sim \frac{d^2\sigma}{d\Omega_1\, d\Omega_2}(\Theta_1^0 = 90°, \Phi_1^0 = 150°, \Theta_2^0 = 90°, \Phi_2^0 = \text{variable}). \quad (5.72b)$$

The A- and B-coefficients depend on the Stokes parameters of the incident light, on the attenuation factors from the large solid angles of the electron spectrometers, and on the dynamical parameters, i.e., on the dipole matrix elements D_γ and relative phases $\Delta_{\gamma\nu}$ of the photoionization process. Therefore, the fit of this theoretical angle dependence to the experimental values shown in Fig. 5.32 provides, in the ratios A_i/A_0 and B_i/A_0, additional information on the photoprocess. Using this in combination with other observables for this photoprocess it is then possible to find a solution for the unknowns D_γ and $\Delta_{\gamma\nu}$. This aspect of a perfect experiment will, however, not be considered here further (for details see [KSc93]). Instead, another advantage of a complete experiment will be illustrated. As the name *complete* implies, all other observables of the selected photoionization (and decay) process can be evaluated. Of special interest in the present context of coincident xenon $4d_{5/2}$ photo- and N_5–$O_{2,3}O_{2,3}$ $^1S_0$ Auger electrons are angular correlation patterns which come from different experimental situations such as other light polarizations and/or other geometries. A selection of these for completely linearly polarized light is shown as spatial views of angle-dependent coincidence intensity in Fig. 5.33. In (a) the photoelectron spectrometer is kept fixed at two selected positions (arrow e_1) whereas the spectrometer for the Auger electron is moved to

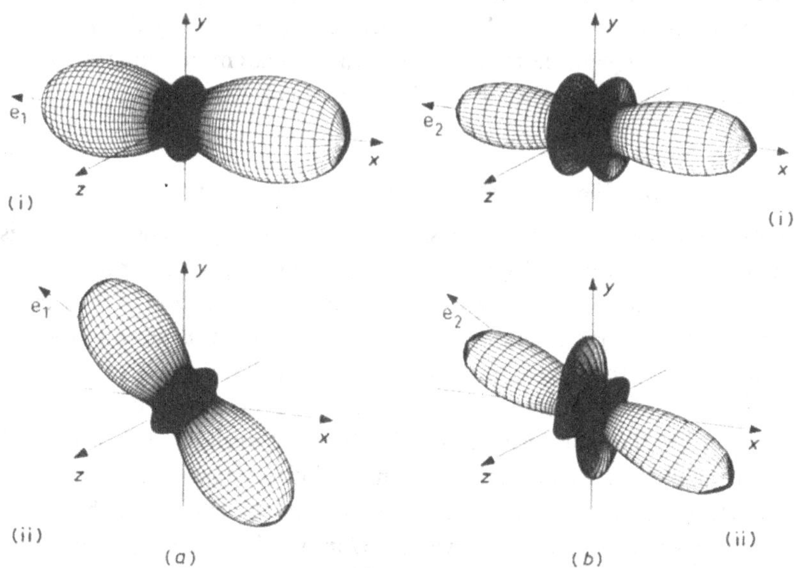

Figure 5.33 Spatial views of angle-resolved intensity patterns for the coincident emission of $4d_{5/2}$ photo- and N_5–$O_{2,3}O_{2,3}$ $^1S_0$ Auger electrons in xenon caused by linearly polarized photons of 94.5 eV (electric field vector along the x-axis). (a) Fixed position of the photoelectron (e_1) with (i) $\Theta_1^0 = 90°$, $\Phi_1^0 = 180°$ and (ii) $\Theta_1^0 = 90°$, $\Phi_1^0 = 150°$, but $\Theta_2^0 = 90°$ and Φ_2^0 = variable for the Auger electron (e_2). (b) Fixed position of the Auger electron (e_2) with (i) $\Theta_2^0 = 90°$, $\Phi_2^0 = 180°$ and (ii) $\Theta_2^0 = 90°$, $\Phi_2^0 = 150°$, but $\Theta_1^0 = 90°$ and Φ_1^0 = variable for the photoelectron (e_1); Θ and Φ are the polar and azimuthal angles in the tilted coordinate frame of Fig. 1.15. From [KSc93].

use the full space; in (*b*) the opposite case is plotted. (The experiment described before then corresponds to a cross-cut in the *xy*-plane for the case shown in Fig. 5.33(*a*)(ii).) Three general statements can be made with respect to these patterns. First, when the observation angle of the fixed electron spectrometer agrees with the direction of the electric field vector of the incident linearly polarized light, one has rotational symmetry around this field vector. Otherwise this cylindrical symmetry is violated, and the pattern is tilted and becomes deformed (see discussion in connection with equ. (4.95c)). Second, due to the general parametrization formula, equ. (4.68), and to the connection of the summation indices k to the orbital angular momenta involved, equ. (4.71b), the pattern contains more structures for higher orbital angular momenta. In the present example one has for the photoelectron, with εp and εf partial waves, summation indices up to $k = 6$, but for the Auger electron, due to its εd partial wave, summation indices up to only $k = 4$. Then, when the direction of the photoelectron is kept fixed, $k = 4$ is responsible for the angle dependences, but when the direction of the Auger electron is kept fixed $k = 6$ is responsible for the angle dependences. Hence, in the latter case the angular correlation pattern has additional lobes. Third, due to the underlying two-step double-ionization process, all angular correlation patterns are less subject to the restrictions of direct double-photoionization processes which affect the probabilities for emission into opposite directions (see Fig. 4.43). (For equal energies of the photo- and Auger electrons the two-step description fails, and the coincident angular pattern changes and becomes more similar to a pattern of direct double photoionization (see [Sch94, VMa94, KSc95]).)

5.7 Threshold double photoionization in argon

Double photoionization in the outer shell of rare gases by a single photon is an important manifestation of electron correlations. One specific aspect which has received much attention over the years is double photoionization in the vicinity of the double-ionization threshold. On the theoretical side, this attention is due to the possibility of deriving certain threshold laws without a full solution of the complicated three-body problem of two electrons escaping the field of the remaining ion. On the experimental side, the study of threshold phenomena always provides the challenge for mastering extremely difficult experiments.

After a short discussion of the underlying theoretical aspects for the threshold double-ionization cross section as given by Wannier theory [Wan53, Wan55], an experimental study for state-dependent double photoionization in the 3p shell of argon leading to the final ionic states $3p^4\ {}^1D^e$, $3p^4\ {}^3P^e$ and $3p^4\ {}^1S^e$ will be presented. The special flavour of this experiment comes from the fact that towards the threshold of double photoionization the cross section approaches zero and the two ejected electrons have zero kinetic energy. Because the doubly-charged photoions do not contain information on the final ionic state, an e/m analysis of the ions can yield only a total, state-insensitive signal. However, if both electrons are energy-analysed and measured in coincidence, a certain state can be selected using the value of the excess energy E_{exc} (see equ. (4.63)). Performing such an experiment close to threshold, one is faced then with the problem of measuring an extremely low intensity (the cross section approaches zero) of all coincident electron pairs (the angle-integrated cross section is of interest) where each electron has practically zero kinetic energy (due to $E_{exc} \to 0$, the individual energies of both electrons, E_a and E_b, also go to zero). Such an experiment seems to be nearly impossible, but two different ways of attacking the problem have been found, and successful experiments have been performed.

5.7.1 Cross section for double photoionization at threshold

Two ingredients are needed to derive theoretically the energy dependence of the double-photoionization cross section σ^{++} at threshold, the so-called *Wannier threshold law*. (It should be noted that a different threshold law has been proposed in which the assumption is made that threshold ionization takes place predominantly in that part of space for which the electrons are at different distances from the nucleus [Tem74, THa74, Tem82]. However, experimental investigations have not yet been able to observe statistically significant oscillations as predicted from this model.) First, appropriate coordinates have to be introduced which allow a transparent description of the correlated motion of the two ejected electrons. This is understandable, because in the threshold region the electrons have low velocities, i.e., there are long interaction times for the evolution of mutual correlation effects. Second, a fundamental result of Wigner's study of threshold properties with short-range interactions [Wig48] is transferred to the present case. Wigner has shown that the energy dependence of relative cross sections in the threshold region arises from the features of the escape process. Therefore a full treatment of the ionization process is only required if the absolute magnitude of the cross section is also of interest. Following Wannier [Wan53], *hyperspherical* coordinates are introduced, and two-electron escape will be discussed in the relevant *Coulomb zone* on the basis of the hyperspherical potential. For simplicity, double photoionization in helium is treated first; the generalizations for state-dependent double photoionization in the 3p shell of argon will be presented afterwards.

Six coordinates are necessary in order to describe the positions $\mathbf{r}_1$ and $\mathbf{r}_2$ of the two helium electrons in a coordinate frame fixed at the nucleus. One possible choice is three hyperspherical coordinates and three Euler angles. The hyperspherical coordinates determine the triangle given by the positions of the three charged particles. They are defined by

(i) the hyperradius R,

$$R = \sqrt{(r_1^2 + r_2^2)}, \tag{5.73a}$$

which gives the general size of the triangle,

(ii) the angle α,

$$\tan \alpha = r_2/r_1, \tag{5.73b}$$

which describes the influence of radial correlation (see Fig. 1.2),

(iii) and the relative angle ϑ_{12} between both electrons,

$$\vartheta_{12} = \angle(\mathbf{r}_1, \mathbf{r}_2), \tag{5.73c}$$

which is responsible for the angular correlation (see Fig. 1.2).

The orientation of this triangle in space is then taken into account by the Euler angles (see discussion of equ. (8.93)). Since the ground state of helium with $^1S_0^e$ symmetry has no orientation in space, these Euler angles are insignificant in this case. However, double photoionization then leads to a $^1P_1^o$ symmetry of the electron pair wavefunction, and the influence of the angular momenta on the threshold law must be studied carefully. In the present case it turns out that the corresponding $L = 1$ equations are equal to the $L = 0$ equations except for additional centrifugal terms (in hyperspherical coordinates) which, however, vanish at threshold [Rot72, KSc76]. Therefore, it is sufficient to consider in the following discussion the simpler $L = 0$ case only.

If the hyperspherical coordinates are inserted into the Schrödinger equation (see equ. (1.1)) one obtains as a differential equation for the wavefunction $\Psi(R; \alpha, \vartheta_{12})$ (see [Fan83])

$$\left(\frac{1}{2}\frac{d^2}{dR^2} + E - U\right)[R^{5/2} \sin \alpha \cos \alpha\, \Psi(R; \alpha, \vartheta_{12})] = 0. \tag{5.74}$$

Here the quantity U is an effective potential that contains three contributions: the kinetic energy for the radial movement of the electrons (in the coordinate α), a centrifugal potential energy, and the Coulomb potential energy $-C(\alpha, \vartheta_{12})/R$ of the system. In the present context of double photoionization it is this Coulomb energy which determines the features of two-electron emission (in atomic units):

$$-\frac{C(\alpha, \vartheta_{12})}{R} = -\left[\frac{Z}{\cos \alpha} + \frac{Z}{\sin \alpha} - \frac{1}{\sqrt{(1 - \sin 2\alpha \cos \vartheta_{12})}}\right]\Big/ R. \tag{5.75}$$

In general, three zones with different implications can be distinguished in double photoionization: the process starts in a small *reaction* zone which is extremely difficult to handle theoretically, but which is left behind rather quickly; then for a much longer time the particles traverse the *Coulomb* zone, in which the Coulomb energy of the system given by equ. (5.75) is comparable to the instantaneous kinetic energy of the system; finally an *outer* zone is reached where the two escaping electrons can be treated as free particles. A closer look at the potential plotted in Fig. 5.34 explains why it is the Coulomb zone which decides whether two-electron escape in the threshold region is possible or not. This figure shows the potential $-C(\alpha, \vartheta_{12})$ for $R = 1$ and $Z = 1$. However, the picture remains qualitatively the same for helium with $Z = 2$. Characteristic features of this potential are the strong Coulomb repulsion between the electrons for $\alpha = 45°$ and $\vartheta_{12} = 0°$ or $360°$ where both electrons would be at the same spatial location, the broad and rather flat

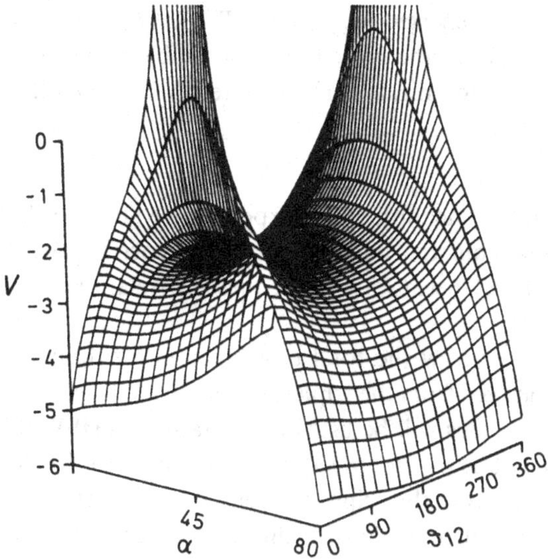

Figure 5.34 Relief plot of the Coulomb potential energy term $V = -C(\alpha, \vartheta_{12})/R$ for a two-electron–one-ion system, given in the hyperspherical coordinates α, ϑ_{12}, and R. The potential is shown for $Z = 1$ and $R = 1$. The centre of the saddle-shaped potential is the Wannier point located at $\alpha = 45°$, $\vartheta_{12} = 180°$ and $V = -2.12$ au. Other characteristic features are the Wannier ridge ($\alpha = 45°$), the two infinitely deep troughs at $\alpha = 0°$ and $\alpha = 90°$, and the two infinitely high towers at $\vartheta_{12} = 0°$ and $\vartheta_{12} = 360°$. From [Eic90], see also [The75].

saddle with the *Wannier point* at $\alpha = 45°$ and $\vartheta_{12} = 180°$, the *Wannier ridge* for $\alpha = 45°$, and the rapid drop to minus infinity for $\alpha \to 0°$ and $\alpha \to 90°$. The crucial influence of this potential on the process of two-electron emission can then be seen by examining how the system evolves for increasing values of the hyperradius R which corresponds in a time-dependent picture to evolution in time. Obviously, *two*-electron escape at threshold is possible only if the system avoids falling into one of the deep troughs in the potential, because otherwise one bounded and one continuum electron would be the result, instead of two continuum electrons. In other words, double ionization occurs only for processes in which the system stays in the vicinity of the Wannier point until R has increased to a value beyond which the two electrons are essentially independent of each other. (In principle, processes at the Wannier ridge can also lead to two-electron emission. However, for $E_{exc} \to 0$ all processes originate asymptotically from the Wannier point only.) Imposing this condition for two-electron escape and assuming that the phase space distribution of the system at the boundary between the reaction and the Coulomb zone is quasi-ergodic (finite and uniform), the Wannier threshold law for the energy dependence of the cross section σ^{++} for double-electron escape can be derived on classical grounds [Wan53] as well as with semiclassical methods [Pet71, Rau71] and gives:

$$\sigma^{++} \sim E_{exc}^{n}, \tag{5.76}$$

where E_{exc} is the excess energy, $E_{exc} = h\nu - E_I^{++} = E_a + E_b$ (equ. (4.63)), and n is the *Wannier exponent*. In the present case of double photoionization in helium, and also for double photoionization in the outer shell of the other rare gases, this exponent has the value

$$n = 1.056. \tag{5.77}$$

(For an experimental verification of this exponent see [KSA88, LIM90, HAD91].) This n value results from the influence of radial correlations described by the hyperspherical angle α;† because without these correlations a linear power law results (see [Rau71]).

In order to elucidate how the total cross section for double photoionization, equ. (5.76), can be derived from the triple-differential cross section, equ. (4.84b), the necessary integration steps will be listed (for details see [HSW91]). Assuming for simplicity completely linearly polarized incident light with the electric field vector defining the reference axis, the triple-differential cross section from equ. (4.84b) including also a constant of proportionality can be reproduced here:

$$\frac{d^3\sigma}{d\Omega_a\, d\Omega_b\, dE} = 2\,|a(^1P^o)|^2\, E_{exc}^{n-2}(\cos\vartheta_a + \cos\vartheta_b)^2\, G(180^\circ - \vartheta_{ab}), \tag{5.78}$$

where the coefficient $a(^1P^o)$ takes into account all phenomena occurring in the reaction zone and provides for the correct strength of double photoionization. If only one of the two ejected electrons is detected, one obtains

$$\frac{d^2\sigma}{d\Omega\, dE} = \frac{1}{4\pi}\frac{d\sigma}{dE}[1 + \beta P_2(\cos\vartheta)], \tag{5.79}$$

where the energy distribution function $d\sigma/d\varepsilon$ is for the production of two electrons, one within the energy range between E and $E + dE$, the other in the range between $E_{exc} - E$ and $E_{exc} - E - dE$, and the angular distribution parameter β is for one of these electrons. Assuming a flat energy distribution (for a more detailed and correct discussion see, for example, [Rea85]), one obtains from the condition that the area of this distribution must equal the total cross section, the relation

$$\frac{d\sigma}{dE} = |a(^1P^o)|^2\, I_{ang}(^1P^o)\, E_{exc}^{n-2}. \tag{5.80}$$

The new quantities I_{ang} and β in equs. (5.79) and (5.80) are fixed by the cumbersome angle integrations. Using for the correlation factor the Gaussian form of equ. (4.83) with the parameter $\Theta_0 = 91^\circ\,eV^{-1/4}$, and keeping for the photon energies near the double ionization threshold the leading terms only, one gets [HSW91]

$$I_{ang}(^1P^o) \rightarrow 14.2\, E_{exc} \tag{5.81}$$

and

$$\beta \rightarrow -1. \tag{5.82}$$

† The radial correlation leads to a dynamical screening of the nuclear charge, imposed by one electron on the other. This allows the electrons to start at the inner boundary of the Coulomb zone with $r_1 \approx r_2$, but to leave the Coulomb zone with $r_1 > r_2$ (or $r_1 < r_2$).

Hence, integration of equ. (5.79) gives for the total cross section

$$\sigma^{++} = 14.2\,|a(^1\mathrm{P^o})|^2\,E^n_{\mathrm{exc}}, \tag{5.83a}$$

abbreviated as

$$\sigma^{++} = \sigma_0 E^n_{\mathrm{exc}}, \qquad \text{where } \sigma_0 = 14.2\,|a(^1\mathrm{P^o})|^2. \tag{5.83b}$$

As was said before, the constant of proportionality σ_0 cannot be predicted by a threshold theory which avoids the complicated treatment of the three-body problem in the reaction zone. However, the numerical value of σ_0 is an interesting quantity, because it determines the strength of the double photoionization. From equ. (5.83) it can be seen that σ_0 depends on both the process in the reaction zone and the integrations over the solid angles of the ejected electrons. It is this which now brings in interest on state-dependent studies of σ^{++}. Using argon as an example, double photoionization in the outer shell can yield three final ionic states with corresponding cross sections

$$\sigma^{++}(3\mathrm{p}^4\,{}^1\mathrm{S^e}\text{ ion}) = \sigma_0(^1\mathrm{S^e})E^n_{\mathrm{exc}}(^1\mathrm{S^e}), \tag{5.84a}$$

$$\sigma^{++}(3\mathrm{p}^4\,{}^1\mathrm{D^e}\text{ ion}) = \sigma_0(^1\mathrm{D^e})E^n_{\mathrm{exc}}(^1\mathrm{D^e}), \tag{5.84b}$$

$$\sigma^{++}(3\mathrm{p}^4\,{}^3\mathrm{P^e}\text{ ion}) = \sigma_0(^3\mathrm{P^e})E^n_{\mathrm{exc}}(^3\mathrm{P^e}). \tag{5.84c}$$

In all cases the same Wannier exponent occurs, and for a selected photon energy the excess energies differ due to the state-dependent ionization energies E_{I}^{++}. However, of importance in the present context are the state-dependent values for the constants of proportionality σ_0. Within the *LS*-coupling scheme the complete final state built from the ion core and the electron-pair wavefunction must have $S_{\mathrm{f}} = 0$ and $L_{\mathrm{f}} = 1$. Therefore one gets the following possibilities:

$$3\mathrm{p}^4\,{}^1\mathrm{D^e}\text{ with the electron pair couplings } (^1\mathrm{F^o},\ {}^1\mathrm{D^o}\text{ or }{}^1\mathrm{P^o}), \tag{5.85a}$$

$$3\mathrm{p}^4\,{}^3\mathrm{P^e}\text{ with the electron pair couplings } (^3\mathrm{D^o}\text{ or }{}^3\mathrm{P^o}), \tag{5.85b}$$

$$3\mathrm{p}^4\,{}^1\mathrm{S^e}\text{ with the electron pair coupling } (^1\mathrm{P^o}). \tag{5.85c}$$

Attached to the special symmetries of the electron pair wavefunctions are certain requirements on their nodal structure in space as well as on the angular functions which describe the emission process. The existence of nodes in these quantities will then decrease the cross section at threshold, i.e., the value of σ_0. A detailed analysis leads in the present example to the result that only the electron-pair wavefunction $^3\mathrm{P^o}$ has no node, all others involved here have a node along the Wannier ridge and/or at the Wannier point. (Compare the classification of these states as *favoured* and *unfavoured*, respectively (see comment following equ. (4.80)).) Hence, one expects at threshold the double-photoionization process to $3\mathrm{p}^4\,{}^3\mathrm{P^e}$ to be stronger than those going to $3\mathrm{p}^4\,{}^1\mathrm{D^e}$ or $^1\mathrm{S^e}$. On the grounds of the differences in the angle-integration factors I_{ang} listed in Table 5.3, a further preference of $3\mathrm{p}^4\,{}^1\mathrm{D^e}$ as compared to $3\mathrm{p}^4\,{}^1\mathrm{S^e}$ may exist (some kind of weak *propensity rule*). However, it is likely that in some cases such a preference is compensated by the coefficients $|a(^{2S+1}L)|^2$ which are due to the actual process in the reaction zone [HSW91].

Table 5.3. *Data relevant for double photoionization in the outer 3p shell of argon.*

Final ionic state	Double-ionization energy E_I^{++}	State of electron-pair function	Constant of proportionality σ_0 at threshold
$^3P^e$	43.457	$^3P^o$	$\lvert a(^3P^o)\rvert^2$ $68.3E_{exc}$
		$^3D^o$	$\lvert a(^3D^o)\rvert^2$ $10.3E_{exc}$
$^1D^e$	45.126	$^1P^o$	$\lvert a(^1P^o)\rvert^2$ $14.2E_{exc}$
		$^1D^o$	$\lvert a(^1D^o)\rvert^2$ $29.2E_{exc}$
		$^1F^o$	$\lvert a(^1F^o)\rvert^2$ const E_{exc} (*)
$^1S^e$	47.514	$^1P^o$	$\lvert a(^1P^o)\rvert^2$ $14.2E_{exc}$

Different electron-pair functions which belong to a selected final ionic state and their individual (incoherent) constants of proportionality σ_0 are shown in different rows. The ionization energies are given in eV, the constants of proportionality in kb(eV)$^{-1.056}$. For the numerical values from the angle-integrated parts in σ_0 see [HSW91]; the asterisk indicates that no numerical value has been given.

5.7.2 Experimental details

A quantitative measurement of the strength of double photoionization in the threshold region means the determination of the state-dependent cross section $\sigma^{++}(^{2S+1}L)$ which is equivalent to the determination of the constant $\sigma_0(^{2S+1}L)$. The intriguing problems of such an experiment have been listed above. The solutions found using two different experimental methods will now be described. The idea underlying both approaches is to extract the threshold electrons created by double photoionization from the source volume, to accelerate these electrons in a certain direction in space, and finally to analyse and detect these fast electrons in coincidence. Of course, to obtain a non-vanishing signal, the photon energy must be slightly above the threshold in order to have a finite cross section. Hence, such 'threshold' electrons typically will have a kinetic energy of several meV. The first experimental method is based on a static electric field penetration into the interaction region which provides the necessary extraction and acceleration of the electrons [CRe74]. The second method makes use of a pulsed photon beam (single-bunch mode of an electron storage ring) and a pulsed and synchronized electric extraction and acceleration field across the interaction region. With a suitable time delay between photoionization and extraction, all electrons with a high kinetic energy leave the source volume before the pulsed electric field is applied and the slow threshold electrons which are still in the source region can be collected with high efficiencies (this approach is often called ZEKE (for 'zero kinetic energy') spectroscopy [MSS84]).

The general experimental set-up of the penetrating-field method is shown schematically in Fig. 5.35. The photon beam is perpendicular to the plane of the drawing and crosses the target gas which emerges from a hypodermic needle. The photon–target interaction region is surrounded by a cage whose walls are at ground potential. This cage has two holes and outside each of these holes is an

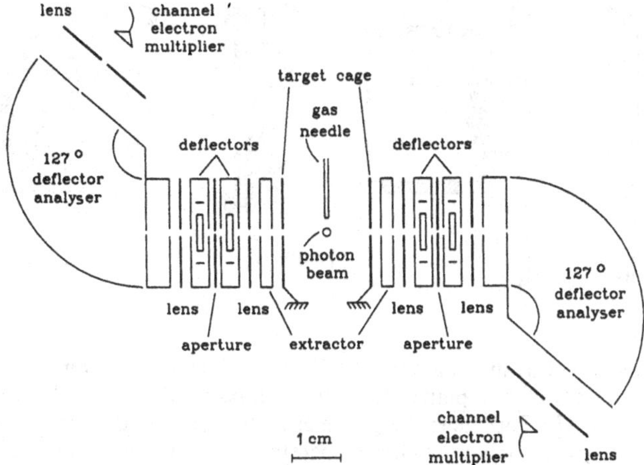

Figure 5.35 Schematic drawing of a threshold-electron coincidence spectrometer based on the penetration-field method. The source volume in the centre of the figure is defined by the intersection of the photon beam and the gas beam. The whole source region is surrounded by a target cage which is kept at ground potential. However, this cage has two holes on opposite sides of the source volume, through which the positive potential from an extracting electrode ('extractor' in the figure) can penetrate from both sides into the source region. Hence, threshold electrons created in the source volume can fall either side of the symmetric saddle-shaped potential in the target cage (for details of the extraction of threshold electrons see also Fig. 4.36). The electrons extracted and accelerated to the right- or left-hand side of the target cage are then fed by a system of two triple-aperture lenses to a 127° cylindrical deflector analyser, and electrons transmitted by these analysers are imaged onto a channeltron detector. From [HDM92].

electrode (again with a hole) to which a positive potential of typically 100 V is applied. This potential then penetrates the interaction region from both sides and produces two troughs the bottoms of which start at the centre with practically zero potential, but which become increasingly positive towards the holes. The resulting electric field has little effect on high kinetic energy electrons, but collects to either side the desired threshold electrons with high efficiency. (It is obvious that the coincident detection of two threshold electrons depends critically on the correct position and shape of the penetration potential at the place of photoionization (for details see [HDM92]).) Furthermore, the penetrating field leads to a cross-over point in the particle trajectories (see Fig. 4.36) which serves as the source point for the subsequent electron optics. Each of the two identical electron optical branches consists of two triple-aperture lens systems (including deflectors for steering the electrons through the apertures), and a 127° cylindrical deflection analyser, followed by a lens to focus the electrons onto the channeltron detector. As was discussed in connection with Fig. 4.36, the whole optical system effectively collects, transmits, and detects threshold electrons from the source region and rejects energetic electrons emitted towards the extraction electrode. Hence, with the output of both spectrometers connected in a coincidence circuit (see Fig. 4.47),

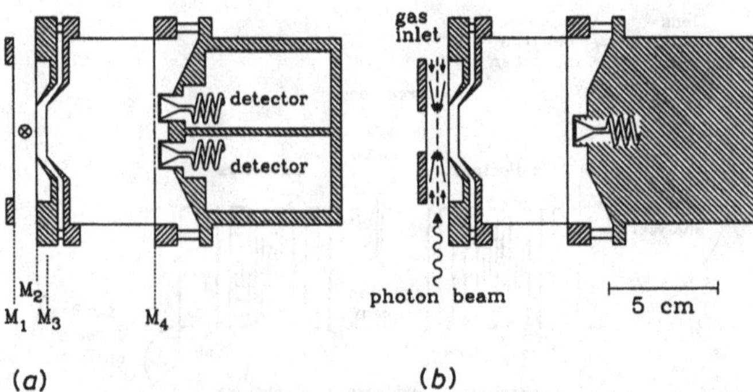

Figure 5.36 Schematic drawing of a double ZEKE coincidence spectrometer. (*a*) Side view: the photon beam intersects the plane of the drawing perpendicular to the point indicated by the crossed circle. (*b*) Top view: the gas-inlet system surrounding the photon beam is indicated by the arrows. The spectrometer operates with a pulsed photon source (single-bunch mode of an electron storage ring) and, synchronized with the light flash, with a pulsed electric field across the source region (between meshes M_1 and M_2). According to the ZEKE principle, the source region is kept field-free during the process of photoionization as well as for an escape interval during which electrons with kinetic energy leave the source region, but before the next light flash arrives, the pulsed electric field is applied across the source region in order to extract and accelerate the slow threshold electrons, which are still in the source region, towards the detector. Since two threshold electrons are detected simultaneously, two separate channeltron detectors are used (see side view). For further details see main text. From [KSc92].

the set-up will register only coincident pairs of threshold electrons, i.e., it operates as a (double) threshold-electron coincidence analyser.

A schematic view of the experimental set-up of the second approach, which might be called a double ZEKE coincidence analyser, is shown in Fig. 5.36. Four different spatial regions with specific electric fields can be distinguished: a source region between two high-transmission gold meshes M_1 and M_2 which carry the voltages V_1 and $V_2(t)$, an acceleration region between the meshes M_2 and M_3 with voltages $V_2(t)$ and V_3, a drift region between the meshes M_3 and M_4 with voltages V_3 and V_4 ($V_4 = V_3$), and a postacceleration region between the mesh M_4 and the detector. The photon beam enters and exits the source region through two opposite 3 mm apertures which limit the length of the viewed source volume to 20 mm. Each of these apertures consists of a set of two concentric cones between which the target gas is directed into the source volume (indicated by the arrows in (*b*), the apertures and the cones themselves are not shown for clarity). According to the ZEKE principle, the source region is kept field-free† except for the duration of the time-dependent electric field used to extract and accelerate the threshold electrons. In particular, photoionization occurs during the field-free intervals, and after a certain time delay T_d the extraction field is switched on by sweeping the potential $V_2(t)$ on mesh M_2 (typically by a voltage pulse with a +1 V swing, 10 ns

† In order to prevent external electrons from entering the spectrometer, the whole system, including the source region, is biased to −100 V against ground.

risetime, and 35 ns duration†). Because of the time delay, the extraction pulse can act only on those electrons which are still in the source region. Hence, the longer the delay, the smaller will be the velocity of the electrons which are collected; in other words, the better will be the energy resolution. V_3 is set to $V_1 + 2$ V to further accelerate the electrons towards the channeltron detector. (The pulse amplitude of $V_2(t)$ and the value of V_3 are additionally optimized in order to reach space focusing for the collected ZEKE electrons [WMc55].) For good detection efficiency there is a +100 V postacceleration voltage between the drift tube and the cones of both channeltrons. Field penetration of this cone voltage into the drift region through the coarse mesh (indicated by dots in the figure) just in front of the channeltron cones results in an enhanced collection efficiency for electrons with a small transverse velocity component. The channeltron pulses are amplified, shaped, and analysed according to their arrival times which are measured with respect to the ionization time given by the pulsed light flash.

A typical time-of-flight spectrum for photoionization processes in argon at 50 eV measured with one of the two channeltron detectors is shown in Fig. 5.37. The

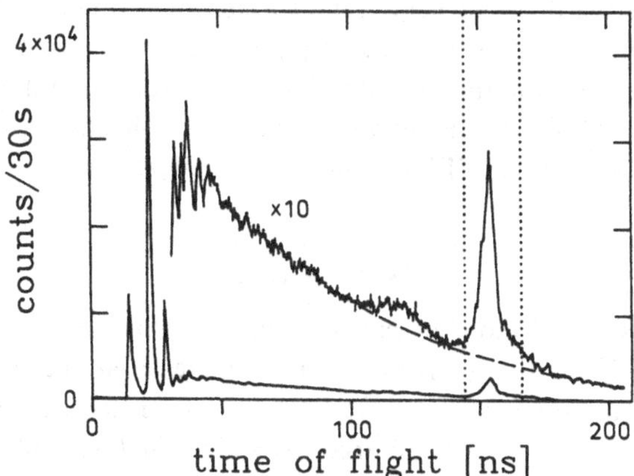

Figure 5.37 Time-of-flight spectrum of one of the two detection branches of the double ZEKE coincidence spectrometer (see Fig. 5.36), for the single-bunch mode of operation of BESSY I (208 ns time period), and for photoprocesses in argon at 50 eV photon energy. The large peaks at short flight times are due to fast electrons from single photoionizations in the 3p and 3s shells (see main text), the smooth continuum following at larger flight times is due to electrons with decreasing kinetic energies. In particular, the dashed curve is a conventional time-of-flight spectrum with time-independent electric fields. However, if the spectrometer is operated as a ZEKE spectrometer, and 35 ns after the light flash a pulsed electric field is applied across the source region, threshold electrons (with kinetic energies lower than approximately 60 meV) are collected, extracted, and detected with high efficiency in the peak seen at approximately 155 ns. The vertical dotted lines around this peak indicate a possible setting for a time-gate which then allows the detection of only threshold electrons. From [Krä94].

† The duration of the extraction pulse is long enough to ensure that the desired threshold electrons have left the source region before this voltage is reset. An alternative way to apply the extracting pulse is to change the voltage $V_1(t)$.

range of flight times covers one period, 208 ns, for the circulation of a single bunch of electrons in the storage ring BESSY I. The largest peak seen in the spectrum at a flight time of 22 ns, and its neighbour at 29 ns, result from 3p and 3s ionizations, respectively. The peak at 14 ns also originates from these ionization processes but is, however, caused by second-order light from the monochromator. (Note the linear scale of flight times which results in a non-linear scale for the kinetic energies. Therefore the second-order 3p and 3s photopeaks are not resolved.) Of interest in the present context is the broad continuum of long flight times which corresponds to low-energy electrons, in particular the region at 155 ns. There the dashed curve indicates the normal time-of-flight spectrum with $V_2(t)$ kept at zero, and nothing special occurs. However, if 35 ns after the light flash the extraction pulse $V_2(t)$ is applied, one gets the solid curve, and the enhanced intensity at 155 ns is due to efficiently collected threshold electrons. (These threshold electrons are released by double photoionization in argon where the energy sharing of the available excess energy, equ. (4.63), also leads to $E_a \approx 0$ and $E_b \approx E_{exc}$, i.e., one threshold electron and one electron with higher kinetic energy. The flight time of these threshold electrons can be calculated following their passage through the different field regions.) Hence, in order to detect threshold electrons only, one can place a time-gate (vertical dotted lines in Fig. 5.37) at this position. Then connecting the outputs of the gates from both detection branches in a coincidence circuit, the set-up will detect only coincident pairs of threshold electrons, i.e., it is a double ZEKE coincidence analyser.

5.7.3 Results

The double threshold-electon coincidence analyser allows one to tune the photon energy and watch for the analyser signal to increase at the double ionization thresholds. As an example, a double ZEKE coincidence spectrum for double photoionization in the 3p shell of argon is shown in Fig. 5.38. The lower part gives an overview on the whole region, and there it can be verified that peaks occur only at the corresponding ionization thresholds (see Table 5.3). These special regions are represented more clearly in the insets which were obtained under slightly different experimental conditions (see figure caption). Since no dependence on the angle between the spectrometer and the electric field vector of the linear polarization was found with this equipment, the spectrum is expected to represent relative intensities of the σ^{++} cross sections. At a first glance there seems to be a confirmation of the propensity rule that $\sigma^{++}(3p^4\ ^3P^e)$ is larger than $\sigma^{++}(3p^4\ ^1D^e)$, which again is larger than $\sigma^{++}(3p^4\ ^1S^e)$. However, a closer look at the observed intensity in the fine-structure components of the $^3P^e_J$ final ionic state reveals that these intensities do not follow the statistical expectation: $^3P^e_2$:$^3P^e_1$:$^3P^e_0$ = 5:3:1. The $J = 1$ component seems to be smaller than the $J = 0$ component, and the $J = 2$ component is much too large. This observation gives a hint that the intensity at the $^3P^e_J$ levels might be severely disturbed by the presence of indirect autoionization

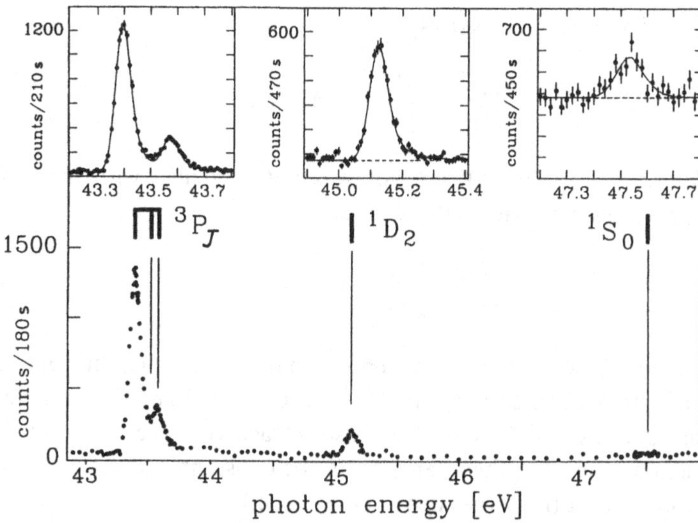

Figure 5.38 Double ZEKE coincidence spectrum of argon in the region of $3p^4\ ^3P_J$, $3p^4\ ^1D_2$, and $3p^4\ ^1S_0$ double-ionization thresholds. The delay of the pulse for the extraction of threshold electrons was 40 ns with respect to the light flash, the monochromator slits were set to 200 μm, and the dwell time for scanning the photon energy was 3 min/channel. Top left and middle insets: Coincidences with a 40 ns delay and improved monochromator resolution (100 μm slits) at the ^3P_J and $^1D_2$ thresholds, respectively. Top right inset: Coincidences at the $^1S_0$ threshold, pulse delay and monochromator slits were set to 22 ns and 200 μm, respectively. From [KSc92]. For an equivalent result obtained by means of a penetration-field threshold-electron coincidence analyser, extended also to the $3s3p^5\ ^3P_J$ and $3s3p^5\ ^1P_1$ double-ionization thresholds in argon, see [HDM92, HME92].

processes which can lead under certain circumstances to two threshold electrons. For example, at a photon energy of 43.41 eV the threshold for photoionization to the $3p^4(^1D^e)6d\ ^2S^e$ final ionic state is reached, and one gets one of the necessary two threshold electrons. This ionic state lies 20 meV above the $3p^4(^3P_2)$ double ionization continuum, and it can decay into it by autoionization (so-called *valence multiplet* Auger transitions [BWH89, ALa91, ALa92, BHL93, BWe94]). Hence, the corresponding autoionization electron has a kinetic energy of 20 meV which places it in the class of 'threshold' electrons detected by the analyser, and both electrons identify a threshold double-ionization process which, however, is a *sequential* two-step process, and not the desired *direct* double-ionization process. There is much detailed complementary information on this problem, see, for example, ion spectrometry [LEN87, HMY88], electron–electron coincidence measurements [PEI89, PEI90], threshold electron and electron–ion spectrometry [HEM92]. Assuming now that the smallest observed component with $J = 1$ is free of disturbances and, furthermore, that all J components are populated according to their statistical weights, an upper limit for the $3p^4\ ^3P^e$ intensity can be derived. From a double ZEKE spectrum which is better resolved than the one shown in Fig. 5.38 one gets [Krä94]

$$\frac{I(3p^4\ ^3P^e_J)}{I(3p^4\ ^1D^e)} = 1.4 \pm 0.3. \tag{5.86a}$$

The ratio of $3p^4\ ^1S^e$ to $3p^4\ ^1D^e$ intensities is not burdened with the problem of sequential processes, and the analysis gives:

$$\frac{I(3p^4\ ^1S^e)}{I(3p^4\ ^1D^e)} = 0.10 \pm 0.02. \tag{5.86b}$$

In order to place these relative intensities on an absolute scale, thus obtaining the constants of proportionality $\sigma_0(3p^4\ ^{2S+1}L^e)$ of equ. (5.84), a double ZEKE coincidence spectrum of helium was also measured. The ratio of observed argon-to-helium intensities then depends on the constants σ_0, the numbers N_{ph} of incident photons, and the target densities n_v. One has

$$\frac{I(\text{Ar};\ 3p^4\ ^{2S+1}L^e_J)}{I(\text{He})} = \frac{\sigma_0(\text{Ar};\ 3p^4\ ^{2S+1}L^e_J)}{\sigma_0(\text{He})}\ \frac{N_{ph}(\text{Ar; at thresh.})}{N_{ph}(\text{He; at thresh.})}\ \frac{n_v(\text{Ar})}{n_v(\text{He})}. \tag{5.87}$$

Measuring the relative photon intensities and the relative target densities, and knowing the numerical value $\sigma_0(\text{He})$ from ion spectrometry [KSA88], one finally

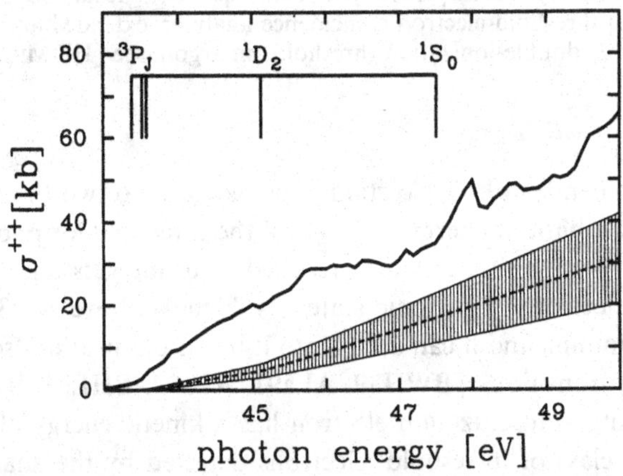

Figure 5.39 Total cross sections for double photoionization in argon in the vicinity of the $3p^4\ ^3P_J$, $3p^4\ ^1D_2$, and $3p^4\ ^1S_0$ thresholds. Solid curve: total cross section for the production of double-charged ions which is the sum of all possibilities, direct and indirect (sequential two-step) double-ionization processes; the relative values are from [LEN87], they have been normalized at a photon energy of 50 eV to the absolute value $\sigma^{++} = 66 \pm 5$ kb from [Kos92]. The dashed curve surrounded by the shaded region (indicating the experimental uncertainties) describes the sum of partial cross sections for direct double photoionization leading to the relevant final ionic states. These partial cross sections are calculated assuming that the Wannier threshold law holds approximately for these larger excess energies, and using in equ. (5.84) the values of equ. (5.88). From [Krä94] with revised data (private communication).

obtains

$$\sigma_0(\mathrm{Ar};\, 3p^4\ {}^3P^e_J) = 2.9 \pm 1.0\ \mathrm{kb}/(\mathrm{eV})^{-1.056}, \tag{5.88a}$$

$$\sigma_0(\mathrm{Ar};\, 3p^4\ {}^1D^e_2) = 2.0 \pm 0.7\ \mathrm{kb}/(\mathrm{eV})^{-1.056}, \tag{5.88b}$$

$$\sigma_0(\mathrm{Ar};\, 3p^4\ {}^1S^e_0) = 0.2 \pm 0.1\ \mathrm{kb}/(\mathrm{eV})^{-1.056}. \tag{5.88c}$$

(These values are from [Krä94]; except for $\sigma_0(\mathrm{Ar};\, 3p^4\ {}^3P^e_J)$ they agree with the estimated values given in [KSc92].)

The large uncertainty in $\sigma_0(\mathrm{Ar};\, 3p^4\ {}^3P^e_J)$, caused by the difficulty in eliminating disturbing indirect double photoionizations, prevents a clear statement concerning the validity of the above-mentioned propensity rule, only $\sigma_0(\mathrm{Ar};\, 3p^4\ {}^1S^e_0)$ as the weakest process is confirmed. Hence, the result will finally be illustrated by looking at the total cross section for the production of doubly-charged argon ions. This cross section is shown in Fig. 5.39 by the solid line. If the Wannier threshold law with $n \approx 1$ is assumed to be approximately valid up to an excess energy of several electron volts, one can use the σ_0 values of equ. (5.88) to predict for the energy range of the figure the partial state-dependent cross sections $\sigma^{++}(3p^4\ {}^3P^e_J)$, $\sigma^{++}(3p^4\ {}^1D^e_2)$, and $\sigma^{++}(3p^4\ {}^1S^e_0)$. Their sum is shown in the figure by the dashed curve, surrounded by a shaded area which takes into account the uncertainties of the experiment. It can be seen that the intensity of direct double photoionization is considerably smaller than the total cross section for the production of doubly charged ions. This means the difference in intensity has to be ascribed to indirect processes of decaying resonances (see also the structures in the observed total cross section). However, because of the difficulties in separating the direct and indirect processes, additional work is needed to clarify the competing roles of these mechanisms better and to derive the correct partition into branches and thereby state-dependent partial cross sections of direct double photoionization in argon.

Part C

Details of specific experimental and theoretical topics

6
Useful reference data

6.1 Atomic units

For all discussions of atomic structure and dynamics caused by some external interaction, the natural measure with which to compare is given by the corresponding quantity of the atom itself. Hence, the hydrogen atom is used as a standard and provides the *atomic units* (au). In addition, if atomic units are used in theoretical expressions, the equations look somewhat simpler. The most important atomic units needed in the present context are (for the numerical values see [CTa87]):

energy E_1 = twice the Rydberg energy = twice the ionization energy of hydrogen with infinite nuclear mass

$$= \frac{m_0 e_0^4}{\hbar^2}\left(\frac{1}{4\pi\varepsilon_0}\right)^2 = \frac{e_0^2}{a_0}\left(\frac{1}{4\pi\varepsilon_0}\right) = 4.35975 \times 10^{-18}\,\mathrm{J} = 27.2116\,\mathrm{eV}$$

length = radius of first Bohr orbit

$$= a_0 = \frac{\hbar^2}{m_0 e_0^2}(4\pi\varepsilon_0) = 5.29177 \times 10^{-11}\,\mathrm{m} = 0.529177\,\text{Å}$$

velocity = classical speed of the electron in the first Bohr orbit

$$= v_0 = \sqrt{(2E_{\mathrm{kin}}/m_0)} = \sqrt{(E_1/m_0)} = \frac{e_0^2}{\hbar}\left(\frac{1}{4\pi\varepsilon_0}\right)$$

$$= \alpha c = 2.18769 \times 10^6\,\mathrm{m/s}$$

time = the time for $(2\pi)^{-1}$ revolutions of the electron in the first Bohr orbit

$$= a_0/v_0 = \frac{\hbar^3}{m_0 e_0^4}(4\pi\varepsilon_0)^2 = 2.41888 \times 10^{-17}\,\mathrm{s}$$

momentum = classical momentum in the first Bohr orbit

$$= m_0 v_0 = \frac{m_0 e_0^2}{\hbar}\left(\frac{1}{4\pi\varepsilon_0}\right) = 1.9929 \times 10^{-24}\,\mathrm{kg\,m/s}$$

angular momentum = $\hbar = 1.05457 \times 10^{-34}\,\mathrm{J\,s}$

mass = electron rest mass

$$= m_0 = 9.10939 \times 10^{-31}\,\mathrm{kg}$$

electric charge = electron charge

$$= e_0 = 1.602177 \times 10^{-19}\,\mathrm{C}$$

electric field = field strength of the hydrogen's proton at a distance of 1 au

$$= \frac{e_0}{a_0^2}\left(\frac{1}{4\pi\varepsilon_0}\right) = 5.1422 \times 10^{11}\ \text{V/m}$$

magnetic moment = Bohr magneton

$$= \mu_B = \frac{e_0\hbar}{2m_0} = 9.2740 \times 10^{-24}\ \text{A m}^2\ (\text{or J/T})$$

magnetic field = field strength at the proton created by the intrinsic magnetic moment of the electron with energy E_1 in the first Bohr orbit

$$= E_1\alpha^2/2\mu_B = 12.52\ \text{T}$$

Miscellaneous:

fine-structure constant $\alpha = \dfrac{e_0^2}{4\pi\varepsilon_0\hbar c} = \dfrac{1}{137.036} = 7.29735 \times 10^{-3}$

permittivity of vacuum = $\varepsilon_0 = 8.854188 \times 10^{-12}$ A s/(V m)

permeability of vacuum $\mu_0 = 4\pi \times 10^{-7}$ V s/(A m)

$$\mu_0\varepsilon_0 = 1/c^2$$

speed of light in vacuum $c = 2.99792458 \times 10^8$ m/s

Planck's constant $\hbar = h/2\pi = 1.054573 \times 10^{-34}$ J s

Boltzmann's constant $k = 1.38066 \times 10^{-23}$ J/K

Electron-volt–wavelength conversion factor and vice versa†

$$1\ \text{eV} = 8065.541\ \text{cm}^{-1},\ 1\ \text{cm}^{-1} = 1.2398424 \times 10^{-4}\ \text{eV}$$

1 eV (electron volt) = 1.602177×10^{-19} J

1 Å (angström) = 10^{-10} m

1 T (tesla) = 1 V s m^{-2} = 10^4 G (gauss)

1 Mb (megabarn) = 10^{-18} cm^2

6.2 Some spectroscopic data for calibration purposes

Energies for single ionization

He	1s	24.5874 eV	[Her58, Moo70]
Ne	$2p_{3/2}$	21.565 eV	[Moo70]
	$2p_{1/2}$	21.661 eV	[Moo70]
	$2s_{1/2}$	48.475 eV	[Moo70, Moo71]
	$1s_{1/2}$	870.21 eV	[PNS82]

† This conversion factor should be used to convert the optical data given by [Moo70, Moo71].

Ar	$3p_{3/2}$	15.760 eV	[Moo70]
	$3p_{1/2}$	15.937 eV	[Moo70]
	$3s_{1/2}$	29.240 eV	[Moo70, Moo71]
	$2p_{3/2}$	248.63 eV	[KTR77, PNS82]
	$2p_{1/2}$	250.78 eV	[KTR77, PNS82]
Kr	$4p_{3/2}$	14.000 eV	[Moo70]
	$4p_{1/2}$	14.666 eV	[Moo70]
	$4s_{1/2}$	27.514 eV	[Moo70, Moo71]
	$3d_{5/2}$	93.79 eV	[KTR77]
	$3d_{3/2}$	95.04 eV	[KTR77]
Xe	$5p_{3/2}$	12.130 eV	[Moo70]
	$5p_{1/2}$	13.436 eV	[Moo70]
	$5s_{1/2}$	23.397 eV	[Moo70, Moo71]
	$4d_{5/2}$	67.548 eV	[KTR77]
	$4d_{3/2}$	69.537 eV	[KTR77]

Energies for double ionization in the outer shell

He	$1s^{-2}$	79.0052 eV	[Her58, Moo70]
Ne	$2p^{-2}\ ^3P^e_2$	62.528 eV	[Moo70]
	$2p^{-2}\ ^3P^e_1$	62.608 eV	[Moo70]
	$2p^{-2}\ ^3P^e_0$	62.642 eV	[Moo70]
	$2p^{-2}\ ^1D^e_2$	65.732 eV	[Moo70]
	$2p^{-2}\ ^1S^e_0$	69.440 eV	[Moo70]
Ar	$3p^{-2}\ ^3P^e_2$	43.389 eV	[Moo70]
	$3p^{-2}\ ^3P^e_1$	43.527 eV	[Moo70]
	$3p^{-2}\ ^3P^e_0$	43.584 eV	[Moo70]
	$3p^{-2}\ ^1D^e_2$	45.126 eV	[Moo70]
	$3p^{-2}\ ^1S^e_0$	47.415 eV	[Moo70]
Xe	$5p^{-2}\ ^3P^e_2$	33.105 eV	[HPe82, PWB88]
	$5p^{-2}\ ^3P^e_1$	34.320 eV	[HPe82, PWB88]
	$5p^{-2}\ ^3P^e_0$	34.113 eV	[HPe82, PWB88]
	$5p^{-2}\ ^1D^e_2$	35.225 eV	[HPe82, PWB88]
	$5p^{-2}\ ^1S^e_0$	37.581 eV	[HPe82, PWB88]

Energies and natural level widths for resonance excitation

Doubly-excited states in He:

He**(2s2p): E_{res} = 60.123 eV [MCo65]; 60.147 eV [DSR96]

Γ = 0.038 eV [MCo65, MEd84, KKS88]; 0.037 eV [DSR96]

He**(3s3p): E_{res} = 69.94 eV [MCo65]; 69.873 eV [DSR96]

Γ = 0.200 eV [KKS88]; 0.181 eV [DSR96]

$4d \rightarrow np$ excitations in xenon (photoabsorption data from [EMa75]; see also electron impact data [KTR77]; for more recent data see [AOM95, MSY95]):

$4d_{5/2} \rightarrow$ 6p:	65.11 eV; $\Gamma = 0.114(8)$ eV
$\rightarrow$ 7p:	66.37 eV; $\Gamma = 0.09(2)$ eV
$\rightarrow$ 8p:	66.84 eV; $\Gamma = 0.09(2)$ eV
$\rightarrow \infty$:	67.55 eV; $\Gamma = 0.11(3)$ eV
$4d_{3/2} \rightarrow$ 6p:	67.04 eV; $\Gamma = 0.13(3)$ eV
$\rightarrow$ 7p:	68.34 eV; $\Gamma = 0.11(3)$ eV
$\rightarrow$ 8p:	68.82 eV; $\Gamma = 0.11(3)$ eV
$\rightarrow \infty$:	69.52 eV; $\Gamma = 0.09(3)$ eV

Auger energies

Ne (for $K\text{–}L_{2,3}L_{2,3}$ transitions: optical data [Moo70] and 1s ionization energy [PNS82]; for other transitions relative values from [ATW90] adapted to $K\text{–}L_{2,3}L_{2,3}\,{}^1D_2$; see also [KMe66, KCM71]):

$K\text{–}L_1L_1$	${}^1S_0$	748.3 eV
$K\text{–}L_1L_{2,3}$	${}^1P_1$	771.9 eV
$K\text{–}L_1L_{2,3}$	${}^3P_J$	782.4 eV
$K\text{–}L_{2,3}L_{2,3}$	${}^1S_0$	800.77 eV
$K\text{–}L_{2,3}L_{2,3}$	${}^1D_2$	804.48 eV

Xe (optical data from [HP382] and 4d ionization energies from [KTR77]; these values correct the ones given by [WBS72], see also [ONS76]):

$N_5\text{–}O_{23}O_{23}\ {}^1S_0$	29.97 eV
$N_4\text{–}O_{23}O_{23}\ {}^1S_0$	31.96 eV
$N_5\text{–}O_{23}O_{23}\ {}^1D_2$	32.32 eV
$N_5\text{–}O_{23}O_{23}\ {}^3P_1$	33.23 eV
$N_5\text{–}O_{23}O_{23}\ {}^3P_0$	33.44 eV
$N_4\text{–}O_{23}O_{23}\ {}^1D_2$	34.31 eV
$N_5\text{–}O_{23}O_{23}\ {}^3P_2$	34.44 eV
$N_4\text{–}O_{23}O_{23}\ {}^3P_1$	35.22 eV
$N_4\text{–}O_{23}O_{23}\ {}^3P_0$	35.43 eV
$N_4\text{–}O_{23}O_{23}\ {}^3P_2$	36.43 eV

Angular distribution parameters β of photoelectrons

1s photoionization in helium: $\beta = 2.0$ (in dipole approximation)

β-parameters with $\beta = 0 \pm 0.05$ [KCF84]:

Ar(3p; $h\nu$ = 16.53 eV)
Ag(4d; $h\nu$ = 19.2 eV)
Ne(2p; $h\nu$ = 26.7 eV)
Xe(5$p_{3/2}$; $h\nu$ = 60.3 eV)

β-parameters for 2p photoionization in neon [Kra80, Sch86]:

$h\nu$ (eV)	β	$h\nu$ (eV)	β
30	0.42 [Kra80]	80	1.31
	0.34 [Sch86]	90	1.33
40	0.79	100	1.34
50	1.03	110	1.36
60	1.18	120	1.38
70	1.27		

Cross sections for 1s photoionization in helium and for 2p photoionization in neon [BWu95]:

$h\nu$ (eV)	σ_{1s}(He) (Mb)	σ_{2p}(Ne) (Mb)	$h\nu$ (eV)	σ_{1s}(He) (Mb)	σ_{2p}(Ne) (Mb)
30	5.38(5)	8.8(3)	80	0.63(1)	4.4(2)
40	3.16(3)	8.4(3)	90	0.468(9)	3.7(2)
50	2.01(3)	7.3(3)	100	0.355(8)	3.2(1)
60	(resonance)	6.2(2)	110	0.276(6)	2.6(1)
70	0.89(2)	5.2(2)	120	0.221(5)	2.2(1)

7
Wavefunctions

7.1 One-electron wavefunctions

The wavefunctions $\varphi(\mathbf{r}, m_s)$ for one electron in the field of a nucleus with charge Z are called *hydrogenic* wavefunctions. In the present context such wavefunctions for discrete orbitals and for the continuum describing the emission of a photo- or Auger electron are of interest.

7.1.1 *Wavefunctions of discrete orbitals*

The wavefunctions of discrete orbitals in a central potential factorize according to

$$\varphi(\mathbf{r}, m_s) = R_{n\ell}(r)\ Y_{\ell m}(\hat{\mathbf{r}})\ \chi_{1/2}^{m_s} \tag{7.1}$$

with

(i) the radial functions $R_{n\ell}(r)$,
(ii) the angular functions $Y_{\ell m}$ (spherical harmonics) where $\hat{\mathbf{r}}$ denotes the polar and azimuthal angles ϑ and φ of the radial coordinate $\mathbf{r}$, and
(iii) the spinor component $\chi_{1/2}^{m_s}$ which characterizes the spin projection 'up' for $m_s = +1/2$ or 'down' for $m_s = -1/2$ measured against a preselected quantization axis; these are abbreviated as $\chi_{1/2}^{+}$ and $\chi_{1/2}^{-}$.

For convenience, the lowest functions are reproduced here using atomic units:

$$R_{10}(r) = R_{1s}(r) = 2\,Z^{3/2}\,\mathrm{e}^{-Zr}, \tag{7.2a}$$

$$R_{20}(r) = R_{2s}(r) = \frac{1}{\sqrt{2}}\,Z^{3/2}\,\mathrm{e}^{-Zr/2}(1 - Zr/2), \tag{7.2b}$$

$$R_{21}(r) = R_{2p}(r) = \frac{1}{\sqrt{(24)}}\,Z^{5/2}\,r\,\mathrm{e}^{-Zr/2}, \tag{7.2c}$$

where, following common practice, the index $\ell = 0, 1, 2, 3, \ldots$ has been replaced by the symbols s, p, d, f, ... Note that instead of the radial functions $R_{n\ell}(r)$ one frequently uses

$$P_{n\ell}(r) = r\,R_{n\ell}(r), \tag{7.3a}$$

because spatial integrations with two such radial functions $R_{n\ell}(r)$ lead to an integrand of the form

$$\int R_{n\ell}(r) \ldots R_{n'\ell'}(r)\,r^2\,\mathrm{d}r = \int P_{n\ell}(r) \ldots P_{n'\ell'}(r)\,\mathrm{d}r, \tag{7.3b}$$

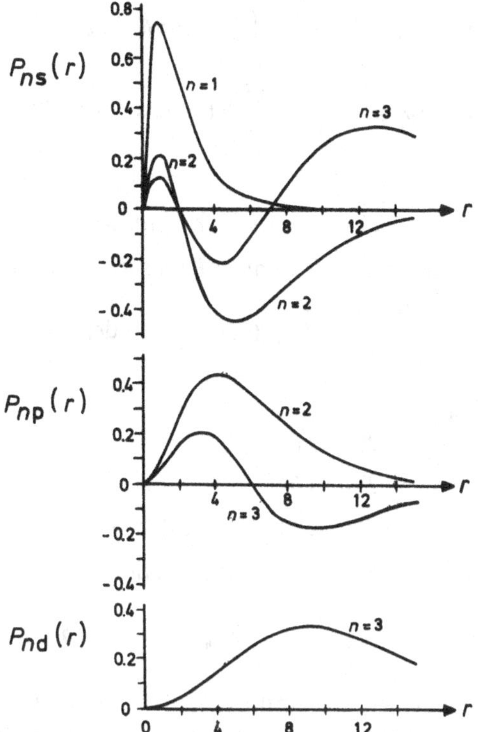

Figure 7.1 Radial eigenfunctions $P_{n\ell}(r) = rR_{n\ell}(r)$ for the electron in the hydrogen atom (in atomic units) where n is the principal quantum number, ℓ the orbital angular momentum. Note that all functions start with a positive slope given by $P_{n\ell}(r) \sim r^{\ell+1}$, have $n - \ell - 1$ zero crossings (nodes), and go outside the atomic region to zero with $P_{n\ell}(r) \sim e^{-r\sqrt{|\varepsilon_{n\ell}|}}$, where $\varepsilon_{n\ell}$ is the single-particle energy of the electron in the orbital $n\ell$. From J. C. Slater, *Quantum theory of atomic structure* (1960) with kind permission of J. F. Slater and The McGraw-Hill Companies.

and the latter expression is easily interpreted as the overlap integral between $P_{n\ell}(r)$ and $P_{n'\ell'}(r)$, possibly modified by an operator which is indicated by the ellipsis. (In particular, $P_{n\ell}(r)P_{n\ell}(r)\,\mathrm{d}r$ describes the radial charge density in the interval $\mathrm{d}r$ at the place r.) In Fig. 7.1 a graphical representation of some radial functions is given.

The lowest spherical harmonics are given by (phase convention of [CSh35])

$$Y_{00}(\vartheta, \varphi) = \frac{1}{\sqrt{(4\pi)}}, \tag{7.4a}$$

$$Y_{10}(\vartheta, \varphi) = \sqrt{\left(\frac{3}{4\pi}\right)} \cos\vartheta, \tag{7.4b}$$

$$Y_{1\pm1}(\vartheta, \varphi) = \mp\sqrt{\left(\frac{3}{8\pi}\right)} \sin\vartheta\, e^{\pm i\varphi}, \tag{7.4c}$$

$$Y_{20}(\vartheta, \varphi) = \frac{1}{4}\sqrt{\left(\frac{5}{\pi}\right)}\,(3\cos^2\vartheta - 1), \tag{7.4d}$$

$$Y_{2\pm1}(\vartheta,\varphi) = \mp\frac{1}{2}\sqrt{\left(\frac{15}{2\pi}\right)}\cos\vartheta\sin\vartheta\,e^{\pm i\varphi}, \tag{7.4e}$$

$$Y_{2\pm2}(\vartheta,\varphi) = \frac{1}{4}\sqrt{\left(\frac{15}{2\pi}\right)}(1-\cos^2\vartheta)\,e^{\pm i2\varphi}, \tag{7.4f}$$

and one has

$$Y^*_{\ell m}(\vartheta,\varphi) = (-)^m\,Y_{\ell -m}(\vartheta,\varphi). \tag{7.5}$$

Special cases are the Legendre polynomials $P_\ell(\cos\vartheta)$, which follow from

$$P_\ell(\cos\vartheta) = \sqrt{\left(\frac{4\pi}{2\ell+1}\right)}\,Y_{\ell 0}(\vartheta,\varphi) \qquad \text{(independent of } \varphi\text{)}, \tag{7.6}$$

i.e.,

$$P_0(\cos\vartheta) = 1, \tag{7.7a}$$

$$P_1(\cos\vartheta) = \cos\vartheta, \tag{7.7b}$$

$$P_2(\cos\vartheta) = \tfrac{1}{2}(3\cos^2\vartheta - 1). \tag{7.7c}$$

For these Legendre polynomials one has the relation

$$P_\ell(\cos\vartheta_{12}) = \frac{4\pi}{2\ell+1}\sum_m Y^*_{\ell m}(\vartheta_1,\varphi_1)\,Y_{\ell m}(\vartheta_2,\varphi_2), \tag{7.8}$$

in which the relative angle ϑ_{12} between the spatial vectors $\mathbf{r}_1$ and $\mathbf{r}_2$ is expressed in terms of the individual polar and azimuthal angles $\hat{\mathbf{r}}_1 = (\vartheta_1, \varphi_1)$ and $\hat{\mathbf{r}}_2 = (\vartheta_2, \varphi_2)$.

7.1.2 Plane wavefunction of an electron

In order to understand the wavefunction of an electron emitted from an atom by a certain ionization process, the wavefunction of a free particle with wavenumber κ travelling along the positive z-axis will first be considered. The space and time dependence of this wavefunction follows from the time-dependent Schrödinger equation with zero potential† and is given by

$$\Psi(z,t) \sim e^{i(\kappa z - \omega t)} \quad \text{with} \quad \hbar\omega = \hbar^2\kappa^2/2m_0, \tag{7.9}$$

where the normalization which fixes the constant of proportionality has not yet been specified. According to the general principles of quantum mechanics, the inherent complex character of this wavefunction only allows an interpretation based on probability densities. Hence, the motion of the free particle described by this wavefunction only can be inferred if a wave packet is constructed and the associated probability flux into certain directions is investigated for increasing times. However, the same result as that from such elaborate treatment can be seen from the simple wavefunction of equ. (7.9) by looking at the phase α: nodes, or

† A constant potential is also allowed because for a particle under the influence of a constant potential the force will vanish, i.e., it is a free particle.

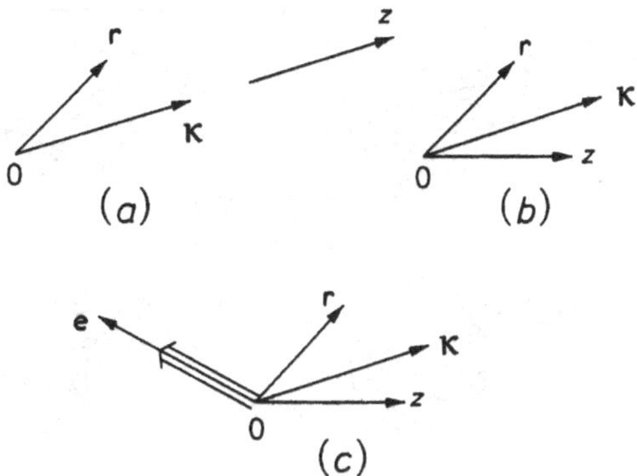

Figure 7.2 Coordinates for a plane wave. Having defined a reference point O (origin), an arbitrary spatial point of the wave travelling with wavenumber vector $\boldsymbol{\kappa}$ into the $\boldsymbol{\kappa}$-direction is indicated by $\mathbf{r}$. Cases (*a*) and (*b*) differ in the direction of the selected reference axis z (the quantization axis) which does or does not coincide with the direction of the wave, respectively. In case (*c*) the spin of an electron wave is also indicated; it is shown as the double arrow pointing into the direction $\mathbf{e}$ against which the spin projection is assumed to be measured.

more generally, points of constant phase will move for increasing times to the right (increasing z values), because in order to keep the phase α constant

$$\alpha = \kappa z - \omega t, \tag{7.10}$$

the whole wavefront characterized by the coordinate z has to increase as the time t increases. (The phase velocity u follows from equ. (7.10) as $u = \omega/\kappa$. For the group velocity V defined by $\mathrm{d}\omega/\mathrm{d}\kappa$ the energy dispersion $\hbar\omega = \hbar^2\kappa^2/2m_0$ leads to the value $V = \hbar\kappa/m_0$ which is identical with the particle velocity v.) As a result, equ. (7.9) represents a plane wave propagating with velocity $v = \hbar\kappa/m_0$ to the right and oscillating with a frequency ω. In the following only the spatial part of this function will be considered because stationary theory is applied most frequently. (It can be shown that stationary theory for photoionization gives the same results as the more elaborate treatment with time-dependent wave packets.)

The expansion into partial waves is of importance for the desired wavefunction of an emitted electron, because it provides a classification into individual angular momenta ℓ which refer to the centre of mass of the atom. As a starting point, the expansion of a plane wave will be considered. If a certain origin is selected and the direction of $\boldsymbol{\kappa}$ is chosen to agree with the z-axis (the quantization axis; see Fig. 7.2(*a*)), one obtains for the spatial part

$$\mathrm{e}^{\mathrm{i}\kappa z} = \sum_{\ell} (2\ell + 1)\, \mathrm{i}^{\ell}\, j_{\ell}(\kappa r)\, P_{\ell}(\cos\vartheta), \tag{7.11}$$

where $j_{\ell}(\kappa r)$ is the spherical Bessel function, for example,

$$j_0(x) = (\sin x)/x, \tag{7.12a}$$

$$j_1(x) = [(\sin x)/x - \cos x]/x, \tag{7.12b}$$

$$j_2(x) = [(3/x^2 - 1)\sin x - (3/x)\cos x]/x, \tag{7.12c}$$

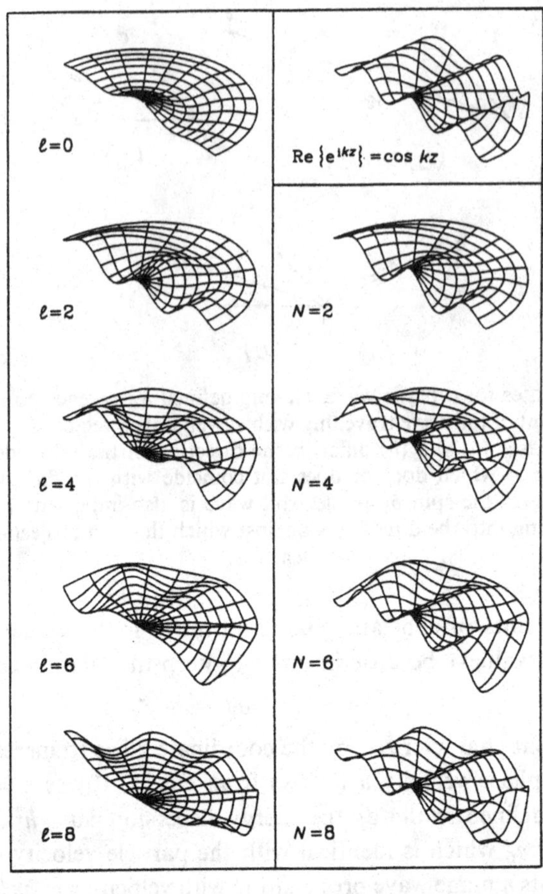

Figure 7.3 Decomposition of a plane wave into spherical waves; (*a*) real part. The plane wave result is shown in the upper right-hand corner, the individual partial wave contributions with a given ℓ value in the left-hand column, the sum of partial wave contributions up to the value $\ell = N$ in the right-hand column. From *The picture book of quantum mechanics*, S. Brandt and H. D. Dahmen, 1st edition 1985, John Wiley & Sons, Inc, © 1985 John Wiley and Sons Inc.

and $P_\ell(\cos\vartheta)$ is a Legendre polynomial with ϑ the polar angle between the z-axis and the spatial coordinate.

In Fig. 7.3 the individual ℓ terms of the given expansion are shown: in Fig. 7.3(*a*) for $\ell = even$ which, because of the phase (i^ℓ) in equ. (7.11), gives real functions; in Fig. 7.3(*b*) for $\ell = odd$ which gives imaginary functions. Each individual contribution, shown on the left-hand side, is centred at the common origin and displays certain symmetry properties with respect to this origin. (This aspect becomes important if the emission of electrons is considered, for example, photoionization of an s-electron leads to only one partial wave with $\ell = 1$, i.e., this pattern provides an approximate view for such an emission process (for the

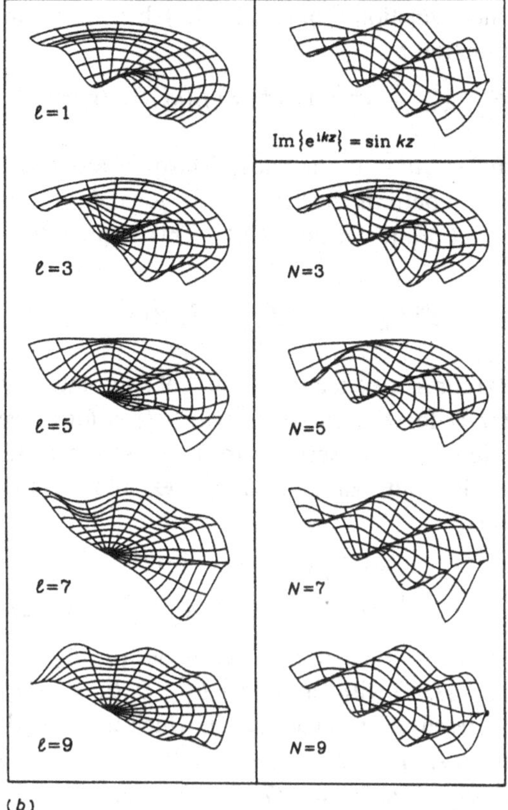

Figure 7.3 (*contd*) Decomposition of a plane wave into spherical waves; (*b*) imaginary part.

necessary modifications due to the underlying plane wave description see below).) The patterns on the right-hand side represent the result of a summation over the individual even/odd contributions, up to the value N. There it can be seen how these sums, centred at the selected origin, progressively form the shape of a plane wave.

In a next step, the plane wave, travelling in the direction of the quantization axis z, is generalized to a plane wave moving along the direction $\boldsymbol{\kappa}$ which differs from the z-direction (see Fig. 7.2(b)). From equ. (7.8)

$$P_\ell(\cos\vartheta) = \frac{4\pi}{2\ell+1}\sum_m Y^*_{\ell m}(\hat{\boldsymbol{\kappa}})\, Y_{\ell m}(\hat{\mathbf{r}}), \tag{7.13}$$

where $\hat{\boldsymbol{\kappa}} = (\Theta, \Phi)$ describes the polar and azimuthal angles for the direction of propagation of the particle, and $\hat{\mathbf{r}} = (\vartheta, \varphi)$ describes the polar and azimuthal angles for a spatial point of the wavefunction. Equ. (7.11) can be replaced by

$$e^{i\boldsymbol{\kappa}\cdot\mathbf{r}} = 4\pi \sum_{\ell m} i^\ell\, Y^*_{\ell m}(\hat{\boldsymbol{\kappa}})\, j_\ell(\kappa r)\, Y_{\ell m}(\hat{\mathbf{r}}). \tag{7.14}$$

If such a wave also contains the electron spin with the projection m_s along the

z-direction, the spinor function from equ. (7.1) has to be included, and one describes the wavefunction by

$$e^{i\boldsymbol{\kappa}\cdot\mathbf{r}}\,\chi_{1/2}^{m_s} = 4\pi \sum_{\ell m} i^{\ell}\, Y_{\ell m}^{*}(\hat{\boldsymbol{\kappa}})\, j_{\ell}(\kappa r)\, Y_{\ell m}(\hat{\mathbf{r}})\, \chi_{1/2}^{m_s}. \tag{7.15}$$

In cases where the polarization of the photoelectron is also to be determined, one needs the spin projection onto three selected orthogonal directions with eigenvalues $\lambda = \pm 1/2$ (see Fig. 7.2(c) and Section 9.2.1). This is provided by the following transformation:

$$e^{i\boldsymbol{\kappa}\cdot\mathbf{r}}\,\chi_{1/2}^{\lambda} = e^{i\boldsymbol{\kappa}\cdot\mathbf{r}} \sum_{m_s} D_{m_s,\lambda}^{1/2}(R)\, \chi_{1/2}^{m_s}. \tag{7.16}$$

The symbol $D_{m_s,\lambda}^{1/2}(R)$ represents a matrix element of the rotation matrix in standard notation and R stands for the rotations which transform the z-axis into the $\hat{\mathbf{e}}$-direction (see Section 8.4.2). Describing in the present case the position of the $\hat{\mathbf{e}}$-direction by the polar and azimuthal angles Θ_s and Φ_s, the rotation matrix elements are given by

$$D_{m_s,\lambda}^{1/2}(R) = \begin{pmatrix} e^{i\Phi_s/2}\cos\Theta_s/2 & e^{-i\Phi_s/2}\sin\Phi_s/2 \\ -e^{i\Phi_s/2}\sin\Theta_s/2 & e^{-i\Phi_s/2}\cos\Theta_s/2 \end{pmatrix}. \tag{7.17}$$

(For a description of the rotation R by Euler angles $(\vartheta_1, \vartheta_2, \vartheta_3)$ see Section 8.4.2, note in particular equ. (8.91) with the values $\vartheta_1 = 0$, $\vartheta_2 = -\Theta_s$, $\vartheta_3 = -\Phi_s$).

Finally, often a representation is required in which the orbital angular momentum ℓ and the electron spin s are coupled to form the total angular momentum j. For the wavefunction given in equ. (7.15) this is provided using the Clebsch–Gordan coupling formula (equ. (7.41b)), giving

$$e^{i\boldsymbol{\kappa}\cdot\mathbf{r}}\chi_{1/2}^{m_s} = 4\pi \sum_{jm_j}\sum_{\ell m} i^{\ell}\, Y_{\ell m}^{*}(\hat{\boldsymbol{\kappa}})\, (\ell m \tfrac{1}{2} m_s | j m_j)\, \varphi_{\ell\frac{1}{2}jm_j}(\mathbf{r}, \mathbf{s}) \tag{7.18}$$

where $\varphi_{\ell\frac{1}{2}jm_j}(\mathbf{r}, \mathbf{s})$ is the spin-orbital function represented in ℓsj coupling.

7.1.3 *Wavefunction of an emitted electron*

A plane wave does not give an accurate representation of the wavefunction of an emitted electron, because it describes a particle passing through the atomic potential without being affected by it. In contrast, a photoelectron is created in the field of the atom and appears only in the distant future as a particle with a well-defined wavenumber $\boldsymbol{\kappa}$. (Due to the long range of the Coulomb potential of the photoion, Coulomb wavefunctions ought to be used, but for simplicity the plane-wave description is kept.) Thus, a solution has to be found which fulfils a certain *boundary condition*. As one might expect naively, the condition for the distant future is simply given by

$$\Psi_{\boldsymbol{\kappa}}^{(-)}(\mathbf{r}, t)\Big|_{\substack{\text{wave}\\\text{packet}}} \xrightarrow[t\to\infty]{} \Phi_{\boldsymbol{\kappa}}(\mathbf{r}, t = +\infty)\Big|_{\substack{\text{planewave}\\\text{packet}}}. \tag{7.19a}$$

In the remote past, before the photon interaction took place, the state $\Psi_{\boldsymbol{\kappa}}^{(-)}(\mathbf{r}, t)$

was subject to the condition

$$\Psi_{\kappa}^{(-)}(\mathbf{r}, t)\Big|_{\substack{\text{wave}\\ \text{packet}}} \xrightarrow[t \to -\infty]{} \Phi_{\kappa}(\mathbf{r}, t = -\infty)\Big|_{\substack{\text{planewave}\\ \text{packet}}} + \text{`scattered' waves}, \quad (7.19b)$$

which reflects that the state is unknown and uncontrolled, except through knowledge of the atomic potential (responsible for the 'scattered' waves) and control over the distant future (responsible for the plane-wave part; for details see [Rom65, Böh79]). The 'scattered' waves are needed to ensure by destructive interference with the plane-wave part that $\Psi_{\kappa}^{(-)}(\mathbf{r}, t)$ describes outgoing radial flux when the photoelectron is released. Therefore, $\Psi_{\kappa}^{(-)}(\mathbf{r}, t)$ is frequently called an 'out'-state. (The 'in'-states (superscript $(+)$) from the time-reversed solution of the scattering problem are suitable for describing a collision experiment in which the incident electron (remote past) is prepared in a well-defined state (with momentum $\boldsymbol{\kappa}$), but the scattering result (distant future) carries information on the scattering potential and the incident electron.)

Instead of the time-dependent description of the photoelectron wavefunction it is easier and correct to use stationary wavefunctions $\Psi_{\kappa}^{(-)}(\mathbf{r})$. The boundary condition for the wave packets then transforms into a boundary condition in the distance coordinate r and requires (for details see [Sta82a])

$$\Psi_{\kappa}^{(-)}(\mathbf{r}) \xrightarrow[r \to \infty]{} \frac{1}{(2\pi)^{3/2}}\left(e^{i\boldsymbol{\kappa}\cdot\mathbf{r}} + f^{(-)}(\vartheta)\frac{e^{-i\kappa r}}{r}\right) \quad (7.20)$$

where $f^{(-)}(\vartheta)$ describes the 'scattering' amplitude and depends on the angle ϑ between $\hat{\boldsymbol{\kappa}}$ and $\hat{\mathbf{r}}$. (The factor $(2\pi)^{-3/2}$ ensures correct normalization in $\boldsymbol{\kappa}$, see equ. (7.28f).) If in this expression one adds the phase factor $e^{-i\omega t}$ of a stationary state, one derives the result that at large distances r the wavefunction has the form of a plane wave travelling in the $\hat{\boldsymbol{\kappa}}$-direction plus incoming spherical waves which contribute with a weight given by the scattering factor. Because, also at large distances r, the plane wave can be represented as a combination of incoming and outgoing spherical waves (see below equ. (7.24)), destructive interference between the two incoming spherical waves (from the plane-wave part and from the boundary condition) ensures that only *outgoing* flux remains as required for an 'out'-state. In the limit of infinite distances, the influence of the spherical wave can be neglected, and it is only the plane wave which remains in equ. (7.20).

Equ. (7.20) describes for an 'out'-state the asymptotic behaviour of the stationary wavefunction. As discussed above, the characteristic property of this state is that the incoming spherical waves $e^{-i\kappa r}/r$ have the scattering amplitude $f^{(-)}(\vartheta)$. It is this minus sign in the exponential term of the incoming spherical waves which is kept as a superscript to characterize the 'out'-state, and the relation described by equ. (7.20) is frequently called the *incoming spherical waves* boundary condition. Hence, one should not mix up the state with the waves.

The boundary condition formulated in equ. (7.20) for the asymptotic behaviour of the photoelectron wavefunction has implications which not only are asymptotic, but also apply to the wavefunction $\Psi_{\kappa}^{(-)}(\mathbf{r})$ represented in full space. In order to work out these implications, one starts with the ansatz

$$\Psi_{\kappa}^{(-)}(\mathbf{r}) = \frac{1}{(2\pi)^{3/2}} \sum_{\ell} \mathrm{i}^{\ell}\,(2\ell+1)\, R_{\kappa\ell}^{(-)}(r)\, P_{\ell}(\cos\vartheta) \tag{7.21a}$$

$$= \sqrt{\left(\frac{2}{\pi}\right)} \sum_{\ell m} \mathrm{i}^{\ell}\, b_{\ell}(\kappa)\, Y_{\ell m}^{*}(\hat{\boldsymbol{\kappa}})\, R_{\kappa\ell}(r)\, Y_{\ell m}(\hat{\mathbf{r}}) \tag{7.21b}$$

with

$$R_{\kappa\ell}^{(-)}(r) = R_{\kappa\ell}(r)\, b_{\ell}(\kappa). \tag{7.21c}$$

This expression differs from the expansion of a plane wave as given in equ. (7.14) in three respects. First, a different overall normalization is used (normalization in $\boldsymbol{\kappa}$-space, see equ. (7.28f)). Second, the radial functions $R_{\kappa\ell}(r)$ are different from the spherical Bessel functions $j_{\ell}(\kappa r)$. Third, the incoming spherical wave boundary condition leads to an additional factor, $b_{\ell}(\kappa)$.

The modification of the radial functions is obvious because the atomic potential $V(r)$ will modify the spherical Bessel functions $j_{\ell}(\kappa r)$ which belong to a free plane wave. Also, the dependence on products of κ and r is lost. The $R_{\kappa\ell}(r)$ functions follow as regular solutions from the time-independent Schrödinger equation†

$$\left[-\frac{1}{2}\frac{\mathrm{d}^2}{\mathrm{d}r^2} + \frac{\ell(\ell+1)}{2r^2} + V(r) - \varepsilon\right][rR_{\kappa\ell}(r)] = 0. \tag{7.22}$$

(Of course, for $V(r) = 0$ this differential equation leads to the spherical Bessel functions.)

The effect of a non-vanishing potential on the radial function is illustrated in Fig. 7.4 for the example of a repulsive potential of rectangular shape. It can be seen that the selected s-wave radial function $R_{\kappa 0}(r)$ strongly differs in the region of potential from the corresponding spherical Bessel function $j_0(\kappa r)$. However, far away from the influence of the potential, the function $R_{\kappa 0}(r)$ behaves like the asymptotic spherical Bessel function $j_0(\kappa r)$, except that it is shifted in phase. A repulsive potential pushes out, and an attractive potential pulls in the radial functions $R_{\kappa\ell}(r)$ as compared to $j_{\ell}(\kappa r)$. This behaviour is expressed in the asymptotic forms of these radial functions (for the general case with ℓ and an attractive potential):

$$j_{\ell}(\kappa r) \xrightarrow[r\to\infty]{} \frac{1}{\kappa r}\sin\left(\kappa r - \frac{\ell\pi}{2}\right), \tag{7.23a}$$

$$R_{\kappa\ell}(r) \xrightarrow[r\to\infty]{} \sqrt{\left(\frac{2}{\pi}\right)}\frac{1}{r}\sin\left(\kappa r - \frac{\ell\pi}{2} + \Delta_{\ell}\right). \tag{7.23b}$$

† This equation can be understood from the Hartree–Fock equation derived for the helium atom, equ. (7.66b)). The discrete negative energy eigenvalue ε_{1s} has to be replaced by the corresponding positive energy $\varepsilon = \kappa^2/2$ (in atomic units) of the continuum electron, the centrifugal potential absent for $\ell = 0$ has to be taken into account for the more general case considered here, and the Hartree–Fock potential $V_{\mathrm{HF}}(r)$ has to be modified to $V(r)$ which in the atomic region is the Hartree–Fock potential, but at large distances is the $(1/r)$ dependent ionic potential (the Latter correction [Lat55]).

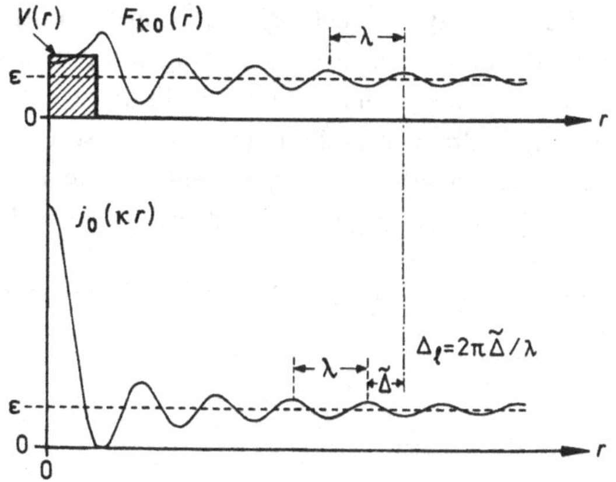

Figure 7.4 Definition of the phase shift Δ_ℓ as introduced by a potential. The solution of the radial function $R_{\kappa\ell}(r)$ of a wave with energy $\varepsilon = \kappa^2/2$ (in atomic units) and with $\ell = 0$ is shown for two situations: under the influence of a repulsive potential $V(r)$ as indicated by the shaded region (top), and for vanishing potential (bottom). In the first case one has $R_{\kappa\ell}(r) = F_{\kappa 0}(r)$, and in the second case the radial function is equal to the spherical Bessel function, i.e., $R_{\kappa\ell}(r) = j_0(\kappa r)$. Asymptotically, both solutions, $F_{\kappa 0}(r)$ and $j_0(\kappa r)$, differ only by a constant distance $\tilde{\Delta}$ in the r coordinate which is related to the phase shift Δ_ℓ as indicated. From *The picture book of quantum mechanics*, S. Brandt and H. D. Dahmen, 1st edition, 1985, John Wiley & Sons Inc., NY. © 1985 John Wiley & Sons Inc.

(These radical functions are normalized according to equ. (7.28c).) Knowing these asymptotic forms, it is now possible to derive the asymptotic limits for the ansatz in equ. (7.21) and the boundary condition in equ. (7.20). A comparison of these expressions will then lead to the unknown coefficients $b_\ell(\kappa)$ in equ. (7.21) and the scattering amplitude $f^{(-)}(\vartheta)$ in equ. (7.20). If one starts with the plane-wave part in equ. (7.11), one obtains with the asymptotic form for $j_\ell(\kappa r)$, equ. (7.23a), and the replacement

$$\sin x = (e^{ix} - e^{-ix})/2i, \tag{7.23c}$$

the result

$$e^{i\kappa z}|_{r\to\infty} = \frac{1}{2\kappa i}\sum_\ell (2\ell+1)\, i^\ell \left(\frac{e^{i(\kappa r - \ell\pi/2)}}{r} - \frac{e^{-i(\kappa r - \ell\pi/2)}}{r}\right) P_\ell(\cos\vartheta). \tag{7.24}$$

Remembering the phase factor $e^{-i\omega t}$, this relation expresses quantitatively the result which was used above, i.e. a plane wave can be described asymptotically as a sum of undisturbed outgoing and incoming spherical waves, which differ only in their relative phases. The boundary condition of equ. (7.20) can then be expressed as

$$\begin{aligned}\Psi_\kappa^{(-)}(r)|_{r\to\infty} &= (\text{incoming spherical waves})|_{\text{undisturbed}}\\ &+ (\text{outgoing spherical waves})|_{\text{undisturbed}}\\ &+ (\text{incoming spherical waves})|_{\text{potential affected}}.\end{aligned} \tag{7.25}$$

(This form of partitioning into incoming and outgoing spherical waves can be generalized to the case of many channels by requiring the outgoing component (flux) to be one specified channel whilst the outgoing components in the admixed channels must vanish.) Similarly, the asymptotic limit of equ. (7.21b) can be worked out using equs. (7.23b) and (7.23c). The comparison of terms belonging to incoming and outgoing spherical waves, respectively, in the two asymptotic expressions of $\Psi^{(-)}_{\kappa}(\mathbf{r})$ then yields (note $e^{i\ell\pi/2} = i^{\ell}$ and $i^{2\ell} = (-1)^{\ell}$)

$$f^{(-)}(\vartheta) = \frac{i}{2\kappa}\sum_{\ell}(-1)^{\ell}(2\ell+1)(e^{-2i\Delta_\ell} - 1)P_\ell(\cos\vartheta) \tag{7.26}$$

and

$$b_\ell(\kappa) = \sqrt{\left(\frac{\pi}{2}\right)}\frac{1}{\kappa}e^{-i\Delta_\ell}, \tag{7.27a}$$

where for a Coulomb wave instead of a plane wave the phase Δ_ℓ contains both δ_ℓ, the phase from the short-range atomic potential, and σ_ℓ, the phase from the long-range Coulomb potential:

$$\Delta_\ell = \delta_\ell + \sigma_\ell, \tag{7.27b}$$

and for single ionization of a neutral atom one has

$$\sigma_\ell = \arg\Gamma(\ell + 1 - i/\kappa), \tag{7.27c}$$

where Γ is the complex gamma function.

In the present context only the factor $b_\ell(\kappa)$ is of further interest. Since it contains the asymptotic phase shift Δ_ℓ and the nominal wavenumber κ which also is defined by its asymptotic value, the expression for $b_\ell(\kappa)$ holds everywhere even though it has been derived in the asymptotic limit (in contrast, the scattering function $f^{(-)}(\vartheta)$ can be defined only in the asymptotic limit). Hence, the correct wavefunction $\Psi^{(-)}_{\kappa}(\mathbf{r})$ for the emission of a photoelectron (or any other electron) is given by (in atomic units)

$$\Psi^{(-)}_{\kappa m_s}(\mathbf{r}) = \frac{1}{\kappa}\sum_{\ell m} i^{\ell}\, e^{-i\Delta_\ell}\, Y^*_{\ell m}(\hat{\mathbf{k}})\, R_{\kappa\ell}(r)\, Y_{\ell m}(\hat{\mathbf{r}})\, \chi^{m_s}_{1/2} \tag{7.28a}$$

or, equivalently,

$$\Psi^{(-)}_{\kappa m_s}(\mathbf{r}) = \frac{1}{\sqrt{\kappa}}\sum_{\ell m} i^{\ell}\, e^{-i\Delta_\ell}\, Y^*_{\ell m}(\hat{\mathbf{k}})\, R_{\varepsilon\ell}(r)\, Y_{\ell m}(\hat{\mathbf{r}})\, \chi^{m_s}_{1/2}, \tag{7.28b}$$

where m_s is the spin projection of the electron along the z-axis, Δ_ℓ the total phase shift (the sum of the phase shifts σ_ℓ and δ_ℓ from the Coulomb and short-range atomic potentials, respectively) and $R_{\kappa\ell}(r)$, or $R_{\varepsilon\ell}(r)$, is the radial part of the wavefunction for different normalizations. The radial functions $R_{\kappa\ell}(r)$ introduced in equ. (7.21) are normalized 'on-the-wavenumber-scale',

$$\int_0^\infty R_{\kappa'\ell}(r)R_{\kappa\ell}(r)r^2\,dr = \delta(\kappa' - \kappa), \tag{7.28c}$$

while the radial functions $R_{\varepsilon\ell}(r)$ are conveniently normalized 'on-the-energy-scale' ε,†

$$\int_0^\infty R_{\varepsilon'\ell}(r) R_{\varepsilon\ell}(r) r^2 \, \mathrm{d}r = \delta(\varepsilon' - \varepsilon). \tag{7.28d}$$

The last normalization means that *one* electron can be found within an interval of *one* energy unit centred around the value ε. For the asymptotic form of $R_{\varepsilon\ell}(r)$ one gets (in atomic units)

$$R_{\varepsilon\ell}(r) \xrightarrow[r\to\infty]{} \sqrt{\left(\frac{2}{\pi\kappa}\right)} \frac{1}{r} \sin\left(\kappa r - \frac{1}{2}\ell\pi + \frac{1}{\kappa}\ln 2\kappa r + \sigma_\ell + \delta_\ell\right), \tag{7.28e}$$

where the logarithmic term is characteristic of the long-range Coulomb potential (this term was missing in the corresponding equ. (7.23b) because there only a plane wave was considered). It should, however, be pointed out that for the *full* wavefunction $\Psi_{\boldsymbol{\kappa}}^{(-)}(\mathbf{r})$, given in equ. (7.28b), the orthonormality condition still holds in **κ**-space, i.e., one has

$$\int \Psi_{\boldsymbol{\kappa}'}^{(-)*}(\mathbf{r}) \Psi_{\boldsymbol{\kappa}}^{(-)}(\mathbf{r}) \, \mathrm{d}\mathbf{r} = \delta(\boldsymbol{\kappa}' - \boldsymbol{\kappa}), \tag{7.28f}$$

and, related to this, the density of states, ρ, in the wavenumber interval dκ with directions of propagation within $\mathrm{d}\Omega_\kappa$ is given by

$$\rho = \mathrm{d}\boldsymbol{\kappa} = \kappa^2 \, \mathrm{d}\kappa \, \mathrm{d}\Omega_\kappa. \tag{7.28g}$$

Finally, the extension of the wavefunction in equ. (7.28b) to the $(\ell + s \to j)$-coupled case will be given. One gets

$$\Psi_{\boldsymbol{\kappa} m_s}^{(-)}(\mathbf{r}) = \frac{1}{\sqrt{\kappa}} \sum_{\ell m_\ell} \sum_{j m_j} \mathrm{i}^\ell \, \mathrm{e}^{-\mathrm{i}(\sigma_{\ell j} + \delta_{\ell j})} \, Y_{\ell m_\ell}^*(\hat{\boldsymbol{\kappa}}) \, (\ell m_\ell \tfrac{1}{2} m_s | j m_j) \, R_{\varepsilon\ell j}(r) \, \mathscr{Y}_{\ell \frac{1}{2} j m_j}(\hat{\mathbf{r}}) \tag{7.29a}$$

with

$$\mathscr{Y}_{\ell \frac{1}{2} j m_j}(\hat{\mathbf{r}}) = \sum_{m_\ell m_s} (\ell m_\ell \tfrac{1}{2} m_s | j m_j) \, Y_{\ell m_\ell}(\hat{\mathbf{r}}) \chi_{1/2}^{m_s}. \tag{7.29b}$$

This form is similar to the one presented in equ. (7.18) for a free plane wave, except for the incorporation of the incoming spherical wave boundary condition, the separate treatment of the radial function, and the normalization of these radial functions on the energy scale. Furthermore, it should be noted that equ. (7.29a) contains a j dependence of the phases and the radial function which can be understood only within a relativistic treatment.

7.1.4 Dimension of wavefunctions

Finally, some remarks will be made concerning the dimension of wavefunctions. The bound-state orbitals are subject to the orthonormality relation

$$\int \varphi_i^*(\mathbf{r}) \varphi_j(\mathbf{r}) \, \mathrm{d}\mathbf{r} = \delta_{ij}, \tag{7.30a}$$

† In the derivation of equ. (8.28) presented below, atomic units are not used. In this case the change from $R_{\kappa\ell}(r)$ to $R_{\varepsilon\ell}(r)$ brings in an additional factor $\hbar m_0^{-1/2}$, because one has $R_{\kappa\ell}(r) = \hbar\sqrt{(\kappa/m_0)} R_{\varepsilon\ell}(r)$.

where δ_{ij} is the Kronecker delta. From this expression one gets

$$dim\{\varphi_i(r)\} = 1/\sqrt{[(\text{length})^3]}. \tag{7.30b}$$

The continuum function is normalized with the δ-function, see equ. (7.28f). Since the δ-function also has a dimension,

$$dim\{\delta(x)\} = 1/dim\{x\}, \tag{7.31a}$$

one obtains

$$dim\{\Psi^{(-)}_{\mathbf{\kappa}}(\mathbf{r})\} = 1/\sqrt{[(\text{length})^3(\text{wavenumber})]} \tag{7.31b}$$

and

$$dim\{R_{\varepsilon\ell}(r)\} = 1/\sqrt{[(\text{length})^3(\text{energy})]}. \tag{7.31c}$$

7.2 Evaluation of determinantal wavefunctions

Determinantal wavefunctions (Slater wavefunctions, [Sla29]) are well suited for representing a special state of a given electron configuration, because the antisymmetrization of the wavefunction with respect to the interchange of any two electrons and the requirement of the Pauli principle that no two electrons agree in all their quantum numbers are automatically fulfilled, due to the properties of determinants. An electron orbital is written in a determinantal wavefunction as $n\ell m_\ell^{m_s}$. A *standard order* is used, i.e., the electron orbitals are arranged according to:

$$\left.\begin{array}{l} n \text{ increasing towards the right direction} \\ \ell \text{ increasing towards the right direction} \\ m_\ell \text{ decreasing towards the right direction} \\ m_s \text{ decreasing towards the right direction} \end{array}\right\} \downarrow \text{ lower priority}. \tag{7.32}$$

The standard order fixes the phase so that the single determinantal wavefunction of a complete shell has the phase $(+1)$. In constructing determinantal wavefunctions, a distinction must be made between *equivalent* and *non-equivalent* electrons, defined by possessing the same value of $n\ell$ or not, respectively, because different construction rules must be applied. (Special treatments also hold if equivalent electrons are described by the $\ell + s \to j$-coupling-scheme quantum numbers (for details see [Cow81]).) For the construction of determinantal wavefunctions some helpful tools are given below (for extensive monographs see [CSh35, Sla60, STa63, Sob72, Cow81]).

7.2.1 Clebsch–Gordan coefficients

The extension of a given determinantal wavefunction (called the *parent* wavefunction, which in the simplest case can be just a one-electron spin-orbital) to include another *non*-equivalent electron (or even a group of non-equivalent electrons) is made with the help of *vector-coupling* or Clebsch–Gordan coefficients

Table 7.1. *Some Clebsch–Gordan coefficients (values which are unity according to equ. (7.36a) or can be calculated from the given values using equs. (7.39) have been omitted).*

1/2 + 1/2 → 0	$(1/2\ 1/2\ 1/2\ -1/2 \mid 0\ 0) = 1/\sqrt{2}$
1/2 + 1/2 → 1	$(1/2\ 1/2\ 1/2\ -1/2 \mid 1\ 0) = 1/\sqrt{2}$
1 + 1/2 → 1/2	$(1\ 1\ 1/2\ -1/2 \mid 1/2\ 1/2) = \sqrt{(2/3)}$ $(1\ 0\ 1/2\ \ 1/2 \mid 1/2\ 1/2) = -1/\sqrt{3}$
1 + 1/2 → 3/2	$(1\ 1\ 1/2\ -1/2 \mid 3/2\ 1/2) = 1/\sqrt{3}$ $(1\ 0\ 1/2\ \ 1/2 \mid 3/2\ 1/2) = \sqrt{(2/3)}$
1 + 1 → 0	$(1\ 1\ 1\ -1 \mid 0\ 0) = 1/\sqrt{3}$ $(1\ 0\ 1\ \ 0 \mid 0\ 0) = -1/\sqrt{3}$
1 + 1 → 1	$(1\ 1\ 1\ \ 0 \mid 1\ 1) = 1/\sqrt{2}$ $(1\ 0\ 1\ \ 1 \mid 1\ 1) = -1/\sqrt{2}$ $(1\ 0\ 1\ \ 0 \mid 1\ 0) = 0$ $(1\ 1\ 1\ -1 \mid 1\ 0) = 1/\sqrt{2}$
1 + 1 → 2	$(1\ 1\ 1\ \ 0 \mid 2\ 1) = 1/\sqrt{2}$ $(1\ 0\ 1\ \ 1 \mid 2\ 1) = 1/\sqrt{2}$ $(1\ 0\ 1\ \ 0 \mid 2\ 0) = \sqrt{(2/3)}$ $(1\ 1\ 1\ -1 \mid 2\ 0) = 1/\sqrt{6}$

[Cle72, Gor75]. Their values follow from the general properties of the vector coupling of angular momentum operators.

For the coupling of two angular momenta **a** and **b** to the resulting value **c** with the corresponding magnetic quantum numbers α, β and γ, the Clebsch–Gordan coefficients $(a\alpha b\beta \mid c\gamma)$ are defined as expansion coefficients in the relation

$$\psi(abc\gamma) = \sum_{\alpha,\beta} (a\alpha b\beta \mid c\gamma)\psi(a\alpha)\psi(b\beta) \tag{7.33a}$$

with the inverse form

$$\psi(a\alpha)\psi(b\beta) = \sum_{c\gamma} (a\alpha b\beta \mid c\gamma)\psi(abc\gamma). \tag{7.33b}$$

(Clebsch–Gordan coefficients are chosen to be real and orthonormal.) The triangle condition from the vector model requires for the angular momenta

$$|a - b| \leq c \leq a + b \qquad \text{and} \qquad \alpha + \beta = \gamma. \tag{7.34}$$

If this condition is not fulfilled, the Clebsch–Gordan coefficients vanish, otherwise they have certain numerical values (see Table 7.1). The Clebsch–Gordan coefficients are related to the *Wigner* coefficients [Wig51], also called *3j symbols* $j_1 = a$, $j_2 = b$, $j_3 = c$, $m_1 = \alpha$, $m_2 = \beta$, $m_3 = \gamma$), defined by

$$\begin{pmatrix} j_1 & j_2 & j_3 \\ m_1 & m_2 & m_3 \end{pmatrix} = \frac{(-1)^{j_1 - j_2 - m_3}}{\sqrt{(2j_3 + 1)}} (j_1 m_1 j_2 m_2 \mid j_3 - m_3). \tag{7.35}$$

These and other j symbols can be found as program packages in many computer libraries.

It is important to note that different authors use different phase conventions. Those of Condon and Shortley [CSh35] will be employed here,† requiring [STa63]

$$(a\,\alpha = a \quad b\,\beta = b \,|\, c = a + b \quad \gamma = \alpha + \beta) = +1 \tag{7.36a}$$

and

$$\sum_{\alpha,\beta} \alpha(a\alpha b\beta \,|\, c\gamma)(a\alpha b\beta \,|\, c-1 \ \gamma) > 0. \tag{7.36b}$$

Besides the phase convention, the following important properties hold for Clebsch–Gordan coefficients:

reality

$$(a\alpha b\beta \,|\, c\gamma)^* = (a\alpha b\beta \,|\, c\gamma); \tag{7.37}$$

orthogonality

$$\sum_{\alpha\beta} (a\alpha b\beta \,|\, c\gamma)(a\alpha b\beta \,|\, c'\gamma') = \delta(c, c')\delta(\gamma, \gamma'), \tag{7.38a}$$

$$\sum_{c\gamma} (a\alpha b\beta \,|\, c\gamma)(a\alpha' b\beta' \,|\, c\gamma) = \delta(\alpha, \alpha')\delta(\beta, \beta'); \tag{7.38b}$$

symmetry

$$(a\alpha b\beta \,|\, c\gamma) = (-)^{a+b-c}(b\beta a\alpha \,|\, c\gamma), \tag{7.39a}$$

$$(a\alpha b\beta \,|\, c\gamma) = (-)^{a-\alpha}\sqrt{\left(\frac{2c+1}{2b+1}\right)}(a\,\alpha\, c-\gamma \,|\, b-\beta), \tag{7.39b}$$

$$(a\alpha b\beta \,|\, c\gamma) = (-)^{b+\beta}\sqrt{\left(\frac{2c+1}{2a+1}\right)}(c-\gamma\; b\beta \,|\, a-\alpha), \tag{7.39c}$$

$$(a\alpha b\beta \,|\, c\gamma) = (-)^{a+b-c}(a-\alpha\; b-\beta \,|\, c-\gamma). \tag{7.39d}$$

Equ. (7.39a) shows that the ordering of coupling is of importance. For example, in LSJ-coupling it matters, whether L plus S or S plus L is coupled to J. Here the coupling of L plus S to J is used (see footnote 14 on p. 57 in [Cow81]).

From the known values of the Clebsch–Gordan coefficients, determinantal wavefunctions with selected angular momenta can easily be evaluated by adding to the parent function the new non-equivalent electron orbital and taking into account the Clebsch–Gordan coefficient:‡

$$\begin{aligned}\tilde{\Psi}(\text{parent}(L_P S_P) n\ell m_\ell^{m_s} L M_L S M_S) = \sum (L_P M_{L_P} \ell m_\ell \,|\, L M_L)(S_P M_{S_P} \tfrac{1}{2} m_s \,|\, S M_S) \\ \times \{\text{DWF}_{\text{parent}}(L_P M_{L_P} S_P M_{S_P}); \varphi(n\ell m_\ell^{m_s})\},\end{aligned} \tag{7.40}$$

† Tabulations with this phase convention can be found in, for example, [CSh35, RBM59, STa63, App68, BSa68, Sob72, Cow81].

‡ In these and subsequent expressions the summation extends over all remaining magnetic quantum numbers.

where $\mathrm{DWF}_{\mathrm{parent}}(L_P M_{L_P} S_P M_{S_P})$ is the determinantal wavefunction of the parent electron configuration in the state $L_P M_{L_P} S_P M_{S_P}$. Using an additional Clebsch–Gordan coefficient, $(LM_L SM_S | JM_J)$, the wavefunction in LSJ-coupling can be obtained:

$$\Psi(LSJM) = \sum (LM_L SM_S | JM) \Psi(LM_L SM_S). \tag{7.41a}$$

A special case of equ. (7.41a) is the ℓsj-coupling of single-particle functions:

$$\varphi_{n\ell jm_j} = \sum (\ell m_\ell \tfrac{1}{2} m_s | jm_j) \varphi_{n\ell m_\ell m_s}, \tag{7.41b}$$

and as an example, the j-coupled single-particle functions for $s_{1/2}$ and p_j electrons will be calculated. One obtains

$$\varphi_{\mathrm{s}1/2 m_j = 1/2} = \varphi_{\mathrm{s}0^+}, \quad \varphi_{\mathrm{s}1/2 m_j = -1/2} = \varphi_{\mathrm{s}0^-}, \tag{7.42}$$

$$\varphi_{\mathrm{p}1/2 m_j = 1/2} = \sqrt{\left(\frac{2}{3}\right)} \varphi_{\mathrm{p}1^-} - \frac{1}{\sqrt{3}} \varphi_{\mathrm{p}0^+}, \tag{7.43a}$$

$$\varphi_{\mathrm{p}1/2 m_j = -1/2} = \frac{1}{\sqrt{3}} \varphi_{\mathrm{p}0^-} - \sqrt{\left(\frac{2}{3}\right)} \varphi_{\mathrm{p}-1^+}. \tag{7.43b}$$

$$\varphi_{\mathrm{p}3/2 m_j = 3/2} = \varphi_{\mathrm{p}1^+}, \quad \varphi_{\mathrm{p}3/2 m_j = -3/2} = \varphi_{\mathrm{p}-1^-}, \tag{7.44a}$$

$$\varphi_{\mathrm{p}3/2 m_j = 1/2} = \sqrt{\left(\frac{2}{3}\right)} \varphi_{\mathrm{p}0^+} + \frac{1}{\sqrt{3}} \varphi_{\mathrm{p}1^-}, \tag{7.44b}$$

$$\varphi_{\mathrm{p}3/2 m_j = -1/2} = \sqrt{\left(\frac{2}{3}\right)} \varphi_{\mathrm{p}0^-} + \frac{1}{\sqrt{3}} \varphi_{\mathrm{p}-1^+}. \tag{7.44c}$$

As another example, the wavefunction for the coupling of a non-equivalent p electron to an $n\mathrm{p}^2(^3\mathrm{P})$ parent state leading to the LSJ final state $^2\mathrm{D}_{5/2}$ will be worked out. For this case one gets

$$\begin{aligned}
\tilde{\Psi}(n\mathrm{p}^2(^3\mathrm{P})n'\mathrm{p}\ ^2\mathrm{D}_{5/2}\ M = 5/2) &= \sum (2M_L \tfrac{1}{2} M_S | \tfrac{5}{2}\tfrac{5}{2})(1M_{L_P} 1m_\ell | 2M_L)(1M_{S_P} \tfrac{1}{2} m_s | \tfrac{1}{2} M_S) \\
&\qquad \times \{\mathrm{DWF}_{\mathrm{parent}}(L_P M_{L_P} S_P M_{S_P});\ n'\mathrm{p}m_\ell^{m_s}\} \\
&= \sum (1M_{L_P} 1m_\ell | 22)(1M_{S_P} \tfrac{1}{2}\, m_s | \tfrac{1}{2}\tfrac{1}{2}) \{\mathrm{DWF}_{\mathrm{parent}}(^3\mathrm{P}M_{L_P} M_{S_P});\ n'\mathrm{p}m_\ell^{m_s}\} \\
&= (11\tfrac{1}{2} - \tfrac{1}{2} | \tfrac{1}{2}\tfrac{1}{2}) \{\mathrm{DWF}_{\mathrm{parent}}(^3\mathrm{P}M_{L_P} = 1 M_{S_P} = 1);\ n'\mathrm{p}1^-\} \\
&\quad + (10\tfrac{1}{2}\tfrac{1}{2} | \tfrac{1}{2}\tfrac{1}{2}) \{\mathrm{DWF}_{\mathrm{parent}}(^3\mathrm{P}M_{L_P} = 1 M_{S_P} = 0);\ n'\mathrm{p}1^+\} \\
&= \frac{2}{\sqrt{6}} \{n\mathrm{p}1^+, n\mathrm{p}0^+, n'\mathrm{p}1^-\} - \frac{1}{\sqrt{6}} \{n\mathrm{p}1^-, n\mathrm{p}0^+, n'\mathrm{p}1^+\} \\
&\quad - \frac{1}{\sqrt{6}} \{n\mathrm{p}1^+, n\mathrm{p}0^-, n'\mathrm{p}1^+\}.
\end{aligned} \tag{7.45}$$

7.2.2 Coefficients of fractional parentage

If an equivalent electron is added to a wavefunction of a given parent electron configuration, *coefficients of fractional parentage* $\{cfp\}$ have to be used instead of Clebsch–Gordan coefficients. These are defined as expansion coefficients of the wavefunction with α equivalent electrons in terms of parent wavefunctions of $\alpha - 1$ equivalent electrons and one 'singled-out' electron. In the LS-coupling case one gets (for jj-coupling, see [STa63])

$$\tilde{\Psi}(n\ell^{\alpha}LM_LSM_S) = \sum \{cfp\}\,\bar{\Psi}(n\ell^{\alpha-1}L_PS_P;\, n\ell LM_LSM_S), \qquad (7.46a)$$

where $\bar{\Psi}$ is defined by

$$\bar{\Psi}(n\ell^{\alpha-1}L_PS_P;\, n\ell LM_LSM_S) = \sum (L_PM_{L_P}\ell m \mid LM_L)(S_PM_{S_P}\tfrac{1}{2}m_s \mid SM_S) \times \tilde{\Psi}(n\ell^{\alpha-1}L_PM_{L_P}S_PM_{S_P})\,\varphi(n\ell m_\ell^{m_s}(\alpha)), \qquad (7.46b)$$

where $\varphi(n\ell m_\ell^{m_s}(\alpha))$ describes the single-particle function of the last (α) electron which is treated *separately*, and the summations run over all remaining quantum numbers. (Note that the wavefunction $\tilde{\Psi}$ is the fully antisymmetric wavefunction even though it is represented as a linear combination of wavefunctions $\bar{\Psi}$ which are antisymmetric only in the remaining $\alpha - 1$ electrons.

The nature of $\{cfp\}$ will be illustrated with the example of coupling three equivalent p-electrons. Assuming, the wavefunction of two equivalent p-electrons is known, the additional coupling of a non-equivalent p′-electron leads to ten different states given by:

$$\left.\begin{array}{ll} p^2(^1S)p': & {}^2P, \\ p^2(^1D)p': & {}^2P,\ {}^2D,\ 2F, \\ p^2(^3P)p': & {}^2S \text{ or } {}^4S,\ {}^2P \text{ or } {}^4P,\ {}^2D \text{ or } {}^4D. \end{array}\right\} \qquad (7.47a)$$

In contrast, only three states are allowed for the coupling of three equivalent electrons:

$$\left.\begin{array}{l} p^3\ {}^2P, \\ p^3\ {}^2D, \\ p^3\ {}^4S. \end{array}\right\} \qquad (7.47b)$$

The {cfp} coefficients then give the corresponding weighting factors (see Table 7.2) for describing these fully antisymmetrized state functions with three equivalent electrons in terms of the antisymmetric two-electron parent functions $p^2\ {}^1S$, $p^2\ {}^1D$, and $p^2\ {}^3P$, and the extra p-electron. Selecting the example $\tilde{\Psi}(p^3\ {}^2D\ M_L = 2\ M_S = \frac{1}{2})$, one can see from equ. (7.47a) that of the three parent configurations only two will contribute when forming $p^3\ {}^2D$; $p^2(^1D)p'$ and $p^2(^3P)p'$. Hence, one gets (see Table 7.2)

$$\tilde{\Psi}(p^3\ {}^2D\ M_L = 2\ M_S = \tfrac{1}{2}) = \frac{1}{\sqrt{2}}\bar{\Psi}(p^2\ {}^3P;p\ {}^2D\ M_L = 2\ M_S = \tfrac{1}{2}) - \frac{1}{\sqrt{2}}\bar{\Psi}(p^2\ {}^1D;p\ {}^2D\ M_L = 2\ M_S = \tfrac{1}{2}) \qquad (7.48a)$$

Table 7.2. *Coefficients of fractional parentage, {cfp}, for p^α electron configurations. From [Cow81].*

Configuration	Parent configuration	Value of *cfp*
$p\ ^2P$	$p^0\ ^1S$	1
$p^2\ ^1S$	$p\ ^2P$	1
$p^2\ ^3P$	$p\ ^2P$	1
$p^2\ ^1D$	$p\ ^2P$	1
$p^3\ ^2D$	$p^2\ ^1S$	0
	$p^2\ ^3P$	$1/\sqrt{2}$
	$p^2\ ^1D$	$-1/\sqrt{2}$
$p^3\ ^4S$	$p^2\ ^1S$	0
	$p^2\ ^3P$	1
	$p^2\ ^1D$	0
$p^3\ ^2P$	$p^2\ ^1S$	$\sqrt{(2/9)}$
	$p^2\ ^3P$	$-1/\sqrt{2}$
	$p^2\ ^1D$	$-\sqrt{(5/18)}$
$p^4\ ^1S$	$p^3\ ^4S$	0
	$p^3\ ^2P$	1
	$p^3\ ^2D$	0
$p^4\ ^3P$	$p^3\ ^4S$	$-1\sqrt{3}$
	$p^3\ ^2P$	$-1/2$
	$p^3\ ^2D$	$\sqrt{(5/12)}$
$p^4\ ^1D$	$p^3\ ^4S$	0
	$p^3\ ^2P$	$-1/2$
	$p^3\ ^2D$	$-\sqrt{(3/4)}$
$p^5\ ^2P$	$p^4\ ^1S$	$1/\sqrt{(15)}$
	$p^4\ ^3P$	$\sqrt{(3/5)}$
	$p^4\ ^1D$	$1/\sqrt{3}$
$p^6\ ^1S$	$p^5\ ^2P$	1

with

$$\begin{aligned}
&\bar{\Psi}(\mathrm{p}^2\ {}^3\mathrm{P};\mathrm{p}\ {}^2\mathrm{D}\ M_L = 2\ M_S = \tfrac{1}{2}) \\
&\quad = \sum (1 M_{L_P}\ 1\ m_\ell \,|\, 22)(1 M_{S_P}\ \tfrac{1}{2}\ m_s|\tfrac{1}{2}\ \tfrac{1}{2})\tilde{\Psi}(\mathrm{p}^2\ {}^3\mathrm{P}\ M_{L_P}\ M_{S_P};\ 1, 2)\varphi(\mathrm{p}m_\ell m_s;\ 3) \\
&\quad = \sqrt{\left(\frac{2}{3}\right)}\tilde{\Psi}(\mathrm{p}^2\ {}^3\mathrm{P}\ M_{L_P} = 1\ M_{S_P} = 1;\ 1, 2)\varphi(\mathrm{p}1^-;\ 3) \\
&\qquad - \frac{1}{\sqrt{3}}\tilde{\Psi}(\mathrm{p}^2\ {}^3\mathrm{P}\ M_{L_P} = 1\ M_{S_P} = 0;\ 1, 2)\varphi(\mathrm{p}1^+;\ 3).
\end{aligned} \tag{7.48b}$$

Introducing the functions $\tilde{\Psi}(\mathrm{p}^2\ {}^3\mathrm{P}\ M_L = 1\ M_S = 1) = \{\mathrm{p}1^+, \mathrm{p}0^+\}$ and $\tilde{\Psi}(\mathrm{p}^2\ {}^3\mathrm{P}$ $M_L = 1\ M_S = 0) = (\{\mathrm{p}1^-, \mathrm{p}0^+\} + \{\mathrm{p}1^+, \mathrm{p}0^-\})/\sqrt{2}$ from Table 3.2, one gets

$$\bar{\Psi}(\mathrm{p}^2\ {}^3\mathrm{P};\mathrm{p}\ {}^2\mathrm{D}\ M_L = 2\ M_S = \tfrac{1}{2})$$
$$= \sqrt{\left(\frac{2}{3}\right)}\{\mathrm{p}1^+, \mathrm{p}0^+\}\varphi(\mathrm{p}1^-; 3) - \frac{1}{\sqrt{6}}\{\mathrm{p}1^-, \mathrm{p}0^+\}\varphi(\mathrm{p}1^+; 3)$$
$$- \frac{1}{\sqrt{6}}\{\mathrm{p}1^+, \mathrm{p}0^-\}\varphi(\mathrm{p}1^+; 3). \qquad (7.48c)$$

A similar calculation leads to

$$\bar{\Psi}(\mathrm{p}^2\ {}^1\mathrm{D};\ \mathrm{p}\ {}^2\mathrm{D}\ M_L = 2\ M_S = \tfrac{1}{2}) = \sqrt{\left(\frac{2}{3}\right)}\{\mathrm{p}1^+, \mathrm{p}1^-\}\varphi(\mathrm{p}0^+; 3)$$
$$- \frac{1}{\sqrt{6}}\{\mathrm{p}1^+, \mathrm{p}0^-\}\varphi(\mathrm{p}1^+; 3) + \frac{1}{\sqrt{6}}\{\mathrm{p}1^-, \mathrm{p}0^+\}\varphi(\mathrm{p}1^+; 3). \qquad (7.48d)$$

With these relations, the desired wavefunction becomes

$$\tilde{\Psi}(\mathrm{p}^3\ {}^2\mathrm{D}\ M_L = 2\ M_S = 1/2) = \frac{1}{\sqrt{3}}(\{\mathrm{p}1^+, \mathrm{p}0^+\}\varphi(\mathrm{p}1^-; 3) + \{\mathrm{p}0^+, \mathrm{p}1^-\}\varphi(\mathrm{p}1^+; 3)$$
$$- \{\mathrm{p}1^+, \mathrm{p}1^-\}\varphi(\mathrm{p}0^+; 3))$$
$$= \{\mathrm{p}1^+, \mathrm{p}0^+, \mathrm{p}1^-\}. \qquad (7.48e)$$

In the last equations, the determinantal two-electron wavefunctions, multiplied by a selected one-particle orbital φ, have been combined using the relation

$$\{a, b\}_{1,2}\varphi(c; 3) + \{b, c\}_{1,2}\varphi(a; 3) - \{a, c\}_{1,2}\varphi(b; 3) = \sqrt{3}\{a, b, c\}. \qquad (7.49)$$

Finally, it should be noted that treating one of the otherwise equivalent electrons in equ. (7.46) individually is frequently used for calculating matrix elements with a one-electron operator (e.g., the photon operator) acting on equivalent electrons. Similarly, if two-electron operators play a role, like in the Coulomb interaction between electrons, then it is convenient to separate *two* electrons from the equivalent electrons. This is done using the *coefficients of fractional grandparentage* (for more details see [Cow81]).

7.2.3 Step-up and step-down operators

Very often a wavefunction is known or can be easily guessed for a state with given magnetic quantum numbers M_L and M_S, but one needs the wavefunction for other M_L and M_S values. Instead of following the coupling procedure, the other functions can be obtained simply by applying *step-up* and *step-down* operators $(L_x \pm \mathrm{i}L_y)_{\mathrm{op}}$ and $(S_x \pm \mathrm{i}S_y)_{\mathrm{op}}$ for the angular momenta L and S, respectively. One has [Sla60]

$$(L_x \pm \mathrm{i}L_y)_{\mathrm{op}}\tilde{\Psi}(\text{el. config. } LM_LSM_S)$$
$$= \sqrt{(L \mp M_L)}\sqrt{(L \pm M_L + 1)}\ \tilde{\Psi}(\text{el. config. } LM_L \pm 1SM_S)$$
$$= \sum_{\text{all possible } (i)} \sqrt{(\ell_i \mp m_{\ell_i})}\sqrt{(\ell_i \pm m_{\ell_i} + 1)}\ \ \{\ldots m_{\ell_i} \pm 1 \ldots\}, \qquad (7.50a)$$

$$(S_x \pm iS_y)_{op}\tilde{\Psi}(\text{el. config. } LM_LSM_S)$$
$$= \sqrt{(S \mp M_S)}\sqrt{(S \pm M_S + 1)}\,\tilde{\Psi}(\text{el. config. } LM_LSM_S \pm 1)$$
$$= \sum_{\text{all possible } (i)} \{\ldots m_{s_i} \pm 1 \ldots\}, \tag{7.50b}$$

i.e., the change in the total quantum numbers M_L or M_S by ± 1 unit is traced back in these expressions to a similar change in the one-electron orbitals m_ℓ or m_s, contained in the Slater determinantal wavefunctions indicated by the symbol $\{\ldots\}$.

As an example, from the wavefunction $\tilde{\Psi}(np^2\ {}^3\mathrm{P}\ M_L = 1\ M_S = 1) = \{np1^+, np0^+\}$ given in Table 3.2 the corresponding function with $M_L = 0$ will be evaluated. Equ. (7.50a) yields

$$\sqrt{2}\,\tilde{\Psi}(np^2\ {}^3\mathrm{P}\ M_L = 0\ M_S = 1) = \sqrt{2}\{np0^+, np0^+\} + \sqrt{2}\{np1^+, np-1^+\}, \tag{7.51a}$$

which is equivalent to

$$\tilde{\Psi}(np^2\ {}^3\mathrm{P}\ M_L = 0\ M_S = 1) = \{np1^+, np-1^+\}, \tag{7.51b}$$

because the determinantal wavefunction with two identical orbitals vanishes.

7.2.4 Recoupling coefficients

For cases in which more than two angular momenta are involved, the coupling can be done in different ways. *Recoupling coefficients* allow the possibility of relating the wavefunctions established in one coupling mode to wavefunctions from the other coupling mode. The most important example is the change between *LSJ*- and *jjJ*-coupling. For a two-electron system one has (summation over all remaining quantum numbers is always required)

$$\Psi(n_a(\ell_a\tfrac{1}{2}j_a)n_b(\ell_b\tfrac{1}{2}j_b)JM_J\,|\,12) = \sum \langle j_aj_b; J\,|\,LS; J\rangle\,\Psi(n_an_b(\ell_a\ell_bL)(\tfrac{1}{2}\tfrac{1}{2}S)JM_J\,|\,12) \tag{7.52a}$$

with the expansion coefficients

$$\langle j_aj_b; J\,|\,LS; J\rangle = \sqrt{(2j_a+1)}\sqrt{(2j_b+1)}\sqrt{(2L+1)}\sqrt{(2S+1)}\begin{Bmatrix} \ell_a & \ell_b & L \\ \frac{1}{2} & \frac{1}{2} & S \\ j_a & j_b & J \end{Bmatrix}. \tag{7.52b}$$

Like the 3*j* symbol introduced in equ. (7.35), the 9*j* symbol defined by the symbol $\{\ldots\}$ in equ. (7.52b) depends only on the coupling of angular momenta and leads to a pure numerical value. (The symmetry properties and numerical values for these 9*j*-coefficients can be found, for example, in [App68, Cow81].) In the present case the 9*j* symbol reflects the two different coupling schemes well: its upper two rows describe the *LSJ*-coupling case, its first two columns the *jjJ*-coupling case, and $L + S$ as well as $j_a + j_b$ gives J. The numbers 1 and 2 in the wavefunction indicate the standard order of the particles 1 and 2 , i.e., the wavefunction is not yet antisymmetrized. When the transformation is applied to an antisymmetrized wavefunction, one has to be aware of the antisymmetrization difficulties for equivalent electrons (see [Cow81]).

7.3 Hartree–Fock approach

The Hartree–Fock (HF) approach [Har27, Foc30, Sla30, Har57, FFi72] is a method for deriving in a self-consistent way optimized single-particle wavefunctions $\varphi(\mathbf{r}, m_s)$. (The relativistic version is called the Dirac–Fock (DF) approach.) The criterion for optimization is minimizing the energy eigenvalue E_g of the many-electron wavefunction calculated with the full Hamiltonian. The quantities to be varied are the radial parts $P(r) = rR(r)$ of the single-particle orbitals, including certain constraints on these orbitals. The HF procedure leads, without any adjustable parameter and within the independent-particle model, to the best solutions for single-particle orbitals.

7.3.1 Derivation of the HF equation for the helium atom

In order to demonstrate the method, the simplest case, the ground state of the helium atom, will be used. Since the two-electron wavefunction is given by $\tilde{\Psi}_0 = \{1s0^+, 1s0^-\}$, one has to find the optimized orbitals $R_{1s}(r)$ which are part of $\varphi_{1s0}(\mathbf{r})$ given by $\varphi_{1s0}(\mathbf{r}) = R_{1s}(r)\, Y_{00}(\hat{\mathbf{r}})$ with $Y_{00}(\hat{\mathbf{r}}) = 1/\sqrt{(4\pi)}$. The starting point is the energy eigenvalue E_g

$$E_g = \langle\tilde{\Psi}_0|H|\tilde{\Psi}_0\rangle = \langle\{1s0^+, 1s0^-\}|H|\{1s0^+, 1s0^-\}\rangle, \tag{7.53a}$$

where the full Hamiltonian H is given in equ. (1.1). Putting the orbitals in the standard order yields

$$\begin{aligned} E_g = \tfrac{1}{2}[&\langle 1s0^+(1), 1s0^-(2)|H|1s0^+(1), 1s0^-(2)\rangle + \langle 1s0^-(1), 1s0^+(2)|H|1s0^-(1), 1s0^+(2)\rangle \\ &- \langle 1s0^+(1), 1s0^-(2)|H|1s0^-(1), 1s0^+(2)\rangle \\ &- \langle 1s0^-(1), 1s0^+(2)|H|1s0^+(1), 1s0^-(2)\rangle]. \end{aligned} \tag{7.53b}$$

In the non-relativistic case H does not act on the spin. Therefore, the spin functions for orbitals are unity if they have the same spin projection on the left- and right-hand sides of H, and zero otherwise. This yields

$$E_g = \langle 1s0(1), 1s0(2)|H|1s0(1), 1s0(2)\rangle. \tag{7.53c}$$

Inserting the operators for the kinetic and potential energies of H, one gets the result

$$E_g = 2\, I(1s) + (1s\,|\,1s). \tag{7.53d}$$

These two contributions represent the value of the 'one-electron integral' $I(1s)$ multiplied by two because there are two electrons in the 1s-orbital, and the 'electron-pair energy' $(1s\,|\,1s)$ between these electrons. Specifically, one has:

(i) For the one-electron integral

$$I(1s) = \langle 1s0| -\tfrac{1}{2}\nabla^2 - \frac{Z}{r} |1s0\rangle\langle 1s0|1s0\rangle$$
$$= \int_0^\infty R_{1s}(r)\frac{1}{r}\left[\left(-\frac{1}{2}\frac{d^2}{dr^2} - \frac{Z}{r}\right)R_{1s}(r)\right]r^2\,dr$$
$$= \int_0^\infty P_{1s}(r)\left(-\frac{1}{2}\frac{d^2}{dr^2} - \frac{Z}{r}\right)P_{1s}(r)\,dr, \tag{7.54a}$$

where the overlap integral $\langle 1s0|1s0\rangle$ is set to unity. The integration over the angular part yields unity, for the radial part note the relation

$$\nabla^2 R_{n\ell}(r)\,Y_{\ell m}(\hat{\mathbf{r}}) = \frac{1}{r}\left[\frac{d^2}{dr^2} - \frac{\ell(\ell+1)}{r^2}\right]P_{n\ell}(r)\,Y_{\ell m}(\hat{\mathbf{r}}). \tag{7.55a}$$

Inserting a hydrogenic wavefunction for the 1s-orbital, one derives

$$I(1s) = 4Z^3\int_0^\infty r\,e^{-Zr}\left(-\frac{1}{2}\frac{d^2}{dr^2} - \frac{Z}{r}\right)r\,e^{-Zr}\,dr$$
$$= 4Z^3\int_0^\infty r\,e^{-Zr}(-\tfrac{1}{2}Z^2 r\,e^{-Zr})\,dr$$
$$= -Z^2/2, \tag{7.54b}$$

in which the following property of the gamma function has been used

$$\int_0^\infty x^{n-1}\,e^{-x}\,dx = \Gamma(n) = (n-1)! \text{ for positive integer } n. \tag{7.55b}$$

(ii) For the electron-pair energy

$$(1s|1s) = \langle 1s0(1)1s0(2)|\frac{1}{r_{12}}|1s0(1)1s0(2)\rangle$$
$$= \int_0^\infty\int_0^\infty P_{1s}(r_1)P_{1s}(r_2)\frac{1}{r_{12}}P_{1s}(r_1)P_{1s}(r_2)\,dr_1\,dr_2. \tag{7.56a}$$

(The integration over the angular part yields unity.) Before this integral is discussed further, it will be put into the more general frame of Coulomb matrix elements with different electron pairs. Using the symbols a, b, c and d to denote the orbitals, one then has to consider

$$\langle\{a,b\}|\frac{1}{r_{12}}|\{c,d\}\rangle, \tag{7.57a}$$

and such matrix elements have been worked out using the relation

$$\frac{1}{r_{12}} = \sum_k \frac{r_<^k}{r_>^{k+1}}P_k(\cos\vartheta_{12}) = \sum_k \gamma_k(r_1,r_2)P_k(\cos\vartheta_{12}), \tag{7.58}$$

where $r_< = \min\{r_1, r_2\}$, $r_> = \max\{r_1, r_2\}$, and $\vartheta_{12} = \text{angle }\{\mathbf{r}_1, \mathbf{r}_2\}$. The result can be expressed as [Sla60]

$$\begin{aligned}\langle\{a,b\}|\frac{1}{r_{12}}|\{c,d\}\rangle &= \delta(m_{s_a},m_{s_c})\delta(m_{s_b},m_{s_d})\delta(m_{\ell_a}+m_{\ell_b},m_{\ell_c}+m_{\ell_d})\\ &\quad\times\sum_k c^k(\ell_a m_{\ell_a};\ell_c m_{\ell_c})c^k(\ell_d m_{\ell_d};\ell_b m_{\ell_b})R^k(n_a\ell_a n_b\ell_b,n_c\ell_c n_d\ell_d)\\ &\quad-\delta(m_{s_a},m_{s_d})\delta(m_{s_b},m_{s_c})\delta(m_{\ell_a}+m_{\ell_b},m_{\ell_c}+m_{\ell_d})\\ &\quad\times\sum_k c^k(\ell_a m_{\ell_a};\ell_d m_{\ell_d})c^k(\ell_c m_{\ell_c};\ell_b m_{\ell_b})R^k(n_a\ell_a n_b\ell_b,n_d\ell_d n_c\ell_c).\end{aligned}\tag{7.57b}$$

The $c^k(\ell m;\ell' m')$ are known coefficients resulting from integration over the angular coordinates $\hat{\mathbf{r}}$;

$$\begin{aligned}c^k(\ell m;\ell' m') &= \int Y^*_{\ell m}(\hat{\mathbf{r}})\sqrt{\left(\frac{4\pi}{2k+1}\right)}Y_{k(m-m')}(\hat{\mathbf{r}})\,Y_{\ell' m'}(\hat{\mathbf{r}})\,\mathrm{d}\Omega\\ &= (-1)^k(\ell' m' k(m-m')\,|\,\ell m)(\ell' 0 k 0\,|\,\ell' 0),\end{aligned}\tag{7.59}$$

i.e., the c^k-coefficients are essentially products of two Clebsch–Gordan coefficients. The $R^k(ab;cd)$ are *generalized Slater integrals* which contain the remaining integration over the radial coordinates r:

$$R^k(ab;cd) = \int_0^\infty\int_0^\infty P_a(r)P_b(r')\gamma_k P_c(r)P_d(r')\,\mathrm{d}r\,\mathrm{d}r'.\tag{7.60a}$$

Special cases are the *direct* Coulomb integral,

$$F^k(ab) = R^k(ab;ab),\tag{7.60b}$$

and the *exchange* Coulomb integral,

$$G^k(ab) = R^k(ab;ba).\tag{7.60c}$$

For helium one simply has

$$(1s\,|\,1s) = F^0(1s\,1s).\tag{7.56b}$$

For a further discussion of this electron-pair energy $(1s\,|\,1s)$, one can introduce into equ. (7.56a) a hydrogenic $P_{1s}(r)$ radial function, and derive

$$\begin{aligned}(1s\,|\,1s) &= 2^4 Z^6\int_0^\infty\int_0^\infty r_1^2\,\mathrm{e}^{-2Zr_1}\,r_2^2\,\mathrm{e}^{-2Zr_2}\frac{1}{r_>}\,\mathrm{d}r_2\,\mathrm{d}r_1\\ &= 2^4 Z^6\int_0^\infty \mathrm{e}^{-2Zr_1}\left(\frac{1}{r_1}\int_0^{r_1}\mathrm{e}^{-2Zr_2}\,r_2^2\,\mathrm{d}r_2 + \int_{r_1}^\infty \mathrm{e}^{-2Zr_2}\,r_2\,\mathrm{d}r_2\right)r_1^2\,\mathrm{d}r_1,\end{aligned}\tag{7.56c}$$

because the integration over r_2 splits into two parts depending on the actual integration range of r_2; for r_2 between zero and r_1 one has $r_> = r_1$, and for r_2 larger than r_1 one has $r_> = r_2$. The integrations can then be performed and result in

$$(1\mathrm{s}\,|\,1\mathrm{s}) = 2^4\,Z^6 \int_0^\infty \mathrm{e}^{-2Zr_1}\left(-\frac{\mathrm{e}^{-2Zr_1}}{4Z^2} - \frac{\mathrm{e}^{-2Zr_1}}{4r_1Z^3} + \frac{1}{4r_1Z^3}\right) r_1^2\,\mathrm{d}r_1, \tag{7.56d}$$

which, together with equ. (7.55b), becomes

$$(1\mathrm{s}\,|\,1\mathrm{s}) = \frac{Z}{16}[-\Gamma(3) - 4\Gamma(2) + 16\Gamma(2)], \tag{7.56e}$$

$$(1\mathrm{s}\,|\,1\mathrm{s}) = \tfrac{5}{8}Z. \tag{7.56f}$$

After these preliminaries, and knowing now the meaning of the individual terms $I(1\mathrm{s})$ and $(1\mathrm{s}\,|\,1\mathrm{s})$ in the ground-state energy E_g of helium, the derivation of the Hartree–Fock equations for this ground state can be begun. This approach requires the optimization of the $P_{1\mathrm{s}}(r)$ in equs. (7.54a) and (7.56a) with respect to their general shape (see equ. (7.71) for this general shape) when seeking a minimum in E_g. The variational principle demands

$$\partial P_{1\mathrm{s}}(r)\,(E_g - \lambda_{1\mathrm{s}1\mathrm{s}}\langle 1\mathrm{s}\,|\,1\mathrm{s}\rangle) \overset{!}{=} 0, \tag{7.61a}$$

where a Lagrange parameter $\lambda_{1\mathrm{s}1\mathrm{s}}$ has been introduced to fulfil the orthonormality condition for the 1s orbitals. (In the general case, all the single-particle functions involved will be orthonormal. Due to the orthogonality of the angular functions, the normalization with Lagrange parameters is required only for radial functions with the same ℓ, and this also applies to the helium example.) Written explicitly, this expression means

$$\partial P_{1\mathrm{s}}(r)\Bigg[2\int_0^\infty P_{1\mathrm{s}}(r)\left(-\frac{1}{2}\frac{\mathrm{d}^2}{\mathrm{d}r^2} - \frac{Z}{r}\right)P_{1\mathrm{s}}(r)\,\mathrm{d}r$$
$$+ \int_0^\infty\int_0^\infty P_{1\mathrm{s}}(r)P_{1\mathrm{s}}(r')\gamma_0(r,r')P_{1\mathrm{s}}(r)P_{1\mathrm{s}}(r')\,\mathrm{d}r\,\mathrm{d}r'$$
$$- \lambda_{1\mathrm{s}1\mathrm{s}}\int_0^\infty P_{1\mathrm{s}}(r)P_{1\mathrm{s}}(r)\,\mathrm{d}r\Bigg] \overset{!}{=} 0. \tag{7.61b}$$

Because the Hamiltonian is an hermitian operator, the two variations

$$\int P_a^* H\,\partial P_a\,\mathrm{d}r \qquad \text{and} \qquad \int \partial P_a H^* P_a^*\,\mathrm{d}r \tag{7.62}$$

are equivalent (one is the complex conjugate of the other). Hence, it is sufficient to vary only the orbitals on the left-hand sides of the respective operators. Noting also that the Coulomb integral can be written equivalently with interchanged indices r and r', one obtains

$$2\int_0^\infty \partial P_{1\mathrm{s}}(r)\left(-\frac{1}{2}\frac{\mathrm{d}^2}{\mathrm{d}r^2} - \frac{Z}{r}\right)P_{1\mathrm{s}}(r)\,\mathrm{d}r + \int_0^\infty\int_0^\infty \partial P_{1\mathrm{s}}(r)P_{1\mathrm{s}}(r')\gamma_0(r,r')P_{1\mathrm{s}}(r)P_{1\mathrm{s}}(r')\,\mathrm{d}r\,\mathrm{d}r'$$
$$+ \int_0^\infty\int_0^\infty P_{1\mathrm{s}}(r')\,\partial P_{1\mathrm{s}}(r)\gamma_0(r,r')P_{1\mathrm{s}}(r')P_{1\mathrm{s}}(r)\,\mathrm{d}r'\,\mathrm{d}r$$
$$- \lambda_{1\mathrm{s}1\mathrm{s}}\int_0^\infty \partial P_{1\mathrm{s}}(r)P_{1\mathrm{s}}(r)\,\mathrm{d}r \overset{!}{=} 0. \tag{7.61c}$$

Since the variations $\partial P_{1s}(r)$ can be selected arbitrarily, it is the remaining quantity which must be zero in order to fulfill equs. (7.61). This condition gives the HF equation for the 1s orbital in helium

$$2\left(-\frac{1}{2}\frac{d^2}{dr^2}-\frac{Z}{r}\right)P_{1s}(r)+2\int_0^\infty P_{1s}(r')\gamma_0(r,r')P_{1s}(r')\,dr'\,P_{1s}(r)=\lambda_{1s1s}P_{1s}(r). \tag{7.63}$$

Introducing [Sob72]†

$$Y^k_{ab}(r)=\int_0^\infty P_a(r')\gamma_k(r,r')P_b(r')\,dr' \tag{7.64}$$

and normalizing λ_{1s1s} against the number of electrons in the 1s orbital, i.e.,

$$\lambda_{1s1s}/2=\varepsilon_{1s}, \tag{7.65}$$

one finally obtains the HF equation for helium in its standard form:

$$\left[-\frac{1}{2}\frac{d^2}{dr^2}-\frac{Z}{r}+Y^0_{1s1s}(r)\right]P_{1s}(r)=\varepsilon_{1s}P_{1s}(r) \tag{7.66a}$$

or, equivalently,

$$\left[-\frac{1}{2}\frac{d^2}{dr^2}+V_{HF}(r)\right]P_{1s}(r)=\varepsilon_{1s}P_{1s}(r), \tag{7.66b}$$

in which the HF potential $V_{HF}(r)$,

$$V_{HF}(r)=-\frac{Z}{r}+Y^0_{1s1s}(r), \tag{7.66c}$$

is introduced.

The quantity ε_{1s}, defined formally in equ. (7.65) as a quantity proportional to the Lagrange parameter λ_{1s1s}, represents the binding energy of a 1s-electron in helium. This can be shown if equ. (7.66a) is multiplied on the left-hand side by $P_{1s}(r)$ and then integrated over dr. This gives

$$\int_0^\infty P_{1s}(r)\left(-\frac{1}{2}\frac{d^2}{dr^2}-\frac{Z}{r}\right)P_{1s}(r)\,dr+\int_0^\infty P_{1s}(r)\,Y^0_{1s1s}(r)P_{1s}(r)=\varepsilon_{1s}, \tag{7.67a}$$

which, according to the equs. (7.54a), (7.56a), and (7.56b), is equivalent to

$$I(1s)+F^0(1s1s)=\varepsilon_{1s}. \tag{7.67b}$$

This expression, however, is equal to the energy difference

$$E_g(1s^2)-E_g^+(1s)=[2I(1s)+F^0(1s1s)]-I(1s) \tag{7.67c}$$

which defines the binding energy E_b of a 1s-electron in the ground state of helium

† This quantity is related to $Y_k(ab/r)$ defined in [Sla60] by

$$rY^k_{ab}(r)=Y_k(ab/r).$$

(Koopman theorem [Koo33]):

$$E_b(1s; \text{He}) = E_g(1s^2) - E_g^+(1s). \tag{7.67d}$$

(Since the same orbitals are used for the ground state and the hole state, no relaxation effects due to the change in the shielded nuclear charges are taken into account, and this model is called the *frozen atomic structure* approximation.) Due to the interpretation of ε_{1s} as the binding energy of one 1s-electron, the differential equation for the $P_{1s}(r)$ orbital, equ. (7.66b), can be interpreted as a one-particle Schrödinger equation for the 1s orbital (see equs. (1.3) and (1.4)):

$$h_i \varphi_i = \varepsilon_i \varphi_i \qquad \text{with} \qquad h_i = -\tfrac{1}{2}\nabla_i + V_{\text{HF}}(r_i). \tag{7.68}$$

According to equ. (7.66c) the HF potential contains the Coulomb interaction between the 1s-electron and the nucleus, and the quantity $Y^0_{1s1s}(r)$ which represents the potential energy of one 1s-electron at a distance r from the origin in the presence of the other 1s-electron, and is given by:

$$\begin{aligned} Y^0_{1s1s}(r) &= \int_0^\infty P_{1s}(r')\gamma_0(r,r')P_{1s}(r')\,\mathrm{d}r' \\ &= \frac{1}{r}\left[\int_0^r P^2_{1s}(r')\,\mathrm{d}r' + r\int_r^\infty \frac{1}{r'}P^2_{1s}(r')\,\mathrm{d}r'\right]. \end{aligned} \tag{7.69}$$

Hence, $Y^0_{1s1s}(r)$, multiplied by r, describes the shielding of the nuclear charge due to the presence of the other electron, thus reducing $Z = 2$ to an effective value Z_{eff}. This potential seen by either one of the two electrons at a distance r can therefore be represented by

$$V_{\text{HF}}(r) = -\frac{Z_{\text{eff}}(r)}{r} = -\frac{Z - \sigma(r)}{r} \tag{7.70a}$$

with a *shielding factor* σ given by

$$\sigma = Z - Z_{\text{eff}} = r\,Y^0_{1s1s}(r). \tag{7.70b}$$

This shielding factor σ is shown in Fig. 7.5. It can be seen that it has the r dependence that one would expect naively: for $r \to 0$ the shielding effect produced by the other electron disappears ($\sigma \to 0$), but for $r \to \infty$ there is full shielding of the nuclear charge by the other electron ($\sigma \to 1$). At the average distance of the electrons from the nucleus $\langle r \rangle = 0.927$ au (the mean radius of the 1s shell in helium) one gets $\sigma = 0.86$, or $Z_{\text{eff}} = 1.14$.

In order to solve the HF equation for helium, one applies an iterative procedure, i.e., one starts with a certain trial function (e.g., a hydrogenic function, see equ. (7.2)), which is then varied, preserving the general shape of the selected orbital. In the present case of a 1s orbital, the general shape has to fulfil the conditions

(i) $P_{1s}(r) \sim r$ for $r \to 0$, (7.71a)
(ii) $P_{1s}(r) \sim \mathrm{e}^{-r}$ for $r \to \infty$, (7.71b)
(iii) $P_{1s}(r)$ has no node between $r = 0$ and $r \to \infty$. (7.71c)

Starting with the trial functions, one first determines the potentials, in the present

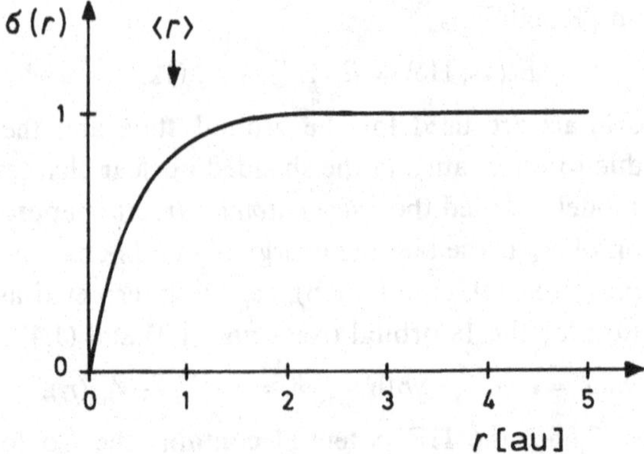

Figure 7.5 Shielding factor $\sigma(r)$ in the ground state of helium as a function of the distance r from the nucleus. $\sigma(r)$ is the difference between the charge $Z = 2$ of the bare nucleus and the effective charge Z_{eff} seen by one of the electrons due to the presence of the other electron. $\langle r \rangle$ indicates the mean distance of the 1s-electron from the nucleus ('radius' of the 1s shell). Data from [WLi35], cf [BSa57] with $Z_{\text{eff}} = Z_p$.

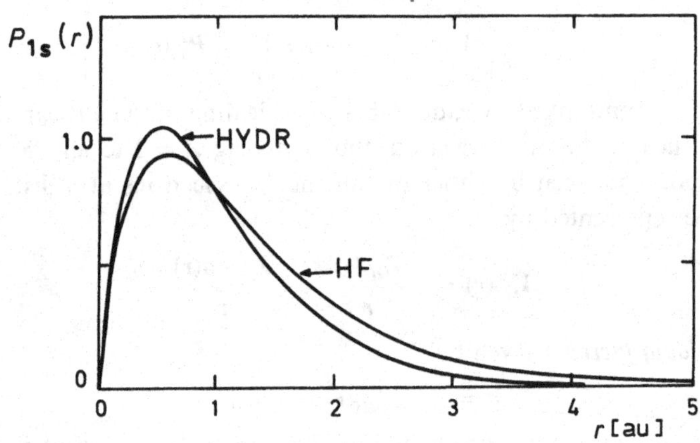

Figure 7.6 Radial wavefunctions $P_{1s}(r) = rR_{1s}(r)$ of helium. HYDR is the hydrogenic wavefunction with $Z = 2$; HF is the Hartree–Fock wavefunction. From [BJo66].

case $Y^0_{1s1s}(r)$, and gets the operators in the HF equation. Since these operators act on the orbitals $P_{n\ell}(r)$, $n\ell = 1\text{s}$ in the present case, the solution of the differential equation then leads to improved orbitals $P_{1s}(r)$, which, in turn, serve to improve the potentials and so on.

The result of such a calculation for $P_{1s}(r)$ of helium and how it compares with the hydrogenic case are shown in Fig. 7.6, and the differences can be seen clearly. For the total energy of the helium ground state the HF approach yields $E_g = -2.862$ au. This is rather close to the experimental value, $E_g(\text{exp.}) = -2.90372$ au, whilst for hydrogenic wavefunctions with $Z = 2$ one gets $E_g = -2.75$ au. (For the energy eigenvalue ε_{1s} one gets $\varepsilon_{1s} = -0.918$ au.)

7.3.2 HF equations for the neon atom

The derivation of the HF equations for atoms other than helium follows along the same lines as presented above, and one gets for closed-shell atoms, in atomic units, [Sla60]

$$\left(-\frac{1}{2}\frac{\mathrm{d}^2}{\mathrm{d}r^2}+\frac{\ell(\ell+1)}{2r^2}-\frac{Z}{r}+\sum_{\text{all }n'\ell'}(4\ell'+2)\,Y^0_{n'\ell'n'\ell'}(r)-\varepsilon_{n\ell}\right)P_{n\ell}(r)$$
$$=\sum_{\text{all }n'\ell'}\sqrt{\left(\frac{2\ell'+1}{2\ell+1}\right)}\sum_k c^k(\ell 0;\ell' 0)\,Y^k_{n\ell n'\ell'}(r)P_{n'\ell'}(r). \quad (7.72)$$

The coefficients $c^k(\ell 0; \ell' 0)$ follow from the integrations over angular functions calculated from equ. (7.59), the potential functions $Y^k_{n\ell n'\ell'}(r)$ have been introduced in equ. (7.64), and the energies $\varepsilon_{n\ell}$ are the straightforward generalization of equ. (7.65). In order to take care of the orthogonality between 1s and 2s orbitals, a Lagrange parameter λ_{1s2s} must also be introduced. However, for closed-shell systems such non-diagonal Lagrange parameters can be eliminated by a unitary transformation.

The following differences from the HF equation for the helium atom can be noted. First, for orbital angular momenta with $\ell > 0$ there is a centrifugal potential. Second, an exchange potential $Y^k_{n\ell n'\ell'}(r)$ appears on the right-hand side of equ. (7.72) in addition to the direct potentials $Y^0_{n'\ell'n'\ell'}(r)$ on the left-hand side. (In helium one has $V_{\text{direct}} = 2Y^0_{1s1s}(r)$ and $V_{\text{exchange}} = -Y^0_{1s1s}(r)$, and their sum can be combined to give $V_{\text{HF}} = -Z/r + V_{\text{direct}} + V_{\text{exchange}} = -Z/r + Y^0_{1s1s}(r)$.)

If this general HF equation is applied to the neon atom, it gives

$$\left[-\frac{1}{2}\frac{\mathrm{d}^2}{\mathrm{d}r^2}-\frac{Z}{r}+Y^0_{1s1s}(r)+2Y^0_{2s2s}(r)+6Y^0_{2p2p}(r)-\varepsilon_{1s}\right]P_{1s}(r)$$
$$=Y^0_{1s2s}(r)P_{2s}(r)+Y^1_{1s2p}(r)P_{2p}(r), \quad (7.73a)$$

$$\left[-\frac{1}{2}\frac{\mathrm{d}^2}{\mathrm{d}r^2}-\frac{Z}{r}+2Y^0_{1s1s}(r)+Y^0_{2s2s}(r)+6Y^0_{2p2p}(r)-\varepsilon_{2s}\right]P_{2s}(r)$$
$$=Y^0_{2s1s}(r)P_{1s}(r)+Y^1_{2s2p}(r)P_{2p}(r), \quad (7.73b)$$

$$\left[-\frac{1}{2}\frac{\mathrm{d}^2}{\mathrm{d}r^2}+\frac{1}{r^2}+\frac{Z}{r}+2Y^0_{1s1s}(r)+2Y^0_{2s2s}(r)+5Y^0_{2p2p}(r)-\tfrac{2}{5}Y^2_{2p2p}(r)-\varepsilon_{2p}\right]P_{2p}(r)$$
$$=\tfrac{1}{3}Y^1_{2p1s}(r)P_{1s}(r)+\tfrac{1}{3}Y^1_{2p2s}(r)P_{2s}(r). \quad (7.73c)$$

From these equations one can see that the radial functions $P_{1s}(r)$, $P_{2s}(r)$, and $P_{2p}(r)$ appear in all three equations in two distinct ways: first as functions to be determined by the solution of the coupled differential equations; and second as functions which determine the direct and exchange parts of the HF potentials. Hence, the desired solution can be found only by an iterative procedure.

Details of the results from HF calculations are given in [FFi72], and for the related relativistic case from DF calculations in [LCM71, Des73, HAC76].

7.4 Mixing of wavefunctions

A stationary state is characterized by all quantum numbers which commute with the time-independent Hamiltonian. For atoms these quantum numbers are the total angular momentum J and its projection M onto a preselected z-axis, the parity π, and the energy E. In order to include electron correlations, such a state can be represented as an expansion in terms of uncorrelated wavefunctions, with different electron configurations which are called *basis* functions, see equ. (1.24). These basis functions have different spatial and angular parts, but they belong to the same symmetry as given by the quantum numbers J, M, and π. The method of finding the stationary solutions within the configuration interaction (CI) approach will be elucidated in the present section. A related problem is the question of how a state described by LSJ-coupled wavefunctions evolves for increasing spin–orbit interactions via *intermediate* coupling towards the jjJ-coupling limit. The CI approach will be explained first and then the fundamental properties of the mixing of wavefunctions will be presented for the simplest example, a two-state system, applying a diagonalization procedure for the Hamiltonian matrix. As a special application, finally the intermediate coupling case with the limits of LSJ- and jjJ-coupling is discussed for the example of (sp $J = 1$)-coupling.

7.4.1 Configuration interaction and stationary wavefunctions

In order to simplify the treatment, a correlated wavefunction $\Psi(\mathbf{r})_{\text{corr}}$ built by CI between two uncorrelated wavefunctions will be considered. (To shorten the notations, the tilde describing the antisymmetric character of the wavefunctions and the spin of the electron have been omitted, and the spatial vectors of all electrons are indicated by the symbol $\mathbf{r}$ only.) One has

$$\Psi(\mathbf{r})_{\text{corr}} = a_1\Psi_1(\mathbf{r}) + a_2\Psi_2(\mathbf{r}). \tag{7.74}$$

The basis functions $\Psi_1(\mathbf{r})$ and $\Psi_2(\mathbf{r})$ are solutions within the independent-particle model (operator H^0, equ. (1.3)):

$$H^0\Psi_j = E_j^0\Psi_j. \tag{7.75a}$$

According to

$$H^0\Psi_j = -\frac{1}{\mathrm{i}}\,\dot{\Psi}_j \tag{7.75b}$$

one can ascribe to these basis functions a time dependence ($E_j^0 = \omega_j$ in atomic units):

$$\Psi_j(\mathbf{r}, t) = \Psi_j(\mathbf{r})\,\mathrm{e}^{-\mathrm{i}\omega_j t}. \tag{7.75c}$$

Because the ω_j values are different for different basis functions j, the time-dependent correlated wavefunction built from such basis functions cannot be a stationary wavefunction of the full Hamiltonian. Instead, the stationary wavefunction must be found using the ansatz

$$\Psi(\mathbf{r}, t)_{\text{corr}} = c_1(t)\Psi_1(\mathbf{r})\,\mathrm{e}^{-\mathrm{i}\omega_1 t} + c_2(t)\Psi_2(\mathbf{r})\,\mathrm{e}^{-\mathrm{i}\omega_2 t}, \tag{7.76}$$

searching for solutions with a common time dependence. If this ansatz is put into the time-dependent Schrödinger equation of the full Hamiltonian, one gets

$$-\frac{1}{\mathrm{i}}(\dot{c}_1(t)\Psi_1(\mathbf{r}, t) + \dot{c}_2(t)\Psi_2(\mathbf{r}, t)) = c_1(t)H_{\text{res}}\Psi_1(\mathbf{r}, t) + c_2(t)H_{\text{res}}\Psi_2(\mathbf{r}, t) \tag{7.77}$$

with the residual Hamiltonian

$$H_{\text{res}} = H - H^0. \tag{7.78}$$

Multiplication of the differential equation (7.77) by $\Psi_1^*(\mathbf{r}, t)$ or $\Psi_2^*(\mathbf{r}, t)$, respectively, and integration over the coordinates $\mathbf{r}$ then leads to the two equations

$$\left.\begin{aligned} -\frac{1}{\mathrm{i}}\dot{c}_1(t) &= c_1(t)H_{11}^{\text{res}} + c_2(t)\,\mathrm{e}^{-\mathrm{i}(\omega_2-\omega_1)t}H_{12}^{\text{res}} \\ -\frac{1}{\mathrm{i}}\dot{c}_2(t) &= c_2(t)\,\mathrm{e}^{-\mathrm{i}(\omega_2-\omega_1)t}H_{21}^{\text{res}} + c_2(t)H_{22}^{\text{res}} \end{aligned}\right\} \tag{7.79}$$

with the matrix elements

$$H_{jk}^{\text{res}} = \langle\Psi_j(\mathbf{r})|H_{\text{res}}|\Psi_k(\mathbf{r})\rangle \tag{7.80a}$$

and the ones needed below

$$H_{jk} = \langle\Psi_j(\mathbf{r})|H|\Psi_k(\mathbf{r})\rangle = \delta_{jk}\omega_j + H_{jk}^{\text{res}} \tag{7.80b}$$

(for simplicity these matrix elements are assumed to be real). The general solution for the coefficients $c_i(t)$, and therefore also for the function $\Psi(\mathbf{r}, t)_{\text{corr}}$, then becomes (see [FMa52, FLS65])

$$\Psi_{\text{corr}}(\mathbf{r}, t) = [a_{\text{I}1}\,\mathrm{e}^{-\mathrm{i}E_{\text{I}}t} + a_{\text{II}1}\,\mathrm{e}^{-\mathrm{i}E_{\text{II}}t}]\Psi_1(\mathbf{r}) + [a_{\text{I}2}\,\mathrm{e}^{-\mathrm{i}E_{\text{I}}t} + a_{\text{II}2}\,\mathrm{e}^{-\mathrm{i}E_{\text{II}}t}]\Psi_2(\mathbf{r}), \tag{7.81a}$$

with

$$E_{\text{I,II}} = \frac{H_{11} + H_{22}}{2} \pm \sqrt{\left(\frac{H_{12}H_{21} + (H_{11} - H_{22})^2}{4}\right)}, \tag{7.81b}$$

and a special relation between the coefficients $a_{\text{I}j}$ and $a_{\text{II}j}$:

$$a_{\text{I}2} = \frac{E_{\text{I}} - H_{11}}{H_{12}}a_{\text{I}1} \qquad \text{and} \qquad a_{\text{II}2} = \frac{E_{\text{II}} - H_{11}}{H_{12}}a_{\text{II}1}. \tag{7.81c}$$

The result can be discussed in two ways:

(i) First, within the basis vectors $\Psi_1(\mathbf{r})$ and $\Psi_2(\mathbf{r})$ of the model Hamiltonian H^0. When the system is assumed to be in the stationary state $\Psi_1(\mathbf{r}, t)$ of H^0 at time $t = 0$, then the residual interaction H_{res} yields oscillations between $\Psi_1(\mathbf{r}, t)$ and $\Psi_2(\mathbf{r}, t)$ described by the absolute value squared of the respective amplitudes $c_1(t)$ and $c_2(t)$:

$$c_1^2(t) = 1 - \frac{H_{12}H_{21}}{\nu^2}\sin^2\nu t, \qquad c_2^2(t) = \frac{H_{12}H_{21}}{\nu^2}\sin^2\nu t, \tag{7.82a}$$

with the frequency ν given by

$$\nu = \sqrt{(H_{12}H_{21} + (H_{22} - H_{11})^2/4)}. \tag{7.82b}$$

(ii) Second, within a new set of basis functions which belong to stationary states of the full Hamiltonian H. In order to find these new basis functions, one has to seek a special solution of $\Psi_{corr}(\mathbf{r}, t)$ where the coefficients $c_1(t)$ and $c_2(t)$ have the same time dependence, namely, e^{-iEt}. Such a solution would then represent the desired stationary state with energy E. From equs. (7.81a)–(7.81c) it can be seen that such a common time dependence can be achieved only for two situations

$$a_{I1} = 0 \text{ (due to its relation with } a_{I2} \text{ then also } a_{I2} = 0), \tag{7.83a}$$

which is connected to $E = E_{II}$, and

$$a_{II1} = 0 \text{ (due to its relation with } a_{II2} \text{ then also } a_{II2} = 0), \tag{7.83b}$$

which is connected to $E = E_I$.

This shows that for the selected two-state system exactly two stationary states exist for the correlated wavefunction, and these are given by

$$\Psi_{corr,I}(\mathbf{r}, t) = \Psi_I(\mathbf{r}, t) = a_{I1}\, e^{-iE_It}\Psi_1(\mathbf{r}) + a_{I2}\, e^{-iE_It}\Psi_2(\mathbf{r}), \tag{7.84a}$$

$$\Psi_{corr,II}(\mathbf{r}, t) = \Psi_{II}(\mathbf{r}, t) = a_{II1}\, e^{-iE_{II}t}\Psi_1(\mathbf{r}) + a_{II2}\, e^{-iE_{II}t}\Psi_2(\mathbf{r}), \tag{7.84b}$$

These stationary states can be written more compactly as

$$\Psi_I(\mathbf{r}, t) = e^{-iE_It}\Psi_I(\mathbf{r}), \tag{7.85a}$$

$$\Psi_{II}(\mathbf{r}, t) = e^{-E_{II}t}\Psi_{II}(\mathbf{r}), \tag{7.85b}$$

with the new basis functions defined by

$$\Psi_I(\mathbf{r}) = a_{I1}\Psi_1(\mathbf{r}) + a_{I2}\Psi_2(\mathbf{r}), \tag{7.86a}$$

$$\Psi_{II}(\mathbf{r}) = a_{II1}\Psi_1(\mathbf{r}) + a_{II2}\Psi_2(\mathbf{r}). \tag{7.86b}$$

The last equation is used in several places in the text, and its physical content is discussed in detail, e.g., in connection with equ. (1.25).

7.4.2 Diagonalization procedure

As has been shown in the foregoing subsection, the stationary states $\Psi_I(\mathbf{r})$ and $\Psi_{II}(\mathbf{r})$ are solutions of the full Hamiltonian H with given energies E_I and E_{II}. This means their Hamilton matrix is diagonal (the subscript I is used now to indicate *all* such stationary states characterized by Roman numbers I, II, III, . . .):

$$\langle \Psi_I(\mathbf{r}) | H | \Psi_{I'}(\mathbf{r}) \rangle = \delta_{II'}. \tag{7.87}$$

It is also possible to derive this result from a different starting point, without explicitly taking into account the time dependences. In this approach one starts with the ansatz

$$\Psi(\mathbf{r})_{corr} = a_1\Psi_1(\mathbf{r}) + a_2\Psi_2(\mathbf{r}) + \cdots, \tag{7.88}$$

which replaces equ. (7.76), and performs on the Schrödinger equation

$$H\Psi(\mathbf{r})_{\text{corr}} = E\Psi(\mathbf{r})_{\text{corr}}, \tag{7.89}$$

the following steps: multiplication by the basis functions $\Psi_i^*(\mathbf{r})$ from the left-hand side and integration over all spatial coordinates. This leads for the unknown coefficients a_k to a system of homogeneous equations given by

$$\sum_k (H_{ik} - E\delta_{ik})a_k = 0 \tag{7.90a}$$

with

$$H_{ik} = \langle \Psi_i(\mathbf{r})|H|\Psi_k(\mathbf{r})\rangle. \tag{7.90b}$$

If equ. (7.90a) is presented in matrix form (H_{ik}) using column vectors (a_k), one gets the equivalent representation

$$(H_{ik})(a_k) = E(\delta_{ik})(a_k). \tag{7.91}$$

The last equation shows that the unknown coefficients are those solutions which diagonalize the Hamilton energy matrix H_{ik}. Hence, they can be found by a diagonalization procedure where for non-trivial solutions the condition

$$\det(H_{ik} - E\delta_{ik}) = 0 \tag{7.92}$$

has to be fulfilled. The roots of this equation with E treated as an unknown parameter then yield the so-called *eigenvalues* E_{I} of equ. (7.91), and for each of these eigenvalues equ. (7.91) can be diagonalized, leading to a set of $a_k(E_{\text{I}}) = a_{\text{I}k}$ coefficients termed *eigenvectors*. These eigenvectors are identical to the stationary states $\Psi_{\text{I}}(\mathbf{r})$, because both fulfil the condition that their matrix elements with the full Hamiltonian H have only diagonal terms.

In order to obtain the eigenvectors, restrictions from the requirement of orthonormality for the wavefunctions $\Psi_{\text{I}}(\mathbf{r})$ have to be also incorporated. This will be demonstrated for the case of a two-state system defined in equ. (7.86). Here these restrictions lead to the additional conditions

$$|a_{\text{I}1}|^2 + |a_{\text{I}2}|^2 = 1, \tag{7.93a}$$

$$|a_{\text{II}1}|^2 + |a_{\text{II}2}|^2 = 1, \tag{7.93b}$$

$$a_{\text{I}1}^* a_{\text{II}1} + a_{\text{I}2}^* a_{\text{II}2} = 0. \tag{7.93c}$$

After some manipulations one then derives for the mixing coefficients the solution (the solution is obtained by fixing a common phase factor $e^{i\delta}$ to unity)

$$a_{\text{I}1} = \frac{1}{\sqrt{(1+\alpha^2)}}, \qquad a_{\text{I}2} = -\frac{\alpha}{\sqrt{(1+\alpha^2)}}, \tag{7.94a}$$

$$a_{\text{II}1} = \frac{\alpha}{\sqrt{(1+\alpha^2)}}, \qquad a_{\text{II}2} = \frac{1}{\sqrt{(1+\alpha^2)}} \tag{7.94b}$$

and

$$\alpha = \frac{H_{12}}{E_{\text{II}} - H_{11}}. \tag{7.94c}$$

These relations will be used in the next section; for an example of a more extended diagonalization procedure see Section 5.3.2.

7.4.3 Wavefunctions in LSJ-, intermediate, and jjJ-coupling

As an application of the diagonalization procedure, the description of two-electron wavefunctions in intermediate coupling will be selected for the special case of (sp $J = 1$ $M = 1$)-coupling. This example is of relevance for K–LL Auger decay leading to the final ionic states $1s^2 2s2p^5\ {}^1P_1$ and $1s^2 2s2p^5\ {}^3P_1$, because the properties of the electron configurations $2s2p^5$, i.e., of the hole configurations $(2s2p)^{-1}$ are related to the electron configuration 2s2p. (Note, however, a change of sign of the spin–orbit parameter $\zeta(2p)$ which then leads to differences in the designation of energy levels in the jjJ-coupling limit (compare Fig. 7.7 with Fig. 3.4 and see the detailed discussion in [Cow81]).)

If spin–orbit effects are negligible, LSJ-coupling can be applied, and the two wavefunctions for $J = 1$ and $M = 1$ are given by

$$\varphi_{LSJ}(\text{sp}\ {}^1\text{P}_1\ J = 1\ M = 1) = \frac{1}{\sqrt{2}}\{\text{s}0^+, \text{p}1^-\} - \frac{1}{\sqrt{2}}\{\text{s}0^-, \text{p}1^+\}, \quad (7.95a)$$

$$\varphi_{LSJ}(\text{sp}\ {}^3\text{P}_1\ J = 1\ M = 1) = \frac{1}{2}\{\text{s}0^+, \text{p}1^-\} + \frac{1}{2}\{\text{s}0^-, \text{p}1^+\} - \frac{1}{\sqrt{2}}\{\text{s}0^+, \text{p}0^+\}. \quad (7.95b)$$

In the limiting case of strong spin–orbit interactions, ℓsj-coupled single-particle orbitals have to be used. With the help of equs. (7.42)–(7.44) they can be traced back to the common spin orbitals, giving

$$\varphi_{jjJ}(\text{s}_{1/2}\text{p}_{1/2}\ J = 1\ M = 1) = \sqrt{\left(\frac{2}{3}\right)}\{\text{s}0^+, \text{p}1^-\} - \frac{1}{\sqrt{3}}\{\text{s}0^+, \text{p}0^+\}, \quad (7.96a)$$

$$\varphi_{jjJ}(\text{s}_{1/2}\text{p}_{3/2}\ J = 1\ M = 1) = \frac{1}{\sqrt{6}}\{\text{s}0^+, \text{p}0^+\} + \frac{1}{2\sqrt{3}}\{\text{s}0^+, \text{p}1^-\} - \frac{\sqrt{3}}{2}\{\text{s}0^-, \text{p}1^+\}. \quad (7.96b)$$

The relation between the wavefunctions φ_{LSJ} and φ_{jjJ} of the two extreme coupling cases can be worked out using the representations directly in the Slater determinantal wavefunctions, equs. (7.95) and (7.96), or alternatively, using the coupling transformation in equ. (7.52). In both cases one gets

$$\varphi_{jjJ}(\text{s}_{1/2}\text{p}_{1/2}\ J = 1\ M) = \frac{1}{\sqrt{3}}\varphi_{LSJ}(\text{sp}\ {}^1\text{P}_1\ M) + \sqrt{\left(\frac{2}{3}\right)}\varphi_{LSJ}(\text{sp}\ {}^3\text{P}_1\ M), \quad (7.97a)$$

$$\varphi_{jjJ}(\text{s}_{1/2}\text{p}_{3/2}\ J = 1\ M) = \sqrt{\left(\frac{2}{3}\right)}\varphi_{LSJ}(\text{sp}\ {}^1\text{P}_1\ M) - \frac{1}{\sqrt{3}}\varphi_{LSJ}(\text{sp}\ {}^3\text{P}_1\ M). \quad (7.97b)$$

For intermediate coupling, however, the wavefunction has to be derived from a diagonalization procedure. With the φ_{LSJ}-functions as the starting point (basis functions), the elements of the Hamilton energy matrix are given by

$$H_{11} = \langle \varphi_{LSJ}(\mathrm{sp}\ ^1\mathrm{P}_1\ M|H|\varphi_{LSJ}(\mathrm{sp}\ ^1\mathrm{P}_1\ M\rangle = E_{\mathrm{av}} + \tfrac{1}{2}G^1(\mathrm{sp}), \tag{7.98a}$$

$$H_{22} = \langle \varphi_{LSJ}(\mathrm{sp}\ ^3\mathrm{P}_1\ M|H|\varphi_{LSJ}(\mathrm{sp}\ ^3\mathrm{P}_1\ M\rangle = E_{\mathrm{av}} - \tfrac{1}{6}G^1(\mathrm{sp}) - \tfrac{1}{2}\zeta(\mathrm{p}), \tag{7.98b}$$

$$H_{12} = H_{21} = \langle \varphi_{LSJ}(\mathrm{sp}\ ^1\mathrm{P}_1\ M|H|\varphi_{LSJ}(\mathrm{sp}\ ^3\mathrm{P}_1\ M\rangle = -\frac{1}{\sqrt{2}}\zeta(\mathrm{p}), \tag{7.98c}$$

where E_{av} is the average energy of these two states, $G^1(\mathrm{sp})$ describes the exchange Coulomb interaction between the electrons, and $\zeta(\mathrm{p})$ is the spin–orbit parameter characterizing the strength of spin–orbit interaction. Note that the sign of the number value in (7.98c) depends on the coupling order $\ell + s \rightarrow j$, or $s + \ell \rightarrow j$ ([Cow81], p. 243), here $\ell + s \rightarrow j$ is always used. According to equs. (7.81b), (7.86), and (7.94), one obtains for the energies E_{I} and E_{II} and the representations of the two states Ψ_{I} and Ψ_{II} in intermediate coupling

$$E_{\mathrm{I,II}} = E_{\mathrm{av}} + \tfrac{1}{6}G^1(\mathrm{sp}) - \tfrac{1}{4}\zeta(\mathrm{p}) \pm \sqrt{\{\tfrac{1}{2}\zeta^2(\mathrm{p}) + [\tfrac{1}{3}G^1(\mathrm{sp}) + \tfrac{1}{4}\zeta(\mathrm{p})]^2\}} \tag{7.99a}$$

and

$$\Psi_{\mathrm{I}}(J = 1\ M) = \frac{1}{\sqrt{(1+\alpha^2)}}\Psi_{LSJ}(\mathrm{sp}\ ^1\mathrm{P}_1\ M) - \frac{\alpha}{\sqrt{(1+\alpha^2)}}\Psi_{LSJ}(\mathrm{sp}\ ^3\mathrm{P}_1\ M), \tag{7.99b}$$

$$\Psi_{\mathrm{II}}(J = 1\ M) = \frac{\alpha}{\sqrt{(1+\alpha^2)}}\Psi_{LSJ}(\mathrm{sp}\ ^1\mathrm{P}_1\ M) + \frac{1}{\sqrt{(1+\alpha^2)}}\Psi_{LSJ}(\mathrm{sp}\ ^3\mathrm{P}_1\ M) \tag{7.99c}$$

and

$$\alpha = \frac{\zeta(\mathrm{p})/\sqrt{2}}{\tfrac{1}{3}G^1(\mathrm{sp}) + \tfrac{1}{4}\zeta(\mathrm{p}) + \sqrt{\{\tfrac{1}{2}\zeta^2(\mathrm{p}) + [\tfrac{1}{3}G^1(\mathrm{sp}) + \tfrac{1}{4}\zeta(\mathrm{p})]^2\}}}. \tag{7.99d}$$

Both the eigenvalues E_{I} and E_{II} and the states Ψ_{I} and Ψ_{II} can then be analysed for the limiting cases of pure *LS*- or pure *jj*-coupling as well as for any intermediate situation. Pure *LS*-coupling is approached for Coulomb interaction energies large compared to spin–orbit interaction energies, pure *jj*-coupling for the opposite, and the spin–orbit parameter $\zeta(2\mathrm{p})$ can be used as a tuning parameter of the actual coupling strength covering the ranges $\zeta(2\mathrm{p}) \rightarrow 0$ and $\zeta(2\mathrm{p}) \rightarrow \infty$, respectively. However, in order to keep the energy range for $E_{\mathrm{I,II}}$ finite and to confine the coupling strength parameter between zero and unity, it is more appropriate to measure the energy difference ($E - E_{\mathrm{average}}$) in units of $[\tfrac{2}{3}G^1(\mathrm{sp}) + \tfrac{3}{2}\zeta(\mathrm{p})]$, and to replace the spin–orbit parameter by the coupling strength parameter $\chi = \tfrac{3}{2}\zeta(\mathrm{p})/[\tfrac{2}{3}G^1(\mathrm{sp}) + \tfrac{3}{2}\zeta(\mathrm{p})]$. A plot of these normalized quantities is shown in Fig. 7.7; the two solid curves show E_{I} and E_{II} of the mixed states, and the dotted straight lines show the pure states (sp $J = 0$) and (sp $J = 2$). In the present context it is

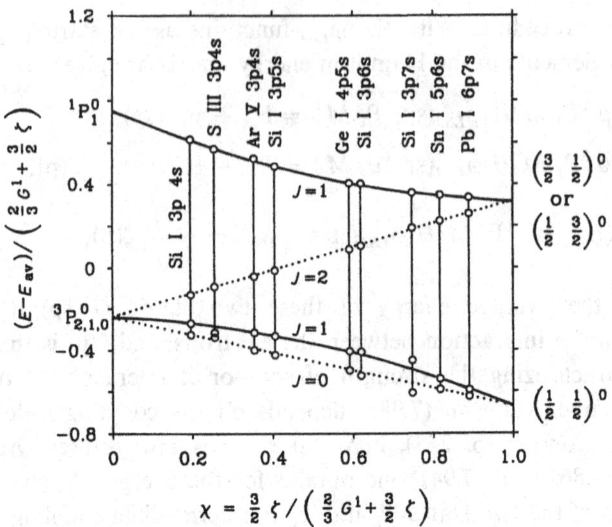

Figure 7.7 Transition from pure *LS*-coupling (left) to pure *jj*-coupling (right) for the electron configurations sp or ps, demonstrated for the energies E of the resulting states with $J = 2$, $J = 1$ (twice), and $J = 0$, shown as a function of the coupling strength parameter χ. Here $\zeta = \zeta(2p)$ is the spin–orbit parameter, and $G^1 = G^1(sp)$ a Slater integral which characterize the strength of spin–orbit and Coulomb interactions, respectively, in the (sp)-electron configuration. In order to keep the energies in a finite range, they are shown as relative values $E - E_{\text{average}}$, normalized against a combination of ζ and G^1 values as indicated on the ordinate. The states with $J = 2$ and $J = 0$ are pure, but the two states with $J = 1$ are mixed and are representative of an important example of intermediate coupling (for details see main text). A short notation is given for the states in the *LS*-coupling limit: $^1P_1^o$ for (sp $^1P_1^o$) or (ps $^1P_1^o$), and $^3P_1^o$ for (sp $^3P_1^o$) or (ps $^3P_1^o$); and in the *jj*-coupling limit: $(\frac{1}{2}\,\frac{3}{2})^o$ or $(\frac{3}{2}\,\frac{1}{2})^o$ for $(s_{1/2}p_{3/2}\;J = 1)^o$ or $(p_{3/2}s_{1/2}\;J = 1)^o$, and $(\frac{1}{2}\,\frac{1}{2})^o$ for $(s_{1/2}p_{1/2}\;J = 1)^o$ or $(p_{1/2}s_{1/2}\;J = 1)^o$. From [Cow81].

the solid curves that are of interest. The *LS*-coupling limit is approached on the left-hand side. Because of $\zeta(2p) \to 0$, i.e., $\chi \to 0$, or $\alpha \to 0$, one gets $\Psi_I \to \varphi_{LSJ}(\text{sp}\;^1P_1\;M)$ with a normalized energy of 0.75, and this point is marked in the figure by the state symbol $^1P_1^o$; similarly, the other state leads to $\Psi_{II} \to \varphi_{LSJ}(\text{sp}\;^3P_1\;M)$ with a reduced energy of -0.25 on the ordinate, which is labelled $^3P^o_{2,1,0}$, which includes the *LS*-coupling limit for (sp $J = 0$) and (sp $J = 2$). The *jj*-coupling limit is approached on the right-hand side where $\zeta(2p) \to \infty$ leads to $\chi \to 1$, $\alpha \to 1/\sqrt{2}$, $\Psi_I \to \varphi_{jjJ}(s_{1/2}p_{3/2}\;J=1\;M)$ with the reduced energy of 1/3, and $\Psi_{II} \to \varphi_{jjJ}(s_{1/2}p_{1/2}\;J=1\;M)$ with the reduced energy $-2/3$, and with the corresponding *jj*-states denoted at these positions by their angular momenta $(j_1 j_2)$ and odd parity. For all four possible states in intermediate coupling, experimental data for the normalized energy splitting of (sp)-electron systems have also been included in the figure. These demonstrate that the actual cases are between the coupling limits: lighter elements are better represented by *LSJ*-coupling, heavier elements by *jjJ*-coupling.

7.5 Recasting of correlated wavefunctions in helium (ground state)

As implied by the name, a *correlated* wavefunction takes into account at least some essential parts of the correlated motion between the electrons which results from their mutual Coulomb interaction. As analysed in Section 1.1.2 for the simplest correlated wavefunction, the helium ground-state function, this correlation imposes a certain spatial structure on the correlated function. In the discussion given there, two correlated functions were selected: a three-parameter Hylleraas function, and a simple CI function. In this section, these two functions will be represented in slightly different forms in order to make their similarities and differences more transparent.

7.5.1 CI approach

When applying the CI model to the ground state of helium, the $2p^2\,{}^1S^e$ basis function makes a significant contribution, and the expansion

$$\tilde{\Psi}_{corr}(\text{`}1s^{2}\text{'}\,{}^1S^e) = A\,\tilde{\Psi}^0(1s^2\,{}^1S^e) + a\,\tilde{\Psi}^0(2p^2\,{}^1S^e) + \cdots \tag{7.100}$$

will be restricted for simplicity to these two terms only. The individual two-electron basis functions are given by Slater determinantal wavefunctions

$$\tilde{\Psi}^0(1s^2\,{}^1S^e) = \{1s0^+, 1s0^-\} \tag{7.101a}$$

and

$$\tilde{\Psi}^0(2p^2\,{}^1S^e) = \frac{1}{\sqrt{3}}(\{2p1^+, 2p-1^-\} - \{2p1^-, 2p-1^+\} - \{2p0^+, 2p0^-\}), \tag{7.101b}$$

with single-particle orbitals $\varphi_{n\ell m_\ell m_s}(\mathbf{r})$ which are assumed to be known. If the two-electron basis function in equ. (7.101a) is worked out explicitly, one obtains

$$\begin{aligned}\tilde{\Psi}^0(1s^2\,{}^1S^e) &= \frac{1}{\sqrt{2}}(\varphi_{1s0^+}(1)\varphi_{1s0^-}(2) - \varphi_{1s0^+}(2)\varphi_{1s0^-}(1))\\ &= \varphi_{1s0}(1)\varphi_{1s0}(2)\,\chi_a\\ &= \frac{1}{4\pi}R_{1s}(r_1)R_{1s}(r_2)\,\chi_a,\end{aligned} \tag{7.102}$$

where χ_a is the antisymmetrized two-electron spinor function given by (note equ. (7.1) for the individual spinor functions)

$$\chi_a = \frac{1}{\sqrt{2}}[\chi^+_{1/2}(1)\chi^-_{1/2}(2) - \chi^+_{1/2}(2)\chi^-_{1/2}(1)]. \tag{7.103}$$

From equ. (7.102) it can be inferred that no spatial structure is imposed on the electrons, i.e., no correlation effects are included.

Similarly, one can calculate the two-electron basis functions contained in equ.

(7.101b)

$$\{2p1^+, 2p-1^-\} = \frac{1}{\sqrt{2}}[\varphi_{2p1}(1)\chi^+_{1/2}(1)\varphi_{2p-1}(2)\chi^-_{1/2}(2)$$
$$- \varphi_{2p1}(2)\chi^+_{1/2}(2)\varphi_{2p-1}(1)\chi^-_{1/2}(1)]$$
$$= \frac{1}{\sqrt{2}} R_{2p}(r_1)R_{2p}(r_2)[Y_{11}(\hat{\mathbf{r}}_1)Y_{1-1}(\hat{\mathbf{r}}_2)\chi^+_{1/2}(1)\chi^-_{1/2}(2)$$
$$- Y_{1-1}(\hat{\mathbf{r}}_1)Y_{11}(\hat{\mathbf{r}}_2)\chi^+_{1/2}(2)\chi^-_{1/2}(1)]$$
$$= \frac{1}{\sqrt{2}} R_{2p}(r_1)R_{2p}(r_2)\frac{3}{8\pi}\sin\vartheta_1\sin\vartheta_2(-1)$$
$$\times [e^{i(\varphi_1-\varphi_2)}\chi^+_{1/2}(1)\chi^-_{1/2}(2) - e^{-i(\varphi_1-\varphi_2)}\chi^+_{1/2}(2)\chi^-_{1/2}(1)], \quad (7.104a)$$

and also

$$\{2p1^-, 2p-1^+\} = \frac{1}{\sqrt{2}} R_{2p}(r_1)R_{2p}(r_2)\frac{3}{8\pi}\sin\vartheta_1\sin\vartheta_2(-1)$$
$$\times [e^{i(\varphi_1-\varphi_2)}\chi^-_{1/2}(1)\chi^+_{1/2}(2) - e^{-i(\varphi_1-\varphi_2)}\chi^+_{1/2}(1)\chi^-_{1/2}(2)], \quad (7.104b)$$

$$\{2p0^+, 2p0^-) = R_{2p}(r_1)R_{2p}(r_2)\frac{3}{4\pi}(\cos\vartheta_1\cos\vartheta_2)\chi_a. \quad (7.104c)$$

Collecting the individual parts gives

$$\tilde{\Psi}^0(2p^2\ {}^1S^e) = \frac{1}{\sqrt{3}} R_{2p}(r_1)R_{2p}(r_2)\frac{3}{4\pi}(-1)$$
$$\times [\sin\vartheta_1\sin\vartheta_2\cos(\varphi_1-\varphi_2) + \cos\vartheta_1\cos\vartheta_2]\chi_a. \quad (7.105)$$

Using spherical trigonometry, the angular functions can be expressed as

$$\cos\vartheta_{12} = \cos\vartheta_1\cos\vartheta_2 + \sin\vartheta_1\sin\vartheta_2\cos(\varphi_1-\varphi_2), \quad (7.106)$$

where ϑ_{12} is the angle between the directions $\hat{\mathbf{r}}_1$ and $\hat{\mathbf{r}}_2$. With this substitution one obtains for the correlated wavefunction defined in equ. (7.100) the equivalent form

$$\tilde{\Psi}_{corr}(\text{`}1s^2\text{'}\ {}^1S^e)$$
$$= \left[A\frac{1}{4\pi}R_{1s}(r_1)R_{1s}(r_2) + a(-1)\frac{\sqrt{3}}{4\pi}R_{2p}(r_1)R_{2p}(r_2)\cos\vartheta_{12} + \cdots\right]\chi_a. \quad (7.107a)$$

Using hydrogenic wavefunctions for the radial functions $R_{n\ell}(r)$ for simplicity, one gets

$$\tilde{\Psi}_{corr}(\text{`}1s^2\text{'}\ {}^1S^e)$$
$$= \left[A\frac{Z^3}{\pi}e^{-Z(r_1+r_2)} + a(-1)\frac{\sqrt{3}}{4\pi}\frac{Z^5}{24}r_1r_2\,e^{-Z(r_1+r_2)/2}\cos\vartheta_{12} + \cdots\right]\chi_a$$
$$= [A2.546\,e^{-2(r_1+r_2)} + a(-0.184)r_1r_2\,e^{-(r_1+r_2)}\cos\vartheta_{12} + \cdots]\,\chi_a. \quad (7.107b)$$

Obviously, the term with the weighting factor $a(-0.184)$ leads to an increase in the amplitude of $\tilde{\Psi}_{corr}$ for $\vartheta_{12} = 180°$, i.e., it describes angular correlation. Further the product r_1r_2, which also appears in this term, gives rise to radial correlations, because if both electrons are on the same radius vector ($\vartheta_{12} = 0°$) and $r_1 + r_2$ is kept constant, the correlated function has a minimum for $r_1 = r_2$, i.e., higher values for the wavefunction are obtained if the electrons tend to be apart. This correlated function is used in equ. (1.27c) for the discussion of electron correlations in the ground state of helium.

7.5.2 *Hylleraas wavefunction*

Due to its importance, the Hylleraas wavefunction introduced in equ. (1.20) will also be discussed in the context of CI. For simplicity, only the three-parameter Hylleraas function is chosen. The essential quantity, responsible for the correlated motion of the electrons, in this function is the distance r_{12} between the electrons. Usually, r_{12} is expressed as

$$r_{12} = \sqrt{(r_1^2 + r_2^2 - 2r_1r_2\cos\vartheta_{12})}, \tag{7.108}$$

where ϑ_{12} is the angle between the radial vectors pointing from the nucleus to the electrons. However, it is also possible to expand r_{12} in terms of Legendre polynomials containing the angle ϑ_{12}. In this case one gets

$$r_{12} = \sum_{\ell} f_\ell(r_1, r_2) P_\ell(\cos\vartheta_{12}), \tag{7.109a}$$

where

$$f_\ell(r_1, r_2) = r_> \left[\frac{(r_</r_>)^2}{2\ell + 3} - \frac{1}{2\ell - 1}\right]\left(\frac{r_<}{r_>}\right)^\ell \tag{7.109b}$$

with

$$r_< = \min\{r_1, r_2\}, r_> = \max\{r_1, r_2\}.$$

If this expansion is applied to the three-parameter Hylleraas function, one obtains [GMM53]

$$\Phi_{\text{Hyll.}}(\mathbf{r}_1, \mathbf{r}_2) = \sum_{\ell} c_\ell F_\ell(r_1, r_2) P_\ell(\cos\vartheta_{12}). \tag{7.110a}$$

The $F_\ell(r_1, r_2)$ are rather cumbersome, but completely known, functions which depend on $r_>$, $r_<$, and ℓ (compare the related case of the functions $f_\ell(r_1, r_2)$ defined in equ. (7.109b)). They are normalized in such a way that the coefficients $|c_\ell|^2$ describe the weight given to the individual ℓ-components. For the three-parameter Hylleraas function, one has [GMM53]:

$$\left.\begin{aligned} c_0 &= 0.997535,\\ c_1 &= 0.069227,\\ c_2 &= 0.010398,\\ c_3 &= 0.003528,\\ \sum c_i^2 &= 0.999989.\end{aligned}\right\} \tag{7.110b}$$

The role of these ℓ-components will now be made more explicit. Using equs. (7.5) and (7.8), together with the property of the special Clebsch–Gordan coefficients,

$$(\ell\, m\, \ell\, -m\,|\,0\,0) = \frac{(-1)^{\ell-m}}{\sqrt{(2\ell+1)}}, \tag{7.111}$$

the expansion of the Legendre polynomials into spherical harmonics can also be written as

$$P_\ell(\cos\vartheta_{12}) = \frac{4\pi(-1)^\ell}{\sqrt{(2\ell+1)}} \sum_m (\ell\, m\, \ell\, -m\,|\,0\,0)\, Y_{\ell m}(\hat{\mathbf{r}}_1)\, Y_{\ell -m}(\hat{\mathbf{r}}_2). \tag{7.112}$$

This result states that the angular momenta ℓ in the expansion of the Hylleraas function impose the following symmetry properties on the individual angular momenta ℓ_1 and ℓ_2 attached to the two electrons with spatial directions $\hat{r}_1$ and $\hat{r}_2$, respectively:

$$\ell_1 = \ell_2 = \ell \qquad \text{and} \qquad \boldsymbol{\ell}_1 + \boldsymbol{\ell}_2 = 0. \tag{7.113}$$

Therefore, the *individual* orbital angular momenta of each electron in the ground state of helium can have *any* value, $\ell = 0, 1, 2, 3, \ldots$, but a certain value for one electron has to be compensated by the same value for the other electron, thus allowing $L = 0$. From this result it becomes understandable that the Hylleraas function can be expressed equivalently in the CI approach using single-particle functions of different electron configurations $(n\ell n'\ell)$, but with the same symmetry $(L = 0, S = 0, \pi = (+))$. (Of course, using a complete set of such basis functions, such an expansion is always possible on the grounds of equ. (1.24).) The quantitative relation can be found if the known radial functions $F_\ell(r_1, r_2)$ are expanded in terms of a complete set of hydrogenic radial functions $R_{n\ell}(r)$ and $R_{m\ell}(r)$:

$$F_\ell(r_1, r_2) = \sum\!\!\!\!\!\!\int d_{n\ell m\ell}\, R_{n\ell}(r_1)\, R_{m\ell}(r_2). \tag{7.114a}$$

(For simplicity, the constant of proportionality which takes care of different normalizations of the radial functions involved has been omitted, the symmetry of the wavefunctions with respect to an interchange of r_1 and r_2 is not incorporated explicitly, and the presence of continuum orbitals is indicated only through the integral symbol.) The $|d_{n\ell m\ell}|^2$ coefficients then describe the quality of this expansion if only certain truncated sets of basis functions are selected, because for a complete basis set

$$\sum\!\!\!\!\!\!\int |d_{n\ell m\ell}|^2 = 1. \tag{7.114b}$$

Using hydrogenic wavefunctions with $Z = 2$, one obtains the following results for the lowest expansion coefficients $F_\ell(r_1, r_2)$ [GMM54]:

(i) The expansion for $F_0(r_1, r_2)$, including the electron configurations $1s^2$, 1s2s, 1s3s, 1s4s, …, 1sεs; $2s^2$, is already nearly complete, because $\sum |d_{nsms}|^2 = 0.99893$. This is because of the shell model which, for the coefficient d_{1s1s} attached to the $1s^2$-electron configuration, leads to the large value $|d_{1s1s}|^2 = 0.92988$. Electron–electron interactions then contribute, for example, 1s2s with $|d_{1s2s}|^2 = 0.04682$, and the continuum with $|d_{1s\varepsilon s}|^2 = 0.01199$.

(ii) In contrast, the expansion for $F_1(r_1, r_2)$ based on a similarly restricted set of electron configurations, namely, $2p^2$, 2p3p, 2p4p, ..., 2pεp; $3p^2$, is rather incomplete. The corresponding coefficients add up to $\sum |d_{npmp}|^2 = 0.662$ only, in which the largest contribution comes from $|d_{2p2p}|^2 = 0.22100$. This low value implies that many more electron configurations than just the selected ones have to be taken into consideration if $F_1(r_1, r_2)$ is to be properly accounted for within such a CI approach, i.e., one has also to include 3p4p, 3p5p, ..., $4p^2$, 4p5p, 4p6p, ..., $5p^2$, and so on.

(iii) Looking at $F_2(r_1, r_2)$, it is clear that it becomes unsuitable for situations with restricted electron configurations based on $3d^2$, 3d4d, 3d5d, ..., $4d^2$. There the dominant contribution of $3d^2$ is only $|d_{3d3d}|^2 = 0.00140$. Therefore, it becomes obvious that any relatively complete expansion of the Hylleraas function in terms of the $F_\ell(r_1, r_2)$ functions and for hydrogenic basis functions with $\ell > 0$ will be a laborious task because so many electron configurations are needed.

Finally a general comment, based on the poor convergence of the expansion of the Hylleraas function in terms of hydrogenic functions, will be added. The problem can be traced back to the different ansatz used for treating electron correlations in the Hylleraas function and in the CI approach; in the Hylleraas function the uncorrelated function $\Phi^0(r_1, r_2)$ is *multiplied* by a correction factor $f_{corr}(r_1, r_2)$ which includes the coordinates $r_1 + r_2$, $r_1 - r_2$ and r_{12} where the ground-state wavefunction has a large amplitude. In contrast, in the CI approach a *summation* is applied over electron configurations built from uncorrelated functions which possess significant amplitudes at radial distances which are large compared to the ground-state wavefunction. Hence, an extremely large set of basis functions is needed to describe adequately correlated motion which is concentrated mainly in the spatial region of the uncorrelated function. (It might be more appropriate to use basis functions which can be adapted to specific spatial regions [RWe60], but the distinction between multiplication and summation still remains.)

8

Special theoretical aspects

8.1 Photon–atom interaction and photoionization matrix elements

8.1.1 Interaction operator

In order to obtain the Hamiltonian for the system of an atom and an electromagnetic wave, the classical Hamilton function H for a free electron in an electromagnetic field will be considered first. Here the mechanical momentum $\mathbf{p}$ of the electron is replaced by the canonical momentum, which includes the vector potential $\mathbf{A}$ of the electromagnetic field, and the scalar potential Φ of the field is added, giving [Sch55]

$$H = \frac{(\mathbf{p} - e_0\mathbf{A})^2}{2m_0} + \Phi. \tag{8.1}$$

The electromagnetic field can be represented in a certain gauge which imposes boundary conditions on the potentials. Using the Coulomb gauge with

$$\nabla\cdot\mathbf{A} = 0 \tag{8.2a}$$

and setting

$$\Phi = 0, \tag{8.2b}$$

these conditions can be fulfilled in a region of space free of sources for the electromagnetic field, i.e., for *external* fields such as monochromatized synchrotron radiation. The simplest formulation follows for a plane monochromatic external field which can be described by the vector potential

$$\mathbf{A}(\mathbf{r}, t) = \tfrac{1}{2}\{\mathbf{A}_0\, \mathrm{e}^{\mathrm{i}(\mathbf{k}\cdot\mathbf{r} - \omega t)} + \mathrm{cc}\}, \tag{8.3}$$

where $\mathbf{A}_0$ is a complex quantity containing field intensity and polarization (see below), $\mathbf{k}$ is the wavenumber vector, $\mathbf{r}$ is the vector to a spatial point of the wave, ω is the angular frequency, t is the time, and cc is the complex conjugate.

The electric field $\mathscr{E}$ is then given by

$$\mathscr{E}(\mathbf{r}, t) = -\frac{\partial\mathbf{A}}{\partial t} = \frac{1}{2}\{\mathscr{E}_0\, \mathrm{e}^{\mathrm{i}(\mathbf{k}\cdot\mathbf{r} - \omega t)} + \mathrm{cc}\}, \tag{8.4a}$$

with

$$\mathscr{E}_0 = \mathrm{i}\omega\mathbf{A}_0 = \mathbf{P}\mathscr{E}_0, \tag{8.4b}$$

where $\mathbf{P}$ is the polarization vector (see equ. (9.26))† and the potential $\mathbf{A}_0$ can be

† A polarization vector $\mathbf{P}_A$ is often introduced by $\mathbf{P}_A = \mathbf{A}_0/|\mathbf{A}_0|$ which yields to the relation $\mathbf{P} = \mathrm{i}\mathbf{P}_A$.

replaced by

$$\mathbf{A}_0 = -\mathrm{i}|\mathbf{A}_0|\mathbf{P} = -\mathrm{i}A_0\mathbf{P}. \tag{8.4c}$$

The result for a free electron in an electromagnetic field can be transferred to the Hamiltonian H of an atom by using the same approach. Because the electromagnetic field depends on time, one starts with the time-dependent Schrödinger equation

$$\mathrm{i}\hbar\frac{\partial\Psi}{\partial t} = \left[\sum_j \frac{(\mathbf{p}_j - e_0\mathbf{A})^2}{2m_0} - \frac{e_0^2 Z}{4\pi\varepsilon_0}\sum_j \frac{1}{r_j} + \frac{e_0^2}{4\pi\varepsilon_0}\sum_{i<j}\frac{1}{r_{ij}}\right]\Psi, \tag{8.5a}$$

in which the canonical momentum has been introduced. Then one has to evaluate the action of the momentum operator $\mathbf{p} = -\mathrm{i}\hbar\nabla$ on the quantities on the right-hand side of equ. (8.5a). This gives (note the conditions from equ. (8.2))

$$\begin{aligned}(\mathbf{p} - e_0\mathbf{A})^2\,\Psi &= (\mathbf{p} - e_0\mathbf{A})\cdot(\mathbf{p} - e_0\mathbf{A})\Psi \\ &= p^2\Psi - e_0\overbrace{\mathbf{p}\cdot\mathbf{A}}\Psi - e_0\overbrace{\mathbf{p}\cdot\mathbf{A}\Psi} - e_0\mathbf{A}\cdot\mathbf{p}\Psi + e_0^2A^2\Psi \\ &= (p^2 - 2e_0\mathbf{A}\cdot\mathbf{p} + e_0^2A^2)\Psi,\end{aligned} \tag{8.6}$$

and finally the result

$$\mathrm{i}\hbar\frac{\partial\Psi}{\partial t} = \left[H_{\text{atom}} + \frac{\mathrm{i}e_0\hbar}{m_0}\sum_j \mathbf{A}(\mathbf{r}_j, t)\cdot\nabla_j + \sum_j \frac{e_0^2}{2m_0}A^2(\mathbf{r}_j, t)\right]\Psi. \tag{8.5b}$$

One can see that the full Hamiltonian consists of three terms, two which describe separately the parts for the atom and the field, and one which represents a coupling between the field (vector potential **A**) and terms from the atom (operator ∇_j). Obviously, it is this mixed term which is responsible for the photon–atom interaction. Provided perturbation theory can be applied, this term then acts as a transition operator between undisturbed initial and final states of the atom. Following this approach, one has to verify whether the disturbance caused by the electromagnetic field in the atom is small enough such that perturbation theory is applicable. Hence, one has to compare the terms which contain the vector potential **A** with an energy E_{ch} that is characteristic for the atomic Hamiltonian:

$$\frac{e_0}{m_0}A_0p/E_{\text{ch}} \qquad \text{and} \qquad \frac{e_0^2}{2m_0}A_0^2/E_{\text{ch}}, \tag{8.7a}$$

with p and E_{ch} taking the typical values of 1 au. For further evaluation of these expressions the strength A_0 of the vector potential of the external electromagnetic field $\mathscr{E}$ is needed. As an example, an energy flux S of 1 W/mm^2 will be assumed for photons of 20 eV. Expressing the magnitude S of the Poynting vector **S** in terms of the electric and magnetic fields,

$$S = |\langle\mathscr{E}\times\mathbf{H}\rangle|_{\text{time averaged}} = \tfrac{1}{2}c\varepsilon_0\,\mathscr{E}_0^2, \tag{8.8a}$$

or, alternatively, in terms of the number n_{ph} of incident photons/(s mm^2), each carrying the energy $\hbar\omega$,

$$S = n_{\text{ph}}\hbar\omega, \tag{8.8b}$$

such a photon beam contains $n_{ph} = 3 \times 10^{17}$ photons/(s mm^2), and has the field parameters $A_0 = 8.9 \times 10^{-13}$ V s/m and $\mathscr{E}_0 = 2.7 \times 10^4$ V/m.

This high photon number n_{ph} cannot be reached with currently available monochromatized synchrotron radiation. Therefore, the estimation based on these data provides an upper estimate. With the numbers given, it follows that

$$\frac{e_0}{m_0} A_0(p = 1 \text{ au})/(E_{ch} = 1 \text{ au}) = 7.1 \times 10^{-8} \tag{8.7b}$$

and

$$\frac{e_0^2}{2m_0} A_0^2/(E_{ch} = 1 \text{ au}) = 2.6 \times 10^{-15}. \tag{8.7c}$$

The result states that it is justified to neglect the term A^2 in equ. (8.5b) and to treat the interaction between an atom and an electromagnetic field by first-order perturbation theory. The interaction operator is then given by

$$\tilde{H}_{int} = \frac{\hbar e_0}{2m_0} \sum_j A_0\{e^{i(\mathbf{k}\cdot\mathbf{r}_j - \omega t)}\mathbf{P}\cdot\mathbf{V}_j + cc\}, \tag{8.9}$$

where the index j runs over all electrons. The justification for the use of first-order perturbation theory in atomic photoionization by monochromatized synchrotron radiation can also be obtained in a different way. If the electric field strength $\mathscr{E}_0$ of the external photon beam is compared with the internal electric field of the atom (again using 1 au as the reference value, see Section 6.1), one has

$$\frac{\mathscr{E}_0(\text{external field})}{\mathscr{E}(1 \text{ au})} = 5.3 \times 10^{-8}. \tag{8.10}$$

The result again states that the external field is small enough to allow the application of first-order perturbation theory.

Because of the time dependence of the vector potential $\mathbf{A}(\mathbf{r}_j, t)$, the photon–atom interaction also depends on time. Hence, time-dependent perturbation theory has to be applied. The *golden rule* (so called by Fermi [Fer50], see also [Dir47, Sch55, LLi58]) for the transition rate w then yields for the change from an initial atomic state $|i\rangle$ to a final atomic state $|f\rangle$

$$w = \frac{2\pi}{\hbar} |\langle f|H_{int}|i\rangle|^2 \, \delta(\text{energy conservation})\, \rho, \tag{8.11}$$

with the time-*independent* interaction

$$H_{int} = \frac{e_0\hbar}{2m_0} A_0 \sum_j e^{i\mathbf{k}\cdot\mathbf{r}_j}\mathbf{P}\cdot\mathbf{V}_j. \tag{8.12}$$

Here the wavefunctions $|i\rangle$ and $|f\rangle$ are eigenfunctions of the stationary atomic Hamiltonian, the δ-function ensures energy conservation, and the quantity ρ describes the density of final states in the photoprocess (see equ. (7.28g)).

It should be noted that equ. (8.11) for the transition rate w has been obtained in a semiclassical approximation, because the motion of the electrons in the atom was treated by quantum mechanics, but a classical description was used for the electromagnetic field. The correct approach would require annihilation (and creation) operators in order to describe the photon interaction. It can be shown, however, that the semiclassical treatment gives the correct answer for absorption and induced emission; it only fails in the case of spontaneous emission (for details, see, for example, [BJa68, Bay69]).

8.1.2 Dipole approximation

The previous formulation for the photoionization process provides the starting point for theoretical calculations. For simplicity, and because the conditions are well fulfilled, in many applications the *dipole approximation* is often used. (For extensions and derivations, relevant in the present context of photoionization studies with synchrotron radiation, see [KJG95] and references therein.) This approximation is based on a special property of the matrix element:

$$\langle \mathrm{f}| \sum_j \mathrm{e}^{\mathbf{k}\cdot\mathbf{r}_j} \mathbf{P}\cdot\nabla_j |\mathrm{i}\rangle . \tag{8.13}$$

Using, for example, Slater wavefunctions for the initial and final states, this matrix element can be evaluated to yield a one-particle matrix element for the active electron and an overlap matrix element for the passive electrons (see equ. (2.4)). Of interest in the present discussion is the one-particle matrix element of the active electron:

$$\langle \boldsymbol{\kappa} m_s^{(-)} | \mathrm{e}^{\mathbf{k}\cdot\mathbf{r}} \mathbf{P}\cdot\nabla | n\ell m_\ell m_s \rangle . \tag{8.14a}$$

The angle brackets indicate an integration over the whole space coordinate $\mathbf{r}$. However, the wavefunction $\varphi_{n\ell m_\ell m_s}$ of the bounded active electron has significant amplitude only within the spatial region of this orbital and approaches zero for larger values of r. Therefore, the restricted range of $\varphi_{n\ell m_\ell m_s}$ will also confine the range in which the photon operator becomes relevant. This suggests an expansion of the exponential function in terms of $\mathbf{k}\cdot\mathbf{r}$, giving in lowest order

$$\langle \boldsymbol{\kappa} m_s^{(-)} | \mathbf{P}\cdot\nabla | n\ell m_\ell m_s \rangle + \mathrm{i} \langle \boldsymbol{\kappa} m_s^{(-)} | \mathbf{k}\cdot\mathbf{r}\, \mathbf{P}\cdot\nabla | n\ell m_\ell m_s \rangle , \tag{8.14b}$$

and this provides an upper boundary for the scalar product $\mathbf{k}\cdot\mathbf{r}$:

$$\mathbf{k}\cdot\mathbf{r} \le k\, a_{n\ell} = \frac{\omega}{c} a_{n\ell} = \frac{a_{n\ell}}{\hbar c}\left(\frac{E_{\mathrm{ph}}}{E_{\mathrm{I}}}\right) E_{\mathrm{I}} \tag{8.15a}$$

which in atomic units gives

$$\mathbf{k}\cdot\mathbf{r} \le \alpha\, a_{n\ell}\left(\frac{E_{\mathrm{ph}}}{E_{\mathrm{I}}}\right) E_{\mathrm{I}}, \tag{8.15b}$$

where $a_{n\ell}$ is the mean orbital radius of the bounded active electron, E_{ph} is the photon energy, E_{I} is the ionization energy of the bounded active electron, and α

Table 8.1. *Hartree–Fock data for neon [FFi72] relevant for a demonstration of the dipole approximation.*

Shell	E_I [eV]	Z_{eff}	$E_{reference}$ [eV]	$\lambda_{reference}$ [nm]	$2\pi a_{n\ell}$ [nm]
1s	892	9.52	26000	0.05	0.05
2s	52.5	6.73	2100	0.60	0.30
2p	23.1	5.18	1200	1.04	0.32

$E_{reference}$ is the reference photon energy defined in equ. (8.15f), with $\lambda_{reference} = hc/E_{reference}$. Note that the E_I values are unrelaxed single-particle energies (see equ. (7.73)).

is the fine-structure constant. Using hydrogenic wavefunctions, one gets

$$a_{n\ell} = n^2/Z_{eff}, \tag{8.15c}$$

$$E_I = Z_{eff}^2/2n^2, \tag{8.15d}$$

and therefore

$$\mathbf{k}\cdot\mathbf{r} \leq \frac{\alpha Z_{eff}}{2}\left(\frac{E_{ph}}{E_I}\right), \tag{8.15e}$$

i.e.,

$$\mathbf{k}\cdot\mathbf{r} \ll 1 \qquad \text{for} \qquad E_{ph} \ll \frac{274}{Z_{eff}} E_I =: E_{reference} \tag{8.15f}$$

If the last condition is fulfilled, the second term in (8.14b) can be neglected, and the approximation is called the *dipole approximation.* In this case the exponential function $e^{\mathbf{k}\cdot\mathbf{r}}$ in (8.13) reduces to unity, and the operator describing the atom–photon interaction in equ. (8.12) to

$$H_{int} = \frac{\hbar e_0 A_0}{2m_0}\mathbf{P}\cdot\sum_j \nabla_j. \tag{8.16}$$

A compilation of data relevant for a discussion of the dipole approximation for photoionization processes in neon is given in Table 8.1. Since E_I should be small compared to $E_{reference}$, it can be seen that the approximation is well justified for excess photon energies of the order of the binding energy of the active $n\ell$-electron, and this statement also holds for other systems (see, however, [KJG95]).

There is another interpretation of the dipole approximation. Because the requirement

$$k\, a_{n\ell} \ll 1 \tag{8.17a}$$

is equivalent to

$$2\pi\, a_{n\ell} \ll \lambda, \tag{8.17b}$$

the last condition states that application of the dipole approximation requires a wavelength λ of the incident electromagnetic field which is large compared to the spatial region $2\pi a_{n\ell}$ in which the active electron is located, i.e., $\lambda \gg \lambda_{reference}$ (or

$\lambda \gg 2\pi a_{n\ell}$, which is the same order of magnitude as $\lambda_{\text{reference}}$; see Table 8.1). In other words, over this region the spatial variation of the field amplitude is small.

8.1.3 Several forms of the dipole matrix element

Within the dipole approximation, one can have different forms for the dipole matrix element (see [BSa57]). The form presented so far is called the *momentum* form (or the *velocity* form) because the relevant operator contains the momentum **p**:

$$\langle \mathrm{f}|\boldsymbol{\nabla}|\mathrm{i}\rangle = \frac{\mathrm{i}}{\hbar}\langle \mathrm{f}|\mathbf{p}|\mathrm{i}\rangle. \tag{8.18}$$

From the definition

$$\mathbf{p} = m_0\dot{\mathbf{r}} \tag{8.19a}$$

and the general commutator relation, which is valid for the time derivative of an arbitrary operator F,

$$\dot{F} = \frac{\mathrm{d}F}{\mathrm{d}t} = \frac{\partial F}{\partial t} + \frac{\mathrm{i}}{\hbar}\{H, F\}_- = \frac{\partial F}{\partial t} + \frac{\mathrm{i}}{\hbar}\{HF - FH\}, \tag{8.20}$$

one finds

$$\begin{aligned}\langle \mathrm{f}|\mathbf{p}|\mathrm{i}\rangle &= m_0\langle \mathrm{f}|\dot{\mathbf{r}}|\mathrm{i}\rangle = m_0\frac{\mathrm{i}}{\hbar}\langle \mathrm{f}|H\mathbf{r} - \mathbf{r}H|\mathrm{i}\rangle \\ &= m_0\frac{\mathrm{i}}{\hbar}(E_\mathrm{f} - E_\mathrm{i})\langle \mathrm{f}|\mathbf{r}|\mathrm{i}\rangle,\end{aligned} \tag{8.19b}$$

in which the energy difference $E_\mathrm{f} - E_\mathrm{i}$, which refers to the eigenvalues of the atomic Hamiltonian, is equal to the photon energy.

The dipole matrix element on the right-hand side of equ. (8.19b) is called the *length* form of the matrix element, because the vector **r** acts as the photon operator (see the discussion of equ. (1.28a) in which the name *dipole* approximation is also explained). Equ. (8.16) can then be replaced by

$$H_{\text{int}} = -\frac{e_0 A_0}{2\hbar} E_{\text{ph}}\,\mathbf{P}\cdot\sum_j \mathbf{r}_j. \tag{8.21}$$

Finally, a third form can be derived by applying equ. (8.20) to the time derivative of the momentum **p**, which gives the *acceleration* form of the dipole matrix element:

$$\langle \mathrm{f}|\mathbf{a}|\mathrm{i}\rangle = \frac{1}{m_0}\langle \mathrm{f}|\dot{\mathbf{p}}|\mathrm{i}\rangle = \frac{1}{m_0}\frac{\mathrm{i}}{\hbar}(E_\mathrm{f} - E_\mathrm{i})\langle \mathrm{f}|\mathbf{p}|\mathrm{i}\rangle. \tag{8.22a}$$

Using the operator relation

$$\{\mathbf{p}, H\}_- = \{\mathbf{p}, V_{\text{atom}}\}_- \qquad \text{with} \qquad \mathbf{p} = -\mathrm{i}\hbar\boldsymbol{\nabla} \tag{8.22b}$$

between the initial and final states, one obtains

$$\langle \mathrm{f}|\mathbf{p}|\mathrm{i}\rangle(E_\mathrm{f} - E_\mathrm{i}) = \mathrm{i}\hbar\langle \mathrm{f}|\boldsymbol{\nabla}V_{\text{atom}}|\mathrm{i}\rangle = \mathrm{i}\hbar\frac{Ze_0^2}{4\pi\varepsilon_0}\langle \mathrm{f}|\frac{\mathbf{r}}{r^3}|\mathrm{i}\rangle. \tag{8.22c}$$

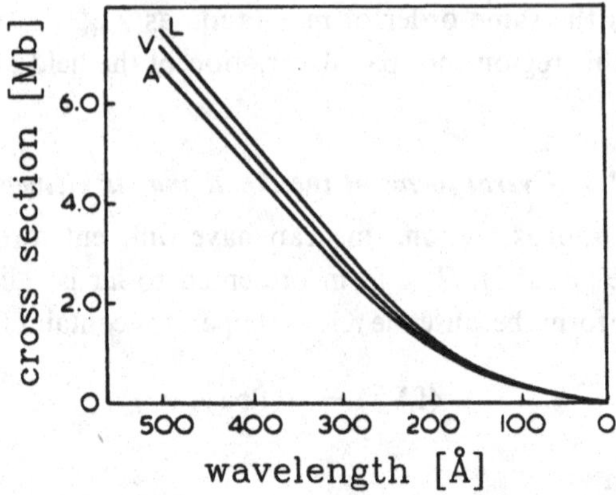

Figure 8.1 Theoretical values for the 1s photoionization cross section of helium. L, V, and A indicate the length, velocity, and acceleration forms of the dipole matrix element, respectively. From [Mar67], using data from [SWe63].

Hence, the acceleration form can be expressed as

$$\langle \mathrm{f}|\mathbf{a}|\mathrm{i}\rangle = -\frac{1}{m_0}\frac{Ze_0^2}{4\pi\varepsilon_0}\langle \mathrm{f}|\frac{\mathbf{r}}{r^3}|\mathrm{i}\rangle. \tag{8.22d}$$

All three forms of the dipole matrix element are equivalent because they can be transformed into each other. However, this equivalence is valid only for exact initial- and final-state wavefunctions. Since the Coulomb interaction between the electrons is responsible for many-body effects (except in the hydrogen atom), and the many-body problem can only be solved approximately, the three different forms of the matrix element will, in general, yield different results. The reason for this can be seen by comparing for the individual matrix elements how the transition operator weights the radial parts $R_i(r)$ and $R_f(r)$ of the single-particle wavefunction differently:

$$\text{the length form } \langle R_f(r)|r|R_i(r)\rangle \text{ weights } \textit{large } r \text{ values;} \tag{8.23a}$$

$$\text{the velocity form } \langle R_f(r)|\frac{\partial}{\partial r}|R_i(r)\rangle \text{ weights } \textit{changes} \text{ in } r; \tag{8.23b}$$

$$\text{the acceleration form } \langle R_f(r)|\frac{1}{r^2}|R_i(r)\rangle \text{ weights } \textit{small } r. \tag{8.23c}$$

As an example, the 1s photoionization cross section in helium calculated for the three forms of the dipole matrix element is shown in Fig. 8.1, and the deviations can be clearly seen.

In the context of the different forms of the dipole matrix elements a comment should be added. A good calculation for observables of photoionization which takes into account most of the important electron correlation effects should yield approximately the same values, whatever the form used for the transition matrix element. However, this necessary condition must not be a sufficient one. The most striking example is the use of hydrogenic wavefunctions where one gets exactly the same results for all three forms of the matrix element, but when calculated for elements other than hydrogen these values deviate from the experimental data. Better agreement is obtained if HF wavefunctions are used. However, there the non-locality of the exchange potential leads to different results, but combining the HF approach with the mean-field correlations as taken into account by a *random phase approximation* (RPA *including exchange*, and its relativistic version RRPA; see Section 5.2.3), the equivalence of the length and velocity form results is restored [ACh75].

Finally, the differential cross section for photoionization, $\mathrm{d}\sigma/\mathrm{d}\Omega$, will be given explicitly for the dipole approximation and the length form of the matrix element by collecting all the individual steps. This cross section is related to the transition rate w by

$$\frac{\mathrm{d}\sigma}{\mathrm{d}\Omega} = \frac{w}{n_{\mathrm{ph}}}, \tag{8.24}$$

and the transition rate w follows from Fermi's golden rule, equ. (8.11), with the interaction operator from equ. (8.21). Using the density ρ of final states as given for the photoelectrons wavefunction in equ. (7.28g)

$$\rho = \mathrm{d}\boldsymbol{\kappa} = \kappa^2\, \mathrm{d}\kappa\, \mathrm{d}\Omega_\kappa = \kappa \frac{m_0}{\hbar^2}\, \mathrm{d}\varepsilon\, \mathrm{d}\Omega_\kappa, \tag{8.25}$$

the δ-function in equ. (8.11) can be eliminated by an integration over the energy parameter ε (not yet fixed by energy conservation), which leads to the transition rate w:

$$w\, \mathrm{d}\Omega_\kappa = \frac{2\pi}{\hbar} |\langle \mathrm{f}|H_{\mathrm{int}}|i\rangle|^2\, \kappa \frac{m_0}{\hbar^2}\, \mathrm{d}\Omega_\kappa, \tag{8.26}$$

where the matrix element has to be understood as 'on-the-energy-shell', i.e., κ and ε are fixed and follow from

$$\kappa^2 = 2m_0\varepsilon/\hbar^2 \qquad \text{and} \qquad \varepsilon = E_{\mathrm{ph}} - E_{\mathrm{I}}. \tag{8.27}$$

In order finally to derive the differential cross section of photoionization one inserts equ. (8.26) in equ. (8.24) and replaces the number n_{ph} of incident photons by $n_{\mathrm{ph}} = c\varepsilon_0 A_0^2\omega/2\hbar$ (see equs. (8.4b) and (8.8a) and (8.8b)) and the interaction operator by equ. (8.21). Then one removes the factor $\hbar^2/m_0$ resulting from the normalization of the continuum function from the matrix element and incorporates it in the final prefactor (see footnote concerning equ. (7.28d)), and one introduces the fine structure constant α using $\alpha = e_0^2/4\pi\varepsilon_0\hbar c$. This leads to (for the summations over magnetic quantum numbers see below)

$$\frac{\mathrm{d}\sigma}{\mathrm{d}\Omega}(\Theta, \Phi) = 4\pi^2\alpha E_{\mathrm{ph}}\kappa \frac{1}{2J_{\mathrm{i}}+1} \sum_{M_{\mathrm{i}}}\sum_{M_{\mathrm{I}}}\sum_{m_s} |\langle J_{\mathrm{I}} M_{\mathrm{I}}, \boldsymbol{\kappa} m_s^{(-)}|\mathbf{P}\cdot\sum_j \mathbf{r}_j|J_{\mathrm{i}} M_{\mathrm{i}}\rangle|^2, \tag{8.28}$$

where for the continuum function the form in equ. (7.28b) has been employed. It should be noted that the wavenumber κ of the emitted photoelectron in front of the summation symbols arises from the definition of the continuum function which contains $1/\sqrt{\kappa}$ if the radial functions $R_{\varepsilon\ell}(r)$ are normalized on the energy scale (equ. (7.28d)). Hence, after taking the absolute squared value of the matrix element, these terms cancel and are, therefore, often omitted. The summations take into account an averaging over the magnetic quantum numbers M_i of the initial state, because one does not know from which level the system starts, as well as a summation over the unobserved magnetic quantum numbers in the final state, M_I and m_s, because one does not know to which level the system proceeds. Equ. (8.28) is then the starting point for all further discussions.

8.2 Different formulations and approximations for photoionization of *n*p-electrons (σ and β)

Photoionization of an np-electron leads to an ion which can be in either of the two fine-structure components belonging to $n\mathrm{p}^{-1}\,{}^2\mathrm{P}_{3/2}$ or $n\mathrm{p}^{-1}\,{}^2\mathrm{P}_{1/2}$. (For short, one might say an $n\mathrm{p}_j$-electron was photoionized with $j = J$. However, this special view from the single-electron picture is not needed here and will be presented only at the end of this section.) If the spacing between these components is large with respect to the natural level width of the photoionized state, the J_c value together with some characteristic angular momentum of the escaping photoelectron then defines a photoionization channel. In the *jj*-coupling limit such a channel will be characterized by the index $\gamma = \{J_I(\ell\frac{1}{2})jJ\}$, where ℓ, $\frac{1}{2}$ and j are the angular momenta of the photoelectron in one channel, and J is the total angular momentum for the complete final state ($J = 1$ for photoionization in closed-shell atoms). If spin–orbit effects in the wavefunction of the photoelectron are negligible, only its orbital angular momentum is involved, i.e., the relevant channel index is given by $\gamma = \{J_I\ell J\}$, and this can be reduced even further to $\gamma = \{L_I\ell L\}$ with $L = 1$ (for photoionization in closed-shell atoms). In reality, intermediate coupling which lies between these two limiting cases has to be applied to the photoelectron wavefunction. Starting with the most general formulation which is well suited for intermediate coupling, the more conventional treatment within *jjJ*-coupling and the necessary approximations towards the Cooper–Zare model will now be explained step by step.

8.2.1 Formulation in the approach of angular momentum transfer

Amongst other advantages, the theoretical approach of *angular momentum transfer* allows a treatment within intermediate coupling of the photoelectron wavefunction. This approach focuses on the important role of the angular momentum j_t transferred during the photoprocess (see [FDi72, DFa72, Dil73, DSM75]). The possible values of j_t follow from two relations which must be fulfilled simultaneously:

$$\mathbf{j}_t = \mathbf{j}_{\text{photon}} - \boldsymbol{\ell} \qquad \text{and} \qquad \mathbf{j}_t = \mathbf{J}_I + \tfrac{1}{2} - \mathbf{J}_i. \tag{8.29}$$

Within the dipole approximation ($j_{\text{photon}} = 1$) and for the example of photoionizing an np-electron from a closed-shell atom ($J^i = 0$), the photoelectron angular momentum is either $\ell = 0$ or 2 and, hence,

$$\left.\begin{aligned} \text{for } J_I = \tfrac{3}{2} \text{ and } \varepsilon\text{d:} \quad & j_t = 2 \text{ or } 1, \\ \text{and } \varepsilon\text{s:} \quad & j_t = 1; \end{aligned}\right\} \tag{8.30a}$$

$$\left.\begin{aligned} \text{for } J_I = \tfrac{1}{2} \text{ and } \varepsilon\text{d:} \quad & j_t = 1, \\ \text{and } \varepsilon\text{s:} \quad & j_t = 1. \end{aligned}\right\} \tag{8.30b}$$

To each ionic state J_I and angular momentum transfer j_t belongs a *reduced scattering amplitude* $S_\ell(j_t)$ – the name for which originates from the formulation of photoionization as a 'half-scattering' process. (In a scattering process with real particles the scattered particle still exists after the interaction, but in photoionization the primary incident particle, the photon, is annihilated.) The angular momentum transfer formulation allows a partition of the amplitudes into two classes, *parity-favoured* and *parity-unfavoured*, where

$$S_\ell(j_t) \text{ is parity-favoured for } \ell = j_t \pm 1, \tag{8.31a}$$

$$S_\ell(j_t) \text{ is parity-unfavoured for } \ell = j_t. \tag{8.31b}$$

(The names parity-favoured and parity-unfavoured should not be confused with parity conservation; in the present case they just describe the partition into the classes by comparing $(-1)^{j_t}$ with the parity $(-1)^\ell$ of the photoelectron (see equ. (8.31)).

Hence, the photoionization process under consideration is determined

$$\text{for } J_I = \tfrac{3}{2} \quad \text{by } S_2(2) \text{ unfav., } S_2(1) \text{ fav., } S_0(1) \text{ fav.;} \tag{8.32a}$$

$$\text{for } J_I = \tfrac{1}{2} \quad \text{by } \tilde{S}_2(1) \text{ fav., } \tilde{S}_0(1) \text{ fav.} \tag{8.32b}$$

If these scattering amplitudes are inserted in the general expressions for σ and β one obtains

$$\sigma(n\text{p}_{3/2}) = \frac{4\pi^2}{3} \alpha E_{\text{ph}} \left[5|S_2(2)|^2 + 3|S_2(1)|^2 + 3|S_0(1)|^2\right] \tag{8.33a}$$

$$\begin{aligned} &\beta(n\text{p}_{3/2})\left[5|S_2(2)|^2 + 3|S_2(1)|^2 + 3|S_0(1)|^2\right] \\ &\qquad = 3|S_2(1)|^2 - 3\sqrt{2}\left[S_2(1)S_0(1)^* + \text{cc}\right] - 5|S_2(2)|^2 \end{aligned} \tag{8.34a}$$

$$\sigma(n\text{p}_{1/2}) = \frac{4\pi^2}{3} \alpha E_{\text{ph}} \left[3|\tilde{S}_2(1)|^2 + 3|\tilde{S}_0(1)|^2\right] \tag{8.33b}$$

$$\beta(n\text{p}_{1/2})\left[|\tilde{S}_2(1)|^2 + |\tilde{S}_0(1)|^2\right] = |\tilde{S}_2(1)|^2 - \sqrt{2}\left[\tilde{S}_2(1)\tilde{S}_0(1)^* + \text{cc}\right]. \tag{8.34b}$$

In this general formulation no information is required concerning the underlying coupling scheme used for the description of the photoionization channels.

8.2.2 Formulation using jjJ-coupled dipole matrix elements

The scattering amplitudes S_γ can also be related to dipole matrix elements D_γ which are defined on the basis of jjJ-coupling and classified by the channel index $\gamma = \{J_I(\ell\frac{1}{2})jJ\}$. For these dipole matrix elements one has [Hua80]

$$D_\gamma = \mathrm{i}^{-\ell_\gamma}\,\mathrm{e}^{\mathrm{i}\sigma_\gamma}\langle\alpha_I\gamma = \{J_I(\ell\tfrac{1}{2})jJ\}^{(-)}\|Op\|\alpha_i J_i\rangle, \tag{8.35}$$

i.e., the phases of the continuum wavefunction (see equ. (7.28b)) are treated differently (the Coulomb phase shift σ_γ is taken out, the short-range phase shift δ_γ remains in the final channel function (symbol $(-)$), and α_I and α_i are additional labels for the final ionic and initial states, respectively. Often the channel index γ is abbreviated by the symbols $\{+,\ 0,\ \text{and}\ -\}$ because in a single-particle approach these symbols reflect the change in the angular momentum j_i of the bounded active electron to that of the photoelectron, j; namely, '+' for increasing, '0' for no change, and '−' for decreasing values. (Note, however, that the formulation is still valid for a many-electron system.) In the present case of np photoionization one has

for $J_I = 3/2$ three dipole matrix elements belonging to the channels

$$\left.\begin{array}{ll} np^{-1}\,{}^2\mathrm{P}_{3/2}\ \varepsilon\mathrm{d}_{5/2}\ J=1 & \text{with } D_+,\\ np^{-1}\,{}^2\mathrm{P}_{3/2}\ \varepsilon\mathrm{d}_{3/2}\ J=1 & \text{with } D_0,\\ np^{-1}\,{}^2\mathrm{P}_{3/2}\ \varepsilon\mathrm{s}_{1/2}\ J=1 & \text{with } D_-; \end{array}\right\} \tag{8.36a}$$

for $J_I = 1/2$ two dipole matrix elements attached to the channels

$$\left.\begin{array}{ll} np^{-1}\,{}^2\mathrm{P}_{1/2}\ \varepsilon\mathrm{d}_{3/2}\ J=1 & \text{with } \tilde{D}_+,\\ np^{-1}\,{}^2\mathrm{P}_{1/2}\ \varepsilon\mathrm{s}_{1/2}\ J=1 & \text{with } \tilde{D}_0; \end{array}\right\} \tag{8.36b}$$

and these quantities are related to the scattering amplitudes by

$$\left.\begin{array}{r} \sqrt{(50)}\,S_2(2) = -D_+ + 3D_0,\\ \sqrt{(30)}\,S_2(1) = 3D_+ + 1D_0,\\ \sqrt{3}\,S_0(1) = 1D_- \end{array}\right\} \tag{8.38a}$$

and

$$\left.\begin{array}{r} \sqrt{3}\tilde{S}_2(1) = -\tilde{D}_+,\\ \sqrt{3}\tilde{S}_0(1) = +\tilde{D}_0. \end{array}\right\} \tag{8.38b}$$

For the general relation valid for closed-shell atoms see, for example, [Sch92a]:

$$S_\ell(j_t) = \sum_j (-1)^{J_I-j+j_t+\ell+1}\sqrt{(2j+1)}\begin{Bmatrix} j_t & 0 & 1\\ j & J_I & 1/2 \end{Bmatrix} D_\gamma. \tag{8.37}$$

One then obtains the result (see [HJC81], note that the length form of the dipole matrix element is employed here)

$$\sigma(np_{3/2}) = \frac{4\pi^2}{3}\,\alpha E_{\mathrm{ph}}\,(|D_+|^2 + |D_0|^2 + |D_-|^2), \tag{8.39a}$$

$$\beta(np_{3/2}) = [0.8|D_+|^2 - 0.8|D_0|^2 + 0.6(D_+D_0^* + cc) - \frac{3}{\sqrt{5}}(D_+D_-^* + cc)$$

$$- \frac{1}{\sqrt{5}}(D_0D_-^* + cc)]/(|D_+|^2 + |D_0|^2 + |D_-|^2), \quad (8.40a)$$

$$\sigma(np_{1/2}) = \frac{4\pi^2}{3}\alpha E_{ph}(|\tilde{D}_+|^2 + |\tilde{D}_0|^2), \quad (8.39b)$$

$$\beta(np_{1/2}) = [|\tilde{D}_+|^2 + \sqrt{2}(\tilde{D}_+\tilde{D}_0^* + cc)]/(|\tilde{D}_+|^2 + |\tilde{D}_0|^2). \quad (8.40b)$$

At this point it is important to note that even though the dipole amplitudes D_γ are formulated in J_1jJ-coupling, they can also be used to describe intermediate coupling equally well as the scattering amplitudes $S_\ell(j_t)$. This is perhaps not so surprising if one follows the above transformation between $S_\ell(j_t)$ and D_γ, but not as evident when starting directly with the channels quoted above for D_γ. In the latter case only the dominant channel is used for the classification, i.e., in reality one has to include mixings between continuum channels. For example, D_+ then stands for the pure J_1jJ channel (superscript 0), but modified by the presence of other channels, i.e.,

$$D_+ = D_+^0(np^{-1}\,{}^2P_{3/2}\,\varepsilon d_{5/2}\,J = 1)$$
$$\times [1 + \cdots D_0^0(np^{-1}\,{}^2P_{3/2}\,\varepsilon d_{3/2}\,J = 1) + \cdots D_-^0(np^{-1}\,{}^2P_{3/2}\,\varepsilon s_{1/2}\,J = 1) + \cdots]. \quad (8.41a)$$

The contribution of $D_0^0(np^{-1}\,{}^2P_{3/2}\,\varepsilon d_{3/2}\,J = 1)$ then leads to intermediate coupling. In addition, $D_-^0(np^{-1}\,{}^2P_{3/2}\,\varepsilon s_{1/2}\,J = 1)$ and the dots in equ. (8.41a) bring in further couplings in the continuum. In particular, the dots might stand for $D_\gamma^0(np^{-1}\,{}^2P_{1/2}\,\varepsilon\ell j\,J = 1)$ and even for photoionization channels with active electrons from other shells where these contributions describe the important class of intershell couplings (see Section 5.2.4). Hence, equ. (8.41a) can also be written as (note $i^{-\ell_\gamma} = e^{-i\ell_\gamma\pi/2}$)

$$D_\gamma = |D_\gamma|\,e^{i(\sigma_\gamma + \delta'_\gamma + \mu_\gamma - \ell_\gamma\pi/2)}, \quad (8.41b)$$

and this form shows that the couplings in the continuum modify not only the value of the amplitude calculated without these couplings ($|D_\gamma^0|$ becomes $|D_\gamma|$), but also the phase through the additional multichannel phase shift μ_γ which leads to $\delta_\gamma = \delta'_\gamma + \mu_\gamma$. (The phase μ_γ is due to complex mixing coefficients in equ. (8.41a) which have been indicated there only by dots.)

8.2.3 Approximations towards the Cooper–Zare model

The formulations so far have been rather general, and in the following they will be reduced to the formulations in the independent-particle approximation. First, the D_γ are replaced by D_γ^0 which leads to the formulation of σ and β in the J_1jJ-coupling case (see [WWa73]). Because only one electron configuration is present in D_γ^0, such a dipole matrix element can be reduced further into the overlap matrix element of the passive electrons and the dipole integral with the active electron $n\ell_i j_i$. The one-electron dipole matrix element leads to numerical factors, phases (note $\Delta_\gamma = \sigma_\gamma + \delta_\gamma$) and a radial dipole integral R_γ. This can be seen from the general relation valid for closed-shell atoms (see, for example, [Sch92a])

$$D_\gamma = \sqrt{(2j_i+1)}\, \mathrm{i}^{-\ell_\gamma}(-1)^{j_i-1/2}\, \mathrm{e}^{\mathrm{i}\Delta_{\varepsilon\ell_\gamma j_\gamma, n\ell_i j_i}} R_{\varepsilon\ell_\gamma j_\gamma, n\ell_i j_i} \sqrt{(2j_\gamma+1)} \begin{pmatrix} j_\gamma & 1 & j_i \\ 1/2 & 0 & -1/2 \end{pmatrix} \tag{8.42a}$$

and also for the present example, where one has

$$\left.\begin{aligned} D_+^0 &= -2\sqrt{\left(\frac{3}{5}\right)}\, \mathrm{e}^{\mathrm{i}\Delta_{\varepsilon d5/2,np3/2}} R_{\varepsilon d5/2,np3/2}, \\ D_0^0 &= -\frac{2}{\sqrt{15}}\, \mathrm{e}^{\mathrm{i}\Delta_{\varepsilon d3/2,np3/2}} R_{\varepsilon d3/2,np3/2}, \\ D_-^0 &= -\frac{2}{\sqrt{3}}\, \mathrm{e}^{\mathrm{i}\Delta_{\varepsilon s1/2,np3/2}} R_{\varepsilon s1/2,np3/2}, \end{aligned}\right\} \tag{8.42b}$$

$$\left.\begin{aligned} \tilde{D}_+^0 &= -\frac{2}{\sqrt{3}}\, \mathrm{e}^{\mathrm{i}\Delta_{\varepsilon d3/2,np1/2}} R_{\varepsilon d3/2,np1/2}, \\ \tilde{D}_0^0 &= +\sqrt{\left(\frac{2}{3}\right)}\, \mathrm{e}^{\mathrm{i}\Delta_{\varepsilon s1/2,np1/2}} R_{\varepsilon s3/2,np1/2}. \end{aligned}\right\} \tag{8.42c}$$

In this single-particle approximation and *jjJ*-coupling scheme one then obtains

$$\sigma(n\mathrm{p}_{3/2}) = \frac{4\pi^2}{3}\alpha E_{\mathrm{ph}}\, \tfrac{1}{15}\,(36R^2_{\varepsilon d5/2,np3/2} + 4R^2_{\varepsilon d3/2,np3/2} + 20R^2_{\varepsilon s1/2,np3/2}), \tag{8.43a}$$

$$\begin{aligned} \beta(n\mathrm{p}_{3/2}) = [&36R^2_{\varepsilon d5/2,np3/2} - 4R^2_{\varepsilon d3/2,np3/2} \\ &- 10R_{\varepsilon d3/2,np3/2}R_{\varepsilon s1/2,np3/2}\cos(\Delta_{\varepsilon d3/2,np3/2} - \Delta_{\varepsilon s1/2,np3/2}) \\ &- 90R_{\varepsilon d5/2,np3/2}R_{\varepsilon s1/2,np3/2}\cos(\Delta_{\varepsilon d5/2,np3/2} - \Delta_{\varepsilon s1/2,np3/2}) \\ &+ 18R_{\varepsilon d5/2,np3/2}R_{\varepsilon d3/2,np3/2}\cos(\Delta_{\varepsilon d5/2,np3/2} - \Delta_{\varepsilon d3/2,np3/2})]/ \\ &(45R^2_{\varepsilon d5/2,np3/2} + 5R^2_{\varepsilon d3/2,np3/2} + 25R^2_{\varepsilon s1/2,np3/2}), \end{aligned} \tag{8.44a}$$

$$\sigma(n\mathrm{p}_{1/2}) = \frac{4\pi^2}{3}\alpha E_{\mathrm{ph}}\, \tfrac{2}{3}\,(2R^2_{\varepsilon d3/2,np1/2} + R^2_{\varepsilon s1/2,np1/2}), \tag{8.43b}$$

$$\begin{aligned} \beta(n\mathrm{p}_{1/2}) = [2R^2_{\varepsilon d3/2,np1/2} - 4R_{\varepsilon d3/2,np1/2}R_{\varepsilon s1/2,np1/2}\cos(\Delta_{\varepsilon s1/2,np1/2} - \Delta_{\varepsilon d3/2,np1/2})]/ \\ (2R^2_{\varepsilon d3/2,np1/2} + R^2_{\varepsilon s1/2,np1/2}). \end{aligned} \tag{8.44b}$$

If in these expressions one neglects the differences in the fine-structure components of the photoelectron wavefunction, one can set

$$R_{\varepsilon dj,npj_i} = R_{\varepsilon d,npj_i}, \tag{8.45a}$$

$$R_{\varepsilon sj,npj_i} = R_{\varepsilon s,npj_i}, \tag{8.45b}$$

$$\Delta_{\varepsilon dj,npj_i} = \Delta_{\varepsilon d,npj_i}, \tag{8.45c}$$

$$\Delta_{\varepsilon sj,npj_i} = \Delta_{\varepsilon s,npj_i}, \tag{8.45d}$$

and then derive

$$\sigma(n\mathrm{p}_{3/2}) = \frac{4\pi^2}{3}\,\alpha E_{\mathrm{ph}}\,\tfrac{4}{3}\,(2R^2_{\varepsilon d,n\mathrm{p}3/2} + R^2_{\varepsilon s,n\mathrm{p}3/2}), \tag{8.46a}$$

$$\beta(n\mathrm{p}_{3/2}) = [2R^2_{\varepsilon d,n\mathrm{p}3/2} - 4R_{\varepsilon d,n\mathrm{p}3/2}R_{\varepsilon s,n\mathrm{p}3/2}\cos(\Delta_{\varepsilon s,n\mathrm{p}3/2} - \Delta_{\varepsilon d,n\mathrm{p}3/2})]/(2R^2_{\varepsilon d,n\mathrm{p}3/2} + R^2_{\varepsilon s,n\mathrm{p}3/2}), \tag{8.47a}$$

$$\sigma(n\mathrm{p}_{1/2}) = \frac{4\pi^2}{3}\,\alpha E_{\mathrm{ph}}\,\tfrac{2}{3}\,(2R^2_{\varepsilon d,n\mathrm{p}1/2} + R^2_{\varepsilon s,n\mathrm{p}1/2}), \tag{8.46b}$$

$$\beta(n\mathrm{p}_{1/2}) = [2R^2_{\varepsilon d,n\mathrm{p}1/2} - 4R_{\varepsilon d,n\mathrm{p}1/2}R_{\varepsilon s,n\mathrm{p}1/2}\cos(\Delta_{\varepsilon s,n\mathrm{p}1/2} - \Delta_{\varepsilon d,n\mathrm{p}1/2})]/(2R^2_{\varepsilon d,n\mathrm{p}1/2} + R^2_{\varepsilon s,n\mathrm{p}1/2}). \tag{8.47b}$$

As a last step towards a simplifying approximation one also might neglect the influence of the fine-structure resolved ion core on the radial integrals and phases. In this case one gets the *LS*-coupling result from the *Cooper–Zare* model [CZa69]:

$$\sigma(2\mathrm{p}) = \sigma(2\mathrm{p}_{3/2}) + \sigma(2\mathrm{p}_{1/2}) = \frac{4\pi^2}{3}\,\alpha E_{\mathrm{ph}}\,2\,(2R^2_{\varepsilon d,n\mathrm{p}} + R^2_{\varepsilon s,n\mathrm{p}}), \tag{8.48}$$

$$\beta(2\mathrm{p}) = \beta(2\mathrm{p}_{3/2}) = \beta(2\mathrm{p}_{1/2}) = \frac{2R^2_{\varepsilon d,n\mathrm{p}} - 4R_{\varepsilon d,n\mathrm{p}}R_{\varepsilon s,n\mathrm{p}}\cos(\Delta_{\varepsilon s} - \Delta_{\varepsilon d})}{R^2_{\varepsilon s,n\mathrm{p}} + 2R^2_{\varepsilon d,n\mathrm{p}}}, \tag{8.49}$$

which are the forms given and discussed in Part A in connection with equ. (2.15).

8.3 Photon-induced L_3–M_1M_1 Auger decay in magnesium described as a one- and a two-step process

Auger electron emission can be described in many cases as a two-step process in which the first step is an inner-shell ionization event and the second step is Auger decay, with the intermediate hole-state JM_J making the link between the steps. The striking result of the two-step formulation is that for non-coincident observation of Auger electron emission the link depends on M_J-dependent *intensities* (population and transition rates) and, for coincident observation of the Auger electron with its preceding photoelectron, on the corresponding M_J-dependent *amplitudes*. However, the correct and more general formulation of Auger decay requires treating the process as a one-step resonance scattering embedded in the double ionization continuum [Abe80, AHo82], called for short the *one-step* description.

In the following, L_3–M_1M_1 Auger decay in magnesium induced by linearly polarized photons is chosen as an illustrative example for both descriptions. First, the one-step formulation is presented for Auger electrons observed in coincidence with the preceding photoelectron. In successive approximations the two-step formulation is then derived and finally applied to the non-coincident emission of Auger electrons.

8.3.1 General formulation as a one-step process

The interaction of a photon with a magnesium atom can lead to the emission of two electrons with wavenumbers $\hat{\boldsymbol{\kappa}}_a$ and $\hat{\boldsymbol{\kappa}}_b$, provided the photon energy is larger than the threshold energy E_I^{++} for double photoionization ($E_I^{++} =$ 22.68 eV for the ejection of the two outer 3s-electrons). Energy conservation then requires

$$\hbar\omega = \varepsilon_a + \varepsilon_b + E_I^{++} \text{ or, alternatively, } \delta(\hbar\omega - \varepsilon_a - \varepsilon_b - E_I^{++}). \qquad (8.50)$$

For photon energies just above this value the underlying process is direct double photoionization. However, at specific higher photon energies (resonance energies) the features typical of direct double ionization (see Section 4.6.1) may change dramatically; for example, the number of emitted electrons as well as their energy and angle distributions. In magnesium such phenomena occur at photon energies above 57.54 eV and 57.82 eV which are the inner-shell $2p^{-1}\,{}^2P_{3/2}$ and $2p^{-1}\,{}^2P_{1/2}$ ionization thresholds, respectively, and the two emitted electrons are usually called the photoelectron and Auger electron (diagram lines).

For a correct treatment of Auger electron emission as a resonance in the double ionization continuum, one has to start with the transition rate P:

$$P \sim \frac{1}{2J_i + 1} \sum_{M_iM_f} \sum_{m_{s_a}m_{s_b}} |T_{fi}(\boldsymbol{\kappa}_a m_{s_a}, \boldsymbol{\kappa}_b m_{s_b}; \hbar\omega)|^2\, \delta(\hbar\omega - \varepsilon_a - \varepsilon_b - E_I^{++}), \quad (8.51)$$

which is equivalent to the differential cross section for two-electron emission (see Section 4.6.1):

$$\frac{d^4\sigma}{d\Omega_1\, d\Omega_2\, d\varepsilon_1\, d\varepsilon_2}(\boldsymbol{\kappa}_a, \boldsymbol{\kappa}_b) = P. \qquad (8.52)$$

In equ. (8.51) the summation over the magnetic quantum numbers takes care of the unobserved substates, and the δ-function ensures energy conservation. The essential part which is of interest in the present context is the transition matrix element $T_{fi}(\boldsymbol{\kappa}_a m_{s_a}, \boldsymbol{\kappa}_b m_{s_b}; \hbar\omega)$ whose dependence on the wavevectors $\boldsymbol{\kappa}_a$ and $\boldsymbol{\kappa}_b$ of the emitted electrons with spin projections m_{s_a} and m_{s_b} and on the photon energy $\hbar\omega$ is indicated explicitly. Following the detailed discussion in [TAA87] this matrix element is given for linearly polarized incident light by two contributions:

$$T_{\text{fi}}^{\text{lin. pol.}}(\boldsymbol{\kappa}_{\text{a}} m_{s_\text{a}}, \boldsymbol{\kappa}_{\text{b}} m_{s_\text{b}}; \hbar\omega) = \langle \text{f}; \chi^{(-)}(\boldsymbol{\kappa}_{\text{a}} m_{s_\text{a}}, \boldsymbol{\kappa}_{\text{b}} m_{s_\text{b}}) | \textstyle\sum z | \text{i} \rangle$$
$$+ \sum_{\nu} \int_0^{\infty} \frac{\langle \text{f}; \chi^{(-)}(\boldsymbol{\kappa}_{\text{a}} m_{s_\text{a}}, \boldsymbol{\kappa}_{\text{b}} m_{s_\text{b}}) | V_{\text{c}} | \chi_\nu(\tau) \rangle \langle \chi_\nu(\tau) | \sum z | \text{i} \rangle}{\varepsilon_{\text{a}} + \varepsilon_{\text{b}} + E_{\text{I}}^{++} - \mathbb{E}_\nu(\tau)} \, \text{d}\tau. \tag{8.53}$$

The first contribution describes *direct double* photoionization. The photon operator (abbreviated to $\sum z$) acts on the initial state $|\text{i}\rangle$, causing a transition to the final doubly-charged state $|\text{f}\rangle$ and two emitted electrons (the superscript minus sign indicates the necessary asymptotic boundary condition for two-electron emission). The second contribution to the full transition amplitude takes into account 'resonances' in the ionization continuum described by the intermediate states $\chi_\nu(\tau)$. The word resonances is in quotation marks to indicate that these resonances are not only produced by discrete excitations, but also by the interacting continua, i.e., each intermediate state $\chi_\nu(\tau)$ characterized by the index ν stands for a full series of single-electron excitations including the continuum. In the example of 2p_j photoionization in magnesium, one such series would be $2\text{p}^5 3\text{s}^2(^2\text{P}_J)n\ell$ excitations and $2\text{p}^5 3\text{s}^2(^2\text{P}_J)\,\varepsilon\ell$ ionizations, and another one processes like $2\text{s}2\text{p}^6(^2\text{S}_{1/2})n\ell$ and $2\text{s}2\text{p}^6\,(^2\text{S}_{1/2})\varepsilon\ell$. Each intermediate state $\chi_\nu(\tau)$, together with the characteristic (resonance) energy denominator, links the photoionization amplitude with the operator $\sum z$, and the non-radiative transition (Auger) amplitude with the Coulomb interaction V_{c}. The subscript ν which characterizes a full series of single-electron excitations/ionizations also stands for a summation over the angular momenta J_ν and the magnetic quantum numbers M_ν of each state. The integration over τ exhausts the whole ionization continua of the intermediate states. Since excitation processes are also involved, it has to be understood that this integration also includes a summation over negative τ values (this region is associated with inner-shell excitation and subsequent 'spectator' Auger decay). The many possibilities for intermediate states $\chi_\nu(\tau)$ and the use of the full two-electron pair wavefunction $\chi^{(-)}(\boldsymbol{\kappa}_{\text{a}}, m_{s_\text{a}}, \boldsymbol{\kappa}_{\text{b}} m_{s_\text{b}})$ in the final channel are typical ingredients for treating the Auger decay as a one-step process.

8.3.2 One-step formula for a well-defined intermediate state

Several approximations are necessary in making the general one-step formula more tractable for applications to specific aspects and, in particular, to simple cases where the two-step formulation can be applied. Again the example of L_3–M_1M_1 Auger decay will considered, and the discussion will closely follow the detailed treatment in [TAA87].

As a first approximation, direct double photoionization will be neglected. This is often justified because the cross section for double photoionization in outer shells, and hence also the corresponding amplitude, is much smaller than the cross section for single photoionization in an inner shell. Therefore, the Auger decay has been classified in Section 1.1.3 as being a 'main' process and not a satellite, even though it originates from electron–electron interactions.

Very frequently the summation over ν can be limited to *one* state only, because a well-defined inner-shell hole-state exists (this is indicated by the subscript p). The requirements for such a case are well-defined values for the properties characterizing the intermediate state, i.e., the energy E_p^+, the angular momentum J_p with projection M_p, and the parity π_p. In addition, the natural level width Γ_p must not be too large, because the lifetime of this intermediate state must be long enough to establish an 'intermediate state'. Finally, the energy separation (including the respective level widths) with neighbouring states must be large compared to the respective level widths Γ in order to avoid overlapping states. For photoionization of a 2p-electron in magnesium, one has $2p^{-1}\,{}^2P^o_{3/2}$ with $E_I^+ = 57.54$ eV and $2p^{-1}\,{}^2P^o_{1/2}$ with $E_I^+ = 57.82$ eV, and $\Gamma \approx 1$ meV. Hence, the states are well separated from each other and have a lifetime long enough to separate photoionization from Auger decay. In contrast, 4p photoionization in xenon is an example of where these conditions are not met. (In the case of 4p photoionization in xenon, extremely strong Coster–Kronig transitions $4p_j \rightarrow 4d^{-2}k\ell$, in particular with $k\ell = 4f$, lead to such strong electron correlations that the single-particle model for $4p_j$ photoionization with a well-defined core-hole breaks down completely. As a consequence, instead of the expected doublet of $4p_{1/2}$, $4p_{3/2}$ photolines there is only one sharp line at a binding energy of 145.5 eV with some internal structure and a weaker, but broad, intensity distribution up to the $4s_{1/2}$ photoline at 213.3 eV binding energy (for the missing $4p_{1/2}$ photoline see Fig. 2.4; for a theoretical interpretation see [LWe74, WOh76]).)

If a well-defined hole-state with quantum numbers J_pM_p exists, for example, the $2p^{-1}\,{}^2P_{3/2}$ hole-state in magnesium, the complex energy $\mathbb{E}_p(\tau)$ can be approximated by

$$\mathbb{E}_p(\tau) \approx E_p^+ + \tau - i\Gamma_p/2. \tag{8.54}$$

Introducing the excess energy $E_{exc,p}$,

$$E_{exc,p} = \hbar\omega - E_p^+, \tag{8.55}$$

the denominator in equ. (8.53) can then be replaced by

$$E_{exc,p} - \tau + i\Gamma_p/2, \tag{8.56}$$

and the transition amplitude T_{fi} reduces to

$$T_{fi}^{\text{lin. pol.}} = \sum_{M_p} \int_0^\infty \frac{\langle f; \chi^{(-)}(\mathbf{\kappa}_a m_{s_a}, \mathbf{\kappa}_b m_{s_b})|V_c|\chi_p(M_p,\tau)\rangle\langle\chi_p(M_p,\tau)|\sum z|i\rangle}{E_{exc,p} - \tau + i\Gamma_p/2}\, d\tau, \tag{8.57}$$

where the dependence of the selected inner-shell hole-state $|\chi_p\rangle$ on the magnetic quantum number M_p and on the integration variable τ is expressed explicitly.

The next step concerns the evaluation of the involved matrix elements. Within the lowest approximation, final-state channel interactions are neglected, and the many-electron wavefunctions are expressed as superpositions of Slater determinantal wavefunctions with the correct symmetry and parity. In the present case of the intermediate $2p^{-1}\,{}^2P_{3/2}$ hole-state in magnesium one has

$$|\chi_p(M_p = 3/2, \tau)\rangle = \{2p1^+, 2p1^-, 2p0^+, 2p0^-, 2p-1^+, 3s0^+, 3s0^-, \tau\ell m_\ell^{m_s}\},$$

$$|\chi_p(M_p = 1/2, \tau)\rangle = \frac{1}{\sqrt{3}}\{2p1^+, 2p1^-, 2p0^+, 2p0^-, 2p-1^-, 3s0^+, 3s0^-, \tau\ell m_\ell^{m_s}\}$$

$$-\sqrt{\left(\frac{2}{3}\right)}\{2p1^+, 2p1^-, 2p0^+, 2p-1^+, 2p-1^-, 3s0^+, 3s0^-, \tau\ell m_\ell^{m_s}\},$$

$$|\chi_p(M_p = -1/2, \tau)\rangle = \frac{1}{\sqrt{3}}\{2p1^+, 2p0^+, 2p0^-, 2p-1^+, 2p-1^-, 3s0^+, 3s0^-, \tau\ell m_\ell^{m_s}\}$$

$$-\sqrt{\left(\frac{2}{3}\right)}\{2p1^+, 2p1^-, 2p0^-, 2p-1^+, 2p-1^-, 3s0^+, 3s0^-, \tau\ell m_\ell^{m_s}\},$$

$$|\chi_p(M_p = -3/2, \tau)\rangle = \{2p1^-, 2p0^+, 2p0^-, 2p-1^+, 2p-1^-, 3s0^+, 3s0^-, \tau\ell m_\ell^{m_s}\}. \tag{8.58}$$

(The 1s and 2s orbitals which are affected by neither the photoionization nor the Auger process are omitted for simplicity.) If these wavefunctions are constructed from single-electron orbitals of a common basis set (the *frozen* atomic structure approximation), the photon operator as a *one*-particle operator allows a change of only *one* orbital. Hence, the photon operator induces the change 2p to τ in these matrix elements:

$$2pm_{\ell_i}^{m_{s_i}} \rightarrow \tau\ell m_\ell^{m_s}. \tag{8.59}$$

In order to calculate the matrix elements with the Coulomb operator V_c, one again uses Slater determinantal wavefunctions, for the intermediate state $\chi_p(M_p, \tau)$ as well as for the complete final state which contains the doubly charged ion, f, and the two ejected electrons, $\chi^{(-)}(\mathbf{\kappa}_a, \mathbf{\kappa}_b)$. Assuming that there is no correlation between the two escaping electrons and that their common boundary condition applies separately to each single-particle function, the directional emission property is included in the factors $f^{(-)}(\hat{\mathbf{\kappa}}_a)$ and $f^{(-)}(\hat{\mathbf{\kappa}}_b)$, and one gets for this Coulomb matrix element C

$$\begin{aligned} C &= (f; \chi^{(-)}(\mathbf{\kappa}_a m_{s_a}, \mathbf{\kappa}_b m_{s_b})|V_c|\chi_p(M_p, \tau)\rangle \\ &= \langle f; \chi(\varepsilon_a, \varepsilon_b)|V_c|\chi_p(M_p, \tau)\rangle f^{(-)}(\hat{\mathbf{\kappa}}_a) f^{(-)}(\hat{\mathbf{\kappa}}_b), \end{aligned} \tag{8.60}$$

where the orbitals $|\varepsilon_a\rangle = |\varepsilon_a\ell_a m_{\ell_a} m_{s_a}\rangle$ and $|\varepsilon_b\rangle = |\varepsilon_b\ell_b m_{\ell_b} m_{s_b}\rangle$ are incorporated in the Slater determinantal wavefunction of the final state. Using the selected example with $|f\rangle = |1s^2 2s^2 2p^6\rangle$, all functions in this matrix element are known within the single-particle approximation, and one derives for frozen atomic structures the values in eqs. (8.61). (For simplicity the *f*-factors are omitted for the moment. Note that all passive electrons on both sides of the matrix element have to be brought to the same order before the overlap matrix element of these electrons can be taken away.)

$$C(M_p = 3/2) = \langle\{2p-1^-; \varepsilon_a, \varepsilon_b\}|V_c|\{3s0^+, 3s0^-, \tau\}\rangle, \tag{8.61a}$$

$$C(M_p = 1/2) = -\frac{1}{\sqrt{3}}(\langle\{2p-1^+, \varepsilon_a, \varepsilon_b\}|V_c|\{3s0^+, 3s0^-, \tau\}\rangle + \sqrt{2}\langle\{2p0^-, \varepsilon_a, \varepsilon_b\}|V_c|\{3s0^+, 3s0^-, \tau\}\rangle), \tag{8.61b}$$

$$C(M_p = -1/2) = +\frac{1}{\sqrt{3}}(\langle\{2p1^-, \varepsilon_a, \varepsilon_b\}|V_c|\{3s0^+, 3s0^-, \tau\}\rangle + \sqrt{2}\langle\{2p0^+, \varepsilon_a, \varepsilon_b\}|V_c|\{3s0^+, 3s0^-, \tau\}\rangle), \tag{8.61c}$$

$$C(M_p = -3/2) = -\langle\{2p1^+, \varepsilon_a, \varepsilon_b\}|V_c|\{3s0^+, 3s0^-, \tau\}\rangle. \tag{8.61d}$$

The Coulomb operator is a *two*-particle operator, i.e., it describes an interaction between at most *two* different orbitals on each side of its matrix element. Therefore, these matrix elements vanish unless the energies and spatial parts of the wavefunction in the orbitals $|\varepsilon_a\rangle$ or $|\varepsilon_b\rangle$ coincide with $|\tau\rangle$. This gives

$$\langle\{2pm_\ell^{m_s}, \varepsilon_a, \varepsilon_b\}|V_c|\{3s0^+, 3s0^-, \tau\}\rangle = \langle\{2pm_\ell^{m_s}, \varepsilon_a\}|V_c|\{3s0^+, 3s0^-\}\rangle\langle\varepsilon_b|\tau\rangle - \langle\{2pm_\ell^{m_s}, \varepsilon_b\}|V_c|\{3s0^+, 3s0^-\}\rangle\langle\varepsilon_a|\tau\rangle, \tag{8.62}$$

where the overlap integrals $\langle\varepsilon_i|\tau\rangle$ require

$$\langle\varepsilon_a|\tau\rangle = \delta(\varepsilon_a - \tau)\delta_{\ell_a\ell_p}\delta_{m_{\ell_a}m_{\ell_p}}\delta_{m_{s_a}m_{s_p}} \tag{8.63a}$$

and

$$\langle\varepsilon_b|\tau\rangle = \delta(\varepsilon_b - \tau)\delta_{\ell_b\ell_p}\delta_{m_{\ell_b}m_{\ell_p}}\delta_{m_{s_b}m_{s_p}}. \tag{8.63b}$$

(This simple result holds for single-particle orbitals of the common basis-set. If different wavefunctions are used for the initial, intermediate and final electron configurations, these overlap matrix elements are capable of describing the postcollision interaction (see Section 5.5) where the original electron with $|\tau\rangle$ is 'shaken down' to $|\varepsilon_i\rangle$.) Even though this result has been derived for the selected example, it essentially holds in general: the exchange of the electrons 'a' and 'b' in the final state wavefunction $|f; \chi^{(-)}(\mathbf{\kappa}_a m_{s_a}, \mathbf{\kappa}_b m_{s_b})\rangle$ leads for each of the two continuum electrons to a Coulomb matrix element with one electron ('a' or 'b'), multiplied by an overlap matrix element between the other electron and the as yet unspecified orbital τ, and one of these products has an overall phase of (-1). At this level of approximation, the τ integration in equ. (8.57) can be performed, and the τ orbital selected by the photoprocess (equ. (8.59)) can be identified with either the ε_a orbital or the ε_b orbital. Incorporating the directional functions $f^{(-)}(\hat{\mathbf{\kappa}}_a)$ and $f^{(-)}(\hat{\mathbf{\kappa}}_b)$ into the corresponding spatial parts described by $|\varepsilon_a\rangle$ and $|\varepsilon_b\rangle$, these new products can be identified with the continuum functions $|\mathbf{\kappa}_a^{(-)}m_{s_a}\rangle$ and $|\mathbf{\kappa}_b^{(-)}m_{s_b}\rangle$,

$$|\mathbf{\kappa}_a^{(-)}m_{s_a}\rangle = |\varepsilon_a\rangle f(\hat{\mathbf{\kappa}}_a) \quad \text{and} \quad |\mathbf{\kappa}_b^{(-)}m_{s_b}\rangle = |\varepsilon_b\rangle f(\hat{\mathbf{\kappa}}_b), \tag{8.64}$$

which finally appear *separately* in the matrix elements of the photoprocess and the non-radiative decay, but at the expense of two terms (see equ. (8.62)). Collecting the information from all individul steps, the transition amplitude of equ. (8.57) can be replaced finally by the equivalent relation

$$T_{fi}^{\text{lin. pol.}} = (-1)\sum_{M_p} \frac{\langle f; \boldsymbol{\kappa}_b^{(-)} m_{s_b}|V_c|J_p M_p\rangle\langle J_p M_p; \boldsymbol{\kappa}_a^{(-)} m_{s_a}|\sum z|i\rangle}{E_{exc,p} - \varepsilon_a + i\Gamma_p/2}$$
$$+ \sum_{M_p} \frac{\langle f; \boldsymbol{\kappa}_a^{(-)} m_{s_a}|V_c|J_p M_p\rangle\langle J_p M_p; \boldsymbol{\kappa}_b^{(-)} m_{s_b}|\sum z|i\rangle}{E_{exc,p} - \varepsilon_b + i\Gamma_p/2}. \tag{8.65}$$

If this expression is inserted in equ. (8.51a), one gets within the *one-step* formulation, and for a well-defined and isolated intermediate state, the differential cross section for the coincident observation of two electrons. These electrons are still classified as 'a' and 'b', because a distinction into a photo- and an Auger electron is not yet possible. Instead, there are two parts in T_{fi} which reflect that either of the two observed electrons, $\boldsymbol{\kappa}_a$ or $\boldsymbol{\kappa}_b$, can be connected to the photo-process, and the other electron connected to the Auger decay.

8.3.3 Two-step formula for a well-defined intermediate state

From equ. (8.65) it can be seen that the two contributions to T_{fi} resonate at $\varepsilon_a = E_{exc}$ and $\varepsilon_b = E_{exc}$ (omitting the subscript p for simplicity). In the following it will be assumed that these energies are *different* from each other (for equal energies the one-step formulation is essential and leads to a different treatment (see [VMa94, KSc95])) and, therefore, only one resonance denominator will be important. In order to calculate the differential cross section for this case, the transition amplitude T_{fi} is inserted in equ. (8.51a), and the δ-function describing energy conservation has to be eliminated by integrating over one energy variable, e.g., ε_a. The resonance then occurs for $\varepsilon_a^0 = E_{exc}$ and, because of energy conservation, ε_b simultaneously takes the value $\varepsilon_b^0 = \hbar\omega - \varepsilon_a - E_I^{++}$. With these relations it is now possible to treat the two electrons differently and to identify electron 'a' with the photoelectron (nominal kinetic energy ε_a^0) and electron 'b' with the Auger electron (nominal kinetic energy ε_b^0). Furthermore, in the absolute squared value of the matrix element, only the terms with the denominator $(E_{exc} - \varepsilon_a + i\Gamma/2)$ will remain, while terms which contain the denominator $(E_{exc} - \varepsilon_b + i\Gamma/2)$ can be neglected. Hence, one derives for the transition rate $P(\hat{\boldsymbol{\kappa}}_a, \hat{\boldsymbol{\kappa}}_b)$

$$P(\hat{\boldsymbol{\kappa}}_a, \hat{\boldsymbol{\kappa}}_b) = \tilde{\Gamma}(\varepsilon)\frac{1}{2J_i + 1}\sum_{M_i M_f}\sum_{m_{s_a} m_{s_b}}\left|\sum_{M_J}\langle f; \boldsymbol{\kappa}_b^{(-)} m_{s_b}|V_c|JM_J\rangle\langle JM_J; \boldsymbol{\kappa}_a^{(-)} m_{s_a}|\sum z|i\rangle\right|^2, \tag{8.66a}$$

where the matrix elements are understood to be 'on-the-energy-shell', i.e.,

$$\varepsilon_a^0 = \varepsilon_{ph}^0 = E_{exc} = \hbar\omega - E_I^+ \tag{8.66b}$$

and

$$\varepsilon_b^0 = \varepsilon_A^0 = E_I^+ - E_I^{++}, \tag{8.66c}$$

and the prefactor $\tilde{\Gamma}(\varepsilon)$ given by

$$\tilde{\Gamma}(\varepsilon) = \left| \frac{1}{E_{\text{exc}} - \varepsilon_a + i\Gamma/2} \right|^2 = \frac{1}{(\varepsilon_a^0 - \varepsilon_a)^2 + \Gamma^2/4} \tag{8.66d}$$

or, using energy conservation between photoelectron and the Auger electron,

$$\tilde{\Gamma}(\varepsilon) = \frac{1}{(\varepsilon_b^0 - \varepsilon_b)^2 + \Gamma^2/4} \tag{8.66e}$$

is proportional to the Lorentzian function.

In analogy to a two-stage cascade of γ-rays, equ. (8.66a) describes the *two-step* formulation for photoionization and subsequent Auger decay, connected via a well-defined and isolated intermediate hole-state. Formally, the two-step result of equ. (8.66a) can be derived from equ. (8.65) by simply neglecting one of the two terms, i.e., the exchange effects between the electrons are thrown away due to their large energy difference. In all discussions of photon-induced Auger decay presented here, such a two-step approach will be used.

As a special application of the two-step model the non-coincident observation of photon-induced Auger electron emission will be considered further. In this case one has to integrate the transition rate P of equ. (8.66a) over $\mathrm{d}\hat{\boldsymbol{\kappa}}_a$, because the photoelectron is not observed, i.e.,

$$\frac{\mathrm{d}P}{\mathrm{d}\Omega_b}(\varepsilon_b, \hat{\boldsymbol{\kappa}}_b) = \frac{1}{(\varepsilon_b^0 - \varepsilon_b)^2 + \Gamma^2/4} \frac{1}{2J_i + 1} \sum_{M_i} \sum_{M_f} \sum_{m_{s_a}} \sum_{m_{s_b}} \times \int \left| \sum_{M_J} \langle \mathrm{f}; \boldsymbol{\kappa}_b^{(-)} m_{s_b} | V_c | JM_J \rangle \langle JM_J; \boldsymbol{\kappa}_a^{(-)} m_{s_a} | \Sigma\, z | \mathrm{i} \rangle \right|^2 \mathrm{d}\Omega_a. \tag{8.67}$$

(Since the energy of the photoelectron is fixed, it is only its direction which counts.) If the absolute squared values of the matrix elements are calculated one gets diagonal terms which can be collected in the expression

$$\sum_{M_J} |\langle \mathrm{f}; \boldsymbol{\kappa}_b^{(-)} m_{s_b} | V_c | JM_J \rangle|^2 \, |\langle JM_J; \boldsymbol{\kappa}_a^{(-)} m_{s_a} | \Sigma\, z | \mathrm{i} \rangle|^2, \tag{8.68a}$$

and cross terms of the form

$$\sum_{M_J \neq M_{J'}} \sum \langle \mathrm{f}; \boldsymbol{\kappa}_b^{(-)} m_{s_b} | V_c | JM_J \rangle \langle JM_J; \boldsymbol{\kappa}_a^{(-)} m_{s_a} | \Sigma\, z | \mathrm{i} \rangle \times \langle \mathrm{f}; \boldsymbol{\kappa}_b^{(-)} m_{s_b} | V_c | JM_{J'} \rangle^* \langle JM_{J'} \boldsymbol{\kappa}_a^{(-)} m_{s_a} | \Sigma\, z | \mathrm{i} \rangle^*. \tag{8.68b}$$

However, because of the integration over all directions of the emitted photoelectron, the cross terms must vanish, and only the diagonal terms remain due to the following symmetry arguments. Within the dipole approximation and for incident linearly polarized light, the convenient quantization axis is the direction of the electric field vector. For randomly oriented atoms in the initial state this electric field vector is then the only direction of preference in the initial system (atom plus photon). Since the observation of the final system (ion, photoelectron and Auger electron) is made only for *one* constituent (the ejected Auger electron), the observed angular distribution must retain the axial symmetry brought in by the electric field vector [Yan48], i.e., a ϑ-dependence exists. Since cross terms in the above summation are connected with a phase factor

$$e^{i(M_J - M_{J'})\phi}, \tag{8.69}$$

axial symmetry can persist only if these ϕ-dependent terms vanish. Then the diagonal terms remain, and for them the summation over M_J is placed outside of the absolute squared matrix element. The result is the two-step formula for non-coincident emission of Auger electrons (subscript 'b' is replaced by 'A' for Auger if necessary [FMS72, CMe74]):

$$I_A(\vartheta, \varepsilon) = \frac{dP}{d\Omega}(\varepsilon, \hat{\boldsymbol{\kappa}}) = \frac{1}{2\pi} \frac{1}{(\varepsilon_A^0 - \varepsilon)^2 + \Gamma^2/4} \sum_{M_J} P_A(JM_J, \vartheta) a_{ph}(JM_J). \tag{8.70}$$

$I_A(\vartheta, \varepsilon)$ describes the angle-dependent intensity of Auger electrons emitted into a solid angle $d\Omega$ in the direction $\hat{\boldsymbol{\kappa}} = (\vartheta, \varphi)$ with respect to the quantization axis defined by the electric field vector of the incident linearly polarized light. The factor in front of the summation is the Lorentzian factor taking care of the inherent linewidth Γ for energies ε of the Auger electrons which differ from the nominal value ε_A^0. The factor $a_{ph}(JM_J)$ gives the population number for the intermediate state (see Sections 3.5 and 8.5), and the factor $P_A(JM_J, \vartheta)$ the M_J-dependent Auger transition rate, with these quantities defined by (in atomic units)

$$a_{ph}(JM_J) = \frac{1}{2J_i + 1} \sum_{M_i} \sum_{m_s} \int |\langle JM_J; \boldsymbol{\kappa}_{ph}^{(-)} m_s | \textstyle\sum z | i \rangle|^2 \, d\hat{\boldsymbol{\kappa}}_{ph}, \tag{8.71}$$

$$P_A(JM_J, \vartheta) = 2\pi \sum_{M_f} \sum_{m_s} |\langle \text{final ion } J_f M_f; \boldsymbol{\kappa}_A^{(-)} m_s | V_c | JM_J \rangle|^2. \tag{8.72}$$

The summations over M_f and m_s are needed because no observation is made with respect to these final-state quantum numbers. $\boldsymbol{\kappa}_A$ and $\boldsymbol{\kappa}_{ph}$ are the wavenumber vectors of the Auger electron and the photoelectron, the minus sign indicates the correct asymptotic boundary condition for the wavefunctions, V_c is the Coulomb interaction between the electrons causing the Auger transition, and $\sum z$ is the dipole operator causing the photoionization process.

Finally, the conditions necessary for deriving the given two-step formula for Auger electron emission will be summarized. Equ. (8.66) describes the coincident observation of the Auger electron with its preceding photoelectron for:

(i) randomly oriented atoms in the initial state;
(ii) a well-defined isolated inner-hole state (also expressed by the condition that the fine-structure splitting must be large compared to the width of the inner-shell hole-state);
(iii) neglect of direct double photoionization;
(iv) expansion of the many-electron wavefunction in terms of Slater determinants with the correct symmetry and parity properties (neglect of final-state channel interactions);

(v) neglect of postcollision interaction;
(vi) different energies ε^0_{ph} and ε^0_A for the photoelectron and the Auger electron;
(vii) no perturbation of the intermediate state by internal/external magnetic fields and/or by ion collision processes (this point was not considered above).

In the case of non-coincident observation of Auger electrons where equ. (8.70) is applied, one has to add:

(viii) no observation of the photoelectron (or ion);
(ix) the quantization axis in the direction of incidence for unpolarized or circularly polarized incident light, but in the direction of the electric field vector if linearly polarized light is considered.

8.4 Derivation of angular distributions by means of statistical tensors

In the differential cross section $d\sigma/d\Omega$ for photoionization (equ. (2.3)) and in the transition rate P of Auger electrons (equ. (3.7)) a summation over magnetic quantum numbers occurs which makes the calculation of these quantities rather cumbersome. There is, however, a very elegant, transparent, and efficient method for deriving these quantities which is based on the algebra of statistical tensors [BBi51, Fan51, Rac51, Fan53, FRa59]. The general idea of this approach along with some recipes and simple examples will be presented here (only a preliminary sketch can be given, for details see [DGo57, Fer65, FSt65, KSa76, Blu81]).

8.4.1 General idea

An operator Op is assumed to cause a transition from a state with angular momentum J_i and magnetic quantum number M_i to a state with quantum numbers $J_f M_f$. The ensemble is assumed to be randomly oriented in the initial state, and no observation with respect to the quantum number M_f is made in the final state. Under these conditions the transition rate P can be calculated from

$$P = \frac{1}{2J_i + 1} \sum_{M_i M_f} |\langle J_f M_f | Op | J_i M_i \rangle|^2. \tag{8.73}$$

The necessary summations reflect the fact that no complete information exists with respect to these magnetic quantum numbers: *statistically* significant information can be derived from initial states with all quantum numbers M_i represented with equal probability, $1/(2J_i + 1)$, the detection of final states with quantum numbers M_f being independent of the actual M_f value, and the summations over M_i and M_f taking care of all possible combinations of matrix elements leading from M_i to M_f substates. The appropriate formalism for such statistical information is that of *density matrices*. In the special representation in which the basic states for defining the density matrix coincide with the actual states of the ensemble, one obtains forms for the density matrices which are easy to interpret: the density matrix $\langle \rho_i \rangle$ attached to the initial, randomly oriented state has the following diagonal form:

$$\langle \rho_i \rangle = \langle J_i M_i | \rho | J_i M_i' \rangle = \frac{1}{2J_i + 1} \delta_{M_i M_i'}. \tag{8.74}$$

In the final state the property of the detector is important. Since the relevant density matrix depends upon this detection efficiency, it is called the *efficiency* matrix $\langle \varepsilon_f \rangle$. In the present example the detection efficiency is independent of M_f, and one gets

$$\langle \varepsilon_f \rangle = \langle J_f M_f | \varepsilon | J_f M_f' \rangle = \text{const}\ \delta_{M_f M_f'}. \tag{8.75}$$

The two matrices $\langle \rho_i \rangle$ and $\langle \varepsilon_f \rangle$ are easy to assess, but they refer to different properties of the ensemble related to the initial and final states, respectively. The link between the states can be established following two rules. First, the transition operator Op transforms the density matrix $\langle \rho_i \rangle$ to a density matrix $\langle \rho_f \rangle$ according to

$$\begin{aligned} \langle \rho_f \rangle &= \langle J_f M_f | \rho | J_f M_f' \rangle \\ &= \sum_{M_i M_i'} \langle J_f M_f | Op | J_i M_i \rangle \langle J_i M_i | \rho | J_i M_i' \rangle \langle J_f M_f' | Op | J_i M_i' \rangle^*. \end{aligned} \tag{8.76}$$

Second, the transition rate P follows from the trace of the matrix built from the density and the efficiency matrix of the final state:

$$P = \text{trace}(\langle \rho_f \rangle \langle \varepsilon_f \rangle). \tag{8.77a}$$

In the present example, applying both rules yields

$$\begin{aligned} P &= \text{trace}\,(\langle J_f M_f | \rho | J_f M_f' \rangle \langle J_f M_f | \varepsilon | J_f M_f' \rangle) \\ &= \text{trace}\,(\langle J_f M_f | \rho | J_f M_f' \rangle\ \text{const}\ \delta_{M_f M_f'}) \\ &= \sum_{M_f} \langle J_f M_f | \rho | J_f M_f \rangle \\ &= \sum_{M_f} \sum_{M_i M_i'} \langle J_f M_f | Op | J_i M_i \rangle \frac{1}{2J_i + 1} \delta_{M_i M_i'} \langle J_f M_f | Op | J_i M_i' \rangle^* \\ &= \frac{1}{2J_i + 1} \sum_{M_f} \sum_{M_i} |\langle J_f M_f | Op | J_i M_i \rangle|^2, \end{aligned} \tag{8.77b}$$

i.e., one arrives back at the formulation given at the beginning in equ. (8.73).

The interpretation of the density (and efficiency) matrices is rather obvious: the diagonal elements represent the probabilities of finding a certain magnetic quantum number of the ensemble, and the non-diagonal elements contain the phase information of the system for the corresponding different magnetic quantum numbers. However, in calculations with these matrices one still has to work out the cumbersome summations over these quantum numbers. It is more convenient to replace the density (and efficiency) matrices by *statistical* and *efficiency* tensors, also called *state multipoles*.

Starting with a density matrix generalized to different values J and J', the statistical tensors $\rho_{k\kappa}(JJ')$ are defined by [FRa59]

$$\rho_{k\kappa}(JJ') = \sum_{MM'} (-1)^{J'-M'}(JMJ' - M' \mid k\kappa)\langle JM|\rho|J'M'\rangle, \tag{8.78a}$$

where $(JMJ' - M' \mid k\kappa)$ is a Clebsch–Gordan coefficient. (The relation for the efficiency tensor is defined analogously.) The inverse relation is given by

$$\langle JM|\rho|J'M'\rangle = \sum_{k\kappa} (-1)^{J'-M'}(JMJ' - M' \mid k\kappa)\rho_{k\kappa}(JJ'). \tag{8.78b}$$

Substitution of the density and efficiency matrices in the expression for the transition rate P in equ. (8.77a) then leads to the result

$$P = \sum_{JJ'k\kappa} \rho_{k\kappa}(JJ')\varepsilon^{*}_{k\kappa}(JJ'). \tag{8.79}$$

At first glance one might think nothing has been gained concerning the summation over magnetic quantum numbers, because even for a well-defined state with $J' = J$ the summation in equ. (8.79) now runs over the *rank parameter* k and its *components* κ of the statistical tensors (with $\kappa = -k$ in steps of unity up to $\kappa = +k$). However, the statistical tensors fulfil certain transformation properties (see below, part (i)) which can be used in connection with the results from the *Wigner–Eckart theorem* (see below, part (ii)) to work out all angular momentum couplings in the magnetic quantum numbers and to derive a closed form for the final expression of the transition rate P in which 'geometry' and 'physics' ('dynamics' in the case of a transition) are separated. The two necessary ingredients for this algebra are:

(i) The transformation properties of statistical tensors. In order to represent a wavefunction $|JM\rangle$ or a statistical tensor $\rho_{k\kappa}(J, J')$, a coordinate frame (x, y, z) must first be defined. If this coordinate frame is subject to a rotation $R(x, y, z \rightarrow x'', y'', z'') = (R\text{: old} \rightarrow \text{new})$, the wavefunction with quantum numbers JM transforms to a wavefunction $|JM''\rangle$ with quantum numbers JM'' in the x'', y'', z'' system according to

$$|JM''\rangle_{\text{new}} = \sum_{M} D^{J}_{MM''}(R\text{: old} \rightarrow \text{new})\, |JM\rangle_{\text{old}}, \tag{8.80a}$$

which can be expressed as

$$|JM''\rangle = \sum_{M} D^{J}_{MM''}(R)\, |JM\rangle, \tag{8.80b}$$

where $D^{J}_{MM''}(R)$ are rotation matrix elements (see below). Note that the angles are also measured with respect to the different coordinate frames. For example, equ. (8.80a) applied to spherical harmonics means the following:

$$Y_{\ell m''}(\vartheta'', \varphi'')|_{\text{new}} = \sum_{m} D^{\ell}_{mm''}(R\text{: old} \rightarrow \text{new})\, Y_{\ell m}(\vartheta, \varphi)|_{\text{old}}. \tag{8.80c}$$

The inverse transformation of equ. (8.80b) is given by

$$|JM\rangle_{\text{old}} = \sum_{M''} D^{J*}_{MM''}(R)\, |JM''\rangle_{\text{new}} = \sum_{M''} D^{J}_{M''M}(R^{-1})\, |JM''\rangle_{\text{new}}. \tag{8.81}$$

where R^{-1} describes the rotation from the new to the old coordinate frame. With these definitions the statistical tensors transform as

$$\rho_{k\kappa''}(JJ') = \sum_{\kappa} D^{k*}_{\kappa\kappa''}(R)\rho_{k\kappa}(JJ') = \sum_{\kappa} D^{k}_{\kappa''\kappa}(R^{-1})\rho_{k\kappa}(J,J') \quad (8.82)$$

The last equation shows that the statistical tensors obey well-defined rotational properties which are similar to those of the angular momentum functions, except for the complex-conjugated rotation matrix elements which describe the inverse rotation $x'', y'', z'' \rightarrow x, y, z$, and that no physical significance is attached to the coefficients k and κ (they are convenient summation indices only).

(ii) The Wigner–Eckart theorem. In the most general case, the transition matrix element in equ. (8.73) connects for $M_f = M_i$ different orientations of the states involved. The Wigner–Eckart theorem [Wig27, Eck30] then provides for an operator Op which can be represented as an *irreducible* (spherical) *tensor operator* $T^{[k]}_{\kappa}$ a simple relationship between these orientations. Such tensor operators are defined to have transformation properties with respect to rotations R of the coordinate frame identical to those given for the angular momentum eigenstates in equs. (8.80) and (8.81) with J replaced by k. (One must distinguish between a tensor operator, $T^{[k]}_{\kappa}$, and a statistical tensor operator, $\rho_{k\kappa}(J, J')$, because they have different transformation properties; it is $\rho^{*}_{k\kappa}(J, J')$ which transforms like an irreducible tensor operator, and not $\rho_{k\kappa}(J, J')$.) As in the case of statistical tensors, k describes the rank of the tensor operator and κ a component $(-\kappa \leq k \leq +\kappa)$. The Wigner–Eckart theorem then states that the transition matrix element with a tensor operator $T^{[k]}_{\kappa}$ can be written in the form

$$\langle J_f M_f | T^{[k]}_{\kappa} | J_i M_i \rangle = (J_i M_i k\kappa | J_f M_f) \langle J_f || T^{[k]} || J_i \rangle, \quad (8.83)$$

where $(J_i M_i k\kappa | J_f M_f)$ is a Clebsch–Gordan coefficient and the factor $\langle J_f || T^{[k]} || J_i \rangle$ is a numerical value called the *reduced* matrix element. (Note that there are different definitions of the Wigner–Eckart theorem in the literature with respect to phases and additional numerical factors.) In this way the matrix element splits into parts which describe the 'geometry', given by the Clebsch–Gordan coefficient which depends on the components of angular momenta along some arbitrary direction, and the 'physics', given by the value of the reduced matrix element which is independent of M_f, M_i, and κ, as the notation implies.

In the present context the two transition operators of relevance are that for photoionization which is given in the dipole approximation, and within the length form, by (see equ. (1.28a))

$$Op_1 = \mathbf{P} \cdot \sum_j \mathbf{r}_j, \quad (8.84a)$$

where $\mathbf{P}$ is the polarization vector of the incident light (equ. (9.26)), and that for

Auger decay given by (equ. (1.28b))

$$Op_2 = \frac{e_0^2}{4\pi\varepsilon_0} \sum_{i<j} \frac{1}{r_{ij}}. \tag{8.84b}$$

Hence, for photoionization one has to consider

$$(Op_1) = \langle J_1 j_p JM | Op_1 | J_i M_i \rangle, \tag{8.85a}$$

where the matrix element describes the transition from the initial state $J_i M_i$ to the final channel state ($J_1 = J_{\text{ion}}$, $j_p = j_{\text{photoelectron}}$, $J = J_{\text{final}}$, $M = M_{\text{final}}$). (In principle, one has to formulate the process in second quantization in terms of a *reaction* matrix element

$$\langle J_1, j_p\text{: } J_{\text{final}} | Op_{\text{reaction}} | j_{\text{atom}}, \text{photon: } J_{\text{initial}} \rangle, \tag{8.85b}$$

where Op_{reaction} is a scalar operator which connects the full initial state with the full final state. However, this form leads to the one given in equ. (8.87a) where the special property of the light polarization is also worked out as indicated in the subsequent relations.) Even though the operator Op_1 is a scalar, a directional property is introduced through the polarization vector since one is interested in calculating the matrix element for a particular kind of light polarization. According to Section 9.2.2 the polarization vector can be represented in different basis systems which can be related to each other, and examples for the important cases of linear and circular polarization are given there. It is also possible to describe the **r** vector in the (x, y, z) basis system using the polar angles ϑ and φ:

$$\mathbf{r} = (x, y, z) = (r \sin\vartheta \cos\varphi, r \sin\vartheta \sin\varphi, r \cos\vartheta), \tag{8.86a}$$

and with this information the scalar product of the photoionization operator can be worked out. For linearly polarized light, with the electric field vector defining the quantization axis, this gives

$$Op_1(\text{lin.}) = \hat{\mathbf{e}}_z \cdot \mathbf{r} = z = r\cos\vartheta = r\sqrt{\left(\frac{4\pi}{3}\right)} Y_{10}(r) = r\, C_0^{[1]}, \tag{8.86b}$$

and for circularly polarized light, with the beam direction defining the quantization axis,

$$\begin{aligned} Op_1(\text{circ.})|_{\text{rcp}} &= \hat{\mathbf{e}}_r \cdot \mathbf{r} = +\frac{1}{\sqrt{2}}(x - \mathrm{i}y) = +\frac{1}{\sqrt{2}} r \sin\vartheta\, \mathrm{e}^{-\mathrm{i}\varphi} \\ &= r\sqrt{\left(\frac{4\pi}{3}\right)} Y_{1-1}(\hat{\mathbf{r}}) = rC_{-1}^{[1]}, \end{aligned} \tag{8.86c}$$

and similarly

$$Op_1(\text{circ.})|_{\ell\text{cp}} = \hat{\mathbf{e}}_\ell \cdot \mathbf{r} = rC_{+1}^{[1]}. \tag{8.86d}$$

In these expressions the quantity $C_\mu^{[1]}$ is proportional to the spherical harmonics $Y_{1\mu}(\hat{\mathbf{r}})$ and represents a tensor operator of rank 1 and component μ. Results for both linear and circular polarizations can then be comprised in

$$\langle Op_1 \rangle = \langle J_1 j_p JM | rC_\mu^{[1]} | J_i M_i \rangle. \tag{8.87a}$$

Application of the Wigner–Eckart theorem yields

$$\langle Op_1 \rangle = (J_i M_i 1\mu \,|\, JM)\,\langle J_I j_p J \| rC^{[1]} \| J_i \rangle \tag{8.87b}$$

with the reduced dipole matrix element

$$\langle J_I j_p J \| rC^{[1]} \| J_i \rangle = D_\gamma . \tag{8.87c}$$

(Note that the Clebsch–Gordan coefficient describes the selection rules of the angular momenta.) However, this quantity does not include the phases of the continuum wavefunction and, therefore, has to be replaced for a real photoionization process by (see equ. (8.35))†

$$D_\gamma = \mathrm{i}^{-\ell_\gamma}\,\mathrm{e}^{\mathrm{i}\sigma_\gamma}\langle \alpha_c \gamma = \{J_I j_p J\}^{(-)} \| Op_1 \| \alpha_i J_i \rangle . \tag{8.87d}$$

For the transition matrix element for a non-radiative Auger decay from the ionic state J_I to J_f one gets

$$\langle Op_2 \rangle = \langle J_f j_A JM | Op_2 | J_I M_I \rangle = (J_I M_I 00 \,|\, JM)\,\langle J_f j_A J \| Op_2 \| J_I \rangle , \tag{8.88a}$$

which gives

$$\langle Op_2 \rangle = \langle J_f j_A J_I \| Op_2 \| J_I \rangle = C_{\gamma_A} , \tag{8.88b}$$

and the reduced Coulomb element C_{γ_A} becomes, in analogy with D_γ:†

$$C_{\gamma_A} = \mathrm{i}^{-\ell_{\gamma_A}}\,\mathrm{e}^{\mathrm{i}\sigma_{\gamma_A}}\,\langle \alpha_f \gamma_A = \{J_f j_A J_I\}^{(-)} \| Op_2 \| \alpha_I J_I \rangle . \tag{8.88c}$$

As a consequence of the Wigner–Eckart theorem, relations between statistical tensors which can be derived from purely vector coupling procedures will be supplemented for *transitions* by introducing the corresponding reduced matrix elements D_γ or C_γ for the process of photoionization or Auger decay, respectively (see equs. (8.102) and (8.103b)).

8.4.2 Some recipes

From the abundance of relations which have been worked out for statistical tensors, only those needed to derive expressions for photon-induced electron emission from closed-shell atoms will be quoted here. (In this context it should be mentioned that many different conventions are in use which are not always compatible and which lead to different phases and/or numerical values. Here the convention of [Fer65] is employed.) The necessary expressions are:

(i) The general formulation of the transition rate. In order to derive the transition rate P of a certain process, one has to write down the reaction for all particles which lead to the final system with non-interacting separated particles (the *dissociated* or *break-up* system). To each of these constituents one can attach a set of statistical tensors $\rho_{k\kappa}(jj')$ and efficiency tensors $\varepsilon_{k\kappa}(jj')$ where the angular momenta j and j' take all possible values. P follows from a

† The operators Op_1 and Op_2 are used in the reduced matrix elements where, however, their tensor character without components μ has to be understood. Op_1 is a tensor of rank 1, Op_2 a tensor of rank 0.

generalization of equ. (8.79) which requires the inclusion of all these statistical and efficiency tensors. In the case of two-particle break-up, i.e., single photoionization with one ion (subscript I) and one electron (subscript j_1) in the final state, this would lead to (all summation indices are omitted for simplicity)

$$P \sim \sum \rho_{k_I\kappa_I}(J_I, J_I')\, \rho_{k_1\kappa_1}(j_1, j_1')\, \varepsilon^*_{k_I\kappa_I}(J_I, J_I')\, \varepsilon^*_{k_1\kappa_1}(j_1, j_1'), \qquad (8.89)$$

where the representations of the statistical and efficiency tensors have to refer to the same coordinate frame (usually the laboratory frame is chosen).

(ii) Efficiency tensors. Since the efficiency tensors $\varepsilon_{k\kappa}(j, j')$ are formulated in terms of the final products, they are related directly to the properties of the corresponding detector. They have two constituents: efficiency tensors defined in the detector frame which are called 'radiation' parameters $c_{k\kappa}(j, j')$, and rotation matrix elements which describe the rotation $R = (\vartheta_1, \vartheta_2, \vartheta_3)$ that moves the axes from the detector frame to the laboratory frame:

$$\varepsilon_{k\kappa}(j, j') = \sum_{\kappa''} D^{k*}_{\kappa''\kappa}(R)\, c_{k\kappa''}(j, j') = \sum_{\kappa''} D^{k}_{\kappa\kappa''}(R^{-1})\, c_{k\kappa''}(j, j'). \qquad (8.90)$$

It is convenient to regard $R^{-1} = (-\vartheta_3, -\vartheta_2, -\vartheta_1)$ as coordinates which specify the detector position in the laboratory frame, and then to replace these Euler angles by the polar and azimuthal angles Θ and Φ which describe the detector axis in the laboratory frame and an angle χ which gives a reorientation of the detector about its figure axis:

$$R^{-1} = (-\vartheta_3, -\vartheta_2, -\vartheta_1) = (\Phi, \Theta, \chi). \qquad (8.91)$$

Hence for point-like detectors one gets

$$\varepsilon_{k\kappa}(jj') = \sum_{\kappa''} c_{k\kappa''}(j, j') D^{k}_{\kappa\kappa''}(\Phi, \Theta, \chi), \qquad (8.92a)$$

if an electron is observed, and trivially

$$\varepsilon_{k\kappa}(jj') = \sqrt{(2j + 1)}\, \delta_{k0}\, \delta_{\kappa 0}\, \delta_{jj'}, \qquad (8.92b)$$

if the particle (electron, ion, ...) is not observed.

(iii) Rotation matrix elements. Many slightly different forms for the rotation matrix elements can be found in the literature. Therefore, the form which will be used here will be reproduced for convenience [Fer65]:†

$$D^{a}_{\alpha\alpha'}(\vartheta_1, \vartheta_2, \vartheta_3) = e^{-i\alpha\vartheta_1}\, d^{a}_{\alpha\alpha'}(\vartheta_2)\, e^{-i\alpha'\vartheta_3}, \qquad (8.93a)$$

with

$$\begin{aligned} d^{a}_{\alpha\alpha'}(\vartheta_2) = {} & \sqrt{[(a+\alpha)!(a-\alpha)!(a+\alpha')!(a-\alpha')!]} \\ & \times \sum_x (-1)^x \cos^{2a+\alpha-\alpha'-2x}(\vartheta_2/2) \sin^{2x-\alpha+\alpha'}(\vartheta_2/2) \\ & \times [(a+\alpha-x)!(a-\alpha'-x)!x!(x-\alpha+\alpha')!]^{-1}, \qquad (8.93b) \end{aligned}$$

where x runs from the larger of the values 0 and $\alpha - \alpha'$ to the smaller of $a + \alpha$

† There is a misprint in the exponent of the sine function quoted in equ. (6.8) of [Fer65]; it should be $\sin^{2x-\alpha+\alpha'}(\vartheta_2/2)$ instead of $\sin^{2x-\alpha-\alpha'}(\vartheta_2/2)$.

and $a - \alpha'$. In these expressions the rotation R which transforms the z-axis into the $\hat{e}$-direction is described by the three Euler angles ϑ_1, ϑ_2, and ϑ_3 defined by

(1) a rotation through ϑ_1 about the z-axis giving the $x', y', z' = z$ frame,
(2) a rotation through ϑ_2 about the new y'-axis giving the $x'', y'' = y', z''$ frame,
(3) a rotation through ϑ_3 about the new z''-axis giving the $x''', y''', z''' = z''$ frame.

Particular values which will be needed below are

$$D^a_{\alpha 0}(\vartheta_1, \vartheta_2, \vartheta_3) = \sqrt{\left(\frac{4\pi}{2a+1}\right)}\, Y^*_{a\alpha}(\vartheta_2, \vartheta_1), \tag{8.94a}$$

$$D^a_{00}(\vartheta_1, \vartheta_2, \vartheta_3) = P_a(\cos\vartheta_2), \tag{8.94b}$$

$$D^a_{\alpha\alpha'}(-\vartheta_3, -\vartheta_2, -\vartheta_1) = D^{a*}_{\alpha'\alpha}(\vartheta_1, \vartheta_2, \vartheta_3). \tag{8.94c}$$

(iv) Radiation parameters of an emitted electron. The radiation parameters of the electron, including the observation of its spin, are given by

$$\begin{aligned} c_{k\kappa}(jj') = {} & \frac{1}{4\pi}\, \rho_{k_s\kappa}(\tfrac{1}{2})\, (\ell 0 \ell' 0|k_\ell 0)\, (k_\ell 0 k_s \kappa|k\kappa)\, (-1)^{\ell'} \\ & \times \sqrt{[(2\ell+1)(2\ell'+1)(2j+1)(2j'+1)(2k_\ell+1)(2k_s+1)]} \\ & \times \begin{Bmatrix} \ell & \frac{1}{2} & j \\ \ell' & \frac{1}{2} & j' \\ k_\ell & k_s & k \end{Bmatrix}. \end{aligned} \tag{8.95}$$

The detector response is comprised in the statistical tensors $\rho_{k\kappa}(\frac{1}{2})$ of the spin subsystem, for which one has

$$\begin{aligned} \rho_{00}(\tfrac{1}{2}) &= \sqrt{(\tfrac{1}{2})}, & \rho_{11}(\tfrac{1}{2}) &= -\tfrac{1}{2}(Q_{x''} - \mathrm{i}Q_{y''}), \\ \rho_{10}(\tfrac{1}{2}) &= \sqrt{(\tfrac{1}{2})}Q_{z''}, & \rho_{1-1}(\tfrac{1}{2}) &= \tfrac{1}{2}(Q_{x''} + \mathrm{i}Q_{y''}). \end{aligned} \tag{8.96}$$

The Q values in these tensors represent the Stokes parameters for the spin sensitivity of the detector in its x'', y'', z'' frame. For example, $Q_{x''}$ describes the detector efficiency for measuring spin projections along $+x''$ and $-x''$, respectively (for the definition of the spin polarization vector see Section 9.2.1).

For a spin-insensitive detector one has $Q_{x''} = Q_{y''} = Q_{z''} = 0$, and the radiation parameters reduce to

$$c_{k0}(jj') = \frac{1}{4\pi}(-1)^{1/2+k-j}\sqrt{[(2j+1)(2j'+1)]}\,(j\tfrac{1}{2}j' - \tfrac{1}{2}|k0) \tag{8.97a}$$

with

$$k \leq 2\ell_{\max} \qquad \text{and} \qquad (\ell + \ell' + k) = \text{even} \tag{8.97b}$$

where the conditions for k follow from the Clebsch–Gordan coefficient in equ. (8.95). (In addition, a factor 2 has been introduced in order to describe unit probability for the electron being detected with either one of the two possible spin projections (for details see [Fer65], p. 35).)

(v) Statistical tensors of the initial state. In addition to the efficiency tensors the corresponding statistical tensors are also needed. These quantities are easily derived for the initial reactants, but not for the final products. In the present context the ones relevant for the initial reactants are given by

$$\rho_{k\kappa}(jj') = 1/\sqrt{(2j+1)}\,\delta_{k0}\,\delta_{\kappa 0}\,\delta_{jj'} \tag{8.98}$$

for a randomly oriented initial state of the atom, and $\rho_{k\kappa}(E1)$ which are the statistical tensors for the incident light in the dipole approximation, given by the components

$$\begin{aligned} &\rho_{00}(E1) = \sqrt{(\tfrac{1}{3})}, \quad &&\rho_{1\pm 1}(E1) = 0, \quad &&\rho_{10}(E1) = -\sqrt{(\tfrac{1}{2})}\widetilde{S}_3, \\ &\rho_{2\pm 2}(E1) = -\tfrac{1}{2}\widetilde{S}_1, \quad &&\rho_{2\pm 1}(E1) = 0, \quad &&\rho_{20}(E1) = \sqrt{(\tfrac{1}{6})}, \end{aligned} \tag{8.99a}$$

where $\widetilde{S}_i$ are the Stokes parameters defined in the tilted collision frame shown in Fig. 1.15 (note $\widetilde{S}_2 = 0$; for the optical definition of $\widetilde{S}_i$ see equ. (9.30); $\widetilde{S}_1 = P_1$ ([Fer65]), $\widetilde{S}_3 = -P_3$ ([Fer65])). If for completely linearly polarized light the reference axis is chosen to coincide with the direction of the electric field vector, one has only $\tilde{\rho}_{00}(E1)$ and $\tilde{\rho}_{20}(E1)$, with

$$\tilde{\rho}_{00}(E1) = 1/\sqrt{3} \quad \text{and} \quad \tilde{\rho}_{20}(E1) = -2/\sqrt{6}. \tag{8.99b}$$

(vi) Coupling of statistical tensors. In order to proceed from the unknown statistical tensors of the final-state system, fragmented into individual components, to the known statistical tensors of the initial reactants, coupling rules for statistical tensors have to be applied. For this coupling it is helpful to look at the corresponding coupling rules of the angular momenta involved. In the present context of photon-induced electron emission the following two relations are needed:

(1) Couplings for the photoionization process: For this case one must investigate how the statistical tensors of the fragmented system, i.e., the photoion, $\rho_{k_I\kappa_I}(J_I, J_I)$, and the photoelectron, $\rho_{k_1\kappa_1}(j_1, j'_1)$, combine to form the statistical tensors in the initial state split into that of the atom, $\rho_{k_a\kappa_a}(J_a, J_a)$, and of the incident photon, $\rho_{k\kappa}(E1)$. Because the common formula for the coupling of statistical tensors applies only to two coupled angular momenta, the photoionization case with four angular momenta involved has to be evaluated by applying the coupling formula twice assuming 'photoexcitation' of a 'resonance' state J_r and 'decay' of this by photoelectron emission. The parentheses are used to express the formal character of this description, because single photoionization is generally not a two-step process. This formal coupling of angular momenta is reflected in the fact that in this coupling procedure reduced matrix elements which describe the actual dynamical process are absent. These reduced matrix elements will be introduced only in the final result of the coupling of angular momenta. (An alternative way of coupling would be: active electron (j_i) and photoionized ion (J_I) to J_a, and j_i

with the photon to j_1.) Because of the relation

$$\mathbf{J}_a + \mathbf{1}\,(\text{photon}) \to \mathbf{J}_r \tag{8.100a}$$

one gets

$$\rho_{k_r\kappa_r}(J_r, J_r) = \sum \rho_{k_a\kappa_a}(J_a, J_a)\,\rho_{k_p\kappa_p}(E1)\,(k_a\kappa_a k_p\kappa_p \mid k_r\kappa_r)$$

$$\times \sqrt{[(2k_a+1)(2k_p+1)]}\,(2J_r+1)\begin{Bmatrix} J_a & 1 & J_r \\ J_a & 1 & J_r \\ k_a & k_p & k_r \end{Bmatrix}, \tag{8.100b}$$

which reduces for closed-shell atoms ($J_a = 0$ and $J_r = 1$) because of

$$\rho_{k_a\kappa_a}(00) = \delta_{k_a0}\,\delta_{\kappa_a0},$$

$$(00k_p\kappa_p \mid k_rk_r) = \delta_{k_pk_r}\,\delta_{\kappa_p\kappa_r},$$

and

$$\begin{Bmatrix} 0 & 1 & 1 \\ 0 & 1 & 1 \\ 0 & k_p & k_r \end{Bmatrix} = \delta_{k_pk_r}/\sqrt{[(2k_r+1)(3)(3)]}$$

to

$$\rho_{k_r\kappa_r}(J_r = 1) = \rho_{k_r\kappa_r}(E1). \tag{8.100c}$$

Because of the relation

$$\mathbf{J}_r \to \mathbf{J}_I + \mathbf{j}_1 \tag{8.101a}$$

one gets

$$\rho_{k_I\kappa_I}(J_I, J_I)\,\rho_{k_1\kappa_1}(j_1, j_1') = \sum \rho_{k_r\kappa_r}(J_r = 1)\,(k_I\kappa_I k_1\kappa_1 \mid k_r\kappa_r)$$

$$\times \sqrt{[(3)(3)(2k_I+1)(2k_1+1)]}\begin{Bmatrix} J_I & j_1 & 1 \\ J_I & j_1' & 1 \\ k_I & k_1 & k_r \end{Bmatrix}. \tag{8.101b}$$

If $\rho_{k_r\kappa_r}(J_r = 1)$ is replaced by $\rho_{k\kappa}(E1)$ and the photoionization matrix elements D_γ from equ. (8.87d), abbreviated to $D_\gamma = D_{j_1}$, are introduced, one finally obtains

$$\rho_{k_I\kappa_I}(J_I, J_I)\,\rho_{k_I\kappa_I}(j_1, j_1') = \sum \rho_{k\kappa}(E1)\,(k_I\kappa_I k_1\kappa_1 \mid k\kappa)\,3\sqrt{[(2k_I+1)(2k_1+1)]}$$

$$\times \begin{Bmatrix} J_I & j_1 & 1 \\ J_I & j_1' & 1 \\ k_I & k_1 & k \end{Bmatrix} D_{j_1} D_{j_1'}^{*}. \tag{8.102}$$

In this expression the role of the summation over the angular momenta j_1 and j_1' of the photoelectron (in *jj*-coupling) is apparent; due to the different partial waves of the photoelectron, interference terms of the matrix elements with j_1 and j_1' occur. In contrast, the ionic state with sharp angular momentum

J_1 has only one value (generalizations to situations with more than one angular momentum are straightforward).

(2) Couplings for Auger electron emission: In this case one has to consider the decay of the intermediate photoionized state (J_1) to the final ionic state J_f by emission of the Auger electron (j_2) taking care also of the Coulomb matrix elements (operator Op_2):

$$\mathbf{J}_1 \to \mathbf{J}_f + \mathbf{j}_2. \tag{8.103a}$$

This leads to (note that J_f is sharp, but for j_2 several partial waves exist)

$$\rho_{k_f\kappa_f}(J_f J_f)\rho_{k_2\kappa_2}(j_2 j_2') = \sum \rho_{k_1\kappa_1}(J_1, J_1)\,(k_f\kappa_f k_2\kappa_2 \mid k_1\kappa_1) \times \sqrt{[(2J_1+1)(2J_1+1)(2k_f+1)(2k_2+1)]} \times \begin{Bmatrix} J_f & j_2 & J_1 \\ J_f & j_2' & J_1 \\ k_f & k_2 & k_1 \end{Bmatrix} C_{j_2} C^*_{j_2'}. \tag{8.103b}$$

8.4.3 Examples

The power of the method of Fano–Racah algebra lies in the fact that the transition rate P, or equivalently the differential cross section, can be worked out for any desired process as a parametrization into 'geometry' and 'dynamics'. (The method was originally applied in nuclear physics; for a first transfer to angular distributions in atomic physics see [KSa76]. It can be extended easily to more complicated cases like photoionization from a laser-polarized initial state, or fluorescence decay following photoionization, and so on.) Three simple examples will be presented which are related to $2p^{-1}\,{}^2P_J$ photoionization in magnesium: the photoprocess itself caused by unpolarized light leading to the ${}^2P_{3/2}$ final ionic state, the non-coincident observation of L_3–M_1M_1 Auger electrons induced by linearly polarized light, and, finally, the angular correlation between coincident photoelectrons and subsequent Auger electrons (arbitrary polarization of the incident light).

8.4.3.1 *Angular distribution of 2p-photoelectrons (unpolarized light)*

$2p^{-1}\,{}^2P_{3/2}$ photoionization in magnesium by unpolarized incident light is described by

$$h\nu + \text{Mg} \to \text{Mg}^+(2p^{-1}\,{}^2P_{3/2}) + e_1(\varepsilon s_{1/2}, \varepsilon d_{3/2}, \varepsilon d_{5/2}). \tag{8.104}$$

In order to obtain the angle-dependent emission of these photoelectrons recorded by a spin-insensitive point-like detector, one has to calculate the transition rate P, where

$$P \sim \sum \rho_{k_1\kappa_1}(J_1, J_1)\,\varepsilon^*_{k_1\kappa_1}(J_1, J_1)\,\rho_{k_1\kappa_1}(j_1, j_1')\,\varepsilon^*_{k_1\kappa_1}(j_1, j_1'). \tag{8.105a}$$

Since the ion is not observed, one has

$$\varepsilon_{k_1\kappa_1}(J_1 = 3/2) = 2\delta_{k_1 0}\,\delta_{\kappa_1 0}.$$

With this information equ. (8.102) leads to

$$\rho_{00}(J_1 = 3/2)\rho_{k_1\kappa_1}(j_1, j'_1) = \sum \rho_{k\kappa}(E1)\,(00k_1\kappa_1 \mid k\kappa)3\sqrt{(2k_1+1)} \times \begin{Bmatrix} \frac{3}{2} & j_1 & 1 \\ \frac{3}{2} & j'_1 & 1 \\ 0 & k_1 & k \end{Bmatrix} D_{j_1}D^*_{j'_1},$$

where the Clebsch–Gordan coefficient requires $k_1 = k$ and $\kappa_1 = \kappa$, and for unpolarized incident light one has (z-axis in photon beam direction)

$$\rho_{k\kappa}(E1) = \begin{cases} \rho_{00}(E1) = 1/\sqrt{3}, \\ \rho_{20}(E1) = 1/\sqrt{6}. \end{cases}$$

Since the electron detector is assumed to be insensitive to the electron spin, and the photoelectron has a well-defined parity (ℓ or ℓ' even), the radiation parameters of the photoelectron are given by ($\kappa_1 = 0$, and $k_1 = 0$ and $k_1 = 2$ only, note $k_1 = k$ in the present case)

$$c_{k0}(j_1, j'_1) = (-1)^{1/2-j_1}\sqrt{[(2j_1+1)(2j'_1+1)]}\,(j\tfrac{1}{2}j' - \tfrac{1}{2}|k0)/4\pi.$$

and for the efficiency tensor of the photoelectron one gets, using equs. (8.92a) and (8.94b)

$$\varepsilon_{k0}(j_1, j'_1) = c_{k0}(j_1, j'_1)D^k_{0,0}(\Phi, \Theta, \chi) = c_{k0}(j_1, j'_1)P_k(\cos\Theta).$$

Collecting all individual terms, one obtains

$$P \sim \sum \frac{6}{4\pi}\sqrt{(2k+1)}\,(-1)^{1/2-j_1}\rho_{k0}(E1)\sqrt{[(2j_1+1)(2j'+1)]}\,(j_1\tfrac{1}{2}j'_1 - \tfrac{1}{2}|k0) \times \begin{Bmatrix} \frac{3}{2} & j_1 & 1 \\ \frac{3}{2} & j'_1 & 1 \\ 0 & k & k \end{Bmatrix} D_{j_1}D^*_{j'_1}\,P_k(\cos\Theta). \quad (8.105b)$$

The terms with $k = 0$ yield

$$P(k=0) = \text{const}(|D_-|^2 + |D_0|^2 + |D_+|^2)/4\pi,$$

with the replacements $D_{1/2} \to D_-$, $D_{3/2} \to D_0$, and $D_{5/2} \to D_+$ for the dipole matrix elements (see equ. (8.35)), and the terms with $k = 2$ give

$$P(k=2) = \text{const}[-\tfrac{2}{5}|D_+|^2 + \tfrac{2}{5}|D_0|^2 - \tfrac{3}{10}(D_+D^*_0 + \text{cc}) + \tfrac{3}{10}\sqrt{5}(D_+D^*_- + \text{cc}) + \tfrac{1}{10}\sqrt{5}(D_-D^*_0 + \text{cc})]P_2(\cos\Theta)/4\pi. \quad (8.105c)$$

If the total result P is recast in the general form for the angular distribution of photoelectrons, one obtains

$$P \sim P(k=0)[1 + P(k=2)/P(k=0)] \sim \frac{\sigma}{4\pi}\left[1 - \frac{\beta}{2}P_2(\cos\Theta)\right] \quad (8.106a)$$

with

$$\sigma \sim (|D_-|^2 + |D_0|^2 + |D_+|^2), \quad (8.106b)$$

$$\beta = \frac{0.8|D_+|^2 - 0.8|D_0|^2 + 0.6(D_+D_0^* + \mathrm{cc}) - (3/\sqrt{5})(D_+D_-^* + \mathrm{cc}) - (1/\sqrt{5})(D_0D_-^* + \mathrm{cc})}{|D_-|^2 + |D_0|^2 + |D_+|^2}. \tag{8.106c}$$

This result is identical to equs. (8.39a) and (8.40a), respectively, except for an overall normalization factor in the cross section σ (which can be included in the matrix elements D_γ).

8.4.3.2 *Angular distribution of non-coincident* L_3–M_1M_1 *Auger electrons (linearly polarized light)*

Within the two-step model the process is characterized by

$$\begin{aligned} h\nu + \mathrm{Mg} \rightarrow\ & \mathrm{Mg}^+(2\mathrm{p}^{-1}\,{}^2\mathrm{P}_{3/2}) + \mathrm{e}_1(\varepsilon\mathrm{s}_{1/2}, \varepsilon\mathrm{d}_{3/2}, \varepsilon\mathrm{d}_{5/2}) \\ & \downarrow \\ & \mathrm{Mg}^{++}(3\mathrm{s}^{-2}) + \mathrm{e}_2(\varepsilon\mathrm{p}_{3/2}). \end{aligned} \tag{8.107}$$

The transition rate P then follows from the statistical and efficiency tensors of the three-particle system in the final state – ion (J_f), Auger electron (j_2), and photoelectron (j_1) – as

$$P \sim \sum \rho_{k_f\kappa_f}(J_f = 0)\varepsilon^*_{k_f\kappa_f}(J_f = 0)\,\rho_{k_2\kappa_2}(\varepsilon\mathrm{p}_{3/2})\varepsilon^*_{k_2\kappa_2}(\varepsilon\mathrm{p}_{3/2})\,\rho_{k_1\kappa_1}(j_1j_1')\varepsilon^*_{k_1\kappa_1}(j_1j_1'). \tag{8.108a}$$

In this expression the well-defined angular momenta are indicated in the respective tensors, and only for the photoelectron described by j_1 do different partial waves exist. Thc three efficiency tensors of the final system are easy to write down:

because the ion is not observed

$$\varepsilon_{k_f\kappa_f}(J_f = 0) = \delta_{k_f0}\,\delta_{\kappa_f0};$$

because the photoelectron is not observed

$$\varepsilon_{k_1\kappa_j}(j_1j_1') = \sqrt{(2j_1 + 1)}\,\delta_{k_10}\,\delta_{\kappa_10}\,\delta_{j_1j_1'}$$

(the non-observation of the photoelectron leads to an incoherent summation over the contributing partial waves);
because the Auger electron is observed with a spin-insensitive detector

$$\varepsilon_{k_2\kappa_2}(\varepsilon\mathrm{p}_{3/2}) = c_{k_20}(\varepsilon\mathrm{p}_{3/2})D^{k_2}_{\kappa_20}(\Phi, \Theta, \chi).$$

For the radiation parameters of the Auger electron without spin observation, one gets only the two possibilities $k_2 = 0$ and $k_2 = 2$ with

$$c_{k_20}(\varepsilon\mathrm{p}_{3/2}) = \frac{(-1)}{\pi}(\tfrac{3}{2}\tfrac{1}{2}\tfrac{3}{2} -\tfrac{1}{2}\,|\,k_20).$$

With these efficiency tensors and the corresponding statistical tensors the transition rate P then follows as

$$P \sim \sum \rho_{00}(J_f = 0)\rho_{00}(j_1j_1)\sqrt{(2j_1 + 1)}\rho_{k_2\kappa_2}(\varepsilon\mathrm{p}_{3/2})\,c_{k_20}(\varepsilon\mathrm{p}_{3/2})\,D^{2*}_{\kappa_20}(\Phi, \Theta, \chi). \tag{8.108b}$$

Replacement of $\rho_{00}(J_f = 0)$ and $\rho_{k_2\kappa_2}(j_2, j_2')$ according to equ. (8.103b) leads to

$$\rho_{00}(J_f = 0)\rho_{k_2\kappa_2}(j_2 = \tfrac{3}{2}) = \sum \rho_{k_I\kappa_I}(J_I = \tfrac{3}{2})(0\,0\,k_2\,\kappa_2\,|\,k_I\,\kappa_I)\sqrt{[(2J_I+1)(2J_I+1)(2k_2+1)]} \times \begin{Bmatrix} 0 & \frac{3}{2} & \frac{3}{2} \\ 0 & \frac{3}{2} & \frac{3}{2} \\ 0 & k_2 & k_I \end{Bmatrix} |C_{j_2}|^2,$$

where the Clebsch–Gordan coefficient requires $k_2 = k_I$ and $\kappa_2 = \kappa_I$.

Next, $\rho_{k_I\kappa_I}(J_I = \frac{3}{2})$ and $\rho_{00}(j_1, j_1)$ are replaced according to equ. (8.102) giving

$$\rho_{k_I\kappa_I}(J_I = \tfrac{3}{2})\rho_{00}(j_1, j_1) = \sum \rho_{k\kappa}(E1)(k_I\kappa_I 0\,0\,|\,k\kappa)3\sqrt{(2k_I+1)} \begin{Bmatrix} \frac{3}{2} & j_1 & 1 \\ \frac{3}{2} & j_1 & 1 \\ k_I & 0 & k \end{Bmatrix} |D_{j_1}|^2,$$

where the Clebsch–Gordan coefficient requires $k_I = k$ and $\kappa_I = \kappa$.

The collection of all terms then yields

$$P \sim \sum 12(2k+1)\sqrt{(2j_1+1)}\rho_{k\kappa}(E1)c_{k0}(\varepsilon p_{3/2})D^{k*}_{\kappa 0}(\Phi, \Theta, \chi) \times |D_{j_1}|^2 \begin{Bmatrix} 0 & \frac{3}{2} & \frac{3}{2} \\ 0 & \frac{3}{2} & \frac{3}{2} \\ 0 & k & k \end{Bmatrix} \begin{Bmatrix} \frac{3}{2} & j_1 & 1 \\ \frac{3}{2} & j_1 & 1 \\ k & 0 & k \end{Bmatrix} |C_{j_2}|^2. \quad (8.108c)$$

Using the tensor operators $\tilde{\rho}_{k\kappa}(E1)$ of equ. (8.99b) defined in the coordinate frame where the electric field vector coincides with the reference axis, one gets

$$P(k=0) = \text{const}\,\frac{1}{4\pi}\frac{6}{\sqrt{3}}|C_{j_2}|^2 \sum \sqrt{(2j_1+1)} \begin{Bmatrix} \frac{3}{2} & j_1 & 1 \\ \frac{3}{2} & j_1 & 1 \\ 0 & 0 & 0 \end{Bmatrix} |D_{j_1}|^2$$

$$= \text{const}\,\frac{1}{4\pi}|C_{j_2}|^2(|D_-|^2 + |D_0|^2 + |D_+|^2)$$

$$= \text{const}\,\frac{1}{4\pi}\omega_A\sigma_{ph}, \quad (8.109a)$$

where ω_A is the Auger yield and σ_{ph} the partial photoionization cross section.

Similarly, one obtains (in order to make the reference axis more explicit, the angles (Φ, Θ, χ) in the rotation matrix elements of equ. (8.108c) have been replaced by $(\varphi, \vartheta, \chi)$)

$$P(k=2) = \text{const}\,\frac{2}{4\pi}\sqrt{(30)}\,P_2(\cos\vartheta)\,|C_{j_2}|^2 \sum \sqrt{(2j_1+1)} \begin{Bmatrix} \frac{3}{2} & j_1 & 1 \\ \frac{3}{2} & j_1 & 1 \\ 2 & 0 & 2 \end{Bmatrix} |D_{j_1}|^2$$

$$= \text{const}\,\frac{2}{4\pi}P_2(\cos\vartheta)\,|C_{j_2}|^2\,(0.5|D_-|^2 - 0.4|D_0|^2 + 0.1|D_+|^2). \quad (8.109b)$$

With $P(0)$ and $P(2)$ one finally gets

$$P \sim P(k=0)[1 + P(k=2)/P(k=0)] \sim \frac{\sigma_{\mathrm{ph}}\omega_{\mathrm{A}}}{4\pi}[1 + \beta_{\mathrm{A}} P_2(\cos\vartheta)] \quad (8.110a)$$

with

$$\beta_{\mathrm{A}} = \frac{0.2|D_+|^2 - 0.8|D_0|^2 + |D_-|^2}{|D_+|^2 + |D_0|^2 + |D_-|^2}. \quad (8.110b)$$

This result is derived in jjJ-coupling. With the following replacements of the dipole matrix elements (see equs. (8.42b) and (5.11a)),

$$\left.\begin{aligned} |D_+|^2 &= \tfrac{36}{15} R^2_{\varepsilon\mathrm{d},2\mathrm{p}_{3/2}} = \tfrac{9}{15}|D_{\mathrm{d}}|^2, \\ |D_0|^2 &= \tfrac{4}{15} R^2_{\varepsilon\mathrm{d},2\mathrm{p}_{3/2}} = \tfrac{1}{15}|D_{\mathrm{d}}|^2, \\ |D_-|^2 &= \tfrac{20}{15} R^2_{\varepsilon\mathrm{s},2\mathrm{p}_{3/2}} = \tfrac{10}{15}|D_{\mathrm{s}}|^2, \end{aligned}\right\} \quad (8.111)$$

one obtains

$$\beta_{\mathrm{A}} = \frac{R^2_{\varepsilon\mathrm{s},2\mathrm{p}_{3/2}} + 0.2R^2_{\varepsilon\mathrm{d},2\mathrm{p}_{3/2}}}{R^2_{\varepsilon\mathrm{s},2\mathrm{p}_{3/2}} + R^2_{\varepsilon\mathrm{d},2\mathrm{p}_{3/2}}} \quad (8.110c)$$

and

$$\beta_{\mathrm{A}} = \frac{|D_{\mathrm{s}}|^2 + 0.1|D_{\mathrm{d}}|^2}{|D_{\mathrm{s}}|^2 + |D_{\mathrm{d}}|^2}, \quad (8.110d)$$

and these expressions are identical to equs. (8.128) and (5.13b) with (5.16a), respectively.

8.4.3.3 *Angular correlation of photoelectrons observed in coincidence with subsequent Auger electrons*

The simultaneous observation of the photoelectron and the subsequent Auger electron, as indicated in equ. (8.107) for the example of $2\mathrm{p}^{-1}\,{}^2\mathrm{P}_{3/2}$ photoionization in magnesium with an $\mathrm{L}_3\mathrm{-M}_1\mathrm{M}_1$ Auger transition, modifies the transition rate P given by equ. (8.108a) via the photoelectron detection efficiency tensor. In order to conform with the general formulation for closed-shell atoms given in [Kab92], the slightly more general case of arbitrary polarization of the incident light and the as yet unspecified angular momenta J_{i} and J_{f} will be considered here (J_{i} is kept at zero). Its application to the example of magnesium will be presented in Section 8.5.2.

The three efficiency tensors of the final system are easy to derive. Because the ion is not observed, one has

$$\varepsilon_{k_{\mathrm{f}}\kappa_{\mathrm{f}}}(J_{\mathrm{f}}) = \sqrt{(2J_{\mathrm{f}}+1)}\,\delta_{k_{\mathrm{f}}0}\,\delta_{\kappa_{\mathrm{f}}0}.$$

For the spin-insensitive and point-like detectors of the observed electrons one gets for the Auger electron

$$\varepsilon_{k_2\kappa_2}(j_2, j_2') = c_{k_20}(j_2, j_2') D^{k_2}_{\kappa_20}(\Phi_2, \Theta_2, \chi_2),$$

and for the photoelectron

$$\varepsilon_{k_1\kappa_1}(j_1, j'_1) = c_{k_1 0}(j_1, j'_1) D^{k_1}_{\kappa_1 0}(\Phi_1, \Theta_1, \chi_1).$$

In these expressions the indices k_2 and k_1 are restricted to even values (cf. equ. (8.97b)), because in the two-step model single photoionization and Auger decay are treated separately.

Taking into account the conditions $k_f = 0$ and $\kappa_f = 0$ from the efficiency tensor $\varepsilon_{k_f\kappa_f}(J_f)$, one can first couple $\rho_{00}(J_f)$ with $\rho_{k_2\kappa_2}(j_2, j'_2)$ to form $\rho_{k_I\kappa_I}(J_I)$. The attached Clebsch–Gordan coefficient brings in the condition $k_I = k_2$ and $\kappa_I = \kappa_2$, and allows the replacement of the coefficients k_I and κ_I which link the first and second steps via the intermediate state. (Only the coefficients k_1, κ_1 and k_2, κ_2 which refer to the first and second steps, respectively, are retained.)

Introducing the couplings of the statistical tensors towards the initial state also, one derives for the transition rate P:

$$\begin{aligned} P \sim \sum 12\pi\sqrt{[(2J_f+1)(2J_I+1)(2J_I+1)(2k_2+1)]}\; c_{k_2 0}(j_2, j'_2)\; Y_{k_2\kappa_2}(\Theta_2, \Phi_2) \\ \times \begin{Bmatrix} J_f & j_2 & J_I \\ J_f & j'_2 & J_I \\ 0 & k_2 & k_2 \end{Bmatrix} C_{j_2} C^*_{j'_2} \\ \times\; \rho_{k_E\kappa_E}(E1)\,(k_2\kappa_2 k_1\kappa_1 \,|\, k_E\kappa_E)\; c_{k_1 0}(j_1, j'_1)\; Y_{k_1\kappa_1}(\Theta_1, \Phi_1) \\ \times \begin{Bmatrix} J_I & j_1 & 1 \\ J_I & j'_1 & 1 \\ k_2 & k_1 & k_E \end{Bmatrix} D_{j_1} D^*_{j'_1}. \end{aligned} \tag{8.112}$$

This result can be recast into different forms. Using the spherical biharmonics defined in equ. (4.67) one gets ($k_1 \le 2\ell_{1\max}$ and even, $k_2 \le 2\ell_{2\max}$ and even)

$$P = \sum \sqrt{[4\pi/(2k_1+1)(2k_2+1)]}\,(-1)^k A(k_1, k_2, k)\, B^{k_1k_2}_{k\kappa}(\Theta_1, \Phi_1; \Theta_2\Phi_2)\, \rho_{k_E\kappa_E}(E1), \tag{8.113a}$$

where, as a consequence of the formulation within the two-step model, the coefficients $A(k_1, k_2, k)$ factorize into

$$A(k_1, k_2, k) = \alpha(k_2)\,\gamma(k_1, k_2, k) \tag{8.113b}$$

with

$$\alpha(k_2) = \sum \sqrt{[(2J_f+1)(2J_I+1)(2k_2+1)]}\; c_{k_2 0}(j_2, j'_2) \begin{Bmatrix} J_f & j_2 & J_I \\ J_f & j'_2 & J_I \\ 0 & k_2 & k_2 \end{Bmatrix} C_{j_2} C^*_{j'_2} \tag{8.113c}$$

and

$$\gamma(k_1, k_2, k_E) = \sum 3\sqrt{[(2J_I+1)(2k_1+1)(2k_2+1)]}\; c_{k_1 0}(j_1, j'_1) \begin{Bmatrix} J_I & j_1 & 1 \\ J_I & j'_1 & 1 \\ k_2 & k_1 & k_E \end{Bmatrix} D_{j_1} D^*_{j'_1}. \tag{8.113d}$$

(Note that the link between the steps goes over k_I, which, however, is identical to k_2.)

It is also possible to introduce the angle-dependent statistical tensor $\rho_{k\kappa}(J_1, J_\mathrm{I}; \Theta_1, \Phi_1)$ of the photoionized state which is defined for $k \leq 2\ell_{2\max}$, $k_1 \leq 2\ell_{1\max}$, with both k and k_1 even, and is given by [Kab92]

$$\rho_{k\kappa}(J_1, J_\mathrm{I}; \Theta_1, \Phi_1) = \sum \sqrt{\left(\frac{4\pi}{2k_1 + 1}\right)} (k\kappa\, k_1\kappa_1 \,|\, k_E\kappa_E) \times Y_{k_1\kappa_1}(\Theta_1, \Phi_1) \rho_{k_E\kappa_E}(E1) B(k_1, k, k_E) \tag{8.114a}$$

with

$$B(k_1, k, k_E) = 3\sqrt{[(2k_1 + 1)(2k + 1)]} \sum c_{k_1 0}(j_1, j_1') \begin{Bmatrix} J_\mathrm{I} & j_1 & 1 \\ J_\mathrm{I} & j_1' & 1 \\ k & k_1 & k_E \end{Bmatrix} D_{j_1} D^*_{j_1'}. \tag{8.114b}$$

(Note $k_2 = k_\mathrm{I} = k$ and $\kappa_2 = \kappa_\mathrm{I} = \kappa$, and furthermore that this expression ought to be multiplied by $\sqrt{(2J_\mathrm{I} + 1)}$, which, however, is not essential here.) Using these definitions, the angle-dependent transition rate P can be recast as

$$P \sim \sum \alpha(k)\, \rho_{k\kappa}(J_1, J_\mathrm{I}; \Theta_1, \Phi_1) \sqrt{\left(\frac{4\pi}{2k + 1}\right)}\, Y_{k\kappa}(\Theta_2, \Phi_2). \tag{8.115a}$$

Introducing normalized parameters, namely, the *Auger decay parameter* α_k and the *angle-dependent alignment tensor* $\mathcal{A}_{k\kappa}(J_1; \Theta_1, \Phi_1)$ of the photoionized intermediate hole-state J_I, both defined for $k =$ even and $k \leq 2\ell_{2\max}$ by

$$\alpha_k = \alpha(k)/\alpha(k = 0), \tag{8.115b}$$

$$\mathcal{A}_{k\kappa}(J_\mathrm{I}; \Theta_1, \Phi_1) = \rho_{k\kappa}(J_1, J_\mathrm{I}; \Theta_1, \Phi_1)/\rho_{00}(J_1, J_\mathrm{I}; \Theta_1, \Phi_1), \tag{8.115c}$$

one then obtains the final result [BKS78, Kab92]

$$P \sim \rho_{00}(J_1, J_\mathrm{I}; \Theta_1, \Phi_1)\, \alpha(0) \left[1 + \sum_{\substack{k>0 \\ \text{even}}}^{2\ell_{2\max}} \alpha_k \sum_{\kappa} \mathcal{A}_{k\kappa}(J_\mathrm{I}; \Theta_1, \Phi_1) \sqrt{\left(\frac{4\pi}{2k+1}\right)}\, Y_{k\kappa}(\Theta_2, \Phi_2) \right], \tag{8.116}$$

where $\rho_{00}(J_1, J_\mathrm{I}; \Theta_1, \Phi_1)$ describes the angle-dependent intensity for non-coincident observation of the photoelectron. Equ. (8.116) has an apparent similarity to the general formulation of non-coincident observation of the Auger electron [BKa77, BKR78]. If the photoelectron is not observed, one has to integrate over its directions giving

$$\int \rho_{k\kappa}(J_1, J_\mathrm{I}; \Theta_1 \Phi_1)\, d\Omega_1 = 4\pi \rho_{k_E\kappa_E}(E1) B(0, k, k)\, \delta_{kk_E}\, \delta_{\kappa\kappa_E}, \tag{8.117a}$$

and because $k_E \leq 2$ one gets for the non-coincident observation of Auger electrons

$$P \sim \text{const} \left[1 + \alpha_2 \sum_{\alpha} \mathcal{A}_{2\kappa}(J_\mathrm{I}) \sqrt{\left(\frac{4\pi}{5}\right)}\, Y_{2\kappa}(\Theta_2, \Phi_2) \right], \tag{8.117b}$$

where the angle-independent alignment tensor $\mathscr{A}_{2\kappa}(J_1)$ is given by

$$\mathscr{A}_{2\kappa}(J_1) = \frac{\rho_{2\kappa}(E1)B(0,2,2)}{\rho_{00}(E1)B(0,0,0)}, \tag{8.117c}$$

and the $\kappa = 0$ component of this quantity is identical to the alignment parameter $\mathscr{A}_{20}(J_1)$ introduced in equ. (3.30). The application of equ. (8.116) for $2p^{-1}\ {}^2P_{3/2}$ photoionization with subsequent L_3–M_1M_1 Auger decay in magnesium is worked out in Section 8.5.2.

8.5 Angular distribution and alignment of L_3–M_1M_1 Auger electrons in magnesium

The two Auger transitions L_2–M_1M_1 and L_3–M_1M_1 in atomic magnesium at kinetic energies of 35.15 eV and 34.87 eV provide an illustrative example for the discussion of the general properties related to the angular distribution of Auger electrons, whether as a measurement of the Auger electron alone, or in coincidence with the initially ejected photoelectron.

8.5.1 Non-coincident case

In the case of a non-coincident angular distribution of Auger electrons it is the angular distribution parameter β_A, introduced in equ. (3.29) as

$$\beta_A = (-2)\mathscr{A}_{20}\alpha_2, \tag{8.118}$$

which is of interest, together with its two constituents, the alignment parameter $\mathscr{A}_{20}$ of the photoionized state J and the Auger decay parameter α_2. Because the alignment parameter $\mathscr{A}_{20}(J)$ is zero for $J \le 1/2$, the L_2–M_1M_1 transition is isotropic,

$$\mathscr{A}_{20}(2p^5 3s^2\ {}^2P_{1/2}) = 0. \tag{8.119a}$$

In contrast, the neighbouring L_3–M_1M_1 transition has a non-vanishing β_A parameter because an alignment exists (and the Auger decay parameter α_2 has a non-vanishing value). Using equ. (3.30d) its alignment parameter is given by

$$\mathscr{A}_{20}(2p^5 3s^2\ {}^2P_{3/2}) = \frac{a(\frac{3}{2}\frac{3}{2}) - a(\frac{3}{2}\frac{1}{2}) - a(\frac{3}{2}\,-\frac{1}{2}) + a(\frac{3}{2}\,-\frac{3}{2})}{a(\frac{3}{2}\frac{3}{2}) + a(\frac{3}{2}\frac{1}{2}) + a(\frac{3}{2}\,-\frac{1}{2}) + a(\frac{3}{2}\,-\frac{3}{2})}, \tag{8.119b}$$

where the $a(JM_J)$ are population numbers for the intermediate hole-state. (Note that equ. (3.30d) holds for normalized population probabilities whilst the values $a(JM_J)$ given here are unnormalized population numbers.) For the Auger decay parameter α_2 defined in equ. (8.115b) one obtains

$$\alpha_2(L_3\text{–}M_1M_1) = -1, \tag{8.120}$$

which is a trivial and constant factor in the present case because only one partial wave, $\varepsilon p_{3/2}$, is allowed for this Auger transition (see [BKa77]).

The population numbers $a(JM_J)$ can be related directly to the photoionization process, because they are equal to the partial cross section which depends on the magnetic quantum number M_J (see equ. (2.3)):

$$a(JM_J) = 4\pi^2\alpha E_{\text{ph}}\kappa \sum_{m_s} \int |M_{\text{fi}}(M_J, m_s)|^2 \, d\Omega_\kappa. \tag{8.121}$$

The population numbers, and hence the alignment, refer to a preselected quantization axis. As a consequence, different values for the alignment are obtained for different polarizations of the incident light. This will be demonstrated for the example of 2p photoionization in magnesium leading to the $^2P_{3/2}$ ionic state. In the simplest approximation of the Cooper–Zare model (see Section 8.2.3), but still distinguishing the fine-structure components $^2P_{3/2}$ and $^2P_{1/2}$ of the photoionized state, one gets (where const $= 4\pi^2\alpha E_{\text{ph}}/3$):

(i) for linearly polarized light with the quantization axis along the electric field vector

$$\left.\begin{aligned} a(\tfrac{3}{2}\tfrac{3}{2})|_{\text{lin}} &= a(\tfrac{3}{2}\,-\tfrac{3}{2})|_{\text{lin}} = \text{const}\ 9R^2_{\varepsilon d,\,2p}/15, \\ a(\tfrac{3}{2}\tfrac{1}{2})|_{\text{lin}} &= a(\tfrac{3}{2}\,-\tfrac{1}{2})|_{\text{lin}} = \text{const}(10R^2_{\varepsilon s,\,2p} + 11R^2_{\varepsilon d,\,2p})/15, \end{aligned}\right\} \tag{8.122a}$$

and therefore

$$\mathcal{A}_{20}(^2P_{3/2})|_{\text{lin}} = -\frac{R^2_{\varepsilon s,\,2p} + 0.2R^2_{\varepsilon d,\,2p}}{R^2_{\varepsilon s,\,2p} + 2R^2_{\varepsilon d,\,2p}}; \tag{8.123a}$$

(ii) for unpolarized light with the quantization axis along the direction of the photon beam

$$\left.\begin{aligned} a(\tfrac{3}{2}\tfrac{3}{2})|_{\text{unpol}} &= a(\tfrac{3}{2}\,-\tfrac{3}{2})|_{\text{unpol}} = \text{const}(15R^2_{\varepsilon s,\,2p} + 21R^2_{\varepsilon d,\,2p})/30, \\ a(\tfrac{3}{2}\tfrac{1}{2})|_{\text{unpol}} &= a(\tfrac{3}{2}\,-\tfrac{1}{2})|_{\text{unpol}} = \text{const}(5R^2_{\varepsilon s,\,2p} + 19R^2_{\varepsilon d,\,2p})/30, \end{aligned}\right\} \tag{8.122b}$$

and therefore

$$\mathcal{A}_{20}(^2P_{3/2})|_{\text{unpol}} = +\frac{1}{2}\frac{R^2_{\varepsilon s,\,2p} + 0.2R^2_{\varepsilon d,\,2p}}{R^2_{\varepsilon s,\,2p} + 2R^2_{\varepsilon d,\,2p}}; \tag{8.123b}$$

(iii) for right circularly polarized light with the quantization axis along the direction of the photon beam

$$\left.\begin{aligned} a(\tfrac{3}{2}\tfrac{3}{2})|_{\text{rcp}} &= \text{const}\ 36R^2_{\varepsilon d,\,2p}/30, \\ a(\tfrac{3}{2}\tfrac{1}{2})|_{\text{rcp}} &= \text{const}\ 24R^2_{\varepsilon d,\,2p}/30, \\ a(\tfrac{3}{2}\,-\tfrac{1}{2})|_{\text{rcp}} &= \text{const}(10R^2_{\varepsilon s,\,2p} + 14R^2_{\varepsilon d,\,2p})/30, \\ a(\tfrac{3}{2}\,-\tfrac{3}{2})|_{\text{rcp}} &= \text{const}(30R^2_{\varepsilon s,\,2p} + 6R^2_{\varepsilon d,\,2p})/30, \end{aligned}\right\} \tag{8.122c}$$

where the population numbers differ for different signs of the M_J value. Therefore, the intermediate state not only possesses alignment as expressed by $\mathcal{A}_{20}$, but also orientation described by $\mathcal{O}_{10}$. The orientation plays a role if circularly polarized light is used and the spin of the emitted Auger electron is also analysed. Using equ. (3.31d) one gets in the present case

$$\mathcal{O}_{10}(2p^5 3s^2\ ^2P_{3/2}) = \frac{\sqrt{5}}{2}\frac{R^2_{\varepsilon d,\,2p} - R^2_{\varepsilon s,\,2p}}{R^2_{\varepsilon s,\,2p} + 2R^2_{\varepsilon d,\,2p}}. \tag{8.124}$$

For the alignment, which is of interest in the present context, one obtains

$$\mathscr{A}_{20}(^2\mathrm{P}_{3/2})|_{\mathrm{rcp}} = +\frac{1}{2}\frac{R^2_{\varepsilon\mathrm{s},\,2\mathrm{p}} + 0.2R^2_{\varepsilon\mathrm{d},\,2\mathrm{p}}}{R^2_{\varepsilon\mathrm{s},\,2\mathrm{p}} + 2R^2_{\varepsilon\mathrm{d},\,2\mathrm{p}}}. \tag{8.123c}$$

Obviously, different expressions are obtained for the alignment parameters, because these quantities are defined with respect to different quantization axes. However, it is possible to express the alignment properties in a unified manner, e.g., $\mathscr{A}_{20}|_{\mathrm{lin}}$ in the coordinate frame with the quantization axis along the photon beam direction. Since the alignment tensor $\mathscr{A}_{2\kappa}$ is defined in connection with statistical tensors, equ. (8.115c), one can use the rotation properties in equ. (8.82) to change the reference axis for the representation from the z-axis to the x-axis:

$$\mathscr{A}_{2\kappa}(J)|_{z\text{-axis}} = \sum D^2_{\kappa 0}(0, \Theta = 90°, 0)\, \mathscr{A}_{20}(J)|_{x\text{-axis}}. \tag{8.125a}$$

This representation gives the following components of the alignment tensor (the index 'z-axis' will be omitted from now on, and quantities referred to the x-axis will be characterized by a tilde)

$$\left.\begin{aligned} \mathscr{A}_{22}(J) &= \mathscr{A}_{2-2}(J) = \sqrt{(3/8)}\tilde{\mathscr{A}}_{20}, \\ \mathscr{A}_{21}(J) &= \mathscr{A}_{2-1} = 0, \\ \mathscr{A}_{20}(J) &= -0.5\tilde{\mathscr{A}}_{20}. \end{aligned}\right\} \tag{8.125b}$$

The symmetry properties of this alignment tensor $\mathscr{A}_{2\kappa}(J)$ express the reflection symmetry with respect to the xz-plane.

From the general form of non-coincident observation of angle-dependent Auger electron emission, equ. (8.117b),

$$W(\Theta, \Phi) = \frac{\sigma\omega_{\mathrm{A}}}{4\pi}\left[1 + \alpha_2 \sum_{\kappa} \mathscr{A}_{2\kappa}\sqrt{\left(\frac{4\pi}{5}\right)}\, Y_{2\kappa}(\Theta, \Phi)\right], \tag{8.126a}$$

one gets in the present case

$$\begin{aligned} W(\Theta, \Phi) = \frac{\sigma\omega_{\mathrm{A}}}{4\pi}\Bigg[1 &+ \alpha_2\, \mathscr{A}_{22}\sqrt{\left(\frac{4\pi}{5}\right)}\, Y_{22}(\Theta, \Phi) + \alpha_2\, \mathscr{A}_{20}\sqrt{\left(\frac{4\pi}{5}\right)}\, Y_{20}(\Theta, \Phi) \\ &+ \alpha_2\, \mathscr{A}_{2-2}\sqrt{\left(\frac{4\pi}{5}\right)}\, Y_{2-2}(\Theta, \Phi)\Bigg], \end{aligned} \tag{8.126b}$$

which gives, using equs. (7.4) and (8.125b),

$$W(\Theta, \Phi) = \frac{\sigma\omega_{\mathrm{A}}}{4\pi}\{1 - \tfrac{1}{2}\alpha_2\, \tilde{\mathscr{A}}_{20}\, [P_2(\cos\Theta) - \tfrac{3}{2}\sin^2\Theta\cos 2\Phi]\}. \tag{8.126c}$$

This result agrees with the general formulation for completely linearly polarized light (equ. (3.28) with $\tilde{S}_1 = 1$) if the angular distribution parameter β_{A} is defined as

$$\beta_{\mathrm{A}} = \alpha_2\, \tilde{\mathscr{A}}_{20}. \tag{8.127a}$$

The last relation differs from the one introduced in equ. (3.29) in which the z-axis along the photon beam is used as the reference axis, i.e.,

$$\beta_A = (-2)\alpha_2 \mathscr{A}_{20}, \tag{8.127b}$$

but in both cases the β_A value of L_3–M_1M_1 Auger electrons following $2p_{3/2}$ photoionization is described by

$$\beta_A = \frac{R^2_{\varepsilon s,\,2p} + 0.2R^2_{\varepsilon d,\,2p}}{R^2_{\varepsilon s,\,2p} + 2R^2_{\varepsilon d,\,2p}}. \tag{8.128}$$

This discussion underlines the importance of specifying the chosen reference axis when referring to the alignment parameter. However, in an actual experiment with arbitrary polarization it is the β_A-parameter which is unique, as follows from equ. (3.28):

$$W(\Theta, \Phi) = \frac{\sigma\omega_A}{4\pi}\{1 - \tfrac{1}{2}\beta_A[P_2(\cos\Theta) - \tfrac{3}{2}\tilde{S}_1 \sin^2\Theta \cos 2\tilde{\Phi}]\}. \tag{8.126d}$$

From the β_A value, one can then derive and discuss the alignment parameter for a conveniently selected type of polarization, using either equ. (8.127a) or equ. (8.127b).

8.5.2 *Auger electrons measured in coincidence with the photoelectron*

Finally, the expression for the angular distribution of L_3–M_1M_1 Auger electrons measured in coincidence with the initially ejected photoelectron in the $2p^{-1}\,{}^2P_{3/2}$ ionization process will be worked out for linearly polarized light (quantization axis along the electric field vector; quantities for which this axis is important will be characterized by a tilde). The starting point is rather similar to the non-coincident angular distribution described by equ. (8.126), but now the full alignment tensor and its dependence on the polar and azimuthal angles (ϑ_1, φ_1) of the detected photoelectron play a role. Using equ. (8.116), and taking into account the restrictions in the summation indices (k = even and $k \le 2\ell_{2\max} = 2$ in the present example with the $\varepsilon_A p_{3/2}$ partial wave of the Auger electron), one has

$$W(\vartheta_1, \varphi_1; \vartheta_2, \varphi_2) = \tilde{\rho}_{00}(J, J; \vartheta_1, \varphi_1)$$

$$\times\left[1 + \alpha_2 \sum_\kappa \tilde{\mathscr{A}}_{2\kappa}(J; \vartheta_1, \varphi_1)\sqrt{\left(\frac{4\pi}{5}\right)}\, Y_{2\kappa}(\vartheta_2, \varphi_2)\right], \tag{8.129a}$$

with

$$\tilde{\rho}_{00}(J, J; \vartheta_1, \varphi_1) = \frac{\sigma_{ph}}{4\pi}[1 + \beta_{ph} P_2(\cos\vartheta_1)], \tag{8.129b}$$

$$\alpha_2 = -1 \text{ (magnesium)}, \tag{8.129c}$$

and

$$\tilde{\mathscr{A}}_{2\kappa}(J, J; \vartheta_1, \varphi_1) = \tilde{\rho}_{2\kappa}(J, J; \vartheta_1, \varphi_1)/\tilde{\rho}_{00}(J, J; \vartheta_1, \varphi_1). \tag{8.129d}$$

For the statistical tensors $\tilde{\rho}_{k\kappa}(J, J; \vartheta_1, \varphi_1)$ one derives

$$\tilde{\rho}_{00}(J = 3/2; \vartheta_1, \varphi_1) = \frac{1}{\sqrt{3}} B(0, 0, 0) - \frac{2}{\sqrt{6}} B(2, 0, 2) P_2(\cos\vartheta_1), \tag{8.130a}$$

$$\begin{aligned}\tilde{\rho}_{20}(J = 3/2; \vartheta_1, \varphi_1) = {}& \frac{1}{\sqrt{15}} B(2, 2, 0) P_2(\cos\vartheta_1) - \frac{2}{\sqrt{6}} B(0, 2, 2)\\ &+ \frac{2}{\sqrt{(21)}} B(2, 2, 2) P_2(\cos\vartheta_1) - \frac{2}{\sqrt{(21)}} B(4, 2, 2) P_4(\cos\vartheta_1),\end{aligned} \tag{8.130b}$$

$$\begin{aligned}\tilde{\rho}_{21}(J = 3/2; \vartheta_1, \varphi_1) = {}& -\frac{1}{\sqrt{(15)}}\sqrt{\left(\frac{4\pi}{5}\right)}\, Y_{2-1}(\vartheta_1, \varphi_1) B(2, 2, 0)\\ &- \frac{1}{\sqrt{(21)}}\sqrt{\left(\frac{4\pi}{5}\right)}\, Y_{2-1}(\vartheta_1, \varphi_1) B(2, 2, 2)\\ &+ \sqrt{\left(\frac{2}{3}\right)}\sqrt{\left(\frac{5}{21}\right)}\sqrt{\left(\frac{4\pi}{9}\right)}\, Y_{4-1}(\vartheta_1, \varphi_1) B(4, 2, 2),\end{aligned}$$

$$\tilde{\rho}_{2-1}(J = 3/2; \vartheta_1, \varphi_1) = -\tilde{\rho}_{21}(J = 3/2; \vartheta_1\varphi_1), \tag{8.130c}$$

$$\begin{aligned}\tilde{\rho}_{22}(J = 3/2; \vartheta_1, \varphi_1) = {}& \frac{1}{\sqrt{(15)}}\sqrt{\left(\frac{4\pi}{5}\right)}\, Y_{2-2}(\vartheta_1, \varphi_1) B(2, 2, 0)\\ &- \frac{2}{\sqrt{(21)}}\sqrt{\left(\frac{4\pi}{5}\right)}\, Y_{2-2}(\vartheta_1, \varphi_1) B(2, 2, 2)\\ &- \frac{1}{3}\sqrt{\left(\frac{5}{7}\right)}\sqrt{\left(\frac{4\pi}{9}\right)}\, Y_{4-2}(\vartheta_1, \varphi_1) B(4, 2, 2),\end{aligned}$$

$$\tilde{\rho}_{2-2}(J = 3/2; \vartheta_1, \varphi_1) = +\tilde{\rho}_{22}(J = 3/2; \vartheta_1\varphi_1). \tag{8.130d}$$

The dynamical parameters $B(k_1, k_2, k_3)$ are fixed by the photoionization matrix elements, e.g., in the *LSJ*-coupling approximation by the radial dipole integrals $R_{\varepsilon s, 2p}$ and $R_{\varepsilon d, 2p}$ and the relative phase Δ, as follows:

$$B(0, 0, 0) = \frac{1}{4\pi}\frac{2}{\sqrt{3}}(R^2_{\varepsilon s, 2p} + 2R^2_{\varepsilon d, 2p}), \tag{8.131a}$$

$$B(2, 2, 0) = \frac{1}{4\pi}\left[-2\sqrt{\left(\frac{5}{3}\right)}\right](R^2_{\varepsilon d, 2p} - 2R_{\varepsilon s, 2p}R_{\varepsilon d, 2p}\cos\Delta), \tag{8.131b}$$

$$B(2, 0, 2) = \frac{1}{4\pi}\left[-2\sqrt{\left(\frac{2}{3}\right)}\right](R^2_{\varepsilon d, 2p} - 2R_{\varepsilon s, 2p}R_{\varepsilon d, 2p}\cos\Delta), \tag{8.131c}$$

$$B(0, 2, 2) = \frac{1}{4\pi}\frac{2}{\sqrt{6}}(R^2_{\varepsilon s, 2p} + \tfrac{1}{5}R^2_{\varepsilon d, 2p}), \tag{8.131d}$$

$$B(2,2,2) = \frac{1}{4\pi}\frac{2}{\sqrt{(21)}}(R^2_{\varepsilon d,2p} + 7R_{\varepsilon s,2p}R_{\varepsilon d,2p}\cos\Delta), \tag{8.131e}$$

$$B(4,2,2) = \frac{1}{4\pi}\frac{108}{5\sqrt{(21)}}R^2_{\varepsilon d,2p}. \tag{8.131f}$$

With the help of these relations, the alignment tensor $\tilde{\mathcal{A}}_{2q}(J = 3/2; \hat{\kappa}_a)$ was calculated and discussed in detail in Section 4.6.1.3, using numerical values for the radial dipole integrals $R_{\varepsilon s,2p}$ and $R_{\varepsilon d,2p}$ and the relative phase Δ.

Alternatively, these B-coefficients are also connected to the A-coefficients used in the general parametrization formula, equ. (4.86a). In the present case one has

$$A_{000} = \frac{1}{\sqrt{3}}B(0,0,0), \tag{8.132a}$$

$$A_{202} = -\frac{8\pi}{\sqrt{(30)}}B(2,0,2), \tag{8.132b}$$

$$A_{022} = \frac{8\pi}{\sqrt{(30)}}B(0,2,2), \tag{8.132c}$$

$$A_{220} = -\frac{1}{\sqrt{(15)}}B(2,2,0), \tag{8.132d}$$

$$A_{222} = \frac{4\pi}{5}\sqrt{\left(\frac{2}{3}\right)}B(2,2,2), \tag{8.132e}$$

$$A_{422} = \frac{8\pi}{3\sqrt{(30)}}B(4,2,2). \tag{8.132f}$$

The correct normalization from the photoionization process gives a factor $(4\pi)^2\alpha E_{ph}/3$ and that from the Auger decay a factor $\omega_A/4\pi$, which finally leads to

$$A_{000} = \frac{\sigma_{ph}(2p_{3/2})}{4\pi}\frac{\omega_A(L_3\text{–}M_1M_1)}{4\pi}, \tag{8.133a}$$

$$A_{202} = \beta_{ph}(2p_{3/2})\frac{4\pi}{\sqrt{5}}A_{000}, \tag{8.133b}$$

$$A_{022} = \beta_A(L_3\text{–}M_1M_1)\frac{4\pi}{\sqrt{5}}A_{000}, \tag{8.133c}$$

$$A_{220} = \tfrac{1}{2}\beta_{ph}(2p_{3/2})A_{000}. \tag{8.133d}$$

$$A_{222} = \frac{\pi}{5}\sqrt{\left(\frac{2}{7}\right)}[18r - 7\beta_{ph}(2p_{3/2})]A_{000}, \tag{8.133e}$$

$$A_{422} = \frac{144\pi}{5\sqrt{(70)}}rA_{000}, \tag{8.133f}$$

where the ratio r is defined in the LSJ-coupling approximation by

$$r = \frac{R^2_{\varepsilon d, 2p}}{R^2_{\varepsilon s, 2p} + 2R^2_{\varepsilon d, 2p}} \tag{8.133g}$$

and can be recast, using equ. (8.128), as

$$r = [1 - \beta_A(L_3\text{-}M_1M_1)]/1.80. \tag{8.133g$'$}$$

The dynamical parameters $A_{k_1k_2k}$, together with the corresponding angular factors, then define the angular distribution pattern represented in equ. (4.86b).

9
Polarization properties

9.1 Angular distributions for different light polarizations

9.1.1 Individual forms for specific polarizations

The angular distribution of photoelectrons (and also of non-coincident Auger electrons) depends on the polarization of the incident light. For completely linearly polarized light where the electric field vector along $\hat{\mathbf{x}}$ also defines the quantization axis against which the polar angle ϑ of the ejected electron is measured, one gets (see equ. (2.13b))

$$\left.\frac{\mathrm{d}\sigma}{\mathrm{d}\Omega}\right|_{\text{lin pol }\hat{\mathbf{x}}} = \frac{\sigma}{4\pi}[1 + \beta P_2(\cos\vartheta)]. \tag{9.1}$$

It is possible to transform this expression into a form which contains the polar and azimuthal angles Θ and $\tilde{\Phi}$ as defined in the tilted coordinate frame of Fig. 1.15. Applying spherical trigonometry, one then has the following relation between these angles

$$\cos\vartheta = \cos(90° - \Theta)\cos\tilde{\Phi} = \sin\Theta\cos\tilde{\Phi}, \tag{9.2a}$$

i.e.,

$$\cos 2\tilde{\Phi} = 2\cos^2\tilde{\Phi} - 1 = 2\frac{\cos^2\vartheta}{\sin^2\Theta} - 1, \tag{9.2b}$$

and it follows that

$$\begin{aligned} P_2(\cos\Theta) - \tfrac{3}{2}\cos 2\tilde{\Phi}\sin^2\Theta &= -3\cos^2\vartheta + 1 \\ &= -2\,P_2(\cos\vartheta). \end{aligned} \tag{9.2c}$$

Therefore, the partial cross section can be written

$$\left.\frac{\mathrm{d}\sigma}{\mathrm{d}\Omega}\right|_{\text{lin pol }\hat{\mathbf{x}}} = \frac{\sigma}{4\pi}\left\{1 - \frac{\beta}{2}[P_2(\cos\Theta) - \tfrac{3}{2}\cos 2\tilde{\Phi}\sin^2\Theta]\right\}. \tag{9.3}$$

Similarly, one obtains for completely linearly polarized light with the electric field vector along the y-direction (note $\cos\vartheta' = \sin\Theta\sin\tilde{\Phi}$)

$$\begin{aligned} \left.\frac{\mathrm{d}\sigma}{\mathrm{d}\Omega}\right|_{\text{lin pol }\hat{\mathbf{y}}} &= \frac{\sigma}{4\pi}[1 + \beta P_2(\cos\vartheta')] \\ &= \frac{\sigma}{4\pi}\left\{1 - \frac{\beta}{2}[P_2(\cos\Theta) + \tfrac{3}{2}\cos 2\tilde{\Phi}\sin^2\Theta]\right\}. \end{aligned} \tag{9.4}$$

The angular distribution pattern for completely right- or left-circularly polarized light is usually calculated using the direction of the incident light as the quantization axis against which the polar angle Θ of the ejected electron is measured. For both right- and left-circular polarization one obtains the result:

$$\left.\frac{d\sigma}{d\Omega}\right|_{\text{circ pol}} = \frac{\sigma}{4\pi}\left[1 - \frac{\beta}{2} P_2(\cos\Theta)\right]. \tag{9.5}$$

For the calculation of the angular distribution of unpolarized light one can use either the form for linearly or the form for circularly polarized light, because unpolarized light can be represented as the incoherent sum of two orthogonal basis states of the light weighted with the factor 0.5. One then gets

$$\left.\frac{d\sigma}{d\Omega}\right|_{\text{unpol}} = \frac{\sigma}{4\pi}\left[1 - \frac{\beta}{2} P_2(\cos\Theta)\right], \tag{9.6}$$

which has the same form as the equation for circularly polarized light.

9.1.2 Common form for arbitrary polarization

Using the framework of the discussion of monochromatized synchrotron radiation described by the Stokes vector **S** (see equ. (1.44b)), the incident light which causes a photoprocess with an ensemble of atoms can be described in the tilted coordinate frame of Fig. 1.15 by the following incoherent contributions: completely polarized wavelets with intensities $I(\text{lin } x)$ and $I(\text{lin } y)$, responsible for the Stokes parameter $\tilde{S}_1$; completely polarized wavelets with intensities $I(\text{right circ})$ and $I(\text{left circ})$, responsible for the Stokes parameter S_3; and completely unpolarized wavelets with intensity $I(\text{unpol})$. Combining these intensities with their corresponding angular distributions, the following angular distribution is derived:

$$\begin{aligned}\frac{d\sigma}{d\Omega}(\Theta, \tilde{\Phi}) = \frac{\sigma}{4\pi}\Bigg(& I(\text{unpol})\left[1 - \frac{\beta}{2} P_2(\cos\Theta)\right] + I(\text{right circ})\left[1 - \frac{\beta}{2} P_2(\cos\Theta)\right] \\ & + I(\text{left circ})\left[1 - \frac{\beta}{2} P_2(\cos\Theta)\right] \\ & + I(\text{lin } x)\left\{1 - \frac{\beta}{2}\left[P_2(\cos\Theta) - \tfrac{3}{2}\cos 2\tilde{\Phi}\sin^2\Theta\right]\right\} \\ & + I(\text{lin } y)\left\{1 - \frac{\beta}{2}\left[P_2(\cos\Theta) + \tfrac{3}{2}\cos 2\tilde{\Phi}\sin^2\Theta\right]\right\}\Bigg),\end{aligned} \tag{9.7a}$$

which is equivalent to

$$\begin{aligned}\frac{d\sigma}{d\Omega}(\Theta, \tilde{\Phi}) = & \frac{\sigma}{4\pi}\left[I(\text{unpol}) + I(\text{right circ}) + I(\text{left circ}) + I(\text{lin } x) + I(\text{lin } y)\right] \\ & \times\left[1 - \frac{\beta}{2} P_2(\cos\Theta)\right] + \frac{\sigma}{4\pi}\left[I(\text{lin } x) - \text{I}(\text{lin } y)\right]\frac{\beta}{2}\frac{3}{2}\cos 2\tilde{\Phi}\sin^2\Theta\end{aligned} \tag{9.7b}$$

and

$$\frac{d\sigma}{d\Omega}(\Theta, \tilde{\Phi}) = \frac{\sigma}{4\pi} I_0 \left\{1 - \frac{\beta}{2}[P_2(\cos\Theta) - \tilde{S}_1 \tfrac{3}{2}\cos 2\tilde{\Phi}\sin^2\Theta]\right\} \tag{9.7c}$$

if the total light intensity I_0,

$$I_0 = I(\text{unpol}) + I(\text{right circ}) + I(\text{left circ}) + I(\text{lin } x) + I(\text{lin } y), \tag{9.7d}$$

and the Stokes parameter $\tilde{S}_1$,

$$\tilde{S}_1 = [I(\text{lin } x) - I(\text{lin } y)]/I_0, \tag{9.7e}$$

are introduced. The form of equ. (9.7c) is used in equ. (1.53). This equation indicates that the angular distribution of photoelectrons (and also of non-coincident Auger electrons) depends only on the Stokes parameter $\tilde{S}_1$ which describes the degree of linear polarization of the incident light, see Fig. 1.15.

It should be pointed out that the expression for $d\sigma/d\Omega$ given in equ. (9.7c) can also be derived if one does not consider the specifically selected kinds of polarization, but stays within one polarization basis only. Selecting the linear basis (see Section 9.2.2), one then has to add in equal portions the contributions from unpolarized and circularly polarized light to $I(\text{lin } x)$ and $I(\text{lin } y)$. This increases the values for the linearly polarized components to $\tilde{I}_x$ and $\tilde{I}_y$ given by

$$\tilde{I}_x = \tfrac{1}{2}I(\text{unpol}) + \tfrac{1}{2}I(\text{right circ}) + \tfrac{1}{2}I(\text{left circ}) + I(\text{lin } x), \tag{9.8a}$$

$$\tilde{I}_y = \tfrac{1}{2}I(\text{unpol}) + \tfrac{1}{2}I(\text{right circ}) + \tfrac{1}{2}I(\text{left circ}) + I(\text{lin } y). \tag{9.8b}$$

With these quantities one obtains

$$\tilde{I}_x = \frac{1 + \tilde{S}_1}{2} I_0 \quad \text{and} \quad \tilde{I}_y = \frac{1 - \tilde{S}_1}{2} I_0, \tag{9.9a}$$

and it follows that

$$\begin{aligned}\frac{d\sigma}{d\Omega}(\Theta, \tilde{\Phi}) = \tilde{I}_x &\left\{1 - \frac{\beta}{2}[P_2(\cos\Theta) - \tfrac{3}{2}\cos 2\tilde{\Phi}\sin^2\Theta]\right\} \\ &+ \tilde{I}_y\left\{1 - \frac{\beta}{2}[P_2(\cos\Theta) + \tfrac{3}{2}\cos 2\tilde{\Phi}\sin^2\Theta]\right\}.\end{aligned} \tag{9.9b}$$

This result shows that the angular distribution $d\sigma/d\Omega$ of non-coincident electrons can be described as an incoherent sum of two contributions, one for completely linearly polarized light with the electric field vector oscillating along the major axis of the polarization ellipse and one oscillating along the minor axis, and the weight for each contribution follows from equ. (9.9a). In these weights it is the value of the Stokes parameter $\tilde{S}_1$ which is important, and for $\tilde{S}_1$ approaching $+1$ or -1, respectively, one is brought back to equs. (9.3) and (9.4).

9.2 Description of electron and photon polarization

Starting from the vector model of a precessing spin, the polarization properties of an electron beam with electrons of spin $(1/2)\hbar$ are explained and then formulated by introducing the Stokes parameters. *A priori*, photon polarization is related to the photon spin by $1\hbar$, but due to the transverse nature of light, the photon polarization obeys the same rules of a two-state system as the electron spin. Hence, the polarization of a photon beam is discussed along the same lines.

9.2.1 Polarization of an electron beam

The Stokes parameters for the polarization of an electron beam can be represented in a Cartesian basis which also provides a convenient pictorial view for the polarization state of an electron beam. Since the polarization of an ensemble of electrons requires the determination of spin projections along preselected directions, the classical vector model of a precessing spin will first be discussed. Here the spin is represented by a vector **s** of length $\sqrt{3}/2$ (in atomic units) which precesses around a preselected direction, yielding as expectation values the projections (in atomic units, see Fig. 9.1)

$$\langle s_z \rangle = m_z \tag{9.10a}$$

with

$$m_z = +1/2 \quad \text{or} \quad m_z = -1/2. \tag{9.10b}$$

This means, the spin projection along the preselected axis (the z-axis) can be

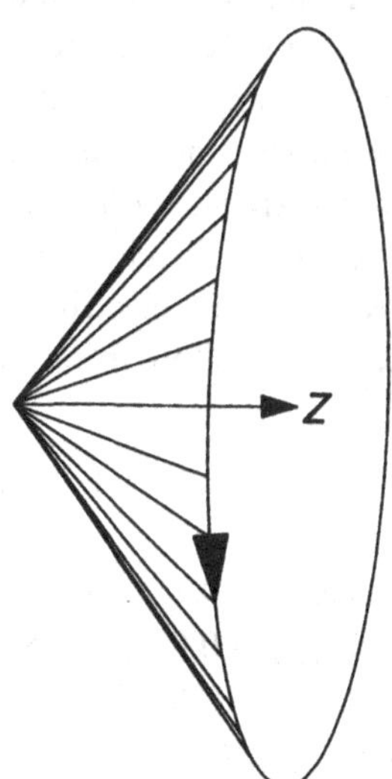

Figure 9.1 Vector model of the electron spin. Using atomic units, the spin is represented by a vector **s** of length $\sqrt{3}/2$ which precesses around the z-axis. By looking at the respective projections of the precessing spin vector, the model provides two important properties: the projection onto the z-axis leads to a sharp value, $m_s = 1/2$ in the case shown ($m_s = -1/2$ for a precession around the negative z-axis), but no sharp values exist for the projections in the xy-plane, i.e., for the projections onto the x- or y-axis one finds with equal probability either $+1/2$ or $-1/2$.

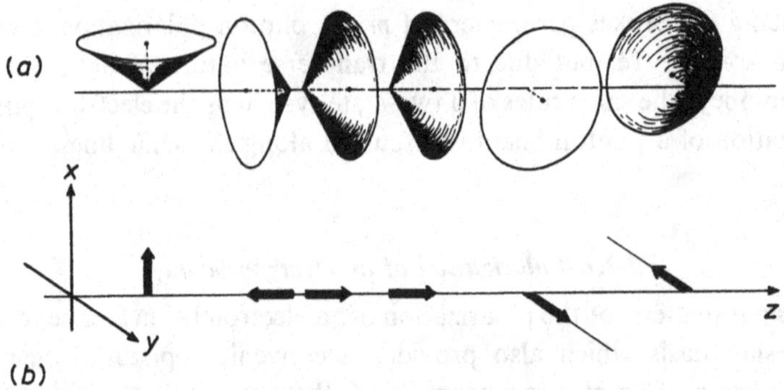

Figure 9.2 Illustration of an electron beam of six electrons with different spin polarization components. It is assumed that the electrons travel along the z-axis and the spin vectors shown are constant in a statistical average, i.e., they are repeated when the beam passes by. (a) Vector model of precessing spins; (b) short notation showing only the corresponding spin projections.

measured unambiguously. For the other components this is not the case. Because

$$|\mathbf{s}|^2 = s_x^2 + s_y^2 + m_z^2 = \tfrac{3}{4} \quad \text{and} \quad m_z^2 = \tfrac{1}{4}. \tag{9.10c}$$

one can conclude only that their average values can be $+1/2$ or $-1/2$:

$$\langle s_x \rangle = \langle s_y \rangle = \pm\tfrac{1}{2}. \tag{9.10d}$$

i.e., one obtains a 50% probability of finding the spin projection along the $+\hat{\mathbf{x}}$ or $-\hat{\mathbf{x}}$ direction (and similarly for finding it along the $+\hat{\mathbf{y}}$ or $-\hat{\mathbf{y}}$ direction).

If the individual electrons of a beam have different spin directions which remain constant in a statistical average, one can describe the polarization of this beam by the result of three successive measurements of the relative excess P_i of spin projections into three preselected Cartesian direction $i = \hat{\mathbf{x}}$, $\hat{\mathbf{y}}$, and $\hat{\mathbf{z}}$. This is demonstrated for an electron beam of six electrons with precessing spins as indicated in Fig. 9.2. The intensities N_i for electrons with $(\pm)$ spin projections along the selected coordinate axes then follow from individual measurements for each of the six precessing spins:

$$\left.\begin{aligned}
N_{+\hat{z}} &= 0.5|_{+\hat{x}} + 0|_{-\hat{z}} + 1|_{+\hat{z}} + 1|_{+\hat{z}} + 0.5|_{+\hat{y}} + 0.5|_{-\hat{y}} = 3.5,\\
N_{-\hat{z}} &= 0.5|_{+\hat{x}} + 1|_{-\hat{z}} + 0|_{+\hat{z}} + 0|_{+\hat{z}} + 0.5|_{+\hat{y}} + 0.5|_{-\hat{y}} = 2.5,\\
N_{+\hat{x}} &= 1|_{+\hat{x}} + 0.5|_{-\hat{z}} + 0.5|_{+\hat{z}} + 0.5|_{+\hat{z}} + 0.5|_{+\hat{y}} + 0.5|_{-\hat{y}} = 3.5,\\
N_{-\hat{x}} &= 0|_{+\hat{x}} + 0.5|_{-\hat{z}} + 0.5|_{+\hat{z}} + 0.5|_{+\hat{z}} + 0.5|_{+\hat{y}} + 0.5|_{-\hat{y}} = 2.5,\\
N_{+\hat{y}} &= 0.5|_{+\hat{x}} + 0.5|_{-\hat{z}} + 0.5|_{+\hat{z}} + 0.5|_{+\hat{z}} + 1|_{+\hat{y}} + 0|_{-\hat{y}} = 3.0,\\
N_{-\hat{y}} &= 0.5|_{+\hat{x}} + 0.5|_{-\hat{z}} + 0.5|_{+\hat{z}} + 0.5|_{+\hat{z}} + 0|_{+\hat{y}} + 1|_{-\hat{y}} = 3.0,
\end{aligned}\right\} \tag{9.11}$$

where the subscripts on the vertical bars indicate the spin direction of the selected electron, and the factor in front of the bar gives the corresponding value measured for this electron. From the numbers N_i, the polarization vector $\mathbf{P}$ can be calculated. It has the components

$$P_x = \frac{N_{+\hat{x}} - N_{-\hat{x}}}{N_{+\hat{x}} + N_{-\hat{x}}} = \frac{1}{6}, \tag{9.12a}$$

$$P_y = \frac{N_{+\hat{y}} - N_{-\hat{y}}}{N_{+\hat{y}} + N_{-\hat{y}}} = 0, \tag{9.12b}$$

$$P_z = \frac{N_{+\hat{z}} - N_{-\hat{z}}}{N_{+\hat{z}} + N_{-\hat{z}}} = \frac{1}{6}. \tag{9.12c}$$

The result reflects what one would have expected naively from looking at Fig. 9.2, in particular from Fig. 9.2(*b*): the spins in the $+y$- and $-y$-directions compensate, hence, there is no net effect for P_y: similarly, one spin along $+\hat{\mathbf{z}}$ compensates for one spin along $-\hat{\mathbf{z}}$. As a result one has effectively one spin component along $+\hat{\mathbf{z}}$, out of six possibilities, i.e., $P_z = +1/6$; and one component along $+\hat{\mathbf{x}}$ out of six possibilities, i.e., $P_x = +1/6$. The beam is *partially* polarized.

In quantum mechanics, spin is described by an operator which acts on a spin wavefunction of the electron. In the present case this operator describes an angular momentum with two possible eigenvalues along a reference axis. The first requirement fixes commutation rules for the spin components, and the second one leads to a representation of the spin operator by 2×2 matrices (*Pauli matrices* [Pau27]). One has

$$\mathbf{s} = \tfrac{1}{2}\boldsymbol{\sigma} = \tfrac{1}{2}(\sigma_x, \sigma_y, \sigma_z) \tag{9.13a}$$

$$\sigma_x\sigma_y = -\sigma_y\sigma_x, \quad \sigma_y\sigma_z = -\sigma_z\sigma_y, \quad \sigma_z\sigma_x = -\sigma_x\sigma_z. \tag{9.13b}$$

$$\sigma_x = \begin{pmatrix} 0 & 1 \\ 1 & 0 \end{pmatrix}, \sigma_y = \begin{pmatrix} 0 & -\mathrm{i} \\ \mathrm{i} & 0 \end{pmatrix}, \sigma_z = \begin{pmatrix} 1 & 0 \\ 0 & -1 \end{pmatrix}, \tag{9.13c}$$

with

$$s_x^2 = s_y^2 = s_z^2 = (\tfrac{1}{2})^2 \begin{pmatrix} 1 & 0 \\ 0 & 1 \end{pmatrix} \quad \text{and} \quad |\mathbf{s}|^2 = \tfrac{3}{4}\begin{pmatrix} 1 & 0 \\ 0 & 1 \end{pmatrix}. \tag{9.13d}$$

The 'wavefunctions' representing the eigenstates of the electron's spin are then two-component vectors (*Pauli spinors*)

$$\chi_{1/2}^{+} = \begin{pmatrix} 1 \\ 0 \end{pmatrix} = \text{'up' which means } m_z = +1/2 \tag{9.14a}$$

and

$$\chi_{1/2}^{-} = \begin{pmatrix} 0 \\ 1 \end{pmatrix} = \text{'down' which means } m_z = -1/2, \tag{9.14b}$$

and the wavefunction

$$\chi = \begin{pmatrix} a_1 \\ a_2 \end{pmatrix} \quad \text{with } |a_1|^2 + |a_2|^2 = 1 \tag{9.14c}$$

represents a state which has a probability $|a_1|^2$ of finding the spin 'up' and a probability $|a_2|^2$ of finding the spin 'down'.

With these properties of the electron's spin, the polarization vector $\mathbf{P}$ for an electron beam can be defined taking into account the contributions from the individual electrons j (for simplicity the subscript j is not always shown in the expressions given):

$$P_x = \sum_j \langle\chi(j)|\sigma_x(j)|\chi(j)\rangle / \sum_j \langle\chi(j)|\chi(j)\rangle$$

$$= \sum (a_1^*, a_2^*)\begin{pmatrix}0 & 1\\ 1 & 0\end{pmatrix}\begin{pmatrix}a_1\\ a_2\end{pmatrix}\Big/\sum (|a_1|^2 + |a_2|^2)$$

$$= \sum (a_1 a_2^* + a_1^* a_2)/\sum (|a_1|^2 + |a_2|^2), \tag{9.15a}$$

$$P_y = \sum_j \langle\chi(j)|\sigma_y(j)|\chi(j)\rangle / \sum_j \langle\chi(j)|\chi(j)\rangle$$

$$= \sum (a_1^*, a_2^*)\begin{pmatrix}0 & -\mathrm{i}\\ \mathrm{i} & 0\end{pmatrix}\begin{pmatrix}a_1\\ a_2\end{pmatrix}\Big/\sum (|a_1|^2 + |a_2|^2)$$

$$= \sum \mathrm{i}(a_1 a_2^* - a_1^* a_2)/\sum (|a_1|^2 + |a_2|^2), \tag{9.15b}$$

$$P_z = \sum_j \langle\chi(j)|\sigma_z(j)|\chi(j)\rangle / \sum_j \langle\chi(j)|\chi(j)\rangle$$

$$= \sum (a_1^*, a_2^*)\begin{pmatrix}1 & 0\\ 0 & -1\end{pmatrix}\begin{pmatrix}a_1\\ a_2\end{pmatrix}\Big/\sum (|a_1|^2 + |a_2|^2)$$

$$= \sum (|a_1|^2 - |a_2|^2)/\sum (|a_1|^2 + |a_2|^2)$$

$$= \frac{N_{+\hat{\mathbf{z}}} - N_{-\hat{\mathbf{z}}}}{N_{+\hat{\mathbf{z}}} + N_{-\hat{\mathbf{z}}}}. \tag{9.15c}$$

Of these components, only the last one is related directly to an observable, in the present case to the relative excess of particles with spin component $+1/2$ along $\hat{\mathbf{z}}$ as compared to those with spin component $-1/2$ along $\hat{\mathbf{z}}$. The other components, P_x and P_y, cannot be interpreted as easily at this stage. However, each measurement of the spin components along the $\hat{\mathbf{x}}$- or $\hat{\mathbf{y}}$-axis means a transformation of the quantization axis into this direction. Such measurements can be described using the general property that the spin state χ specified in the representation of equ. (9.14c) is connected to a representation at an arbitrary direction $\hat{\mathbf{e}}$

$$\chi = a_{+e}\begin{pmatrix}1\\ 0\end{pmatrix}_{+\hat{\mathbf{e}}} + a_{-e}\begin{pmatrix}0\\ 1\end{pmatrix}_{-\hat{\mathbf{e}}} \tag{9.16a}$$

by the rotation matrix given in equ. (7.17). One then gets the coefficients

$$a_{+e} = \mathrm{e}^{\mathrm{i}\Phi/2}\cos\left(\frac{\Theta}{2}\right)a_1 + \mathrm{e}^{-\mathrm{i}\Phi/2}\sin\left(\frac{\Theta}{2}\right)a_2, \tag{9.16b}$$

$$a_{-e} = -\mathrm{e}^{\mathrm{i}\Phi/2}\sin\left(\frac{\Theta}{2}\right)a_1 + \mathrm{e}^{-\mathrm{i}\Phi/2}\cos\left(\frac{\Theta}{2}\right)a_2, \tag{9.16c}$$

where a_1 and a_2 refer to the amplitudes with spin projections $+1/2$ and $-1/2$ along the original z-axis. From these relations one can now calculate the special cases of $\mathbf{e}$ pointing in the x- and y-directions ($\hat{\mathbf{e}} = \hat{\mathbf{x}}$ means $\Theta = 90°$, $\Phi = 0°$; $\hat{\mathbf{e}} = \hat{\mathbf{y}}$ means $\Theta = 90°$ and $\Phi = 90°$), giving

$$a_1 a_2^* + a_1^* a_2 = |a_{+\hat{\mathbf{x}}}|^2 - |a_{-\hat{\mathbf{x}}}|^2 \tag{9.17a}$$

and

$$i(a_1 a_2^* - a_1^* a_2) = |a_{+\hat{y}}|^2 - |a_{-\hat{y}}|^2. \qquad (9.17b)$$

These expressions finally allow an interpretation of the other two components of the spin polarization vector in terms of relative excess of particles with spin projections in the respective directions:

$$P_x = \sum (|a_{x1}|^2 - |a_{x2}|^2) / \sum (|a_{x1}|^2 + |a_{x2}|^2)$$
$$= \frac{N_{+\hat{x}} - N_{-\hat{x}}}{N_{+\hat{x}} + N_{-\hat{x}}}, \qquad (9.18a)$$
$$P_y = \sum (|a_{y1}|^2 - |a_{y2}|^2) / \sum (|a_{y1}|^2 + |a_{y2}|^2)$$
$$= \frac{N_{+\hat{y}} - N_{-\hat{y}}}{N_{+\hat{y}} + N_{-\hat{y}}}. \qquad (9.18b)$$

These quantum mechanical results are identical to the ones derived before based on the vector model (equs. (9.12)), which were used to interpret Fig. 9.2.

9.2.2 Polarization of a photon beam

Photons have the spin value 1 (in atomic units). Therefore one can expect their polarization properties to differ from those of electrons. However, due to the transverse nature of light, only two projections of the photon spin along the propagation direction of the light are allowed, i.e., the polarization of light is also a two-state system, and this property makes the polarization properties of photons and electrons rather similar. In contrast to the spin polarization of electrons which is represented conveniently in a Cartesian coordinate system, for photons the circular basis is more appropriate, in particular for circularly polarized photons related to the *helicity* states (states with given projection of the photon spin along the propagation direction, see below). In order to derive these different representations and their relations, the fact that any vector **b** can be expressed in different basis systems is employed:

(i) In the Cartesian basis:

$$\mathbf{b} = \sum_i \alpha_i \hat{\mathbf{e}}_i = \alpha_x \hat{\mathbf{e}}_x + \alpha_y \hat{\mathbf{e}}_y + \alpha_z \hat{\mathbf{e}}_z \qquad (9.19a)$$

with the unit vectors $\hat{\mathbf{e}}_x$, $\hat{\mathbf{e}}_y$ and $\hat{\mathbf{e}}_z$ in the directions $\hat{\mathbf{x}}$, $\hat{\mathbf{y}}$ and $\hat{\mathbf{z}}$. There are different conventions for defining the polarization of light, and two are used here: the *optical* definition (see Section 1.3) and the definition based on *helicity* states (see below). The latter will be characterized by the superscript h, and equ. (9.19a) will be then written as

$$\mathbf{b} = \sum \alpha_i^{\mathrm{h}} \hat{\mathbf{e}}_i = \alpha_x^{\mathrm{h}} \hat{\mathbf{e}}_x + \alpha_y^{\mathrm{h}} \hat{\mathbf{e}}_y + \alpha_z^{\mathrm{h}} \hat{\mathbf{e}}_z. \qquad (9.19b)$$

(ii) In the spherical basis ($\mu = -1, 0, +1$):

$$\mathbf{b} = \sum_\mu \alpha^\mu \hat{\mathbf{e}}_\mu = \alpha^+ \hat{\mathbf{e}}_+ + \alpha^- \hat{\mathbf{e}}_- + \alpha^0 \hat{\mathbf{e}}_0 \qquad (9.20a)$$

with the covariant spherical basis vectors

$$\hat{\mathbf{e}}_+ = -\frac{1}{\sqrt{2}}(\hat{\mathbf{e}}_x + \mathrm{i}\hat{\mathbf{e}}_y), \quad \hat{\mathbf{e}}_- = +\frac{1}{\sqrt{2}}(\hat{\mathbf{e}}_x - \mathrm{i}\hat{\mathbf{e}}_y), \quad \hat{\mathbf{e}}_0 = \hat{\mathbf{e}}_z \tag{9.20b}$$

and the contravariant components

$$\alpha^+ = -\frac{1}{\sqrt{2}}(\alpha_x^{\mathrm{h}} - \mathrm{i}\alpha_y^{\mathrm{h}}), \quad \alpha^- = +\frac{1}{\sqrt{2}}(\alpha_x^{\mathrm{h}} + \mathrm{i}\alpha_y^{\mathrm{h}}), \quad \alpha^0 = \alpha_z. \tag{9.20c}$$

The different basis systems will be illustrated for two examples, linear polarization along the x-direction and right circular polarization. According to equ. (8.4) the electric field vector $\mathscr{E}$ is represented as

$$\mathscr{E}(\mathbf{r}, t) = \tfrac{1}{2}\mathscr{E}_0(\mathbf{P}\,\mathrm{e}^{\mathrm{i}(\mathbf{k}\cdot\mathbf{r}-\omega t)} + \mathrm{cc}) \tag{9.21}$$

with $\mathbf{P}$ describing the polarization. For

$$\mathbf{P} = \hat{\mathbf{e}}_x \tag{9.22a}$$

one derives linear polarized light along $\hat{\mathbf{x}}$:

$$\begin{aligned}\mathscr{E}_{\text{lin pol}\,\hat{\mathbf{x}}} &= \tfrac{1}{2}\mathscr{E}_0(\hat{\mathbf{e}}_x\,\mathrm{e}^{\mathrm{i}(\mathbf{k}\cdot\mathbf{r}-\omega t)} + \hat{\mathbf{e}}_x\,\mathrm{e}^{-\mathrm{i}(\mathbf{k}\cdot\mathbf{r}-\omega t)}) \\ &= \mathscr{E}_0 \cos(k\cdot r - \omega t)\hat{\mathbf{e}}_x,\end{aligned} \tag{9.22b}$$

because the electric field vector oscillates with angular frequency ω along the x-axis. For

$$\mathbf{P} = \hat{\mathbf{e}}_{\mathrm{r}} = \hat{\mathbf{e}}_- \tag{9.23a}$$

one obtains

$$\begin{aligned}\mathscr{E}_{\mathrm{rcp}}(\mathbf{r}, t) &= \tfrac{1}{2}\mathscr{E}_0(\hat{\mathbf{e}}_-\,\mathrm{e}^{\mathrm{i}(\mathbf{k}\cdot\mathbf{r}-\omega t)} + \hat{\mathbf{e}}_-^*\,\mathrm{e}^{-\mathrm{i}(\mathbf{k}\cdot\mathbf{r}-\omega t)}) \\ &= \frac{1}{\sqrt{2}}\mathscr{E}_0(\cos(\mathbf{k}\cdot\mathbf{r} - \omega t)\hat{\mathbf{e}}_x + \sin(\mathbf{k}\cdot\mathbf{r} - \omega t)\hat{\mathbf{e}}_y),\end{aligned} \tag{9.23b}$$

i.e., in this case the electric field vector has a constant magnitude, but for an observer facing the incident light the field vector rotates clockwise around the propagation direction of the light. (This can easily be seen by setting $\mathbf{k}\cdot\mathbf{r}$ to zero and following the expression for increasing phase ωt.) This kind of polarization is called in optics *right*-circularly polarized light (base vector $\hat{\mathbf{e}}_{\mathrm{r}}$, component α_{r}). In quantum mechanics, the photon is said to be in a state of *negative helicity* because the photon has a negative projection of its spin onto its propagation direction (base vector $\hat{\mathbf{e}}_-$, component α^-). Similarly, *positive* helicity then corresponds to *left*-circularly polarized light. Using the representation of the polarization ellipse given in equ. (1.40), it is the angle γ which changes sign between the representations. If one keeps the definition that $\gamma > 0$ will describe right-elliptical polarization, but also positive helicity, one has to adapt the α_x^{h} and α_y^{h} components correspondingly, giving (λ is the tilt angle in Fig. 1.15)

$$\alpha_x^{\mathrm{h}} = \cos\gamma\cos\lambda - \mathrm{i}\sin\gamma\sin\lambda, \tag{9.24a}$$

$$\alpha_y^{\mathrm{h}} = \cos\gamma\sin\lambda + \mathrm{i}\sin\gamma\cos\lambda. \tag{9.24b}$$

In the discussion of light polarization so far the Cartesian basis and spherical basis have been considered. Because the linear polarization might be tilted with respect to the $(\hat{\mathbf{e}}_x, \hat{\mathbf{e}}_y)$-basis, a third basis system has to be introduced against which such a tilted polarization state can be measured via its non-vanishing components. This coordinate system is called $(\hat{\mathbf{e}}_1, \hat{\mathbf{e}}_2)$ and its axes are rotated by $+45°$ with respect to the previous ones. This leads to a third representation of the arbitrary vector $\mathbf{b}$:

(iii) In the $(\hat{\mathbf{e}}_1, \hat{\mathbf{e}}_2)$ basis:

$$\mathbf{b} = \alpha_1\hat{\mathbf{e}}_1 + \alpha_0\hat{\mathbf{e}}_0 + \alpha_2\hat{\mathbf{e}}_2 \tag{9.25a}$$

with the unit vectors

$$\hat{\mathbf{e}}_1 = +\frac{1}{\sqrt{2}}(\hat{\mathbf{e}}_x + \hat{\mathbf{e}}_y), \quad \hat{\mathbf{e}}_2 = -\frac{1}{\sqrt{2}}(\hat{\mathbf{e}}_x - \hat{\mathbf{e}}_y), \quad \hat{\mathbf{e}}_0 = \hat{\mathbf{e}}_z \tag{9.25b}$$

and the components

$$\alpha_1^{\mathrm{h}} = +\frac{1}{\sqrt{2}}(\alpha_x^{\mathrm{h}} + \alpha_y^{\mathrm{h}}), \quad \alpha_2^{\mathrm{h}} = -\frac{1}{\sqrt{2}}(\alpha_x^{\mathrm{h}} - \alpha_y^{\mathrm{h}}), \quad \alpha_0 = \alpha_z. \tag{9.25c}$$

An arbitrary polarization $\mathbf{P}$ of a photon beam can then be represented in either basis system: in the helicity formulation by

$$\mathbf{P} = \alpha_x^{\mathrm{h}}\hat{\mathbf{e}}_x + \alpha_y^{\mathrm{h}}\hat{\mathbf{e}}_y = \alpha_1^{\mathrm{h}}\hat{\mathbf{e}}_1 + \alpha_2^{\mathrm{h}}\hat{\mathbf{e}}_2 = \alpha^+\hat{\mathbf{e}}_+ + \alpha^-\hat{\mathbf{e}}_-; \tag{9.26a}$$

in the optical formulation by

$$\mathbf{P} = \alpha_x\hat{\mathbf{e}}_x + \alpha_y\hat{\mathbf{e}}_y = \alpha_1\hat{\mathbf{e}}_1 + \alpha_2\hat{\mathbf{e}}_2 = \alpha_\ell\hat{\mathbf{e}}_\ell + \alpha_r\hat{\mathbf{e}}_r; \tag{9.26b}$$

and the polarization properties outlined above for an electron beam can be transferred. The only difference is in the basis sytems: Cartesian basis for the electron beam (Pauli matrices $\sigma_x, \sigma_y, \sigma_z$, and Stokes parameters P_x, P_y, P_z); special basis for the photon beam (Pauli matrices $\sigma_1, \sigma_2, \sigma_3$, and Stokes parameters S_1, S_2, S_3). Within the circular basis, a photon state in the helicity system is described by

$$\chi_{\mathrm{photon}} = \begin{pmatrix}\alpha^+\\ \alpha^-\end{pmatrix} \quad \text{with } |\alpha^+|^2 + |\alpha^-|^2 = 1, \tag{9.27}$$

and the polarization properties of an ensemble follow from the expectation values with the Pauli matrices

$$\sigma_1 = \begin{pmatrix}0 & 1\\ 1 & 0\end{pmatrix}, \sigma_2 = \begin{pmatrix}0 & -\mathrm{i}\\ \mathrm{i} & 0\end{pmatrix}, \sigma_3 = \begin{pmatrix}1 & 0\\ 0 & -1\end{pmatrix}. \tag{9.28}$$

One then derives for the Stokes parameters of the photon beam with individual wavelets (j) in the helicity basis (superscript h):

$$\begin{aligned} S_3^{\mathrm{h}} &= \sum_j \langle\chi_{\mathrm{photon}}(j)|\sigma_3(j)|\chi_{\mathrm{photon}}(j)\rangle / \sum_j \langle\chi_{\mathrm{photon}}(j)\,|\,\chi_{\mathrm{photon}}(j)\rangle \\ &= \sum_j [|\alpha^+(j)|^2 - |\alpha^-(j)|^2] / \sum_j [|\alpha^+(j)|^2 + |\alpha^-(j)|^2] \\ &= \frac{N_{\hat{\mathbf{e}}^+} - N_{\hat{\mathbf{e}}^-}}{N_{\hat{\mathbf{e}}^+} + N_{\hat{\mathbf{e}}^-}} = \left\{\begin{matrix}\text{relative excess of photon intensities with}\\ \text{positive and negative helicities,}\end{matrix}\right\} \end{aligned} \tag{9.29a}$$

$$S_2^{\rm h} = \sum_j \langle \chi_{\rm photon}(j)|\sigma_2(j)|\chi_{\rm photon}(j)|/\sum_j \langle \chi_{\rm photon}(j)\,|\,\chi_{\rm photon}(j)\rangle$$
$$= \sum_j {\rm i}[\alpha^{-*}({\rm j})\alpha^{+}({\rm j}) - \alpha^{-}(j)\alpha^{+*}(j)]/\sum_j [|\alpha^{+}(j)|^2 + |\alpha^{-}(j)|^2]$$
$$= \sum_j [|\alpha_2^{\rm h}(j)|^2 - |\alpha_1^{\rm h}(j)^2]/\sum_j [|\alpha_2^{\rm h}(j)|^2 + |\alpha_1^{\rm h}(j)|^2]$$
$$= \frac{N_{\hat{\mathbf{e}}_2} - N_{\hat{\mathbf{e}}_1}}{N_{\hat{\mathbf{e}}_2} + N_{\hat{\mathbf{e}}_1}} = \left\{\begin{array}{l}\text{relative excess of photon intensities with}\\ \text{linear polarization along } \hat{\mathbf{e}}_2 \text{ and linear}\\ \text{polarization along } \hat{\mathbf{e}}_1,\end{array}\right\} \tag{9.29b}$$

$$S_1^{\rm h} = \sum_j \langle \chi_{\rm photon}(j)|\sigma_1(j)|\chi_{\rm photon}(j)\rangle/\sum_j \langle \chi_{\rm photon}(j)\,|\,\chi_{\rm photon}(j)\rangle$$
$$= \sum_j [\alpha^{-}(j)\alpha^{+*}(j) + \alpha^{-*}(j)\alpha^{+}(j)]/\sum_j [|\alpha^{+}(j)|^2 + |\alpha^{-}(j)|^2]$$
$$= \sum_j [|\alpha_y^{\rm h}(j)|^2 - |\alpha_x^{\rm h}(j)|^2]/\sum_j [|\alpha_y^{\rm h}(j)|^2 + |\alpha_x^{\rm h}(j)|^2]$$
$$= \frac{N_{\hat{\mathbf{e}}_y} - N_{\hat{\mathbf{e}}_x}}{N_{\hat{\mathbf{e}}_y} + N_{\hat{\mathbf{e}}_x}} = \left\{\begin{array}{l}\text{relative excess of photon intensities with}\\ \text{linear polarization along } \hat{\mathbf{e}}_y \text{ and linear}\\ \text{polarization along } \hat{\mathbf{e}}_x.\end{array}\right\} \tag{9.29c}$$

The polarization properties are often discussed for different definitions of the Stokes parameters. For the optical convention used in Parts A and B, one has

$$S_3 = -S_3^{\rm h}, \quad S_2 = -S_2^{\rm h} \quad \text{and} \quad S_1 = -S_1^{\rm h}. \tag{9.30}$$

10
Special instrumental aspects

10.1 Time-of-flight analysis of electrons

A time-of-flight (TOF) analyser measures the time t required for a particle to travel a fixed distance d. If applied to electron spectrometry, non-relativistic electrons with kinetic energy E_{kin} have a velocity v

$$v = 0.5931\sqrt{(E_{kin}/\text{eV})}\ \text{mm/ns} \tag{10.1}$$

and travel a distance d in a field-free region† in the time t

$$t = d/v. \tag{10.2}$$

Therefore, electrons emerging from atoms at the same time, but with different kinetic energies can be distinguished through their corresponding flight times t. These times can be measured using a time-to-digital converter which determines the time difference between its two inputs (START, STOP; see Fig. 4.47), provided that the creation time of the continuum electrons is known and that it is a short time interval. The latter condition is fulfilled when working at a large electron storage ring that operates in the single-bunch mode. In this case the creation time of electrons is set by each short light flash which passes the interaction region. In addition, the circulation time T_{circ} of the electron bunch in the storage ring is then large enough to complete a TOF analysis of the emitted electrons, before the next flash occurs.

An example of the general lay-out for electron spectrometry using TOF analysis of the electrons is shown in Fig. 10.1. After monochromatization ('mono' in the figure) the light flashes from the synchrotron radiation in the single-bunch mode enter the apparatus where they interact with the sample. These light flashes provide the necessary reference signal for the TOF analysis which can be derived either from the high-frequency cavity of the electron storage ring or from the measured current of a photodiode. Whether electrons created in the reaction volume enter the TOF analyser depends on their direction of emission with respect to the analyser. In Fig. 10.1 an angle of 54.7° with respect to the electric field vector of

† Such a TOF analysis with only one field-free region is in contrast to the TOF analysers frequently used for ion spectrometry. There the ions travel through different regions, supplied by zero and/or non-vanishing constant and/or time-dependent electrical fields, in order to achieve good spatial and energy focusing (see [WMc55, Ste74]). For electrons, retardation is also applied in order to expand the transit time of fast electrons and in this way improve the resolution (for the spectrometer in [BSK89a] see, for example, [Lan92]). For efficient high-resolution spectrometry of electrons with (practically) zero kinetic energy the technique of applying a pulsed electrical field across the region of the source volume, synchronized with the light flash of the incident photon beam, is also used (*ZEKE-spectroscopy* [MSS84]; see also Section 5.7.2).

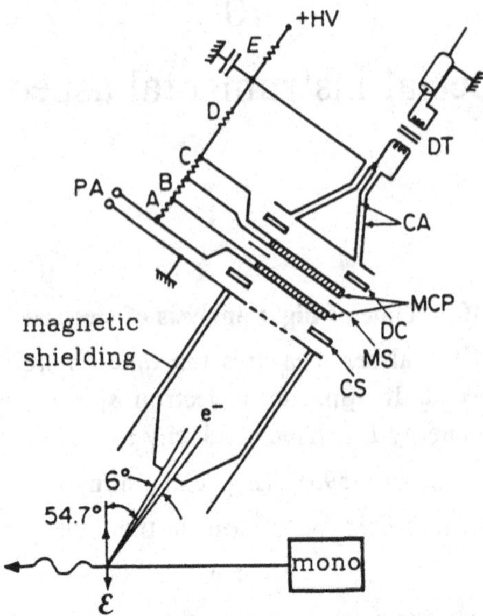

Figure 10.1 Schematic set-up (not to scale) of a TOF electron energy analyser. The monochromatized photon beam coming from the monochromator (mono) is indicated at the target by a wavy line with an electric field vector $\mathscr{E}$. Electrons ejected at a mean angle of 54.7° with respect to this field vector can enter the TOF analyser and travel towards the channelplate detector. Before they hit the first channelplate the electrons are postaccelerated (PA) by 100 V. A–E indicate a set of voltage dividers connected to the high-voltage source (HV). The detector is a Chevron mounting of two micro-channelplates (MCPs) with a ceramic spacer (CS), and a Mylar spacer (MS). Using a decoupling capacitor (DC), a conical coaxial anode (CA) and a decoupling transformer (DT), an electronic signal is produced and fed into the electronics held at ground potential. From [WRG79]; see also [BSK89a].

the nearly completely linearly polarized light was chosen. The solid angle accepted by the analyser was defined by an entrance hole of 2.4 mm diameter at a distance of 23 mm from the source volume (this corresponds to a cone angle of $\pm 3°$). The selected electrons travel in a field-free region a distance $d = 285$ mm. After passing a mesh which ends this field-free region, the electrons are postaccelerated (100 V, PA in the figure) and detected with a set of two microchannelplates (MCP) of 40 mm diameter. These channelplates amplify the incident electron, and the resulting charge cloud strikes a conical anode (CA) of 50 Ω impedance and is finally fed as a voltage pulse to the time-measuring device (time-to-amplitude converter with a multichannel analyser or time-to-digital converter with a histogramming memory and a computer). The lower rate of recorded electrons is used as a START signal, the higher rate of delayed light pulses is used as the STOP signal. (The time-measuring device has a dead time if there is no or one STOP signal, and dead-time losses can be reduced if unnecessary STARTs are avoided.) The resulting TOF spectrum then appears as a function of $T_{\text{circ}} - t$. Typical flight times in the spectrometer shown in Fig. 10.1 are 481 ns for 1 eV electrons, 96 ns for 25 eV electrons and 48 ns for 100 eV electrons.

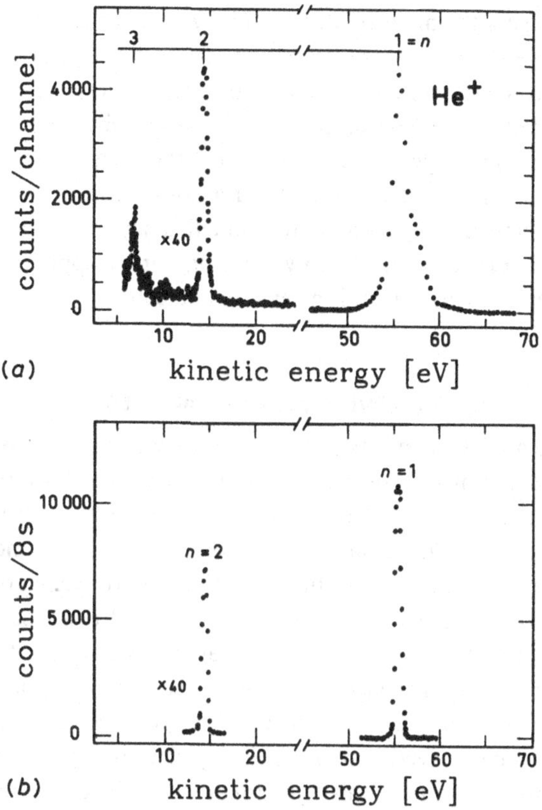

Figure 10.2 Photoelectron spectra of helium at 80 eV photon energy obtained under comparable experimental conditions taken with: (*a*) a TOF analyser (from [Lin83]); (*b*) a sector CMA (see [SDM82]). For details see main text.

The main advantages of the TOF analysers are the recording of the whole electron spectrum at once (i.e., not via a scanning procedure) and the good resolution at low electron kinetic energies. In Fig. 10.2 two electron spectra are shown for photoionization at 80 eV in helium; Fig. 10.2(*a*) was recorded with a time-of-flight analyser, and Fig. 10.2(*b*) was recorded with a sector CMA. Both spectra were taken under approximately the same experimental conditions with respect to photon intensity and bandpass, target pressure and total data collection time. The spectra show the main photoline for 1s photoionization in helium and, in addition, some satellite lines which belong to final excited ions $He^{*+}(n)$, where n is the principal quantum number. When comparing the spectra, three main differences can be noted. First, in the TOF spectrum one can see an increasing density of data points towards lower kinetic energy. This is the result of plotting the TOF spectrum on a linear energy axis, even though the natural axis of a TOF spectrum is the flight-time axis. These axes are related to each other by equ. (10.1). Second, the relative intensities (line areas) look different in the spectra, as can be noticed by comparing the main photoline He^+ ($n = 1$) with the satellite line He^{*+} ($n = 2$). In the case of an electrostatic energy analyser operated in the EDC mode where the spectrometer voltage is changed (see Section 4.2.4), the dispersion correction is needed before correct intensity ratios are obtained; such a correction is not required for a TOF spectrum. Third, within the same time for data collection, the TOF measurement yields many more data points. In this context, however, it should be noted that the lower accumulation efficiency of the electrostatic energy analyser can be improved considerably (by a factor of approximately 20) if the channeltron is replaced by a position-sensitive detector.

10.2 Optical properties of CMAs

In order to derive analytical expressions for the optical properties of an electrostatic spectrometer, one assumes a point source and follows the trajectories of electrons passing through the analyser. The feature of interest is the dependence of these trajectories on the geometrical lay-out of the selected spectrometer and on the deflection field in the analyser (produced by the spectrometer voltage U_{sp}). As a measure for such a trajectory one might use the actual distance travelled by the electron, but for a CMA an equivalent and even easier quantity to use follows from the z-component of the trajectory which is the distance z measured along the symmetry axis of the analyser. For the CMA shown schematically in Fig. 10.3, this distance z will be derived first by arbitrary electron trajectories entering the analyser at an angle ϑ. From this functional dependence the general focusing properties and the energy dispersion of the CMA will be calculated.

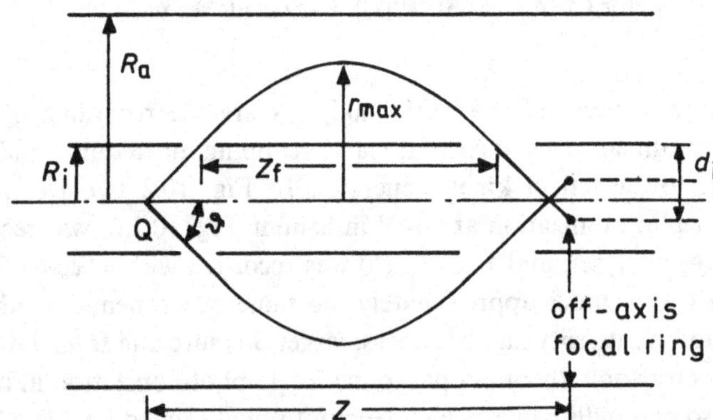

Figure 10.3 Definition of geometrical parameters for a CMA shown for the case of off-axis focusing of a point source (Q). R_i and R_a are the radii of the inner and outer field cylinders, respectively; d_i is the radial image distance to the inner cylinder (in analogy, one can introduce a radial source distance d_s, in the present case one has $d_s = R_i$); z is the total distance between source and image measured along the symmetry axis of the analyser; z_f is the corresponding distance for the field region; ϑ is the entrance angle into the analyser (due to symmetry properties this is equal to the exit angle).

10.2.1 Electron trajectories

According to Fig. 10.3 the distance z of an electron trajectory measured along the symmetry axis of the CMA contains three contributions: z_{field}, the distance the electrons travel in the field region of the analyser, and z_{before} and z_{after}, the distance, the electrons travel in the field-free regions before and after the analyser field:

$$z = z_{\text{before}} + z_{\text{field}} + z_{\text{after}}. \tag{10.3}$$

While the distances in the field-free regions are easy to calculate

$$z_{\text{before}} = \tan^{-1}\vartheta\, d_s \quad \text{and} \quad z_{\text{after}} = \tan^{-1}\vartheta\, d_i, \tag{10.4a}$$

$$z_{\text{before}} + z_{\text{after}} = d_{\text{tot}} \tan^{-1}\vartheta \tag{10.4b}$$

with (for d_s and d_i see the caption of Fig. 10.3)

$$d_{\text{tot}} = d_s + d_i, \tag{10.4c}$$

z_{field} is more difficult to assess. Following the detailed treatment given in [Sar67, Ris72], one starts with the equations of motion for an electron in a cylindrical electric field

z-coordinate: $$m_0\, \ddot{z} = 0, \tag{10.5a}$$

φ-coordinate: $$2m_0\, r\, \dot{r}\, \dot{\varphi} + m_0\, r^2\, \ddot{\varphi} = 0, \tag{10.5b}$$

r-coordinate: $$m_0\, \ddot{r} - m_0\, \dot{\varphi}^2\, r + \frac{e_0 U_{\text{sp}}}{\ln(R_a/R_i)}\frac{1}{r} = 0. \tag{10.5c}$$

From the first two equations one gets conservation of the linear momentum along z:

$$p_z = m_0\, \dot{z} = m_0\, v_0 \cos\vartheta \tag{10.6a}$$

and the orbital angular momentum along z:

$$\ell_z = m_0\, v\, r = m_0\, r^2\, \dot{\varphi} \tag{10.6b}$$

For a point source, ℓ_z is zero, and from equ. (10.6b) it follows that $\varphi = \text{const.}$ This means that the electrons ejected from the source move in a plane which contains the z-axis. Hence, the focal point must also be in this plane, and this fact considerably facilitates the discussion of the optical properties of the spectrometer. Due to the conservation conditions, one is left with only one differential equation which describes the radial movement of the electron, i.e.,

$$m_0\, \ddot{r} + \lambda/r = 0, \tag{10.7a}$$

with

$$\lambda = e_0 U_{\text{sp}}/\ln(R_a/R_i). \tag{10.7b}$$

A first integration leads to the 'radial' energy $E_{\text{rad}}(r)$

$$E_{\text{rad}}(r) = \tfrac{1}{2} m_0\, \dot{r}^2 = -\lambda \ln r + \text{const}, \tag{10.7c}$$

which is the kinetic energy calculated with the velocity in the radial direction. The constant of integration is fixed by the boundary conditions at the entrance to the field region and at the point of maximum deflection:†

$$E_{rad}(r = R_i) = E_0 \sin^2 \vartheta \tag{10.8a}$$

and

$$E_{rad}(r = r_{max}) = 0. \tag{10.8b}$$

Both conditions are fulfilled if the value *const* is chosen such that equ. (10.7c) takes the form

$$\tfrac{1}{2}m_0 \dot{r}^2 = \lambda \ln(r_{max}/r) \tag{10.9a}$$

with

$$r_{max} = R_i \, e^{E_0 \sin^2 \vartheta/\lambda}. \tag{10.9b}$$

Following [Sar67] a parameter k_0 is now introduced, which is defined by

$$k_0 = \frac{E_0}{e_0 U_{sp}} \ln(R_a/R_i) \tag{10.10}$$

and with this quantity one gets for the maximum radial distance r_{max}

$$r_{max} = R_i \, e^{k_0 \sin^2 \vartheta} \tag{10.11}$$

and for the spectrometer factor, defined by $f = E_0/U_{sp}$:

$$f = \frac{e_0 k_0}{\ln(R_a/R_i)}. \tag{10.12}$$

After these preliminaries one has to calculate z_{field}. This quantity follows as the product of the electron velocity v_z in the z-direction and the travelling time T along a selected trajectory in the field region. One has

$$z_{field} = v_z \, T = v_0 \cos\vartheta \, T. \tag{10.13}$$

and T follows if the integral

$$T = 2 \int_{R_i}^{r_{max}} \frac{dr}{v_r}, \tag{10.14a}$$

is calculated. (r_{max} is the maximum radial distance to the deflection point, the factor 2 arises from the symmetry of the trajectory to and away from the deflection point and $v_r = \dot{r}$ is given by equ. (10.9a).) Individual steps along this integration are

$$T = \frac{R_i}{\surd(2\lambda/m_0)} \int_{R_i}^{r_{max}} \{\ln[(R_i/r)\, e^{k_0 \sin^2 \vartheta}]\}^{-1/2} \, dr/R_i, \tag{10.14b}$$

† The electron trajectory within the analyser field follows an arc which is symmetrical with respect to the point of maximum deflection (at r_{max} in Fig. 10.3).

and

$$T = \frac{R_i\sqrt{k_0}}{v_0}\int_{R_i}^{r_{max}} [k_0 \sin^2 \vartheta - \ln(r/R_i)]^{-1/2}\, \mathrm{d}r/R_i. \tag{10.14c}$$

Introducing

$$k^2 = 2k_0 \sin^2 \vartheta \tag{10.15a}$$

and

$$u^2 = k^2 - 2\ln(r/R_i), \tag{10.15b}$$

the following transformations can be performed

$$[k_0 \sin^2 \vartheta - \ln(r/R_i)] = u^2/2, \tag{10.16a}$$

$$\mathrm{d}r/u = -r\,\mathrm{d}u, \tag{10.16b}$$

$$[k_0 \sin^2 \vartheta - \ln(r/R_i)]^{-1/2}\, \mathrm{d}r/R_i = -\sqrt{2}\,(r/R_i)\,\mathrm{d}u, \tag{10.16c}$$

$$\ln(r/R_i) = (k^2 - u^2)/2, \tag{10.16d}$$

$$r/R_i = \mathrm{e}^{0.5k^2}\,\mathrm{e}^{-u^2/2}. \tag{10.16e}$$

to derive the time T

$$T = \frac{R_i\sqrt{(2k_0)}}{v_0}\int_{u_1}^{u_2} (-1)\,\mathrm{e}^{0.5k^2}\,\mathrm{e}^{-u^2/2}\,\mathrm{d}u \tag{10.17a}$$

with the integration limits

$$u_2 = \sqrt{[k^2 - 2\ln(r_{max}/R_i)]}\ (\text{which gives } u_2 = 0), \tag{10.17b}$$

$$u_1 = \sqrt{(k^2 - 2\ln 1)}\ (\text{which gives } u_1 = k), \tag{10.17c}$$

and, finally,

$$T = \frac{R_i\sqrt{(2k_0)}}{v_0}\,\mathrm{e}^{0.5k^2}\int_0^k \mathrm{e}^{-u^2/2}\,\mathrm{d}u \tag{10.18a}$$

or, equivalently,

$$T = \frac{R_i k}{v_0 \sin\vartheta}\,\mathrm{e}^{0.5k^2}\int_0^k \mathrm{e}^{-u^2/2}\,\mathrm{d}u, \tag{10.18b}$$

where the latter integral is known from the normal error distribution.

Collecting the individual contributions to z, one obtains the final result for the distance z:

$$z = d_{tot}\tan^{-1}\vartheta + 2\,k\,R_i\tan^{-1}\vartheta\,\mathrm{e}^{0.5k^2}\int_0^k \mathrm{e}^{-u^2/2}\,\mathrm{d}u. \tag{10.19}$$

This expression describes how z depends, for a fixed analyser geometry and for arbitrarily selected electron trajectories, on the entrance angle ϑ and, through the coefficient k, on the energy.

10.2.2 Focusing properties

The distance z has been derived for any trajectory with a preselected entrance angle ϑ, i.e., a *principal trajectory* as defined in connection with Fig. 4.4 is not yet specified. In order to fix such a principal trajectory, which goes from the source to the image, one has to find the focusing properties of the analyser. In other words, one has to investigate how the distance function $z(\vartheta)$ behaves for small changes in ϑ. For this purpose one starts with a Taylor expansion of z in $\pm\Delta\vartheta$ around a still arbitrarily selected ϑ_0 value:

$$z = z_0 + \sum_{n>0} \frac{1}{n!} \Delta\vartheta^n \frac{\partial^n}{\partial\vartheta^n} z(\vartheta)\bigg|_0 . \tag{10.20}$$

If the first derivative is taken, one gets

$$\begin{aligned}\frac{\partial z}{\partial\vartheta}\bigg|_0 = -\frac{d_{tot}}{\sin^2\vartheta} + 2R_i\bigg(&-\sin^{-2}\vartheta\, k\, e^{k^2/2} \int_0^k e^{-u^2/2}\, du \\ &+ \tan^{-2}\vartheta\, k\, e^{k^2/2} \int_0^k e^{-u^2/2}\, du + \\ &+ \tan^{-2}\vartheta\, k^3\, e^{k^2/2} \int_0^k e^{-u^2/2}\, du + \tan^{-2}\vartheta\, k^2\bigg).\end{aligned} \tag{10.21}$$

From the Taylor expansion it can be seen that electron trajectories deviating $\pm\Delta\vartheta$ around ϑ_0 will have the same z_0 (in nth order) if all derivations up to the order n vanish. Hence, the selected entrance angle ϑ_0 identifies a principal trajectory. If the first derivative is zero, one gets focusing in first order, and the corresponding condition is given by (for a point source the φ angle is of no special relevance, there is focusing in all orders)

$$\sin^2\vartheta\, k\, e^{k^2/2} \int_0^k e^{-u^2/2}\, du = \frac{d_{tot}/(2R_i) - k^2\cos^2\vartheta}{k^2\tan^{-2}\vartheta - 1}. \tag{10.22}$$

This condition for first-order focusing can be fulfilled for a whole set of pairs ϑ and k (or equivalently k_0, see equ. (10.15a)). The interesting combinations are plotted in Fig. 10.4(*a*) with $d = d_{tot}/R_i$ as a parameter. For a given value of d, a certain entrance angle $\vartheta = \vartheta_0$ can be selected, and the k_0 value necessary for first-order focusing follows from the figure. Knowing k_0, one can then calculate the k value from equ. (10.15a), and with this information the total distance z follows from equ. (10.19), and from equ. (10.20) follows a numerical value for the second derivative of the Taylor expansion. These two quantities are shown in Figs. 10.4(*b*) and (*c*) with d as a parameter. From Fig. 10.4(*c*) it can be seen that the second derivative vanishes for a certain angle ϑ. At this angle, focusing in second order exists, and therefore this angle is usually selected as the entrance angle of the principal trajectory into the CMA.

As an application of the preceding discussion, the optical properties of the sector CMA mentioned several times in Part B will be given now, based on the formulas given above. The analyser is the rotatable double-sector CMA shown schematically

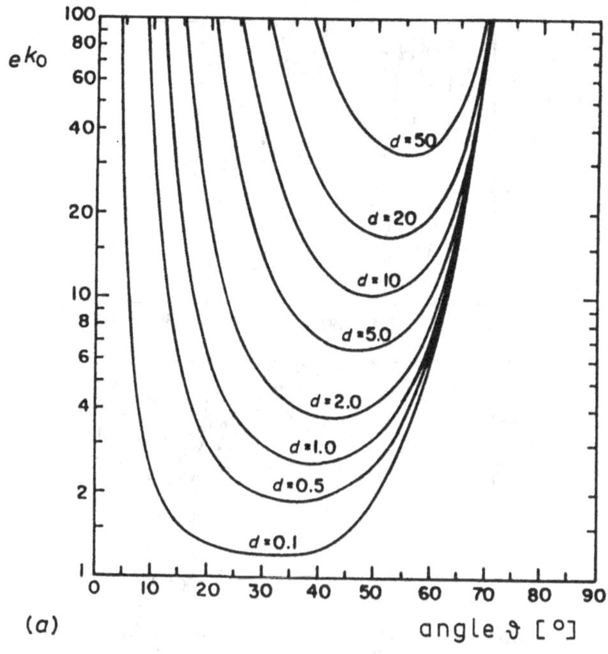

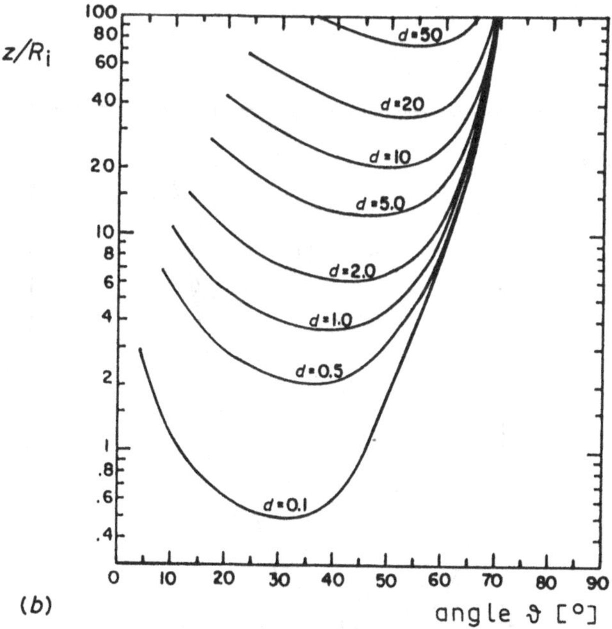

Figure 10.4 Design parameters for a CMA fulfilling first-order focusing: (*a*) e^{k_0} as a function of the entrance angle ϑ; (*b*) total distance z as a function of ϑ. (d_{tot} is the total radial distance of a selected principal ray in the field-free regions, R_i is the inner radius of the field region, and for a given outer radius R_a of the field region the parameter k_0 is related to the spectrometer factor f). From [Ris72].

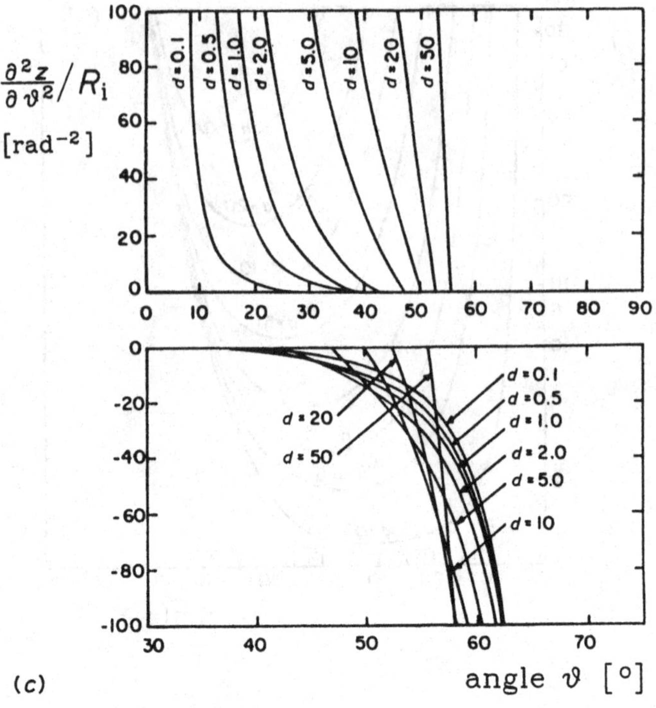

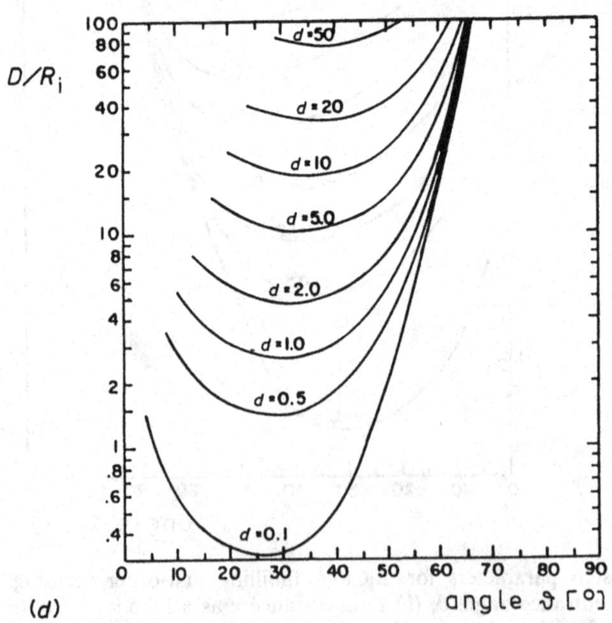

Figure 10.4 (contd) (c) Second derivative of z as a function of ϑ; (d) dispersion as a function of ϑ; in all cases with $d = d_{tot}/R_i$ as a parameter. From [Ris72].

in Fig. 1.17. In order to have space for two separate channeltron detectors, off-axis focusing in second order for ϑ_0 has been selected with $d_s = R_i$ and $d_i = 1.2R_i$, i.e. $d_{tot} = 2.2R_i$. From Fig. 10.4(*c*) one then gets $\vartheta_0 = 42.79°$. Using this value, one obtains from Fig. 10.4(*a*) $e^{k_0} = 3.913$, and from Fig. 10.4(*b*) the total distance $z = 6.586R_i$. For good energy resolution the size of the source has to be small compared to a characteristic dimension of the spectrometer, and for this the radius R_i of the inner field cylinder is a good measure. In the example selected, R_i has been chosen to be 60 mm. From equ. (10.11) one then gets, together with the other fixed parameters, $r_{max} = 113$ mm which is smaller than $R_a = 135$ mm. (For the actual situation it has to be checked that electrons with $\vartheta_0 + \Delta\vartheta$ are still far away from the outer cylinder of the analyser.) With fixed R_a and k_0 one finally obtains the spectrometer factor $f = 1.682$ eV/V from equ. (10.12).

10.2.3 Energy dispersion

When the design parameters of the electron spectrometer are fixed, the variation of the distance z with small changes in the electron energy $\pm\Delta E$ around the value E_0 attached to the principal trajectory can be investigated. A Taylor expansion up to the first order yields in this case

$$z(\Delta E) = z_0 + \Delta E \left.\frac{\partial z(\Delta E)}{\partial E}\right|_0 \tag{10.23a}$$

or, equivalently,

$$\frac{z(\Delta E) - z_0}{\Delta E} = \left.\frac{\partial z(\Delta E)}{\partial E}\right|_0. \tag{10.23b}$$

The quantity was introduced in equ. (4.6) as an energy dispersion D divided by E_0. If the first derivative is to be evaluated, one has to take into account the dependences of k on k_0, and of k_0 on E_0, as given in equs. (10.15a) and (10.10). This gives

$$\left.\frac{\partial z}{\partial E}\right|_0 = 2R_i \tan^{-1}\vartheta \left[k\,(2E_0)^{-1}\, e^{k^2/2} \int_0^k e^{-u^2/2}\, du \right.$$
$$\left. + k^3\,(2E_0)^{-1}\, e^{k^2/2} \int_0^k e^{-u^2/2}\, du + k^2(2E_0)^{-1} \right]. \tag{10.24}$$

In this expression, k depends on the entrance angle ϑ which is usually selected by the condition of second-order focusing. Of course, the condition for first-order focusing is then also fulfilled, and the corresponding equ. (10.22) can be used to replace the integrations over u in equ. (10.24) . This gives

$$\left.\frac{\partial z}{\partial E}\right|_0 = \frac{R_i \tan^{-1}\vartheta_0}{E_0} \left[k^2 + \frac{k^2 \cos^2\vartheta_0 - d_{tot}(2R_i)^{-1}}{\sin^2\vartheta_0 - k^2\cos^2\vartheta_0}(1 + k^2) \right]. \tag{10.25}$$

With this result the energy dispersion D follows as

$$D = R_i \tan^{-1}\vartheta_0 \left[k^2 + \frac{1+k^2}{\sin^2\vartheta_0} \frac{d_{tot}(2R_i)^{-1} - k^2\cos^2\vartheta_0}{k^2\tan^{-2}\vartheta_0 - 1} \right], \tag{10.26}$$

and this quantity is shown in Fig. 10.4(*d*). For the example discussed previously one obtains $D = 6.115R_i$.

10.3 Supplements for electrostatic lenses

10.3.1 Derivation of the fundamental lens equation

In order to derive the optical properties of an electrostatic lens, one has to establish and solve the equations of motion for a transmitted particle of mass m and charge q. Since the particle moves in the lens under the action of an electric field which can be derived from a potential φ, this potential will be considered first. Assuming cylindrical symmetry around the optical axis (z-axis) and treating paraxial rays only, the potential $\varphi(\rho, z)$ depends on the cylindrical coordinates ρ and z. It can be expanded as a power series in ρ with z-dependent coefficients. Due to the rotational symmetry, only even powers of ρ appear in the expansion, and one has the ansatz

$$\varphi(\rho, z) = \varphi(0, z) + \tfrac{1}{2}\rho^2 A_2(z) + \cdots. \tag{10.27}$$

Putting this expansion into the Laplace equation

$$\frac{\partial^2\varphi}{\partial\rho^2} + \frac{1}{\rho}\frac{\partial\varphi}{\partial\rho} + \frac{\partial^2\varphi}{\partial z^2} = 0, \tag{10.28}$$

the coefficients $A_n(z)$ can be determined, giving for the lowest one

$$A_2(z) = -\frac{1}{2}\frac{\partial^2\varphi(\rho, z)}{\partial z^2}\bigg|_{\rho=0} = -\tfrac{1}{2}\Phi''(z). \tag{10.29}$$

Therefore, the potential φ is given by

$$\varphi(\rho, z) = \Phi(z) - \tfrac{1}{4}\rho^2\,\Phi''(z) + \cdots, \tag{10.30a}$$

where the potential $\Phi(z)$ along the z-axis,

$$\Phi(z) = \varphi(\rho = 0, z), \tag{10.30b}$$

has been introduced. (Differentiations with respect to the z coordinate are indicated by primes in order to distinguish them from time derivatives indicated by dots.) This result states that the potential $\varphi(\rho, z)$ and, hence, the electric field vector $\mathcal{E}$, follow uniquely from the values of the potential $\Phi(z)$ along the z-axis. For the electric field components one then obtains

$$\mathcal{E}_\rho = -\frac{\partial\varphi(\rho, z)}{\partial\rho} = \tfrac{1}{2}\,\rho\,\Phi''(z), \tag{10.31a}$$

$$\mathcal{E}_z = -\frac{\partial\varphi(\rho, z)}{\partial z} = -\Phi'(z), \tag{10.31b}$$

and the equations of motion are given by

$$m\ddot{\rho} = q\mathscr{E}_\rho = \tfrac{1}{2}\, q\, \rho\, \Phi''(z), \tag{10.32a}$$

$$m\ddot{z} = q\mathscr{E}_z = -q\, \Phi'(z). \tag{10.32b}$$

For charged particles travelling in a potential, energy conservation requires

$$E = E_{\text{kin}} + E_{\text{pot}} = \tfrac{1}{2}mv^2 + q\varphi(\rho, z) = \text{const.} \tag{10.33a}$$

In the present context the total energy E can be identified with the kinetic energy E^0_{kin} which the particles have before they enter the lens system, and it is convenient also to relate this energy to a voltage by

$$E = E^0_{\text{kin}} = qU = e_0 U_e, \tag{10.33b}$$

where for electrons, both q and U are negative values, but the elementary charge e_0 and U_e are positive.

If the approximations for paraxial rays,

$$\dot{r} \ll \dot{z} \quad \text{and} \quad \varphi(r, z) \approx \Phi(z), \tag{10.34}$$

can be made, one obtains for equ. (10.33)

$$q[U - \Phi(z)] = \tfrac{1}{2}mv^2 = \tfrac{1}{2}m(\dot{r}^2 + \dot{z}^2) \approx \tfrac{1}{2}m\dot{z}^2 = \frac{p_z^2}{2m}. \tag{10.35a}$$

The result means that the longitudinal momentum p_z is given by

$$p_z^2 \approx q2m[U - \Phi(z)] = -q2m\,\tilde{\Phi}(z), \tag{10.35b}$$

with

$$\tilde{\Phi}(z) = \Phi(z) - U = \Phi(z) + U_e. \tag{10.35c}$$

i.e., the longitudinal momentum at each point along the z-axis is determined by the value of the *reduced* potential $\tilde{\Phi}(z)$.

Using the time derivative of the z coordinate, one then has

$$\frac{\mathrm{d}z}{\mathrm{d}t} = \dot{z} = \frac{1}{m}p_z = \sqrt{\left[-\frac{2q}{m}\tilde{\Phi}(z)\right]} \tag{10.36a}$$

giving

$$\frac{\mathrm{d}}{\mathrm{d}t} = \sqrt{\left[-\frac{2q}{m}\tilde{\Phi}(z)\right]}\frac{\mathrm{d}}{\mathrm{d}z}. \tag{10.36b}$$

With the help of this last relation, the time derivatives in equ. (10.32a) can be replaced by derivatives with respect to the z coordinate, thus eliminating the time dependences. Together with equ. (10.32a) (equ. (10.32b) is taken care of by implementing energy conservation which leads to equ. (10.35)) this gives

$$m\ddot{\rho} = m\sqrt{\left[-\frac{2q}{m}\tilde{\Phi}(z)\right]}\frac{\mathrm{d}}{\mathrm{d}z}\left(\sqrt{\left[-\frac{2q}{m}\tilde{\Phi}(z)\right]}\frac{\mathrm{d}\rho}{\mathrm{d}z}\right) = \frac{q}{2}\rho\,\tilde{\Phi}''(z), \tag{10.37}$$

and, after some manipulation, the fundamental differential equation of an electrostatic lens for the coordinate ρ as a function of the potential $\tilde{\Phi}$ and the derivatives in z:

$$2\tilde{\Phi}(z)\rho'' + \tilde{\Phi}'(z)\rho' + \tfrac{1}{2}\tilde{\Phi}''(z)\rho = 0. \tag{10.38}$$

An equivalent form of this differential equation is presented in equ. (4.38), and important aspects which can be deduced from this equation are discussed there.

10.3.2 Liouville theorem and related forms

The Helmholtz–Lagrange relation given in equ. (4.46) is related to many other forms which all state certain conservation laws (the Clausius theorem, Abbe's relation, the Liouville theorem). The most important one in the present context is the *Liouville theorem* [Lio38] which describes the invariance of the volume in phase space. The content of this theorem will be discussed and represented finally in a slightly different form which allows a new access to the luminosity introduced in equ. (4.14).

The applications of the Helmholtz–Lagrange relation given in equ. (4.46) to an arbitrary conjugate object–image pair characterized by left (ℓ) and right (r), respectively, leads to

$$y_\ell \tan \alpha_\ell \sqrt{\tilde{\Phi}_1} = y_r \tan \alpha_r \sqrt{\tilde{\Phi}_r} \tag{10.39a}$$

which is equivalent to (paraxial rays with $\tan \alpha \approx \alpha$)

$$y_\ell \, \alpha_\ell \sqrt{\tilde{\Phi}_1} = y_r \, \alpha_r \sqrt{\tilde{\Phi}_r}. \tag{10.39b}$$

The product $y\,\alpha \sqrt{\tilde{\Phi}}$ is called the *emittance*, and when divided by a factor of π it is termed the *normalized* emittance. Equ. (10.39) states that the (normalized) emittance remains constant, provided an ideal optical system, i.e., without intensity losses, is given.

Introducing the transverse momentum p^{trans}, defined by†

$$p^{\text{trans}} = p \sin \alpha = \sqrt{(2me_0\tilde{\Phi})} \sin \alpha, \tag{10.40}$$

equ. (10.39b) derived for paraxial rays is equivalent to

$$y_\ell \, p_\ell^{\text{trans}} = y_r \, p_r^{\text{trans}}. \tag{10.41}$$

This last form describes the invariance of a two-dimensional volume in phase space. It is this relation which is strictly speaking known as the *Liouville theorem.* Liouville considered a volume V in the six-dimensional phase space (three coordinates and three momenta for each particle) and proved from general principles of particle dynamics that for an ensemble of particles homogeneously filling such a volume V, this volume and, hence, the density of particles remain constant in time, if the forces acting on the particles can be derived from a potential. In the present cases of interest, the motions in the three planes spanned by the conjugated variables $\{x; p_x\}$, $\{y; p_y\}$ and $\{z; p_z\}$ are not coupled, and the six-dimensional phase-space volume V factorizes into the corresponding two-dimensional phase-space areas: $V = V_x \; V_y \; V_z$ with $V_x = xp_x$, $V_y = yp_y$, $V_z = zp_z$. For axial symmetry, only two such areas are of relevance for the longitudinal and

† This equation with $\sin \alpha$ in the transverse momentum is more general than equ. (10.39a) with $\tan \alpha \approx \alpha$ derived in the paraxial approximation.

transverse components, $V_{\text{long}} = V_z$, and V_{trans} which represents V_x and V_y. If there is acceleration/retardation in the lens, the longitudinal components z and p_z do change, but their product remains constant†

$$V_{\text{long}} = zp_z = \text{const}, \tag{10.42a}$$

and the Liouville theorem reduces to

$$V_{\text{trans}} = \rho p^{\text{trans}} = \text{const}, \tag{10.42b}$$

which is identical to equ. (10.41).

In the context of conservation relations another form will be added. If equ. (10.39b) is squared, one can introduce an area A to replace y^2, a solid angle Ω to replace α^2 and a so-called *brightness* B which is proportional to $\tilde{\Phi}$ (see below). This gives

$$A_\ell \Omega_\ell B_\ell = A_r \Omega_r B_r. \tag{10.43}$$

In optical radiometry with plane solid surfaces the product $A_\ell \Omega_\ell$ at the lens input represents the *luminosity* L_{opt}:

$$L_{\text{opt}} = A\,\Omega. \tag{10.44a}$$

(In order to distinguish this luminosity from the quantity introduced in equ. (4.14) which was also called luminosity, the subscript 'opt' is introduced.) The product of L_{opt} and B describes the overall intensity I accepted (I_ℓ) and transmitted (I_r) by an optical system which remains constant for an ideal system:

$$I_\ell = L_{\text{opt},\ell} B_\ell = L_{\text{opt,r}}\, B_r = I_r. \tag{10.44b}$$

The brightness B_ℓ describes the strength of electron emission at the sample, i.e., it is given by the current per area and per unit solid angle. It is an inherent property of the process of electron production, and according to the substitution made when deriving equ. (10.44b) from equ. (10.39b) it depends linearly on the kinetic energy E_{kin} of the electrons. This dependence will be explained with the help of Fig. 10.5, where electrons are emitted from an area 'A' to two sides; into a field-free region on the left-hand side and into a homogeneous electric field which retards the electrons on the right-hand side. It can be seen that the individual electron trajectories which start from unit areas into a certain solid angle become bent in the field-region such that an observer viewing the electron emitter receives (with constant acceptance angle) many fewer electrons on the right-hand side than on the left-hand side. Hence, the source viewed after the retardation of the electrons looks less 'bright' when compared to the case without retardation. Of course, the overall intensities emitted into the *adapted* solid angles, α_ℓ and α_r, remain the same, because one has

$$I_\ell = B_\ell A \int_0^{\alpha_\ell} \cos\vartheta \, d\Omega = B_r A \int_0^{\alpha_r} \cos\vartheta \, d\Omega = I_r. \tag{10.45a}$$

† Assuming the electrons are confined at $t = t_0$ in a two-dimensional area $\{z, p_z\}$ defined by the four points $t_0\{0; 0\}$, $t_0\{z_0; 0\}$, $t_0\{0; p_z^0\}$ and $t_0\{z_0; p_z^0\}$, one gets $V_z(t_0) = z_0 p_z^0$. At a time $t_1 = t_0 + \Delta t$ these points are transferred to $t_1\{\Delta z, \Delta p_z\}$, $t_1\{z_0 + \Delta z; \Delta p_z\}$, $t_1\{\Delta z + p_z^0\,\Delta t; p_z^0 + \Delta p_z\}$ and $t_1\{z_0 + \Delta z + p_z^0\,\Delta t; p_z^0 + \Delta p_z\}$ and these points define a parallelogram whose area is given by $V_z(t_1) = z_0 p_z^0$ which is the same as $V_z(t_0)$, i.e., $V_z = \text{const}$.

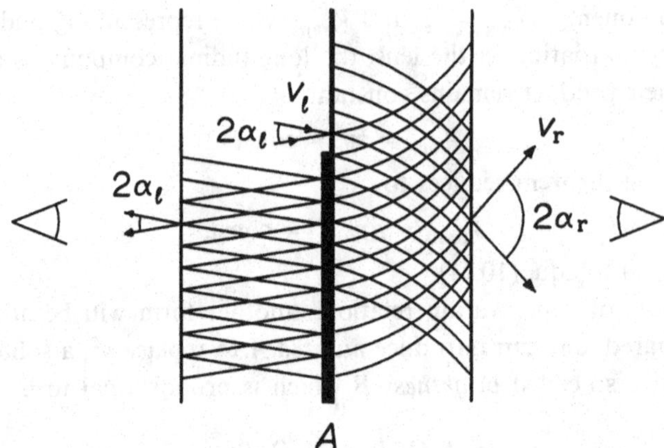

Figure 10.5 Demonstration of the brightness of an electron source placed at the surface A which emits equal numbers of electrons into a field-free region (left) and into a homogeneous electric field causing retardation (right). Representative electron trajectories starting from a unit area and into a unit solid angle (cone angle $2\alpha_\ell$) are shown. The angle symbols on each side indicate the acceptance angle of an observer viewing the electron emitter. Due to the retardation effect on the right-hand side, the angular range accepted by the observer is smaller than the cone angle $2\alpha_r$, and this leads to a reduced brightness. From [HWe68].

(Since the observer views radiation from all parts of the extended area A, but at different angles α, Lambert's cosine law, which states that the intensity in a given direction will change in proportion to the cosine of the angle ϑ from the normal, has to be taken into account.) Equ. (10.45a) is equivalent to

$$I_\ell = B_\ell\, A\, \pi \sin^2\alpha_\ell = B_r\, A\, \pi \sin^2\alpha_r = I_r. \tag{10.45b}$$

Since in the present example the homogeneous electric field affects only the longitudinal velocity, but not the transverse velocity, one has

$$v_\ell^{\text{trans}} = v_\ell \sin\alpha_\ell = v_r \sin\alpha_r = v_r^{\text{trans}}. \tag{10.46}$$

Using this relation to replace $\sin\alpha$ in equ. (10.45b) and further substituting v^2 by $2E_{\text{kin}}/m$, one then obtains

$$B_\ell/E_{\text{kin},\ell} = B_r/E_{\text{kin},r}, \tag{10.47}$$

and this result confirms the linear dependence of the brightness on the kinetic energy as was used to derive equ. (10.43).

After these explanations the content of equ. (10.44b) can be formulated as follows: for an ideal electron-optical system the overall intensity of transmitted electrons remains constant and separates into two factors, one (the luminosity) describing the geometrical parameters which the electrons have on either side of the system, the other (the brightness) describing the kinetic energies.

10.4 Convolution procedures

For a correct analysis of photoionization processes studied by electron spectrometry, convolution procedures are essential because of the combined influence of several distinct energy distribution functions which enter the response signal of the electron spectrometer. In the following such a convolution procedure will be formulated for the general case of photon-induced two-electron emission needed for electron–electron coincidence measurements. As a special application, the convolution results for the non-coincident observation of photoelectrons or Auger electrons, and for photoelectrons in coincidence with subsequent Auger electrons are worked out. Finally, the convolutions of two Gaussian and of two Lorentzian functions are treated.

10.4.1 General formulation for photon-induced two-electron emission

Two limiting cases will be distinguished when formulating the response of two electron spectrometers to photon-induced two-electron emission. These cases are *direct* double photoionization where the two electrons are ejected simultaneously and *two-step* double ionization where photoelectron emission is followed by Auger decay. One is then faced with the following ingredients and constraints:

(i) The distribution function of monochromatized photons,

$$G_B(E_{ph}, E^0_{ph}), \tag{10.48}$$

is approximated by a Gaussian distribution normalized to unit area (see equ. (2.28e)), i.e., the total number N_{ph} of photons incident per second on the sample is treated separately and the factor of proportionality is given by $1/(1.06\,\Delta E_B)$ where ΔE_B is the fwhm. (Note that the Gaussian function depends only on the difference between E and E^0. Therefore, it can also be characterized by $G(E - E^0)$, a form which means it can be characterized in a way similar to that used for the δ-function. This comment is also valid for a Lorentzian distribution.)

(ii) The distribution function related to an intermediate photoionized state with ionization energy E_I^+ is in the case of two-step double photoionization a Lorentzian function (for simplicity the final two-hole state with ionization energy E_I^{++} is assumed to be sharp)

$$L_\Gamma(E_\ell, E_I^+), \tag{10.49a}$$

and in the case of direct double ionization, one has instead a δ-function

$$\delta(E_\ell - E_I^+). \tag{10.49a}$$

(Formally, this δ-function converts the two separate energy relations $E_{ph} = E_{kin\,1} + E_\ell$ and $E_\ell = E_{kin\,2} + E_I^{++}$ from a two-step description to the relation $E_{ph} = E_{kin\,1} + E_{kin\,2} + E_I^{++}$ needed in a one-step description.)

(iii) The spectrometer functions for each of the two analysers is approximated by Gaussian functions. Depending on the normalization, two different forms are used (see equ. (2.28)),

$$G_{sp}(E_{kin}, E_{pass}) \quad \text{with } G_{max} = T, \tag{10.50a}$$

where T is the transmission of the spectrometer defined in equ. (4.16), and

$$\tilde{G}_{sp}(E_{kin}, E_{pass}) \quad \text{with } \tilde{G}_{max} = 1. \tag{10.50b}$$

According to equs. (1.48a) and (1.48b), integration over the spectrometer function can be done in either E_{kin} or E_{pass}. One obtains for the area over $G_{sp}(E_{kin}, E_{pass})$ the value $Tk_1\,\Delta E_{sp}$, and for the area over $\tilde{G}_{sp}(E_{kin}, E_{pass})$ the value $k_1\,\Delta E_{sp}$, where ΔE_{sp} is the fwhm of the spectrometer function, and the constant $k_1 = 1.06$ is often set to unity (see footnote concerning equ. (2.39a)).

(iv) In addition to the given energy distributions one has also to consider the relations which follow from energy conservation:

direct double photoionization: $E_{ph} = E_{kin1} + E_{kin2} + E_{I}^{++}$ (10.51a)

two-step double ionization: $E_{ph} = E_{kin1} + E_{\ell}, E_{\ell} = E_{kin2} + E_{I}^{++}$. (10.51b)

In these expressions, the energies refer to the actual values (the mean values are characterized by a superscript 0).

After these preliminaries, one can introduce the intensity distribution function $I_s(E_{kin1}, E_{kin2})$ for electron pairs created in the source volume (subscript 's') with kinetic energies E_{kin1} and E_{kin2}. For direct double photoionization one obtains

$$I_s(E_{kin1}, E_{kin2})|^{dd} = N_{ph} n_v \,\Delta z \int \frac{d^3\sigma}{d\Omega_1\, d\Omega_2\, dE_1} G_B(E_{ph}, E_{ph}^0) \times \delta(E_{ph} - E_{kin1} - E_{kin2} - E_{I}^{++})\, dE_{ph}, \tag{10.52a}$$

where the differential cross section which is a smooth function of the kinetic energies can be taken out of the integral as an 'on-the-energy-shell' value. Integration over the photon energies then leads to

$$I_s(E_{kin1}, E_{kin2})|^{dd} = N_{ph} n_v \,\Delta z \frac{d^3\sigma}{d\Omega_1\, d\Omega_2\, dE}(E_{kin1}, E_{kin2})\, F_s(E_{kin1}, E_{kin2})|^{dd} \tag{10.52b}$$

with the energy distribution function

$$F_s(E_{kin1}, E_{kin2})|^{dd} = G_B(E_{kin1} + E_{kin2} + E_{I}^{++}, E_{ph}^0). \tag{10.52c}$$

For photon-induced two-step double ionization the treatment is slightly different. As expressed in equ. (4.62c), the Lorentzian function $L_\Gamma(E_\ell, E_I^+)$, which arises from the decay of the intermediate state, appears and energy conservation holds separately for each of the two steps (see equ. 10.51b)). Since the Lorentzian function is not smooth it has to be taken into account explicitly in the evaluation of the energy integrals. Hence, one gets

$$I_s(E_{kin1}, E_{kin2})|^{2\text{-step}} = N_{ph} n_v \,\Delta z \frac{d^2\sigma}{d\Omega_1\, d\Omega_2} \iint G_B(E_{ph}, E_{ph}^0) L_\Gamma(E_\ell, E_I^+) \times \delta(E_{ph} - E_{kin1} - E_\ell)\, \delta(E_\ell - E_{kin2} - E_{I}^{++})\, dE_{ph}\, dE_\ell. \tag{10.53a}$$

The evaluation leads to

$$I_s(E_{\text{kin}1}, E_{\text{kin}2})|^{2\text{-step}} = N_{\text{ph}} n_v \,\Delta z \frac{\mathrm{d}^2\sigma}{\mathrm{d}\Omega_1\,\mathrm{d}\Omega_2}(E^0_{\text{ph}})\, F_s(E_{\text{kin}1}, E_{\text{kin}2})|^{2\text{-step}} \quad (10.53b)$$

with the energy distribution function

$$F_s(E_{\text{kin}1}, E_{\text{kin}2})|^{2\text{-step}} = G_B(E_{\text{kin}1} + E_{\text{kin}2} + E_I^{++}, E^0_{\text{ph}}) L_\Gamma(E_{\text{kin}2} + E_I^{++}, E_I^{+}) \quad (10.53c)$$

and (see equs. (4.86b) and (4.87))

$$\frac{\mathrm{d}^2\sigma}{\mathrm{d}\Omega_1\,\mathrm{d}\Omega_2} = \frac{\sigma_{\text{ph}}}{4\pi}\frac{\omega_A}{4\pi}(1 + \text{angular terms}), \quad (10.53d)$$

where σ_{ph} is the inner-shell photoionization cross section and ω_A the yield of the subsequent Auger transition.

10.4.2 Non-coincident observation of the photoelectron or the Auger electron

In order to derive the relevant energy distribution function for non-coincidence experiments, one has to integrate $F_s(E_{\text{kin}1}, E_{\text{kin}2})$ over the kinetic energies of the unobserved electron, include the spectrometer function for the observed electron, and perform an integration over the kinetic energies offered to the spectrometer, keeping the pass energy E_{pass} fixed. (Note that an integration over the angular range accepted by the analysers also has to be performed. Neglecting the angular terms in equ. (10.53d), one then obtains for each spectrometer a factor $\Omega_{\text{acc}}/4\pi = T$, where Ω_{acc} is the acceptance angle and T the spectrometer transmission. This value T is incorporated in the response function of equ. (10.50a).) In the case of two-step double ionization one then derives for the response function $F(E_{\text{pass}1})$ of the photoelectron

$$F(E_{\text{pass}1}) = \int\!\!\int G_{\text{sp}1}(E_{\text{kin}1}, E_{\text{pass}1}) F_s(E_{\text{kin}1}, E_{\text{kin}2})|^{2\text{-step}}\, \mathrm{d}E_{\text{kin}1}\,\mathrm{d}E_{\text{kin}2}. \quad (10.54a)$$

Because of $E_{\text{kin}2} = E_\ell - E_I^{++}$, the result can be recast in the form given in equ. (2.37)

$$F(E_{\text{pass}1}) = \int\!\!\int G_{\text{sp}1}(E_{\text{kin}1}, E_{\text{pass}1}) G_B(E_{\text{kin}1} + E_\ell, E^0_{\text{ph}}) L_\Gamma(E_\ell, E_I^{+})\, \mathrm{d}E_{\text{kin}1}\,\mathrm{d}E_\ell$$
$$= G_{\text{sp}1} \otimes G_B \otimes L_\Gamma. \quad (10.54b)$$

In particular, one obtains for the area $A_{\text{exp}}(E^0_{\text{kin}})$ of a photoelectron line (see equ. (2.38); the subscript 1 is omitted, and the efficiency ε of the electron detector is included):

$$A_{\text{exp}}(E^0_{\text{kin}}) = N_{\text{ph}}\, \sigma_{\text{ph}}(E^0_{\text{ph}})\, n_v\, \Delta z\, T\varepsilon\, \tilde{F} \quad (10.55a)$$

with

$$\tilde{F} = \int F(E_{\text{pass}})\, \mathrm{d}E_{\text{pass}}$$
$$= \int\!\!\int\!\!\int L_\Gamma(E_\ell, E_I^{+}) G_B(E_{\text{kin}} + E_\ell, E^0_{\text{ph}}) \tilde{G}_{\text{sp}}(E_{\text{kin}}, E_{\text{pass}})\, \mathrm{d}E_\ell\,\mathrm{d}E_{\text{kin}}\,\mathrm{d}E_{\text{pass}}. \quad (10.55b)$$

Because all the integrations cover the whole integration region, one can arbitrarily change the order of integration. Starting with integration over the spectrometer function, one gets

$$\tilde{F} = \Delta E_{sp} \iint L_\Gamma(E_\ell, E_I^+) G_B(E_{kin} + E_\ell, E_{ph}^0)\, dE_\ell\, dE_{kin}, \tag{10.56a}$$

i.e., the coupling between $G_{sp}(\ldots)$ and $G_B(\ldots)$ made by E_{kin} disappears, and this allows in a next step integration over the bandpass function which then gives

$$\tilde{\tilde{F}} = \Delta E_{sp} \int L_\Gamma(E_\ell, E_I^+)\, dE_\ell. \tag{10.56b}$$

Because of the normalization of the Lorentzian function one obtains finally

$$\tilde{\tilde{F}} = \Delta E_{sp}, \tag{10.56c}$$

which for equ. (10.55a) yields

$$A_{exp}(E_{kin}^0) = N_{ph}\, n_v\, \sigma_{ph}(E_{ph}^0)\, \Delta z\, T\varepsilon\, \Delta E_{sp}, \tag{10.57a}$$

or, alternatively, for the dispersion corrected area A_D (see equ. (2.39a))

$$A_D = N_{ph}\, n_v\, \sigma_{ph}(E_{ph}^0)\, \Delta z\, T\varepsilon \left(\frac{\Delta E}{E}\right)_{sp}. \tag{10.57b}$$

Similarly, for the response function $F(E_{pass2})$ of the Auger electron one gets

$$\begin{aligned} F(E_{pass2}) &= \iint G_{sp2}(E_{kin2}, E_{pass2}) F_s(E_{kin1}, E_{kin2})|^{2\text{-step}}\, dE_{kin1}\, dE_{kin2} \\ &= \iint G_{sp2}(E_{kin2}, E_{pass2}) G_B(E_{kin1} + E_{kin2} + E_I^{++}, E_{ph}^0) \\ &\quad \times L_\Gamma(E_{kin2} + E_I^{++}, E_I^{++})\, dE_{kin1}\, dE_{kin2}. \end{aligned} \tag{10.58a}$$

Since the integrations cover the whole range, the bandpass function integrated over E_{kin}, yields the same result as when integrated over E_{ph}, giving unity with the selected normalization. Using $E_{kin2}^0 = E_I^+ - E_I^{++}$ in the Lorentzian function, one finally derives

$$\begin{aligned} F(E_{pass2}) &= \int G_{sp2}(E_{kin2}, E_{pass2}) L_\Gamma(E_{kin2}, E_{kin2}^0)\, dE_{kin2} \\ &= G_{sp2} \otimes L_\Gamma. \end{aligned} \tag{10.58b}$$

This result is used in equ. (3.23) where other factors determining the observed intensity have also been included.

10.4.3 Coincident observation of the photoelectron and its subsequent Auger electron

For electron–electron coincidence measurements both spectrometer functions have to be taken into account, giving as the response function for selected pass energies E_{pass1} and E_{pass2}:

$$F(E_{pass1}, E_{pass2}) = \iint G_{sp2}(E_{kin2}, E_{pass2}) G_{sp1}(E_{kin1}, E_{pass1}) \times F_s(E_{kin1}, E_{kin2}) \, dE_{kin1} \, dE_{kin2}. \tag{10.59}$$

The presence of both spectrometer functions now prevents the easy evaluation of the convolution integral which was possible for the non-coincident case, because the energy balance between the coincident electrons couples the response of the spectrometers. This will be discussed for the case of two-step double photoionization with the help of Fig. 10.6 which is a schematic representation of the individual contributions to equ. (10.59). In Fig. 10.6(*a*) the bandpass function G_B of the monochromatized light and the energy distribution function of the intermediate hole-state L_Γ are shown. Since energy conservation connects the kinetic energies of both the photoelectron and the Auger electron through the energies E_ℓ in the distribution function L_Γ, it is convenient to use these E_ℓ values as labels for corresponding electrons. In the example given, five E_ℓ values, labelled 1–5, are selected for the subsequent discussion.

The spectral distribution of photoelectrons with E_{kin1} and Auger electrons with E_{kin2} follows from the function $F_s(E_{kin1}, E_{kin2})$ defined in equ. (10.53c). Selecting a specific E_ℓ value, one gets a one-to-one correspondence to E_{kin2}, which, therefore, can also be used to label the E_ℓ values. (The final state of Auger decay is assumed to be sharp.) The Auger intensity at certain energy positions n then follows as a sum over all photoionization processes within the given bandpass, weighted by the value of the L_Γ-function at E_ℓ, i.e.,

$$F_A(E_{kin2} = n) = \sum_{E_{kin1}} F_S(E_{kin1}, E_{kin2} = n). \tag{10.60}$$

As can be seen in Fig. 10.6(*b*), the Auger line F_A analysed with respect to its n constituents, i.e., $F_A(E_{kin2} = n)$, is equal to the distribution function $L_\Gamma(E_\ell = n)$, except for a magnification which results from the accumulation of photoionization processes within the full bandpass function.

Having established the Auger line, one can concentrate now on the photoline described by the function F_P, which is similarly analysed for selected E_ℓ or E_{kin2} constituents, i.e.,

$$F_P(E_{kin1}, E_{kin2} = n). \tag{10.61}$$

These functions are shown on the left-hand side of Fig. 10.6(*b*). The sum of all contributions gives the shape of the photoline whose area agrees with the area of the Auger line. (The Auger yield is assumed to be unity.) However, the decomposition of this photoline into contributions numbered 1–5 clearly demonstrates that different parts of this photoline are associated with different parts of the Auger line. This point becomes extremely important in coincidence experiments where the two electron energy analysers are usually set at fixed pass energies E_{pass1} and E_{pass2}. Assuming analyser settings as indicated in 'case 1' of Fig. 10.6(*c*), it can be

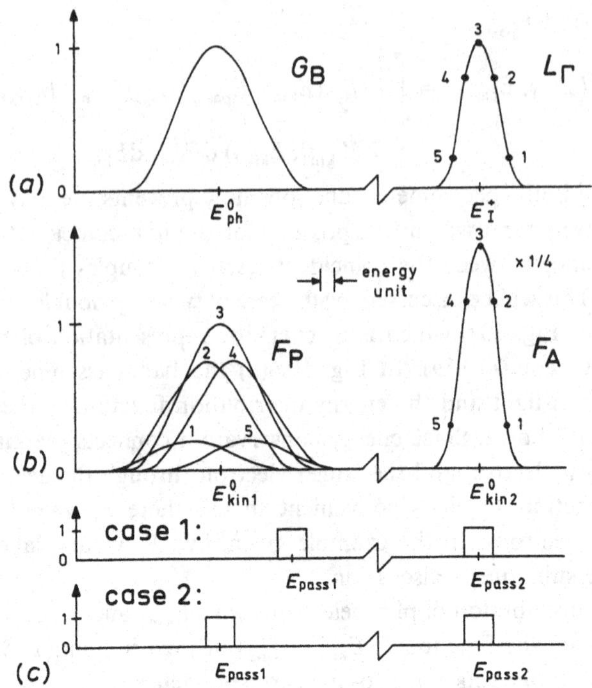

Figure 10.6 Demonstration of the effect the energy conservation has on the coincident emission of a photoelectron and a subsequent Auger electron. (*a*) Bandpass function G_B of the monochromatized incident light (left) and Lorentzian function L_Γ of the inner-shell photoionized state (right). The numbers 1–5 refer to different energies E_ℓ within the distribution function L_Γ. These labels establish the link between the photoelectron and the Auger electron due to energy conservation, $E_{kin1} = h\nu - E_\ell$ for the photoelectron and $E_{kin2} = E_\ell - E_I^{++}$ for the Auger electron. Taking into account the bandpass function, each selected E_ℓ value results from different parts of the G_B distribution. This leads for the kinetic energies of the photoelectrons to the distribution functions F_P labelled with the numbers $E_\ell = 1$–5, and the sum of these individual distributions gives 'the' photoline. For the corresponding Auger transition the area of each $F_P(E_\ell)$ photoline determines the weight of the Auger line at the position E_ℓ. The result is shown as an F_A distribution, again with labels $E_\ell = 1$–5. For the Auger yield $\omega_A = 1$ assumed here, the areas of the full photoline and the full Auger line are equal (note the scaling factor 1/4 at the F_A distribution). The F_P and F_A distributions, as shown in (*b*) are offered to the electron spectrometers. Two cases are selected in (*c*) to demonstrate the influence of different settings of the pass energies E_{pass1} and E_{pass2} (rectangular spectrometer functions). Since energy conservation links photoelectrons and Auger electrons through the label E_ℓ, coincidences will be registered with maximum intensity for 'case 2' but not for 'case 1'.

seen that spectrometer 2 selects $n = 3$ with full intensity, but spectrometer 1 selects $n = 3$ only partially, with contributions from $n = 4$ and 5. Hence, only a weak coincidence signal from $n = 3$ will result. The optimum of the coincidence signal will be reached for 'case 2' where both pass energies are set to the corresponding peak maxima:

$$\left.\begin{aligned} E_{pass1} \text{ set to } E^0_{pass1} &= E^0_{kin1}, \\ E_{pass2} \text{ set to } E^0_{pass2} &= E^0_{kin2}. \end{aligned}\right\} \tag{10.62}$$

Summarizing the discussion of Fig. 10.6, the response function $F(E_{\text{pass}1}, E_{\text{pass}2})$ in electron–electron coincidence experiments depends on the correct settings of the pass energies and on the accepted energy ranges of the spectrometers ($\Delta E_{\text{sp}1}, \Delta E_{\text{sp}2}$) as well as on other broadening effects like ΔE_{B} and Γ. In equ. (4.108) these dependences are included in the factor f(e.c.) where 'e.c.' indicates the implications of energy conservation.

In order to elucidate the foregoing discussions further, two-step double photoionization will be treated more quantitatively. The coincidence counting rate $I(E^0_{\text{pass}1}, E^0_{\text{pass}2})$ at adapted pass energies of the spectrometers is given by

$$I(E^0_{\text{pass}1}, E^0_{\text{pass}2})|^{\text{2-step}} = N_{\text{ph}} n_{\text{v}} \, \Delta z \, \sigma_{\text{ph}} \omega_{\text{A}} \, T_1 \varepsilon_1 \, T_2 \varepsilon_2 \, F(E^0_{\text{pass}1}, E^0_{\text{pass}2})|^{\text{2-step}} \quad (10.63\text{a})$$

with the distribution function (note equ. (10.62))

$$F(E^0_{\text{pass}1}, E^0_{\text{pass}2})|^{\text{2-step}} = \int\!\!\int \tilde{G}_{\text{sp}2}(E_{\text{kin}2}, E^0_{\text{kin}2}) \tilde{G}_{\text{sp}1}(E_{\text{kin}1}, E^0_{\text{kin}1})$$

$$\times \, G_{\text{B}}(E_{\text{kin}1} + E_{\text{kin}2} + E^{++}_{\text{I}}, E^0_{\text{ph}}) L_\Gamma(E_{\text{kin}2} + E^{++}_{\text{I}}, \text{E}^{+}_{\text{I}}) \, \text{d}E_{\text{kin}1} \, \text{d}E_{\text{kin}2}. \quad (10.63\text{b})$$

In order to simplify the integration, the Lorentzian function is replaced by a Gaussian distribution $G_\Gamma(E_\ell, \text{E}^+_{\text{I}})$ normalized to unit area (note the resulting proportionality factor $G_{\text{max}} = 1/1.06 \, \Delta E_\Gamma$). Introducing the short notations

$$\varepsilon_i = E_{\text{kin},i} - E^0_{\text{kin}} \quad (10.63\text{c})$$

and

$$\delta_i = \Delta E_i / 2\sqrt{(\ln 2)}, \quad (10.63\text{d})$$

one gets

$$F(E^0_{\text{pass}1}, E^0_{\text{pass}2})|^{\text{2-step}} = \frac{1}{\pi \delta_{\text{L}} \delta_{\text{B}}} \int \text{e}^{-\varepsilon_2^2/\delta_2^2} \, \text{e}^{-\varepsilon_2^2/\delta_{\text{L}}^2} \int \text{e}^{-\varepsilon_1^2/\delta_1^2} \, \text{e}^{-(\varepsilon_1 + \varepsilon_2)^2/\delta_{\text{B}}^2} \, \text{d}\varepsilon_1 \, \text{d}\varepsilon_2. \quad (10.63\text{e})$$

If the integrations are performed, one obtains

$$F(E^0_{\text{pass}1}, E^0_{\text{pass}2})|^{\text{2-step}} = \sqrt{\left(\frac{\delta_1^2 \delta_2^2}{\delta_1^2 \delta_{\text{L}}^2 + \delta_2^2 \delta_{\text{L}}^2 + \delta_1^2 \delta_2^2 + \delta_2^2 \delta_{\text{B}}^2 + \delta_{\text{L}}^2 \delta_{\text{B}}^2} \right)}. \quad (10.63\text{f})$$

(For direct double photoionizaton see [SSc95]; there one gets a different result with respect to the general form, but also because $F(E^0_{\text{pass}1}, E^0_{\text{pass}2})|^{\text{dd}}$ is given in energy units (it is called the *effective coincidence energy resolution* [LWD85]).)

This result reflects the intricate dependences of the response function $F(E_{\text{pass}1}, E_{\text{pass}2})$, even for the case of properly selected pass energies and for the replacement of $L_\Gamma(E_\ell, E^+_{\text{I}})$ by $G_\Gamma(E_\ell, E^{++}_{\text{I}})$. For $\Delta E_\Gamma = \Gamma \to 0$ which is equivalent to $\delta_L \to 0$ one gets

$$F(E^0_{\text{pass}1}, E^0_{\text{pass}2})|^{\text{2-step}}_{\delta_{\text{L}} \to 0} = \sqrt{\left(\frac{\delta_1^2}{\delta_1^2 + \delta_{\text{B}}^2} \right)}, \quad (10.63\text{g})$$

i.e., the dependence on δ_2, which is proportional to $\Delta E_{\text{sp}2}$, vanishes. This must be so for two reasons: first, the Auger energy $E_{\text{kin}2}$ is always equal to $E^0_{\text{kin}2}$ and second, since the spectrometer is set to $E_{\text{pass}2} = E^0_{\text{kin}2}$, the actual $\Delta E_{\text{sp}2}$ value has no influence on its response. On the contrary, for $\Delta E_{\text{B}} \to 0$ which is equivalent to $\delta_{\text{B}} \to 0$, one obtains

$$F(E^0_{\text{pass}1}, E^0_{\text{pass}2})|^{\text{2-step}}_{\delta_{\text{B}}\to 0} = \sqrt{\left(\frac{\delta_1^2\delta_2^2}{\delta_1^2\delta_2^2 + \delta_1^2\delta_{\text{L}}^2 + \delta_2\delta_{\text{L}}^2}\right)}, \tag{10.63h}$$

i.e., the coincidence signal depends on both spectrometer resolutions, and these quantities are also coupled to δ_{L}.

The result obtained in equ. (10.63f) is valid only for Gaussian distribution functions. In Section 5.6 an example was described in which this condition is not fulfilled due to distortions by postcollision interaction. In such a situation one has to seek a response function which is free of the actual shape of the distribution function $F_{\text{s}}(E_{\text{kin}1}, E_{\text{kin}2})$ offered to the electron spectrometers. Guided by the standard method of non-coincidence spectrometry, a scanning over the pass energy of the analyser points in the right direction. If transferred to electron–electron coincidence experiments, the sum (area) A_{true} of all true coincident events for all combinations of the pass energy settings in both spectrometers has to be taken. Mathematically this can be expressed as

$$A_{\text{true}} = \iiiint G_{\text{sp}2}(E_{\text{kin}2}, E_{\text{pass}2})G_{\text{sp}1}(E_{\text{kin}1}, E_{\text{pass}1}) \times F_{\text{s}}(E_{\text{kin}1}, E_{\text{kin}2})\,\mathrm{d}E_{\text{pass}1}\,\mathrm{d}E_{\text{pass}2}\,\mathrm{d}E_{\text{kin}1}\,\mathrm{d}E_{\text{kin}2}, \tag{10.64a}$$

and the calculation yields

$$A_{\text{true}} = \Delta E_{\text{sp}1}\,\Delta E_{\text{sp}2} \iint F_{\text{s}}(E_{\text{kin}1}, E_{\text{kin}2})\,\mathrm{d}E_{\text{kin}1}\,\mathrm{d}E_{\text{kin}2}, \tag{10.64b}$$

which is equivalent to

$$A_{\text{true}} = \Delta E_{\text{sp}1}\,\Delta E_{\text{sp}2}\,\text{const}, \tag{10.64c}$$

because the integration over all kinetic energies of the electron pairs created in the source volume leads to a constant value (with the normalization described above one has const = 1), independent of the actual form of the distribution function.

In contrast to the simple result of equ. (10.64c), the experimental determination of A_{true} is rather cumbersome. A considerable simplification occurs if one of the analysers (e.g., analyser 1 for the photoelectrons) is capable of recording the transmitted electrons with equal transmission and detection efficiencies, independent of their actual kinetic energies $E_{\text{kin}1}$ (large channelplate detector, see Section 5.6). This means that the whole photoline is recorded at once, and the requirement of energy conservation is always fulfilled. In other words, no scanning over the photoline is necessary, only a scanning over the Auger line. Mathematically, this corresponds to

$$G_{\text{sp}1}(\text{channelplate}) = 1, \tag{10.65a}$$

and equ. (10.59) reduces to

$$F(\text{channelplate}, E_{\text{pass2}})|^{2}_{\text{-step}} = \iint G_{\text{sp2}}(E_{\text{kin2}}, E_{\text{pass2}}) L_{\Gamma}(E_{\text{kin2}} + E_{\text{I}}^{++}, \text{E}_{\text{I}}^{+})$$
$$\times\, G_{\text{B}}(E_{\text{kin1}} + E_{\text{kin2}} + E_{\text{I}}^{++}, E_{\text{ph}}^{0})\, \mathrm{d}E_{\text{kin1}}\, \mathrm{d}E_{\text{kin2}}, \quad (10.65\text{b})$$

Since E_{kin1} now only occurs in the bandpass function, the integration can be performed, giving

$$F(\text{channelplate}, E_{\text{pass2}})|^{\text{2-step}} = \int G_{\text{sp2}}(E_{\text{kin2}}, E_{\text{pass2}}) L_{\Gamma}(E_{\text{kin2}}, E_{\text{kin2}}^{0})\, \mathrm{d}E_{\text{kin2}}$$
$$= G_{\text{sp2}} \otimes L_{\Gamma}. \quad (10.65\text{c})$$

If the coincident Auger line is then scanned, one gets

$$F(\text{channelplate, all } E_{\text{pass2}})|^{\text{2-step}} = \iint G_{\text{sp2}}(E_{\text{kin2}}, E_{\text{pass2}}) L_{\Gamma}(E_{\text{kin2}}, E_{\text{kin2}}^{0})\, \mathrm{d}E_{\text{kin2}}\, \mathrm{d}E_{\text{pass2}}. \quad (10.65\text{d})$$

Performing first the integration over E_{pass2}, one obtains ΔE_{sp2}, and the remaining integration over E_{kin2} then leads to the final result

$$F(\text{channelplate, all } E_{\text{pass2}})|^{\text{2-step}} = \Delta E_{\text{sp2}}. \quad (10.65\text{e})$$

This result shows that for constant resolution of the second analyser the coincidence signal obtained in this way is also a correct measure of the coincidence intensity. This method has been applied in Section 4.6.2 and equ. (5.69a).

10.4.4 Convolution of two Gaussian or two Lorentzian functions

As an example of a convolution procedure which gives a result in closed form, the convolution of two Gaussian functions will be treated first by means of a direct calculation of the convolution integral. Then the more powerful approach of Fourier transformations will be used to derive the same result for the Gaussian functions, but extending the application to the convolution of two Lorentzian functions.

The convolution of two Gaussian functions requires the calculation of

$$\Psi(x) = \int_{-\infty}^{\infty} G(y) G(x-y)\, \mathrm{d}y$$
$$= c_1 c_2 \int_{-\infty}^{\infty} \mathrm{e}^{-y^2/\delta_1^2}\, \mathrm{e}^{-(x-y)^2/\delta_2^2}\, \mathrm{d}y$$
$$= c_1 c_2\, \mathrm{e}^{-x^2/\delta_2^2} \int_{-\infty}^{\infty} \mathrm{e}^{-a[y^2 - (2x/a\delta_2^2)y]}\, \mathrm{d}y, \quad (10.66\text{a})$$

with

$$a = \frac{1}{\delta_1^2} + \frac{1}{\delta_2^2} = \frac{\delta^2}{\delta_1^2 \delta_2^2}; \quad \delta^2 = \delta_1^2 + \delta_2^2. \quad (10.66\text{b})$$

Making the substitutions

$$b = \frac{x}{\delta_2^2 a}, \quad z = y - b, \tag{10.66c}$$

one gets

$$\Psi(x) \sim \mathrm{e}^{-x^2(1/\delta_2^2 - 1/\delta_2^4 a)} \int_{-\infty}^{\infty} \mathrm{e}^{-az^2}\, \mathrm{d}z, \tag{10.66d}$$

where the last integral is known and yields

$$\int_{-\infty}^{\infty} \mathrm{e}^{-az^2}\, \mathrm{d}z = \sqrt{(\pi/a)}. \tag{10.66e}$$

Because of the equality

$$\frac{1}{\delta_2^2} - \frac{1}{\delta_2^4 a} = \frac{1}{\delta^2}, \tag{10.66f}$$

one gets the result

$$\Psi(x) \sim \mathrm{e}^{-x^2/\delta^2} \tag{10.67a}$$

with

$$\delta^2 = \delta_1^2 + \delta_2^2, \tag{10.67b}$$

i.e., the convolution of two Gaussian functions with the width parameters δ_1^2 and δ_2^2 again yields a Gaussian function with $\delta^2 = \delta_1^2 + \delta_2^2$.

Working with Fourier transformed functions allows the use of the *convolution theorem* which states that the Fourier transform $F(\omega)$ of a function $f(t)$ resulting from the convolution of two functions $f_1(t)$ and $f_2(t)$ is equal to the product of the Fourier transforms $F_1(\omega)$ and $F_2(\omega)$ of these two functions.

Because the Fourier transform $F(\omega)$ of a Gaussian function $G(x)$

$$G(x) = \mathrm{e}^{-x^2/\delta^2} \tag{10.68a}$$

is given by

$$F(G(x)) = F(\omega) \sim \mathrm{e}^{-\omega^2\delta^2/4}, \tag{10.68b}$$

one obtains for the Fourier transform $F(\omega)$ resulting from the convolution of two Gaussian functions,

$$F(\omega, \text{convolution}) \sim \mathrm{e}^{-\omega^2\delta_1^2/4}\, \mathrm{e}^{-\omega^2\delta_2^2/4} = \mathrm{e}^{-\omega^2(\delta_1^2+\delta_2^2)/4}, \tag{10.68c}$$

which corresponds to the function $f(x, G_1 \otimes G_2)$

$$f(x, G_1 \otimes G_2) \sim \mathrm{e}^{-x^2/(\delta_1^2+\delta_2^2)}, \tag{10.68d}$$

i.e., a Gaussian function is reproduced with width parameter δ^2 as obtained in equ. (10.67b).

The analogous steps for the convolution of two Lorentzian functions are first the Fourier transform of a Lorentzian function

$$L(x) \sim \frac{1}{x^2 + \delta_1^2}, \tag{10.69a}$$

which is given by

$$F(L(x)) = F(\omega) \sim \mathrm{e}^{-\delta_1 \omega}, \tag{10.69b}$$

and second the Fourier transform $F(\omega, \text{convolution})$ resulting from the convolution of two Lorentzians

$$F(\omega, \text{convolution}) \sim \mathrm{e}^{-\delta_1 \omega}\,\mathrm{e}^{-\delta_2 \omega} = \mathrm{e}^{-(\delta_1+\delta_2)\omega}, \tag{10.69c}$$

which corresponds to the convoluted function $f(x, L_1 \otimes L_2)$

$$f(x, L_1 \otimes L_2) \sim \frac{1}{x^2 + (\delta_1 + \delta_2)^2}. \tag{10.69d}$$

The result states that the convolution of two Lorentzian functions with the width parameters δ_1 and δ_2 yields again a Lorentzian function with $\delta = \delta_1 + \delta_2$.

10.5 Solid-angle corrections

To obtain a non-vanishing signal, all electron spectrometers must accept electrons emitted in a finite solid angle Ω. Due to the relation $T = \Omega/4\pi$ for the spectrometer transmission T, a larger transmission or, equivalently, a larger solid angle is necessary in order to obtain a higher signal. Though a good signal is always advantageous, large acceptance angles can lead to problems if angle-dependent intensities are being investigated, because large acceptance angles allow the recording of electrons at angles differing from the principal ray. The observed angular distribution pattern will then be the result of the weighted contributions from different angles, and it will deviate from the angular distribution patterns 'observed' with point-like detectors. Because only the latter pattern is related directly to the theoretical form of the differential cross section, the influence of finite acceptance angles has to be worked out quantitatively in order to allow a correct interpretation of the experimental data.

The solid-angle effect can be treated easily for a point-like source, because then general transformation properties of the spherical harmonics can be used which lead to a closed form for the angular distribution pattern observed for finite acceptance angles (see [Fra51, LFr53]). The approach of the 'solid-angle' correction will be demonstrated here for the important case of coincidence experiments because in this case high transmission with the necessary large acceptance angles is essential, and the correction for only one spectrometer can be derived from this general formula as a special case.

Starting with the parametrization for two-electron emission given in equ. (4.68), and using equ. (4.67) which relates the bipolar spherical harmonics to spherical harmonics attached to the actual directions $\hat{\boldsymbol{\kappa}}_\mathrm{a} = \{\vartheta_\mathrm{a}, \varphi_\mathrm{a}\}$ and $\hat{\boldsymbol{\kappa}}_\mathrm{b} = \{\vartheta_\mathrm{b}, \varphi_\mathrm{b}\}$ of electron emission, the triple differential cross section for two-electron emission can be rewritten as

$$\frac{\mathrm{d}^3\sigma}{\mathrm{d}\Omega_\mathrm{a}\,\mathrm{d}\Omega_\mathrm{b}\,\mathrm{d}E}(\hat{\boldsymbol{\kappa}}_\mathrm{a}, \hat{\boldsymbol{\kappa}}_\mathrm{b}) = \sum_{k_1 q_1 k_2 q_2} A_{k_1 q_1 k_2 q_2}\, Y_{k_1 q_1}(\hat{\boldsymbol{\kappa}}_\mathrm{a})\, Y_{k_2 q_2}(\hat{\boldsymbol{\kappa}}_\mathrm{b}). \tag{10.70}$$

(Note that the coefficients $A_{k_1q_1k_2q_2}$ differ from, but are related to the coefficients $A_{k_1k_2k}$ used in Section 4.6.1.1.)

The observed intensity $C(\hat{\mathbf{K}}_a, \hat{\mathbf{K}}_b)$ of true coincidences is then characterized by the settings of both spectrometers at $\hat{\mathbf{K}}_a = \{\Theta_a, \Phi_a\}$ and $\hat{\mathbf{K}}_b = \{\Theta_b, \Phi_b\}$, respectively, where these angles refer to the principal rays. Assuming constant detection efficiencies, the intensity $C(\hat{\mathbf{K}}_a, \hat{\mathbf{K}}_b)$ then follows from an integration over the corresponding acceptance angles:

$$C(\hat{\mathbf{K}}_a, \hat{\mathbf{K}}_b) = \sum_{k_1q_1k_2k_2} A_{k_1q_1k_2q_2} \int_{\Omega_a}\int_{\Omega_b} Y_{k_1q_1}(\hat{\mathbf{\kappa}}_a)\, Y_{k_2q_2}(\hat{\mathbf{\kappa}}_b)\, d\Omega_a\, d\Omega_b. \quad (10.71)$$

Each of the two spherical harmonics refers to the laboratory frame, but with the transformations into the detector frames described in equ. (8.92a), one gets†

$$Y_{kq}(\hat{\mathbf{\kappa}}; \text{lab}) = \sum_{q'} D^{k*}_{qq'}(\Phi, \Theta, \chi)\; Y_{kq'}(\hat{\mathbf{\kappa}}', \text{det}), \quad (10.72)$$

where the Euler angles (Θ, Φ, χ) specify the detector in the laboratory frame. Assuming a polarization-insensitive detector, the reorientation angle χ plays no role and can be set to zero. This leads to

$$C(\hat{\mathbf{K}}_a, \hat{\mathbf{K}}_b) = \frac{1}{4\pi} \sum_{k_1q_1q'_1k_2q_2q'_2} A_{k_1q_1k_2q_2} \sqrt{[(2k_1+1)(2k_2+1)]}$$
$$\times\, Q_{k_1q'_1}\, D^{k_1*}_{q_1q'_1}(\Phi_a, \Theta_a, \chi_a = 0)\, Q_{k_2q'_2}\, D^{k_2*}_{q_2q'_2}(\Phi_b, \Theta_b, \chi_b = 0) \quad (10.73)$$

with the solid-angle factors

$$Q_{kq'} = \sqrt{\left(\frac{4\pi}{2k+1}\right)} \int Y_{kq'}(\hat{\mathbf{\kappa}}', \text{det})\, d\Omega_{\text{det}}. \quad (10.74a)$$

These solid-angle factors $Q_{kq'}$ can be calculated easily because all quantities refer to the detector frame alone. In particular, for $k = 0$ one gets

$$Q_{00} = \int_{\vartheta_{\min}}^{\vartheta_{\max}} \sin\vartheta'\, d\vartheta' \int_{\varphi_{\min}}^{\varphi_{\max}} d\varphi' = \Omega_{\text{acc}}, \quad (10.74b)$$

where the integration limits refer to the accepted angular range of the electron spectrometer, i.e., Q_{00} is equal to the accepted solid angle Ω_{acc}.

An interesting result is obtained if the electron spectrometers possess axial symmetry. In this case one has $q' = 0$, equ. (8.94a) means

$$D^{k*}_{q0}(\Phi, \Theta, \chi) = \sqrt{\left(\frac{4\pi}{2k+1}\right)}\, Y_{kq}(\Theta, \Phi) = \sqrt{\left(\frac{4\pi}{2k+1}\right)}\, Y_{kq}(\hat{\mathbf{K}}), \quad (10.75)$$

and one obtains for the coincidence intensity

$$C(\hat{\mathbf{K}}_a, \hat{\mathbf{K}}_b) = \sum_{k_1q_1k_2q_2} A_{k_1q_1k_2q_2}\, Q_{k_10}\, Y_{k_1q_1}(\hat{\mathbf{K}}_a)\, Q_{k_20}\, Y_{k_2q_2}(\hat{\mathbf{K}}_b). \quad (10.76)$$

† Note the difference in the transformation properties of statistical (efficiency) tensors as compared to angular functions (compare equ. (8.82) with equs. (8.80)).

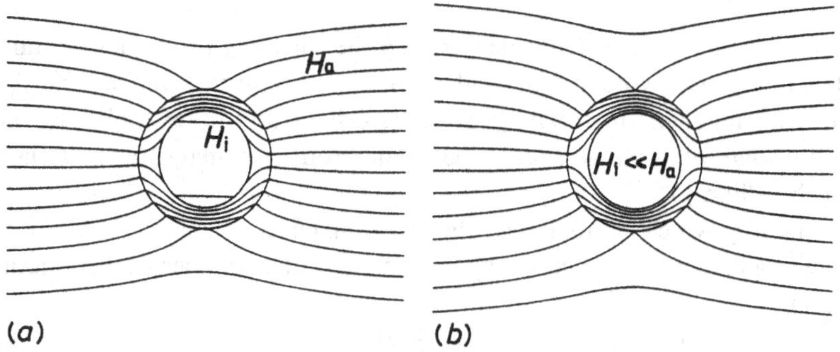

Figure 10.7 Calculated magnetic field of a cylinder placed perpendicular to a formerly homogeneous magnetic field H_a: (*a*) for a permeability $\mu = 20$; (*b*) for $\mu \to \infty$. H_i is the field which penetrated the cylinder. From [VAC75].

Usually, the solid-angle factor Q_{00} is treated separately in this expression by introducing *attenuation factors* defined by

$$\tilde{Q}_{k0} = Q_{k0}/Q_{00}. \tag{10.77}$$

(Note that $Q_{kq} < Q_{00}$ which explains the name 'attenuation' factor for the ratio Q_{kq}/Q_{00}.) One then gets

$$C(\hat{\mathbf{K}}_a, \hat{\mathbf{K}}_b) = Q_{00}(\text{det}, 1) Q_{00}(\text{det}, 2) \sum_{k_1 q_1 k_2 q_2} A_{k_1 q_1 k_2 q_2} \{\tilde{Q}_{k_1 0}\, Y_{k_1 q_1}(\hat{\mathbf{K}}_b)\} \{\tilde{Q}_{k_2 0}\, Y_{k_2 q_2}(\hat{\mathbf{K}}_b)\}. \tag{10.78}$$

This result states that the form of the angular correlation pattern measured with finite acceptance angles differs from that for a point-like detector only through attenuation factors which can be calculated for the given experimental set-up.

10.6 Shielding of the earth's magnetic field

To protect the electron trajectories from disturbance by the earth's magnetic field (see Section 4.5.2), two methods are used: shielding with Helmholtz coils and/or shielding with layers of μ-metal. In the first approach a field-free region in space is obtained by compensating the earth's field with an appropriate magnetic field from a pair of Helmholtz coils (for details see [Fir66]). Since this compensation requires correct balancing, it is rather sensitive to any changes of the earth's field produced, for example, by moving iron in the neighbourhood of the desired field-free region. Due to the great fluctuation of experimental set-ups at neighbouring beam lines, this method is not practicable for electron spectrometry with synchrotron radiation. In contrast, shielding with a μ-metal cage is free of such problems because it prevents the magnetic field penetrating into its interior. This effect is demonstrated in Fig. 10.7 where the shielding effect is shown for two metals with different permeability μ. It can be seen that for $\mu \to \infty$ perfect shielding is achieved. For special alloys called *μ-metal* (e.g., trade name *Hyperm*) one can get very high μ values, typically $\mu = 35\,000$,[†] provided the μ-metal shielding is

[†] This value refers to a field of 5 mOe, the maximum permeability lies at approximately $\mu_{max} = 90\,000$.

annealed after fabrication. (In order to keep the shielding properties of annealed μ-metal, one must avoid mechanical tension, temperatures above 400 °C, and contact with magnetic materials (e.g., screw drivers and other tools used in the neighbourhood of μ-metal should be made from non-magnetic material like copper–beryllium).)

The quantity which characterizes the shielding effect of μ-metal boxes is called the shielding factor S. It is defined by the ratio of outer to inner field strength (see Fig. 10.7),

$$S = H_a/H_i. \tag{10.79}$$

For a static magnetic field and a sphere with inner and outer diameters, D_i and D_o, respectively, this shielding factor can be calculated giving [VAC75]

$$S = \tfrac{2}{9}\,\mu\,(1 - D_i^3/D_o^3) + 1. \tag{10.80a}$$

When d is the thickness of μ-metal, with $d \ll D_o$, this result can be approximated as

$$S \approx \tfrac{4}{3}\,\mu\, d/D_o, \tag{10.80b}$$

and this equation allows a rough estimate of the shielding effect for geometries which differ slightly from a sphere. For example, if a vacuum chamber of 1 m diameter is shielded by a layer of μ-metal with $\mu = 35\,000$, with a thickness $d = 1.5$ mm, one obtains $S \approx 70$. According to the discussion in Section 4.5.2 such a value is not sufficient to perform electron spectrometry at lower kinetic energies.

The effect of μ-metal shielding can be improved by increasing the wall thickness d, e.g., by constructing the whole vacuum chamber in μ-metal. A different possibility is to use two (or even more) layers of μ-metal separated by a distance Δ. If each individual shielding factor of two layers, S_1 and S_2, is larger than 1 and the average diameter D of both layers is large compared to Δ, the total shielding factor S can be evaluated from [VAC75]

$$S = S_1\, S_2\, 4\Delta/D. \tag{10.81}$$

This formula shows that the distance Δ plays an important role. For the example given above, the use of two layers separated by $\Delta = 15$ mm gives $S \approx 294$, and this reduces the earth's field of approximately 500 mOe to a value less than 2 mOe.

This discussion has, however, been rather idealized because in a real experimental set-up a closed cage cannot be used: one needs pumping holes and several feed-throughs. In order to keep the resulting disturbances small, these openings have to be covered with strips or cylinders (see Fig. 10.8). A calculation of the resulting shielding factor is then rather complicated (for details see [VAC75]). However, as a rule of thumb for the construction of the necessary strips and cylinders one can say that their length should be at least equal to the diameter of the attached opening.

Finally, it should be mentioned that small magnetic fields, down to approximately 0.1 mOe, can be measured conveniently with a *Förster* probe [För55]. Its principal of operation will be explained with the help of Fig. 10.9. The probe consists of a pair of ferromagnetic cores (K_1 and K_2), each surrounded by a primary and a secondary coil. A sinoidal alternating current with a frequency of 20 kHz is sent

Figure 10.8 Magnetic shielding of a vacuum chamber with 'holes' for pumping and electrical feed-through. From [VAC75].

through the primary coils, and the resulting magnetic field $H_\sim$ leads to an alternating magnetization $B(t)$ which extends far into the region of saturation. The winding of the two primary coils is such as to produce equal, but opposite magnetizations, $B_1(t)$ and $B_2(t)$, while the winding of the two secondary coils leads to the sum signal of both induced voltages, $U_{sec}(t) = U_1(t) + U_2(t)$. (The relative directions of the currents in the primary and secondary coils can be interchanged. This pair then measures the gradient of a magnetic field.) If both parts of this probe are in a field-free region, one gets the dashed curves B_1 and B_2 shown in Fig. 10.9(*b*)(i). Since $B_1 + B_2$ cancels, the total induced voltage U_{sec} is zero. However, if the probe is placed in a small constant field H_0, this field adds to the primary oscillating magnetic field and shifts the magnetizations B_1 and B_2 as indicated by the solid curves in Fig. 10.9(*b*)(i). The sum of the magnetizations yields a non-vanishing signal proportional to H_0 (Fig. 10.9(*b*)(ii)), and one obtains

$$U_{sec} \sim \frac{\mathrm{d}(B_1 + B_2)}{\mathrm{d}t} = \frac{\mathrm{d}(B_1 + B_2)}{dH_\sim}\frac{\mathrm{d}H_\sim}{\mathrm{d}t}. \tag{10.82}$$

Analysing the two constituents of this equation, one verifies that U_{sec} crosses zero not only at $t = 0, T/2, T, \ldots$ as the primary field does (Fig. 10.9(*b*)(iv)), but also if $\mathrm{d}(B_1 + B_2)/\mathrm{d}H_\sim$ is zero (Fig. 10.9(*b*)(iii)). Since the latter occurs for $t = 0, T/2, T, \ldots$, but also at $t = T/4, 3T/4, \ldots$, U_{sec} appears with twice the frequency of the primary current. Hence, one has to extract the second harmonics from the voltage signal U_{sec} in order to get a measure for the strength of H_0. (There are also odd overtones due to the saturation effects in the magnetization curves and the antisymmetry of these curves with respect to $H_\sim = 0$, but these can be effectively discriminated against the desired second harmonics.)

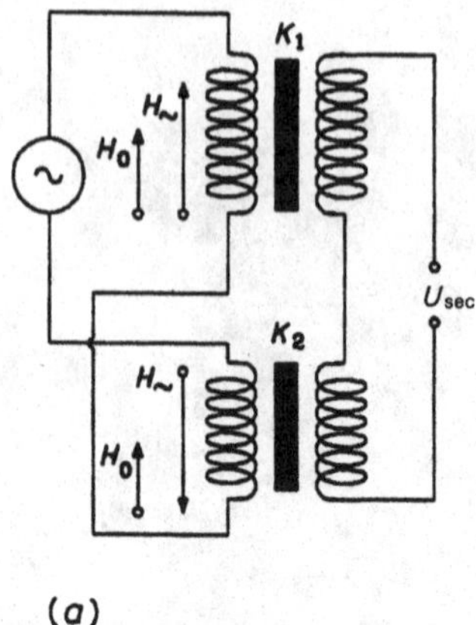

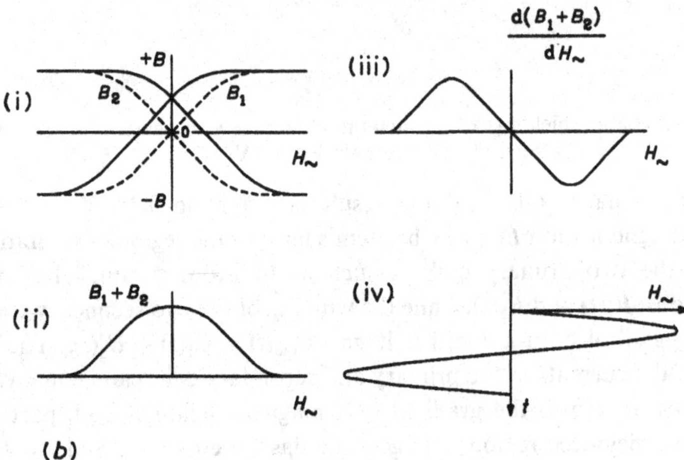

Figure 10.9 Principle of a Förster probe for measuring small magnetic fields. (*a*) Electrical circuit consisting of a pair of primary and secondary coils which surround the two ferromagnetic cores K_1 and K_2. An alternating current applied to the primary circuit produces the magnetic field $H_\sim$ with directions as indicated. The resulting voltage U_{sec} induced in the two coils of the secondary circuit carries information on the strength of a constant magnetic field H_0 superimposed on $H_\sim$. This is explained with the quantities shown in (*b*): (i) individual magnetizations B_1 and B_2 in the ferromagnetic cores as functions of the field $H_\sim$, dashed curves for $H_0 = 0$, solid curves for $H_0 \neq 0$; (ii) sum of both magnetizations for $H_0 \neq 0$; this quantity is responsible for the induced voltage U_{sec}. According to equ. (10.82) U_{sec} is the product of two contributions which are shown in (iii) and (iv), respectively, the derivative of $(B_1 + B_2)$ for $H_0 \neq 0$ with respect to $H_\sim$, and the time-dependence of $H_\sim$. From [För55].

10.7 Formation of a gas beam by capillaries

The counting rate that can be obtained in electron spectrometry with gaseous species is an important figure to be considered. It depends on two competing factors, the target density in the source volume and the limitations set by the scattering losses of the electrons. Therefore, special attention must be given to the gas-inlet system: the aim is to have high pressure in the source region, but low pressure everywhere else (see Section 4.5.1). In the following gas flow through capillaries which results in a collimated beam will be discussed, first for the total rate of flow, second, for the spatial distribution of the beam formed and, third, for the target density that can be achieved in a colliminated beam.

10.7.1 General aspects of flowing gases and total rate of flow

In order to describe the formation of a gas beam by a specially shaped inlet system, the gas flow through a single tube with radius a and length ℓ, caused by a pressure difference on each side (labelled 1 and 2) with $p_1 > p_2$ will be considered first (p_1 is frequently called the *driving pressure*). The throughput Q of such a tube is defined as the volume V_t of gas flowing per unit time through a plane at which the average pressure $\bar{p}$ exists, i.e.,

$$Q = V_t \bar{p} \quad \text{with} \quad \bar{p} = (p_1 + p_2)/2. \tag{10.83}$$

If the conductance L which describes the capability of the tube to transmit the particles is introduced, Q can also be represented as

$$Q = L\,\Delta p \quad \text{with} \quad \Delta p = p_1 - p_2. \tag{10.84}$$

Using the ideal gas equation

$$pV = nkT, \tag{10.85}$$

where n is the number of particles, k is the Boltzmann constant and T is the gas temperature, the pressure p and the particle density n_v can be related to each other by

$$p = n_v kT, \tag{10.86a}$$

which is, in more convenient units, equivalent to

$$p[\text{mbar}] = 1.3807 \times 10^{-19} n_v\,[\text{cm}^{-3}]\,T\,[\text{K}]. \tag{10.86b}$$

Equ. (10.83) can be replaced by

$$Q = NkT, \tag{10.87}$$

where the throughput N of particles per unit time, which is also called the *total rate of flow*, has been introduced. Combining equs. (10.84), (10.87), and (10.86a), one gets

$$N = L\,\Delta n_v \quad \text{with} \quad \Delta n_v = n_{v,1} - n_{v,2}. \tag{10.88}$$

Since the total rate of flow is one characteristic property of any gas inlet system, it becomes obvious from equ. (10.88) that the conductance L plays a key role in describing the flow of particles through a tube. L depends on the geometry of the tube and the properties of the particle flow inside the tube. Two limiting cases for the particle flow can be distinguished:

(i) The *Poisseuille* mode of flow. Here the conductance results from collisions of the particles with each other, i.e., collisions with the tube wall are negligible.

Hence, the viscosity η is important, and a streamline or *laminar* viscous flow of the particles occurs. (For velocities larger than the critical value v_R given by

$$v_R = R\eta/(\rho a) \tag{10.89}$$

turbulent flow will occur (R is the Reynolds number, $R \approx 1000$, and ρ the density).) For the conductance one obtains

$$L(\text{Poiseuille}) = \frac{\pi a^4}{8\eta\ell}\bar{p}, \qquad \bar{p} = (p_1 + p_2)/2, \tag{10.90}$$

i.e., L is a linear function of the average pressure $\bar{p}$ in the tube.

(ii) The *Knudsen* mode of flow. In this region collisions of the particles with each other are negligible in comparison with collisions with the inner walls of the tube. If a particle strikes the surface at some point, it is repelled according to a cosine distribution, and free or molecular flow prevails for these particles leaving the tube. For the conductance one obtains

$$L(\text{Knudsen}) = \frac{2\pi}{3}\bar{v}\,\frac{a^3}{\ell}, \tag{10.91}$$

i.e., for a given gas with a velocity $\bar{v}$ in the tube, L depends only on the geometry of the tube. (The average thermal velocity follows from

$$\bar{v} = \sqrt{\left(\frac{8kT}{\pi m}\right)}, \qquad \text{i.e., } \bar{v}\,[\text{m/s}] = 1.46 \times 10^2 \sqrt{(T[\text{K}]/M)}, \tag{10.92}$$

where m and M are the mass and mass number of the gas, respectively.)

According to the different collision mechanisms in these modes, a distinction between them can be made with the help of the *Knudsen number K* which compares the mean-free-path length λ of the particles with a characteristic dimension d of the tube:

$$K = \lambda/d. \tag{10.93}$$

(Following the kinetic theory of gases, λ can be expressed as

$$\lambda = 1/\sqrt{2}\pi n_v \sigma^2, \tag{10.94}$$

where σ is the particle diameter in the collision.)

λ values for some gases are plotted in Fig. 10.10, from which an important observation can be drawn: since λ depends on the gas selected, the association of the gas flow with a specific mode and the properties of the gas beam formed will depend on the gas, i.e., for a given capillary the formation of a gas beam will differ from gas to gas.

For a simple circular orifice the characteristic dimension d needed in the Knudsen number is clearly the hole diameter, $d = 2a$, and

the Poiseuille region with $K \le 1/100$, (10.95a)

a transition region with $1/100 \le K \le 1/3$, (10.95b)

and the Knudsen region with $K \ge 1/3$ (10.95c)

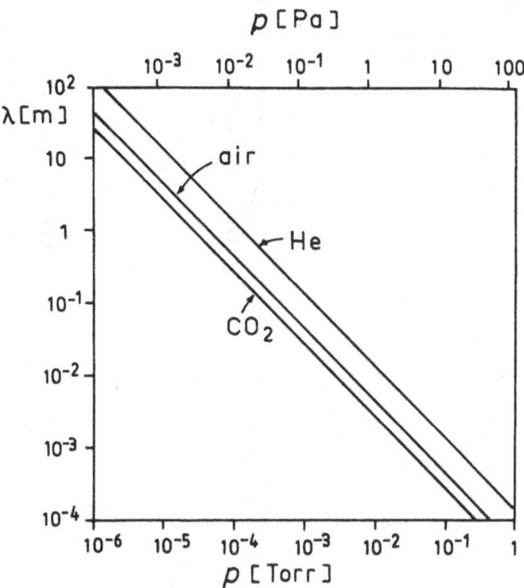

Figure 10.10 Mean-free-path λ of helium, air and carbon dioxide as a function of the pressure p and at the temperature $T = 293$ K. From [Ard64].

can be identified unambiguously. However, for a tube the situation is not so simple, because there is a pressure gradient, and hence a change of λ in the tube. In addition, the position-dependent λ value has to be compared with two characteristic dimensions, $2a$ and ℓ. Only for $\lambda_1 > \ell > a$ with λ_1 the mean-free-path in the source region with pressure p_1 is a clear case of Knudsen flow realized.

This, however, leads to rather low intensities in the beam formed, and for more realistic applications in which such clear conditions do not occur, a different criterion has to be found. Following the detailed discussion in [GWa60] the tube can then be classified as being *transparent* or *opaque* depending on the value of a new parameter x defined by

$$x = \frac{\sigma\ell}{2^{3/4}a^{3/2}}\sqrt{\left(\frac{3N}{\bar{v}}\right)}, \tag{10.96a}$$

with

$$x \ll 1 \quad \text{for a transparent tube} \tag{10.96b}$$

and

$$x \gg 1 \quad \text{for an opaque tube.} \tag{10.96c}$$

(In this context it should be mentioned that other possibilities for distinguishing the two limiting regions of Poiseuille and Knudsen modes have been introduced such as a partition based on equal amounts in the two constituents of equ. (10.97a). However, for gas flow through a long tube the parameter x is appropriate because even for $x \gg 1$ the term belonging to Poiseuille flow still might be negligible (see below).)

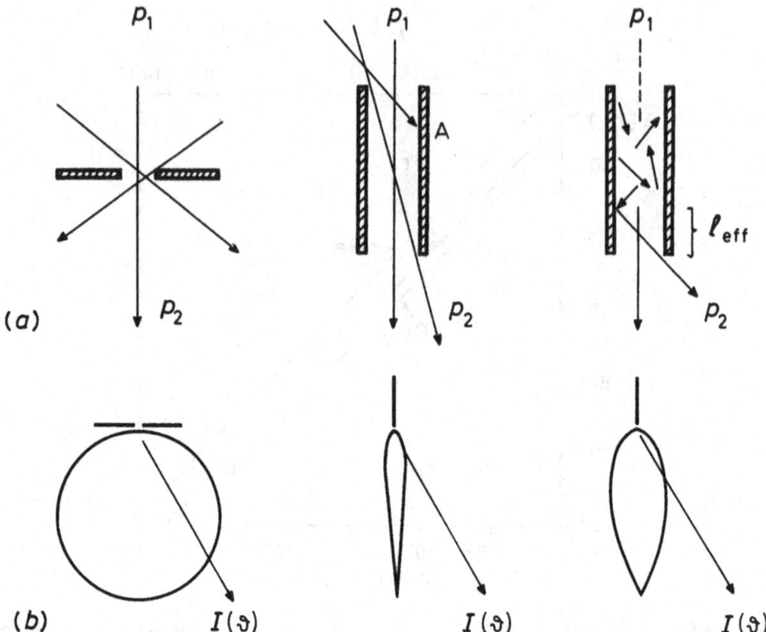

Figure 10.11 Demonstration of the formation of a gas beam by an orifice (left), a transparent tube (middle), and an opaque tube (right). (*a*) Some particle trajectories in the vicinity of the orifice or within the tube, respectively; (*b*) resulting angle-dependent intensities $I(\vartheta)$. The driving pressure p_1 is taken to be large compared to the pressure p_2 where the beam formed is observed. The conditions for Knudsen flow are always fulfilled in the left-hand and middle diagrams but in the right-hand diagram they are only fulfilled in a restricted region (indicated by ℓ_{eff}, note the different lengths of the arrows which indicate the mean-free-path of some particles). If a particle hits a surface, such as at point A, it is assumed to be repelled with a cosine distribution.

The term 'transparent' characterizes a situation in which particles entering the high-pressure end of the tube in the direction of the tube axis pass through the tube without any collision. In contrast, in an 'opaque' tube the density in the tube is so large that the probability of passing through the tube without a collision becomes negligible. Both cases are also shown in Fig. 10.11.

The rate of flow N bridging the whole range between the Poiseuille and the Knudsen regions has not yet been fully worked out. An early attempt based on experimental data is given by [Knu09]

$$N = \left(\frac{\pi a^4}{8\eta\ell}\bar{p} + \frac{1 + K_1 a\bar{p}}{1 + K_2 a\bar{p}}\frac{2\pi}{3}\bar{v}\,\frac{a^3}{\ell}\right)\Delta n_v \tag{10.97a}$$

with

$$K_1 = \frac{4}{\eta}\sqrt{\left(\frac{2}{\pi}\right)}\frac{1}{\bar{v}} \quad \text{and} \quad K_2 = 1.24K_1. \tag{10.97b}$$

In this equation it can be seen that both limits are approached correctly, L(Poiseuille) for large $\bar{p}$, and L(Knudsen) for small $\bar{p}$.

10.7.2 Spatial distribution of the formed beam

The property characterizing the collimation of the beam formed is the spatial distribution $I(\vartheta, \varphi)$ of the emitted particles. Due to the axial symmetry of the tube, the angle dependence is only on ϑ which is the angle of the emitted particles against the tube axis. The connection between $I(\vartheta)$ and the total rate of flow N is given by

$$2\pi \int_0^{\pi/2} I(\vartheta) \sin \vartheta \, \mathrm{d}\vartheta = N. \tag{10.98}$$

Usually, the peak intensity $I(0)$ and the fwhm $\vartheta_{1/2}$ are taken as a rough characterization of the distribution function $I(\vartheta)$.

In order to form a collimated beam by sending the gas through a tube, Knudsen flow is an essential prerequisite. This is demonstrated with the help of Fig. 10.11 where different angular distributions are shown. The left-hand and middle diagram shows two cases where Knudsen flow dominates: a simple orifice and a tube with a diameter-to-length ratio of $\gamma = 0.1$ (a transparent tube); the right-hand diagrams show a situation in which Knudsen flow is restricted to a length ℓ_{eff} at the low-pressure side of the tube only (an opaque tube). It can be seen that in the Knudsen region the orifice leads to a $\cos \vartheta$ distribution, i.e., no collimation effect exists, but for the transparent tube with a small γ value a remarkable peaking effect can be noted. For the opaque tube shown in the right-hand diagrams, Knudsen flow is limited to a length ℓ_{eff} at the low-pressure end of the tube, and this region is responsible for the remaining peaking effect in $I(\vartheta)$. Obviously, this peaking will become less pronounced if the backing pressure p_1 is increased further, and ℓ_{eff} becomes smaller.

The qualitative description of Fig. 10.11 can be put onto a quantitative basis by reproducing the results for the peak intensity $I(0)$, the fwhm $\vartheta_{1/2}$, and the effective length ℓ_{eff}. Depending on the parameter x in equ. (10.96a) one has for long tubes [GWa60, Zug66][†]

$$I(0)|_{\mathrm{transparent}} = \frac{a^2 \bar{v}}{4} \Delta n_v, \tag{10.99a}$$

$$I(0)|_{\mathrm{opaque}} = 0.103 \frac{\sqrt{(a\bar{v})}}{\sigma} \sqrt{N}, \tag{10.99b}$$

$$\hat{\vartheta}_{1/2}|_{\mathrm{transparent}} = 1.68 \frac{a}{\ell}, \tag{10.100a}$$

$$\hat{\vartheta}_{1/2}|_{\mathrm{opaque}} = 1.97 \frac{\sigma}{\sqrt{(a\bar{v})}} \sqrt{N} \bigg/ \left(1 - 3.10 \frac{\sigma}{\sqrt{(a\bar{v})}} \sqrt{N}\right), \tag{10.100b}$$

$$\ell_{\mathrm{eff}}|_{\mathrm{transparent}} = \ell, \tag{10.101a}$$

$$\ell_{\mathrm{eff}}|_{\mathrm{opaque}} \approx 1.94 \frac{a\sqrt{(a\bar{v})}}{\sigma} \frac{1}{\sqrt{N}}. \tag{10.101b}$$

† Note that the treatment in [Zug66] corrects the assumption of zero density at the tube exit made in [GWa60]; see also [JSP66, JOR69]. The case of very high driving pressure with a free gas jet expanding into a vacuum is not treated here.

From these relations it can be seen that for an opaque gas inlet system the parameters which define the beam depend on the gas kinetic diameter σ (for this quantity see equ. (10.94)). Therefore, as was said above, the properties of a certain gas inlet system determined for one kind of gas cannot be transferred to a different gas without great caution.

To show the relations characterizing the gas flow through a transparent or opaque tube, a compilation of relevant data is given in Table 10.1. The data refer to carbon dioxide gas, because for this case comparison with available experimental values is possible. The gas is at 20 °C and flows through a collimated hole structure of 1.3 cm diameter assembled from 1.28×10^4 tubes with $a = 0.00235$ cm and $\ell = 0.31$ cm. When calculating the properties of gas flow though such a collimated hole structure, it is assumed in a first approximation that the collimation properties described by $\vartheta_{1/2}$ for a single tube in equs. (10.100) still hold, but that the peak intensity $I_{tot}(0)$ and the total rate of flow N_{tot} are given by the single-tube values of equs. (10.99) and (10.97), multiplied with the number r of individual tubes assembled in the collimated hole structure. (Such an approximation can be applied only if the distance between the gas outlet and the observation of the emitted particles is large compared to the diameter of the gas outlet. For the data given, this ratio is 40.5 cm to 1.3 cm.)

The three examples in Table 10.1 refer to different driving pressures p_1, the pressure p_2 at the outlet of the tube being much smaller. In all three cases, equ. (10.97a) yields a negligible contribution from the term describing Poiseuille flow, but, as can be seen from the characteristic parameter x and from the effective length ℓ_{eff}, the tube clearly becomes opaque for $p_1 > 0.03$ Torr. If the calculated values of N_{tot} are compared with the experimental data, a systematic discrepancy can be seen which indicates that equ. (10.97a) might not describe this pressure regime correctly. Therefore, all further quantities in this table have been calculated using the experimental values for N. On this basis, good agreement can be noted between calculated and measured values.

10.7.3 Target density of a formed beam

In the context of electron spectrometry with gaseous samples, the target density n_v^{target} achievable in the interaction region with the ionizating radiation is of interest. This quantity is difficult to estimate, because it depends on the type of gas inlet system, and on the geometrical arrangement, in particular, on the distance D between the gas outlet and the ionizing beam and the area F of the opening through which the particles leave the gas outlet. However, some general comments can be made:

(i) Since the total flow emerges from the area F of the gas outlet, the maximum value of n_v^{target} is obtained if all these particles flow through the same area F

Table 10.1. *Beam formation data for carbon dioxide flowing through a collimated-hole structure and observed at a distance of* 40.6 *cm. From* [*GWa*60, *Zan*66].

Pressure $p_1 = 0.003$ Torr $= 0.4$ Pa; $\bar{p} = 0.2$ Pa
$x = 0.4 \ll 1$, i.e., **transparent**

	Calculated	Measured
N [molecules/s] =	3.2×10^{11}	
N_{tot} [molecules/s] =	4.1×10^{15}	
$I(0)$ [molecules/(sterad s)] =	5.1×10^{12}	
$I_{tot}(0)$ [molecules/(sterad s)] =	6.5×10^{16}	
$\vartheta_{1/2}$ [degree] =	0.7	

Pressure $p_1 = 0.03$ Torr $= 4.0$ Pa; $\bar{p} = 2.0$ Pa
$x = 1.2$, i.e., **between transparent and opaque**

	Calculated	Measured
N [molecules/s] =	3.2×10^{12}	
N_{tot} [molecules/s] =	4.1×10^{16}	2.7×10^{17}
$I(0)$ [molecules/(sterad s)] ≈	7.4×10^{13} [a,b]	
$I_{tot}(0)$ [molecules/(sterad s)] ≈	9.4×10^{17} [a,b]	9.2×10^{17}
$\vartheta_{1/2}$ [degree] =	2.7 [a,c]	3.5
ℓ_{eff} [mm] =	2.0 [a,c]	

Pressure $p_1 = 0.44$ Torr $= 58.6$ Pa; $\bar{p} = 29.3$ Pa
$x = 4.5 \ll 1$, i.e., **opaque**

	Calculated	Measured
N [molecules/s] =	4.6×10^{13}	
N_{tot} [molecules/s] =	5.8×10^{17}	4.6×10^{18}
$I(0)$ [molecules/(sterad s)] =	4.0×10^{14} [a]	
$I_{tot}(0)$ [molecules/(sterad s)] =	5.1×10^{18} [a]	5.0×10^{18}
$\vartheta_{1/2}$ [degree] =	14.8 [a]	15.7
ℓ_{eff} [mm] =	0.5 [a]	

The gas inlet is a collimated hole structure of 1.3 cm diameter with 1.28×10^4 capillaries of radius $a = 0.0235$ mm and length $\ell = 3.1$ mm. The other relevant parameters are the temperature $T = 293$ K, the collision diameter $\sigma = 4.6 \times 10^{-10}$ m, the viscosity $\eta = 1.46 \times 10^{-5}$ Pa s and the velocity $\bar{v} = 3.74 \times 10^2$ m/s. The uncertainty in the measured pressure value is ± 0.003 Torr. The numerical value of $I(0)$ is larger than that of N because $I(\vartheta)$ is multiplied by $\sin\vartheta$ when forming the integrand for N (see equ. (10.98)).

[a] Due to the discrepancy between calculated and measured data for N, the experimental value for N has been used for these data.

[b] Average value between the results for a transparent and an opaque tube.

[c] Larger value of the results for a transparent and an opaque tube.

in the interaction region. In this case one gets†

$$n_v^{\text{target}}|_{\max} = N/(\bar{v}F). \tag{10.102}$$

This relation holds independent of the collimation properties of the beam. In particular, it includes the possibility of high N values, far beyond the limit which is set by the requirement of good beam collimation. However, due to the spatial distribution of the outgoing gas, the actual target density will decrease for finite distance D and for target regions larger than that determined by F. Nevertheless, it follows that for high target densities the gas outlet should be placed as close as possible to the ionizing beam ($D \to 0$), and the area F of the opening of the gas outlet should match the diameter of the ionizing beam.

(ii) If a collimated beam is used to provide a good target density at a finite distance D, the beam quality is characterized by the peak intensity $I(0)$, and the fwhm angle $\vartheta_{1/2}$ plays an important role. For a rough estimate of the target density one can consider only those gas particles which emerge at a distance D through an area F which is spanned by a cone with opening angle $\vartheta_{1/2}$, and assume for these particles a constant intensity given by $I(0)$. In this approximation the target density follows as

$$n_v^{\text{target}} < \frac{1}{\bar{v}F} 2\pi \int_0^{\vartheta_{1/2}} I(0) \sin\vartheta \, d\vartheta, \tag{10.103a}$$

which gives

$$n_v^{\text{target}} < \frac{I(0)}{\bar{v}F} \Omega_0, \tag{10.103b}$$

where Ω_0 is the solid angle of the cone with half-angle $\vartheta_{1/2}$. Using equ. (4.11c) one gets

$$n_v^{\text{target}} < \frac{I(0)}{\bar{v}} \frac{1}{D^2}. \tag{10.103c}$$

As an example, this result is now applied to the data in Table 10.1 for a driving pressure p_1 of 0.44 Torr and a single tube. With $\vartheta_{1/2} = 15.7°$ and $I(0) = 3.9 \times 10^{14}$ molecules/(sterad s) on gets for a distance $D = 1$ cm an upper limit of the target density $n_v \approx 1 \times 10^{10}$ cm^{-3}. If many capillaries are stacked together in a collimated hole structure and the distance D is not large compared to the size of this capillary array, the target density is again difficult to estimate. However, advantage of a large number r of capillaries can be made if a focusing collimated hole structure, which confines the forward intensity $I(0)$ of all tubes into one spot, is used.

† In all these discussions it is assumed that the mean-free-path λ_2 is large compared to the distance D.

References

[AAk95] Aksela H and Aksela S 1995 private communication.
[Abe67] Åberg T 1967 *Phys. Rev.* **156** 35.
[Abe69] Åberg T 1969 *Ann. Acad. Sci. Fennicae, Ser. A VI Physica* **308** 1.
[Abe75] Åberg T 1975 in: *Atomic inner-shell processes*, ed. B. Crasemann, Vol. I., *Ionization and transition probabilities* (Academic Press, London) p. 353.
[Abe80] Åberg T 1980 *Phys. Scripta* **21** 495.
[Abe81] Åberg T 1981 *XI Int. Conf. on X-ray processes and inner-shell physics (Stirling, UK 1980)* eds. D. J. Fabian, H. Kleinpoppen and L. M. Watson (Plenum Press, New York) p. 251.
[ABu58] Asaad WN and Burhop EHS 1958 *Proc. Phys. Soc.* **71** 369.
[ACC71] Amusia MYa, Cherepkov NA and Chernysheva LV 1971 *Sov. Phys. JETP* **33** 90.
[ACh75] Amusia MYa and Cherepkov NA 1975 *Case Studies At. Molec. Phys.* **5** 47.
[AHo82] Åberg T and Howat G 1982 Theory of the Auger effect, in: *Corpuscles and radiation in matter*, ed. W Mehlhorn, Encyclopedia of Physics XXXI, ed. in chief S. Flügge (Springer-Verlag, Berlin) p. 469.
[AKh85] Amusia MYa and Kheifets AS 1985 *J. Phys. B.: At. Mol. Phys.* **18** L679.
[AKS79] Amusia MYa, Kuchiev MYu and Sheinerman SA 1979 *Sov. Phys. JETP* **49** 238.
[ALa91] Armen GB and Larkins FP 1991 *J. Phys. B: At. Mol. Opt. Phys.* **24** 741.
[ALa92] Armen GB and Larkins FP 1992 *J. Phys. B: At. Mol. Opt. Phys.* **25** 931.
[ALK92] Amusia MYa, Lee IS and Kilin VA 1992 *Phys. Rev. A* **45** 4576.
[Alt89] Altun Z 1989 *Phys. Rev. A* **40** 4968.
[Amu80] Amusia MYa 1980 *Appl. Optics* **19** 4042.
[Amu85) Amusia MYa 1985 *Comments At. Mol. Phys.* **16** 143.
[Amu90] Amusia MYa 1990 *Atomic Photoeffect* (Plenum Press, New York).
[AMW85] Adam MY, Morin P and Wendin G 1985 *Phys. Rev. A* **31** 1426.
[AOM95] Ausmees A, Osborne, SJ, Moberg R, Svensson S, Aksela S, Sairanen O-P, Kibimäki A, Naves de Brito A, Nõmmiste E, Jauhiainen J and Aksela H 1995 *Phys. Rev. A* **51** 855.
[App68] Appel H 1968 Numerical tables for angular correlation computation in α-, β- and γ-spectroscopy: 3j-, 6j-, 9j-symbols, F- and Γ-coefficients, in: *Landolt–Börnstein, Numerical Data and Functional Relationships in Science and Technology*, New Series, ed. H Schopper, ed in chief KH Hellwege, Vol. 3 (Springer-Verlag, Berlin).
[Ard64] von Ardenne M 1964 *Tabellen zur angewandten Physik*, Vol. 2, (VEB deutscher Verlag der Wissenschaften, Berlin).
[Asa63] Asaad WN 1963 *Nucl. Phys.* **44** 399.
[Asa65] Asaad WN 1965 *Nucl. Phys.* **66** 494.
[ASW87] Armen GB, Sorensen SL, Whitfield SB, Ice GE, Levin JC, Brown GS and Crasemann B 1987 *Phys. Rev. A* **35** 3966.
[ATA87] Armen GB, Tulkki J, Åberg T and Crasemann B 1987 *Phys. Rev. A* **36** 5606.
[ATW90] Albiez A, Thoma M, Weber W and Mehlhorn W 1990 *Z. Physik D – Atoms, Molecules and Clusters* **16** 97.
[Aug95] Auger P 1925 *Compt. Rend.* **180** 65, and 19926 *Ann. Physique (Paris)* **6** 183.
[AWK78] Adam MY, Wuilleumier F, Krummacher S, Schmidt V and Mehlhorn W 1978 *J. Phys. B: At. Mol. Phys.* **11** L413.
[Bar11] Barkla CG 1911 *Philos. Mag.* **22** 396.
[Bay69] Baym G 1969 *Lectures on quantum mechanics* (Benjamin, New York).
[BBe66] Barker RB and Berry HW 1966 *Phys. Rev.* **151** 14.
[BBi51] Blatt JM and Biedenharn LC 1951 *Phys. Rev.* **82** 123.

[BBK89] Brauner M, Briggs JS and Klar H 1989 *J. Phys. B: At. Mol. Opt. Phys.* **22** 2265.
[BBK9] Brauner M, Briggs JS, Klar H, Broad JT, Rösel T, Jung K and Ehrhardt H 1991 *J. Phys. B: At. Mol. Opt. Phys.* **24** 657.
[BBr86] Brauner M and Briggs JS 1986 *J. Phys. B: At. Mol. Phys.* **19** L325.
[BBT87] Brion CE, Bawagan AO and Tan KH 1987 *Chem. Phys. Lett.* **134** 76.
[BBT88] Brion CE, Bawagan AO and Tan KH 1988 *Can. J. Chem.* **66** 1877.
[BCC80] Brown GS, Chen MH, Crasemann B and Ice GE 1980 *Phys. Rev. Lett.* **45** 1937.
[BDa95] Brandt S and Dahmen HD 1995 *The picture book of quantum mechanics*, 2nd edition (Springer-Verlag, New York). (1st edition 1985 by Wiley & Sons, New York.)
[Ber62] Berry HW 1962 *Phys. Rev.* **127** 1634.
[BES79] BESSY, Technische Studie zum Bay von Strahlrohren und Monochromatoren 1979.
[Beu35] Beutler H 1935 *Z. Physik* **93** 177.
[Beu45] Beutler HG 1945 *J. Opt. Soc. Am.* **35** 311.
[[BGW87] Bizau JM, Gérard P, Wuilleumier FJ and Wendin G 1987 *Phys. Rev. A* **36** 1220.
[BHL93] Becker U, Hemmers O, Langer B, Lee I, Menzel A, Wehlitz R and Amusia MYa 1993 *Phys. Rev. A* **47** R767.
[BJa68] Bethe HA and Jackiw RW 1968 *Intermediate quantum mechanics*, 2nd edition (Benjamin, New York).
[BJo66] Byron FW and Joachain CJ 1966 *Phys. Rev.* **146** 1.
[BKa77] Berezhko EG and Kabachnik NM 1977 *J. Phys. B: At. Mol. Phys.* **10** 2467.
[BKH93] Berakdar J, Klar H, Huetz A and Selles P 1993 *J. Phys. B: At. Mol. Opt. Phys.* **26** 1463.
[BKl92]Berakdar J and Klar H 1992 *Phys. Rev. Lett.* **69** 1175.
[BKL93] Baltzer P, Karlsson L, Lundqvist M and Wannberg B 1993 *Rev. Sci. Instrum.* **64** 2179.
[[BKR78] Berezhko EG, Kabachni NM and Rostovsky VS 1978 *J. Phys. B: At. Mol. Phys.* **11** 1749.
[BKS78] Berezhko EG, Kabachnik NM and Sizov VV 1978 *J. Phys. B: At. Mol. Phys.* **11** 1819.
[BLK88] Becker U, Langer B, Kerkhoff HG, Kupsch M, Szostak D, Wehlitz R, Heimann PA, Liu SH, Lindle DW, Ferrett TA and Shirley DA 1988 *Phys. Rev. Lett.* **60** 1490.
[Blu81] Blum K 1981 *Density matrix theory and applications* (Plenum Press, New York).
[Böh79] Böhm A 1979 *Quantum mechanics* (Springer-Verlag, New York).
[Boy41] Boyce JC 1941 *Rev. Mod. Phys.* **13** 1.
[BRA89] Buckley C, Rarback H, Alforque R, Shu D, Ade H, Hellman S, Iskander N, Kirz J, Lindaas S, Mcnulty I, Oversluizen M, Tang E, Attwood D, DiGennaro R, Howells M, Jacobsen C, Vladimirsky Y, Rothman S, Kern D and Sayre D 1989 *Rev. Sci. Instrum.* **60** 2444.
[Bri32] Brillouin L 1932 *J. Physique et Radium* **3** 373, and 1934 *Actualités Scientif. et Industr.* No. 159 (Hermann et Cie, Paris) p. 3.
[Bro33] Brode RB 1933 *Rev. Mod. Phys.* **5** 257.
[Bru83] Bruneau J 1983 *J. Phys. B: At. Mol. Phys.* **16** 4135.
[[BSa57] Bethe HA and Salpeter EE 1957 *Quantum mechanics of one- and two-electron atoms* (Springer-Verlag, Berlin).
[BSa68] Brink DM and Satchler GR 1968 *Angular momentum*, 2nd edition (Clarendon, Oxford).
[BSc74] Breuckmann B and Schmidt V 1974 *Z. Physik* **268** 235.
[BSc86] Borst M and Schmidt V 1986 *Phys. Rev. A* **33** 4456.
[BSh96] Becker U and Shirley DA 1996 *VUV and soft X-ray photoionization* Physics of Atoms and Molecules, eds. PG Burke and H Kleinpoppen (Plenum Press, London).
[BSK89a] Becker U, Szostak D, Kerkhoff HG, Kupsch M, Langer B, Wehlitz R, Yagishita A and Hayaishi Y 1989 *Phys. Rev. A* **39** 3902.
[BSK89b] Becker U, Szostak D, Kupsch M, Kerkhoff HG, Langer B and Wehlitz R 1989 *J. Phys. B: At. Mol. Opt. Phys.* **22** 749.
[BSS76] Breuckmann B, Schmidt V and Schmitz W 1976 *J. Phys. B: At. Mol. Phys.* **9** 3037.
[BTH53] Broyles CD, Thomas DA and Haynes SK 1953 *Phys. Rev.* **89** 715.
[BWe94] Becker U and Wehlitz R 1994 *J. Electron Spectrosc. Relat. Phenom.* **67** 341.
[BWH89] Becker U, Wehlitz R, Hemmers O, Langer B and Menzel A 1989 *Phys. Rev. Lett.* **63** 1054.
[BWu95] Bizau JM and Wuilleumier FJ 1995 *J. Electr. Spectr.* **71** 205.
[BZh85] Best PE and Zhu H 1985 *Rev. Sci. Instrum.* **56** 389.
[CDS80] Cederbaum LS, Domcke W, Schirmer J and von Niessen W 1980 *Phys. Scripta* **21** 481.

[CEK87] Connerade JP, Esteva JM and Karnatak RC (ed) 1987 *Giant resonances in atoms, molecules and solids*, NATO ASI Series B, Physics, Vol. 151 (Plenum Press, New York).
[CFa76] Chang TN and Fano U 1976 *Phys. Rev. A* **13** 263.
[Che79] Cherepkov NA 1979 *J. Phys. B: At. Mol. Phys.* **12** 1279.
[CKr35] Coster D and Kronig R 1935 *Physica* **2** 13.
[CKr65] Carlson TA and Krause MO 1965 *Phys. Rev. Lett.* **14** 390.
[Cle72] Clebsch A 1872 *Theorie der binären algebraischen Formen* (Teubner, Leipzig).
[CMe74] Cleff B and Mehlhorn W 1974 *J. Phys. B: At. Mol. Phys.* **7** 593.
[CNe73] Carlson TA and Nestor CW 1973 *Phys. Res. A* **8** 2887.
[CNT68] Carlson TA, Nestor CW, Tucker TC and Malik FB 1968 *Phys. Rev.* **169** 27.
[Coo62] Cooper JW 1962 *Phys. Rev.* **128** 681.
[Cow81] Cowan RD 1981 *The theory of atomic structure and spectra* (University of California Press, Berkeley, Los Angeles, London).
[Cra87] Crasemann B 1987 *J. Physique* **48(C9)** 389.
[CRe74] Cvejanovic S and Read FH 1974 *J. Phys. B: At. Mol. Phys.* **7** 1841.
[CSh35] Condon EU and Shortley GH 1935 *The theory of atomic spectra* (Cambridge University Press, Cambridge) reprint 1959.
[CTa87] Cohen ER and Taylor BN 1987 *Rev. Mod. Phys.* **59** 1121.
[CWe87] Crljen Z and Wendin G 1987 *Phys. Rev. A* **35** 1555 and 1571.
[CWS93] Cubrić D, Wills, AA, Sokell E, Comer J and MacDonald MA 1993 *J. Phys. B: At. Mol. Opt. Phys.* **26** 4425, and revised data from J Comer 1994, private communication.
[CZa69] Cooper J and Zare RN 1969 *Lecture notes in theoretical physics*, Vol. XIV-C, eds S Geltman, KT Mahanthappa and WE Brittin (Gordon and Breach Science Publ., New York), p. 317.
[Dah73] Dahl P 1973 *Introduction to electron and ion optics* (Academic Press, New York and London).
[DAM95] Dawber G, Avaldi L, McConkey AG, Rojas H, MacDonald MA and King GC 1995 *J. Phys. B: At. Mol. Opt. Phys.* **28** L271.
[DDA90] Dahl DA, Delmore JE and Appelhans AD 1990 *Rev. Sci. Instrum.* **61** 607.
[Der84] Derenbach H 1984 PhD thesis University of Freiburg, Germany.
[Des73] Desclaux JP 1973 *Atomic Data Nucl. Data Tables* **12** 311.
[DFa72] Dill D and Fano U 1972 *Phys. Rev. Lett.* **29** 1203.
[DFM87] Derenbach H, Franke Ch, Malutzki R, Wachter A and Schmidt V 1987 *Nucl. Instr. Meth. A* **260** 258.
[DGo57] Devons S and Goldfarb LJB 1957 Angular Correlations, in: *Encyclopedia of physics* Vol. XLII ed. S. Flügge (Springer-Verlag, Berlin) p. 362.
[Dil73] Dill D 1973 *Phys. Rev. A* **7** 1976.
[Dir47] Dirac PAM 1947 *The principles of quantum mechanics*, 3rd edition (Clarendon, Oxford).
[DLa82] Dyall KG and Larkins FP 1982 *J. Phys. B: At. Mol. Phys.* **15** 219.
[DLe55] Dalgarno A and Lewis JT 1955 *Proc. R. Soc. London A* **233** 70.
[DMa83] Deshmukh PC and Manson ST 1983 *Phys. Rev. A* **28** 209.
[DRK92] Domke M, Remmers G and Kaindl G 1992 *Phys. Rev. Lett.* **69** 1171.
[DSa73] Dehmer JL and Saxon RP 1973, Argonne National Laboratories Report No. 8060 Part I, p. 102.
[DSM75] Dill D, Starace AF and Manson ST 1975 *Phys. Rev. A* **11** 1596.
[DSR96] Domke M, Schulz K, Remmers G, Kaindl G and Wintgen D 1996 *Phys. Rev. A* **53** 1424.
[DVi66] Dalgarno A and Victor GA 1966 *Proc. R. Soc. London A* **291** 291.
[DXP91] Domke M, Xue C, Puschmann A, Mandel T, Hudson E, Shirley DA, Kaindl G, Greene CH, Sadeghpour HR and Petersen H 1991 *Phys. Rev. Lett.* **66** 1306.
[EBM81] Engelhardt HA, Bäck W, Menzel D and Liebl H 1981 *Rev. Sci. Instrum.* **52** 835.
[Eck30] Eckart C 1930 *Rev. Mod. Phys.* **2** 305.
[EDH80] Eastman DE, Donelon JJ, Hien NC and Himpsel FJ 1980 *Nucl. Instr. Meth.* **172** 327.
[EGL47] Elder FR, Gurewitsch AM, Langmuir RV and Pollock HC 1947 *Phys. Rev.* **71** 829.
[Ehr91] Ehret R 1991 Diplom-Thesis University of Freiburg, Germany (unpublished).
[Eic90] Eichmann U 1990 PhD thesis University of Freiburg, Germany (unpublished).
[Elm82] Elmiger J 1992 Diplom-Thesis University of Freiburg, Germany (unpublished).
[ElK70] El-Kareh AB and El-Kareh JCJ 1970 *Electron beams, lenses and optics* (Academic Press, New York and London).

[EMa75] Ederer DL and Manalis M 1975 *J. Opt. Soc. Am.* **64** 634.
[ERe85] Eisberg R and Resnick R. 1985 *Quantum physics of atoms, molecules, solids, nuclei, and particles*, 2nd edition (John Wiley & Sons, New York).
[ESc96] Eland JHD and Schmidt V 1996 Coincidence measurements on ions and electrons, in: *VUV and soft X-ray photoionization studies*, eds. U Becker and DA Shirley (Plenum, New York).
[Fan35] Fano U 1935 *Nuovo Cim.* **12** 154.
[Fan51] Fano U 1951 *Nat. Bur. Stand. Rept.* 1214.
[Fan53] Fano U 1953 *Phys. Rev.* **90** 577.
[Fan61] Fano U 1961 *Phys. Rev.* **124** 1866.
[Fan69] Fano U 1969 *Phys. Rev.* **178** 131 and *Phys. Rev.* **184** 250.
[Fan74] Fano U 1974 *J. Phys. B: At. Mol. Phys.* **7** L401.
[Fan83] Fano U 1983 *Rep. Progr. Phys.* **46** 97.
[FDi72] Fano U and Dill D 1972 *Phys. Rev. A* **6** 185.
[Fer28] Fermi E 1928 *Quantum Theorie und Chemie*, ed. H. Falkenhagen (Hinzel-Verlag, Leipzig).
[Fer50] Fermi E 1950 *Nuclear physics* (Univ. of Chicago Press, Chicago) p. 142.
[Fer65] Ferguson AJ 1965 *Angular correlation methods in gamma-ray spectroscopy* (North-Holland, Amsterdam).
[FFi72] Froese Fischer Ch 1972 *Atomic data* **4** 301.
[FFi77] Froese Fischer Ch 1977 *The Hartree–Fock method for atoms* (John Wiley & Sons, New York).
[FGW74] Fellner-Feldegg H, Gelius U, Wannberg B, Nilsson AG, Basilier E and Siegbahn K 1974 *J. Electron. Spectrosc. Relat. Phenom.* **5** 643.
[Fir66] Firester AH 1966 *Rev. Sci. Instrum.* **37** 1264.
[FJH59] Frey WF, Johnston RE and Hopkins JI 1959 *Phys. Rev.* **113** 1057; *Bull. Am. Phys. Soc.* **3** (1958) No. 4 299.
[FJM66] Fink RW, Jopson RC, Mark H and Swift CD 1966 *Rev. Mod. Phys.* **38** 513.
[FLS65] Feynman RP, Leighton RB and Sands M 1965 *The Feynman lectures on physics, quantum mechanics* (Addison-Wesley, Reading MA).
[FMa52] Flügge S and Marschall H 1952 *Rechenmethoden der Quanten-theorie* (Springer-Verlag, Berlin).
[FMS72] Flügge S, Mehlhorn W and Schmidt V 1972 *Phys. Rev. Lett.* **29** 7, *erratum* 1288.
[Foc30] Fock V 1930 *Z. Physik* **61** 126 and 1930 *Z. Physik* **62** 795.
[För55] Förster F 1955 *Z. Metallkunde* **46** 358.
[FPe66] Frankowski K and Pekeris CL 1966 *Phys. Rev.* **146** 46.
[Fra51] Frankel S 1951 *Phys. Rev.* **83** 673.
[Fra84] Fraser GW 1984 *Nucl. Instr. Meth.* **221** 115.
[FRa59] Fano U and Racah G 1959 *Irreducible tensorial sets* (Academic Press Inc., New York).
[FSt66] Frauenfelder H and Steffen RM 1966 (second printing) Angular correlations, in: *Alpha- beta- and gamma-ray spectroscopy*, Vol. 2, ed. K. Siegbahn (North-Holland, Amsterdam) p. 997.
[Gal77] Galileo Electro-Optics Corp., Data Sheet 4000A (1977).
[GBS73] Gelius U, Basilier E, Svensson S, Bergmark T and Siegbahn K 1973 *J. Electron. Spectrosc. Relat. Phenom.* **2** 405.
[GHe67] Goff RF and Hendee CF 1967 *Report on the 27th Annual Conf. on Physical Electronics* (MIT) p. 231.
[GKB72] Green MI, Kenealy PF and Beard GB 1972 *Nucl. Instr. Meth.* **99** 445.
[GKB75] Green MI, Kenealy PF and Beard GB 1975 *Nucl. Instr. Meth.* **126** 175.
[GKe91] Gellrich A and Kessler J 1991 *Phys. Rev. A* **3** 204.
[Gla52] Glaser W 1952 *Grundlagen der Elektronenoptik* (Springer-Verlag, Vienna).
[GMa41] Goeppert Mayer M 1941 *Phys. Rev.* **60** 184.
[GMi80] Garibotti CR and Miraglia JE 1980 *Phys. Rev. A* **21** 572.
[GMM53] Green LC, Mulder MM and Milner PC 1953 *Phys. Rev.* **91** 35.
[GMM54] Green LC, Mulder MM, Milner PC, Lewis MN, Woll JW, Kolchin EK and Mace D 1954 *Phys. Rev.* **96** 319.
[[GMU52] Green LC, Mulder MM, Ufford CW, Slaymaker E, Krawitz E and Mertz RT 1952 *Phys. Rev.* **85** 65.
[Gor75] Gordan P 1875 *Über das Formensystem binärer Formen* (Teubner, Leipzig).

[GRa82] Greene CH and Rau ARP 1982 *Phys. Rev. Lett.* **48** 533.
[GRa83] Greene CH and Rau ARP 1983 *J. Phys. B: At. Mol. Phys.* **16** 99.
[Gri72] Grivet P 1972 *Electron optics*, 2nd edn (Pergamon Press, Oxford).
[Gro52] de Groot SR 1952 *Physica* **18** 1201.
[GSa75] Gardner JL and Samson JAR 1975 *J. Electron. Spectrosc. Relat. Phenom.* **6** 53.
[GSS74] Gelius U, Svensson S, Siegbahn H, Basilier E, Faxälv Å and Siegbahn K 1974 *Chem. Phys. Lett.* **28** 1.
[GvW83] Granneman EHA and Van der Wiel MJ 1983 Transport, dispersion and detection of electrons, ions, and neutrals, in: *Handbook on Synchrotron Radiation*, Vol. 1A, ed. EE Koch (North-Holland, Amsterdam) p. 367.
[GWa60] Giordmaine JA and Wang TC 1960 *J. Appl. Phys.* **31** 463.
[HAC76] Huang K-N, Aoyagi M, Chen MH, Crasemann B and Mark H 1976 *At. Data Nucl. Data Tables* **18** 243.
[HAD91] Hall RI, Avaldi L, Dawber G, Zubek M, Ellis K and King GC 1991 *J. Phys. B: At. Mol. Opt. Phys.* **24** 115.
[HAG78] Howat G, Åberg T and Goscinski O 1978 *J. Phys. B: At. Mol. Phys.* **11** 1575.
[Han82] Hansen JE 1982 *Comments Atom. Molec. Phys.* **12** 197.
[Har27] Hartree DR 1927 *Proc. Cambridge Phil. Soc.* **24** 89, and 111.
[Har57] Hartree DR 1957 *The calculation of atomic structures* (John Wiley & Sons, New York).
[HBC92] Hatfield JV, Burke SA, Comer J, Currell F, Goldfinch J, York YA and Hicks PJ 1992 *Rev. Sci. Instrum.* **63** 235.
[HBW80] Hink W, Brunner K and Wolf A 1980 *J. Phys. E: Sci. Instrum.* **13** 882.
[HDM92] Hall RI, Dawber G, McConkey AG, MacDonald MA and King GC 1992 *Z. Physik D – Atoms, Molecules and Clusters* **23** 377.
[Hed71] Heddle DWO 1971 *J. Phys. E: Sci. Instrum.* **4** 589.
[Hei27] Heisenberg W 1927 *Z. Physik* **43** 172.
[Hei80] Heinzmann U 1980 *J. Phys. B: At. Mol. Phys.* **13** 4353 and 4367.
[HEM92] Hall RI, Ellis K, McConkey A, Dawber, G, Avaldi L, MacConald MA and King GC 1992 *J. Phys. B: At. Mol. Opt. Phys.* **25** 377.
[Her40] Herzog R 1940 *Physikal. Zeitschr.* **41** 18.
[Her58] Herzberg G 1958 *Proc. Roy. Soc. A* **248** 309.
[HES86] Heckenkamp Ch, Eyers A, Schäfers F, Schönhense G and Heinzmann U 1986 *Nucl. Instr. Meth. A* **246** 500.
[HHa87] Hibbert A and Hansen JE 1987 *J. Phys. B: At. Mol. Phys.* **20** L245.
[HJC81] Huang K-N, Johnson WR and Cheng KT 1981 *At. Data Nucl. Data Tables* **26** 33.
[HKK88] Hausmann A, Kämmerling B, Kossmann H and Schmidt V 1988 *Phys. Rev. Lett.* **61** 2669.
[HLA94] Huetz A, Lablanquie P, Andric L, Selles P and Mazeau J 1994 *J. Phys. B: At. Mol. Opt. Phys.* **27** L13.
[HLT82] Henke BL, Lee P, Tanaka TJ, Shimabukuro RL and Fujikawa BK 1982 *At. Data Nucl. Data Tables* **27** 1.
[HME92] Hall RI, McConkey A, Ellis K, Dawber G, MacDonald MA and King GC 1992 *J. Phys. B: At. Mol. Opt. Phys.* **25** 799.
[HMY88] Hayaishi T, Murakami E, Yagishita A, Koike F, Morioka Y and Hansen JE 1988 *J. Phys. B: At. Mol. Opt. Phys.* **21** 3203.
[Hof58] Hoffmann KW 1958 *Naturwiss.* **45** 377.
[HPe82] Hansen JE and Persson W 1982 *Phys. Scripta* **25** 487.
[HRe76] Harting E and Read FH 1976 *Electrostatic lenses* (Elsevier, Amsterdam).
[HRi70] Huchital DA and Rigden JD 1970 *Appl. Phys. Lett.* **16** 348.
[HSk63] Herman F and Skillman S 1963 *Atomic structure calculations* (Prentice-Hall, Englewood Cliffs, New Jersey).
[HSS84] Heckenkamp Ch, Schäfers F, Schönhense G and Heinzmann U 1984 *Phys. Rev. Lett.* **52** 421.
[HSS86] Heckenkamp Ch, Schäfers F, Schönhense G and Heinzmann U 1986 *Z. Physik D – Atoms, Molecules and Clusters* **2** 257.
[HSW91] Huetz A, Selles P, Waymel D and Mazeau J 1991 *J. Phys. B: At. Mol. Opt. Phys.* **24** 1917.
[Hua80] Huang KN 1980 *Phys. Rev. A* **22** 223.
[Hue93] Huetz A 1993 private communication, see also [HLA94].

[HWe68] Helmer JC and Weichert NH 1968 *Appl. Phys. Lett.* **13** 266.
[Hyl29] Hylleraas EA 1929 *Z. Physik* **54** 347.
[INo74] Inokuti M and Noguchi T 1974 *Am. J. Phys.* **42** 118.
[JCh92] Johnson WR and Cheng KT 1992 *Phys. Rev. A* **46** 2952.
[JOK69] Jones RH, Olander DR and Kruger VR 1969 *J. Appl. Phys.* **40** 4641.
[JSP66] Johnson JC, Stair AT and Pritchard JL 1966 *J. Appl. Phys.* **37** 1551.
[Kab92] Kabachnik NM 1992 *J. Phys. B: At. Mol. Opt. Phys.* **25** L389.
[Kar81] Karaziya RI 1981 *Sov. Phys. Usp.* **24** 775.
[KCF84] Krause MO, Carlson TA and Fahlman A 1984 *Phys. Rev. A* **30** 1316.
[KCM71] Krause MO, Carlson TA and Moddeman WE 1971 *J. Physique (Paris)* **32** C4-139.
[KCW81] Krause MO, Carlson TA and Woodruff PR 1981 *Phys. Rev. A* **24** 1374.
[KCW92] Krause MO, Caldwell CD, Whitfield SB, Shaphorst SJ and Azuma Y 1992 *Synchrotron radiation news* **5** 25 (Gordon and Breach, New York).
[KEF83] Koch E-E, Eastman DE and Farge Y 1983 Synchrotron radiation – A powerful tool in science, in: *Handbook on synchrotron radiation*, Vol. 1A, ed. EE Koch (North-Holland, Amsterdam), p. 1.
[Kel64] Kelly HP 1964 *Phys. Rev.* **136** B896.
[Kel75] Kelly HP 1975 *Phys. Rev. A* **11** 556.
[Kel85] Kelly HP 1985 Many body calculations in atomic physics, in: *Fundamental processes in atomic collision physics (Santa Flavia, Italy, 1984)*, NATO ASI Series B, Physics 134, eds. H Kleinpoppen, JS Briggs and HO Lutz (Plenum Press, New York) p. 239.
[Kes73] Keski-Rahkonen O 1973 *Phys. Scripta* **7** 173.
[Kes85] Kessler J 1985 *Polarized electrons*, 2nd edn (Springer-Verlag, Berlin).
[KFe92] Klar H and Feht M 1992 *Z. Physik D – Atoms, Molecules and Clusters* **23** 295.
[KHA84] Kim KJ, Halbach K and Attwood D 1984 *Laser techniques in the extreme ultraviolet* (AIP Conf. Proc.) Vol. 119 eds. SE Harris and TB Lucatorto (AIP, New York), p. 267.
[KHe85] Kheifets AS 1985 *Sov. Phys. JETP* **62** 260.
[KHL92] Kämmerling B, Hausmann A. Läuger and Schmidt V 1992 *J. Phys. B: At. Mol. Opt. Phys.* **25** 4773.
[Kin57] Kinoshita T 1957 *Phys. Rev.* **105** 1490.
[KJG95] Krässig B, Jung M, Gemmel DS, Kanter EP, LeBrun T, Southworth SH and Young L 1995 *Phys. Rev. Lett.* **15** 4736.
[KKK87] Keller H, Klingelhöfer G and Kankeleit E 1987 *Nucl. Instr. Meth. A* **258** 221.
[KKS87] Kossmann H, Krässig B, Schmidt V and Hansen JE 1987 *Phys. Rev. Lett.* **58** 1620.
[KKS88] Kossmann H, Krässig B and Schmidt V 1988 *J. Phys. B: At. Mol. Opt. Phys.* **21** 1489.
[KKS89] Kämmerling B, Kossmann H and Schmidt V 1989 *J. Phys. B: At. Mol. Opt. Phys.* **22** 841.
[KKS92] Kämmerling B, Krässig B and Schmidt V 1992 *J. Phys. B: At. Mol. Opt. Phys.* **25** 3621.
[KKS93] Kämmerling B, Krässig B and Schmidt V 1993 *J. Phys. B: At. Mol. Opt. Phys.* **26** 261.
[Kli66] van Klinken J 1966 *Nucl. Phys.* **75** 161.
[KLS94] Kämmerling B, Läuger J and Schmidt V 1994 *J. Electron Spectrosc. Relat. Phenom.* **67** 363.
[KMe66] Körber H and Mehlhorn W 1966 *Z. Physik* **191** 217.
[KNA93] Kivimäki A, Naves de Brito A, Aksela S, Aksela H, Sairanen O-P, Ausmees A, Osborne SJ, Dantas LB and Svensson S 1993 *Phys. Rev. Lett.* **71** 4307.
[Knu09] Knudsen M 1909 *Ann. Physik* **28** 75.
[KOl79] Krause MO and Oliver JH 1979 *J. Phys. Chem. Ref. Data* **8** 329.
[Koo33] Koopmans TA 1933 *Physica* **1** 104.
[Kos92] Kossman H 1992 PhD thesis University of Freiburg, Germany (unpublished).
[Kos93] Kossman H 1993 *Meas. Sci. Techn.* **4** 16.
[KOs92] Kazansky AK and Ostrovsky VN 1992 *J. Phys. B: At. Mol. Opt. Phys.* **26** 2231.
[KOs95] Kazansky AK and Ostrovsky VN 1995 *J. Phys. B: At. Mol. Opt. Phys.* **28** 1453, and *Phys. Rev. A* **51** 3712.
[KPW83] Krinsky S, Perlman ML and Watson RE 1983 Characteristics of synchrotron radiation and of its sources, in: *Handbook on synchrotron radiation* Vol. 1A, ed. EE Koch (North-Holland, Amsterdam) p. 65.
[Kra79] Krause MO 1979 *J. Phys. Chem. Ref. Data* **8** 307.

[Kra80] Krause MO 1980 Electron spectrometry of atoms and molecules, in: *Synchrotron radiation research*, eds. H Winick and S Doniach (Plenum Press, New York) p. 101.
[Krä94] Krässig B 1994 PhD thesis University of of Freiburg, Germany (unpublished).
[KSa76] Kabachnik NM and Sazhina IP 1976 *J. Phys. B: At. Mol. Phys.* **9** 1681.
[KSA88] Kossmann H, Schmidt V and Andersen T 1988 *Phys. Rev. Lett.* **60** 1266.
[KSc76] Klar H and Schlecht W 1976 *J. Phys. B: At. Mol. Phys.* **9** 1699.
[KSc91] Kämmerling B and Schmidt V 1991 *Phys. Rev. Lett.* **67** 1848, and 1992 *Phys. Rev. Lett.* **69** 1144.
[KSc92] Krässig B and Schmidt V 1992 *J. Phys. B: At. Mol. Opt. Phys.* **25** L327.
[KSc93] Kämmerling B and Schmidt V 1993 *J. Phys. B: At. Mol. Opt. Phys.* **26** 1441, note misprints in the equation for $\beta(4d_{5/2})$.
[KSc95] Kabachnik NM and Schmidt V 1995 *J. Phys. B: At. Mol. Opt. Phys.* **28** 233.
[KSh89] Kuchiev MYu and Sheinerman SA 1989 *Soc. Phys. Usp.* **32** 569.
[KSK89] Kossmann H, Schwarzkopf O, Kämmerling B, Braun W and Schmidt V 1989 *J. Phys. B: At. Mol. Opt. Phys.* **22** L411.
[KSL70] Krause MO, Stevie FA, Lewis LJ, Carlson TA and Moddeman WE 1970 *Phys. Lett.* **31A** 81.
[KTR77] King GC, Tronc M, Read FH and Bradford RC 1977 *J. Phys. B: At. Mol. Phys.* **10** 2479.
[Kuy68] Kuyatt CE 1968 Measurement of electron scattering from a static gas target, p. 1, in: *Atomic and electron physics, atomic interactions*, Part A, eds. B Bederson and WL Fite, Vol. 7 of the series Methods of Experimental Physics, ed. in chief B Pederson (Academic Press, London).
[KWC92] Krause MO, Whitfield SB, Caldwell CD, Wu J-Z, van der Meulen P, de Lange CA and Hansen RWC 1992 *J. Electron. Spectros. Relat. Phenom.* **58** 79.
[KZR87] King GC, Zubek M, Rutter PM and Read FH 1987 *J. Phys. E: Sci. Instrum.* **20** 440.
[Lan92] Langer B 1992 PhD thesis TU Berlin, Vol. 2 in *Studies of vacuum ultravioet and X-ray processes*, ed. U Becker (AMS Press, New York).
[Lat55] Latter R 1955 *Phys. Rev.* **99** 510.
[LCM71] Lu CC, Carlson TA, Malik FB, Tucker TC and Nestor CW 1971 *Atomic Data* **3** 1.
[Lec87] Leckey RCG 1987 *J. Electron Spectrosc. Relat. Phenom.* **43** 183.
[LEN87] Lablanquie P, Eland JHD, Nenner I, Morin P, Delwiche J and Hubin-Franskin MJ 1987 *Phys. Rev. Lett.* **58** 992.
[LFr53] Lawson JS and Frauenfelder H 1953 *Phys. Rev.* **91** 649.
[LFr94] Lohmann B and Fritzsche S 1994 *J. Phys. B: At. Mol. Opt. Phys.* **27** 2919.
[LIM90] Lablanquie P., Ito K, Morin P, Nenner I and Eland JHD 1990 *Z. Physik D – Atoms, Molecules and Clusters* **16** 77.
[Lin83] Lindle W 1983 PhD thesis University of California (Berkeley), see also Lindle DW, Ferrett TA, Becker U, Kobrin PH, Truesdale CM, Kerkhoff HG and Shirley DA 1985 *Phys. Rev. A* **31** 714.
[Lio38] Liouville J 1838 *Journal de Math* **3** 349.
[LLi58] Landau LD and Lifshitz EM 1958 *Quantum mechanic*, vol. 3 (Pergamon, Oxford).
[LMA95] Lablanquie P, Mazeau J. Andric L, Sellas P and Huetz A 1995 *Phys. Rev. Lett.* **74** 2192.
[LMR79] Lucatorto TB, McIlrath TJ and Roberts JR 1979 *Appl. Optics* **18** 2505.
[LWD85] Lahman-Bennani A, Wellenstein HF, Duguet A and Lecas M 1985 *Rev. Sci. Instrum.* **56** 43.
[LWe74] Lundqvist S and Wendin G 1974 *J. Electron. Spectros. Relat. Phenom.* **5** 513.
[Mal82] Malutzki R 1982 Diplom thesis University of Freiburg, Germany (unpublished).
[Man72] Manson ST 1972 *J. Electron Spectrosc. Relat. Phenom.* **1** 413.
[Mar67] Marr GV 1967 *Photoionization processes in gases* (Academic Press, New York).
[MBB94] Mårtensson N, Baltzer P, Brühwiler PA, Forsell J-O, Nilsson A, Stenborg A and Wannberg B 1994 *J. Electron. Spectrosc. Relat. Phenom.* **70** 117.
[MBr93] Maulbetsch F and Briggs JS 1993 *J. Phys. B: At. Mol. Opt. Phys.* **26** L647 and 1679.
[MBr94] Maulbetsch F and Briggs JS 1994 *J. Phys. B: At. Mol. Opt. Phys.* **27** 4095.
[MBr95] Maulbetsch F and Briggs JS 1995 *J. Phys. B: At. Mol. Opt. Phys.* **28** 551.
[McG69] McGuire EJ 1969 *Phys. Rev.* **185** 1.
[McG72] McGuire EJ 1972 *Phys. Rev. A* **5** 1043.
[MCo65] Madden RP and Codling K 1965 *Astrophys. J.* **141** 364; see also 1963 *Phys. Rev. Lett.* **10** 516.
[MDG75] Moak CD, Datz S, Garcia Santibanez F and Carlson TA 1975 *J. Electron Spectrosc, Relat. Phenom.* **6** 151.
[MEd84] Morgan HD and Ederer DL 1984 *Phys. Rev. A* **29** 1901.

[Meh68a] Mehlhorn W 1968 *Phys. Lett. A* **26** 166.
[Meh68b] Mehlhorn W 1968 *Z. Physik* **208** 1.
[Meh78] Mehlhorn W 1978 Electron Spectroscopy of Auger and autoionizing states: Experiment and theory, Lectures held at the Institute of Physics, University of Aarhus, Denmark.
[Meh85] Mehlhorn W 1985 Auger-electron spectrometry of core levels of atoms, in: *Atomic inner-shell physics*, ed. B Crasemann (Plenum Press, New York).
[Mer77] Merkuriev SP 1977 *Theor. Math. Phys.* **32** 680.
[MJL81] Martin C, Jelinsky P, Lampton M and Malina RF 1981 *Rev. Sci. Instrum.* **52** 1067.
[Moo70] Moore ChE 1970 Ionization potentials and ionization limits derived from the analyses of optical spectra, US Govt. Printing Office, Washington DC; NSRDS-NBS34.
[Moo71] Moore ChE 1971 Atomic energy levels as derived by the analysis of optical spectra, Vol. I, II, III, reissued 1971, US Govt Printing Office, Washington DC.
[Mot29] Mott NF 1929 *Proc. Roy. Soc.* (*London*) *A* **124** 426, and 1932 *Proc. Roy. Soc.* (*London*) *A* **135** 429.
[MPB95] Maulbetsch F, Pont M, Briggs JS and Shakeshaft R 1995 *J. Phys. B: At. Mol. Opt. Phys.* **28** L341.
[MSR83] Martinez G, Sancho M and Read FH 1983 *J. Phys. E: Sci. Instrum.* **16** 631.
[MSS84] Müller-Dethlefs K, Sander M and Schlag EW 1984 *Z. Naturforsch.* **39a** 1089.
[MSV68] Mehlhorn W, Stalherm D and Verbeek H 1968 *Z. Naturf.* **23a** 287.
[MSY95] Masui, S, Shigemasa E, Yagishita A and Sellin IA 1995 *J. Phys. B: At. Mol. Opt. Phys.* **28** 4529.
[New66] Newton RG 1966 *Scattering theory of waves and particles* (McGraw-Hill, New York).
[New71] Newsom GH 1971 *Astrophys. J.* **166** 243.
[Nie77] Niehaus A 1977 *J. Phys. B: At. Mol. Phys.* **10** 1845.
[Nie78] Niehaus A 1979 *Int. conf. on physics of electronic and atomic collisions* (*Paris, France 1977*) ed. G Watel (North-Holland, Amsterdam), p. 185
[Ogu83] Ogurtsov GN 1983 *J. Phys. B: At. Mol. Phys.* **16** L745.
[OKr70] Olander DR and Kruger V 1970 *J. Appl. Phys.* **41** 2769.
[ONS76] Ohtani S, Nishimura H, Suzuki H and Wakiya K 1976 *Phys. Rev. Lett.* **36** 863.
[Ost79] Ostgaard Olsen J 1979 *J. Phys. E.: Sci. Instrum.* **12** 1106.
[Pau25] Pauli W 1925 *Z. Physik* **31** 765.
[Pau27] Pauli W 1927 *Z. Physik* **43** 601.
[PEI89] Price SD and Eland JHD 1989 *J. Phys. B: At. Mol. Opt. Phys.* **22** L153.
[PEI90] Price SD and Eland JHD 1990 *J. Electron. Spectrosc. Relat. Phenom.* **52** 649.
[PEM74] Parkes W, Evans KD and Mathieson E 1974 *Nucl. Instr. Meth.* **121** 151.
[Pet71] Peterkop R 1971 *J. Phys. B: At. Mol. Phys.* **4** 513.
[Pet82] Petrini D 1982 *Canad. J. Phys* **60** 644.
[PGG77] Plummer EW, Gustafsson, T. Gudat W and Eastman DE 1977 *Phys. Rev. A* **15** 2339.
[PNS82] Pettersson L, Hordgren J, Selander L, Nordling C and Siegbahn K 1982 *J. Electron Spectrosc. Relat. Phenom.* **27** 29.
[PSh95] Pont M and Shakeshaft R 1995 *Phys. Rev. A* **51** R2676.
[PSD84] Parr AC, Southworth SH, Dehmer, JL and Holland DMP 1984 *Nucl. Instr. Meth.* **222** 221.
[PTL81] Pollard JE, Trevor DJ, Lee YT and Shirley DA 1981 *Rec. Sci. Inst.* **52** 1837.
[PWB88] Persson W, Wahlström CG, Bertucelli G, Di Rocco HO, Reyna Almandos JG and Gallardo M 1988 *Phys. Scripta* **38** 347.
[Rac51] Racah G 1951 *Phys. Rev.* **84** 910.
[Ram21] Ramsauer C 1921 *Ann. Physik* **66** 545, and 1923 *Ann. Physik.* **72** 345.
[Rau71] Rau ARP 1971 *Phys. Rev. A* **4** 207.
[RBK67] Radeloff J, Buttler N, Kesternich W and Bodensedt E 1967 *Nucl. Instrum. Meth.* **47** 109.
[RBM59] Rotenberg M, Bivins R, Metropolis N and Wooten JK 1059 *The 3j and 6j-symbols* (Technology Press, Massachusetts Institute of Technology).
[Rea70] Read FH 1970 *J. Phys. E: Sci. Instrum.* **3** 127.
[Rea83] Read FH 1983 *J. Phys. E: Sci. Instrum.* **16** 636.
[Rea85] Read FH 1985 Threshold behaviour of ionization cross-sections, in: *Electron impact ionization*, ed. T.D. Märk and G.H. Dunn (Springer-Verlag, Vienna) p. 42.

[Red73] Redmond PJ 1973 (unpublished), presented without derivation by Rosenberg L 1973 *Phys. Rev. D* **8** 1833.
[RHo86] Richter LJ and Ho W 1986 *Rev. Sci. Instrum.* **57** 1469.
[Ris72] Risley JS 1972 *Rev. Sci. Instrum.* **43** 95.
[RMe86] Russek A and Mehlhorn W 1986 *J. Phys. B: At. Mol. Phys.* **19** 911.
[Rom65] Roman P 1965 *Advanced quantum theory, an outline of the fundamental ideas* (Addison-Wesley Publ. Co., Reading, MA).
[Rot72] Roth TA 1972 *Phys. Rev. A* **5** 476.
[RSa25] Russel NH and Saunders FA 1925 *Astrophys. J.* **61** 38.
[RSo95] Rose KJ and Sonntag B 1995 *Rev. Sci. Instrum.* **66** 4409.
[RTr90] Roy D and Tremblay D 1990 *Rep. Progr. Phys.* **53** 1621.
[RWe60] Roothaan CCJ and Weiss AW 1960 *Rev. Mod. Phys.* **32** 194.
[Ryd89] Rydberg JR 1889 *Kgl. Svenska Akad. Handl.* N.F. 23 No 11 1.
[SAA95] Sairanen O-P, Aksela H, Aksela S, Mursu J, Kivimäki A, Naves de Brito A, Nommiste E, Osborne SJ, Ausmees A and Svensson S 1995 *J. Phys. B: At. Mol. Opt. Phys.* **28** 4509.
[Sam67] Samson JAR 1967 *Techniques of vacuum ultraviolet spectroscopy*, John Wiley & Sons, New York.
[Sam79] Samson JAR 1979 *J. Electron Spectrosc. Relat. Phenom.* **15** 257.
[San86] Sandner W 1986 *J. Phys. B: At. Mol. Phys.* **19** L863.
[Sar67] Sar-El HZ 1967 *Rev. Sci. Instrum.* **38** 1210.
[Sch55] Schiff LI 1955 *Quantum mechanics*, 2nd edn (McGraw-Hill, New York).
[Sch73] Schmidt V 1973 *Phys. Lett.* **45A** 63.
[Sch82] Schmidt V 1982 Post-collision interaction in inner-shell ionization, in: *X-ray and atomic inner-shell physics*, ed. B Crasemann, AIP Confer. Proceed. No. 94 (AIP, New York) p. 544.
[Sch86] Schmidt V 1986 *Z. Physik D – Atoms, Molecules and Clusters* **2** 275.
[Sch87] Schmidt V 1987 *J. Physique (Paris)* **48** C9-401.
[Sch92a] Schmidt V 1992 *Rep. Prog. Phys.* **55** 1483.
[Sch92b] Schnetz M 1992 PhD thesis University of Freiburg, Germany (unpublished).
[Sch94] Schmidt V 1994 *Nucl. Instr. Meth. B* **87** 241.
[Sch95] Schaphorst S 1995 private communication.
[SCL88] Samson JAR, Chung Y and Lee EM 1988 *Phys. Lett. A* **127** 171.
[SDe66] Sachenko VP and Demekhin VF 1966 *Sov. Phys. JETP* **22** 532.
[SDM82] Schmidt V, Derenbach H and Malutzki R 1982 *J. Phys. B: At. Mol. Phys.* **15** L523.
[SEM88] Svensson S, Eriksson, B. Mårtssen N, Wendin G and Gelius U 1988 *J. Electron Spectrosc. Relat. Phenom.* **47** 327.
[Sep67] Septier A 1967 *Focusing of charged particles* (Academic Press, New York).
[Sev72] Sevier KD 1972 *Low energy electron spectrometry* (John Wiley & Sons, New York).
[Sev79] Sevier KD 1979 *At. Data Nucl. Data Tables* **24** 323.
[SHa81] Smid H and Hansen JE 1981 *J. Phys. B: At. Mol. Phys.* **14** L811.
[SHa83] Smid H and Hansen JE 1983 *J. Phys. B: At. Mol. Phys.* **16** 3339.
[Sche56] Sherman N 1956 *Phys. Rev.* **103** 1601.
[SHe83] Schönhense G and Heinzmann U 1983 *J. Phys. E: Sci. Instrum.* **16** 74.
[SHC87] Svensson S, Helenelund K and Gelius U 1987 *Phys. Rev. Lett.* **58** 1624.
[Sie66] Siegbahn K 1966 Beta-ray spectrometer theory and design. Magnetic alpha-ray spectroscopy. High resolution spectroscopy, in: *Alpha-, beta- and gamma-ray spectroscopy*, Vol. I, 2nd edn, ed. K Siegbahn (North-Holland, Amsterdam) p. 79.
[SKE93] Schwarzkopf O, Krässig B, Elmiger J and Schmidt V 1993 *Phys. Rev. Lett.* **70** 3008.
[SKS94] Schwarzkopf O, Krässig B, Schmidt V, Maulbetsch F and Briggs JS 1994 *J. Phys. B: At. Mol. Opt. Phys.* **27** L347.
[Sla29] Slater JC 1929 *Phys. Rev.* **34** 1283.
[Sla30] Slater JC 1930 *Phys. Rev.* **35** 210.
[Sla60] Slater JC 1960 *Quantum theory of atomic structure*, Vols. I and II (McGraw-Hill, New York).
[SLP94] Sukhorukov VL, Lagutin BM, Petrov ID, Lavrentiev SV, Schmoranzer H and Schartner K-H 1994 *J. Electron. Spectrosc. Relat. Phenom.* **68** 255.
[SLS92] Sukhorukov VL, Lagutin BM, Schmoranzer H, Petrov ID and Schartner KH 1992 *Phys. Lett. A* **169** 445.

[SMB76] Svensson S, Mårtensson N, Basilier E, Malmquist PA, Gelius U and Siegbahn K 1976 *Phys. Scripta* **14** 141.

[SNF67] Siegbahn K, Nordling C, Fahlman A, Nordberg R, Hamrin K, Hedman Y, Johanssen G, Bergmark T, Karlson S-E, Lindgren I and Lindberg B 1967 *Nova Acta Regiae Societatis Scientiarum Upsaliensis* Ser. IV, Vol. 20.

[SNJ69] Siegbahn K, Nordling C, Johansson G, Hedman H, Hedén PF, Hamrin K, Gelius U, Bergmark T, Werme LO, Manne R and Baer Y 1969 *ESCA applied to free molecules* (North-Holland, Amsterdam).

[Sob72] Sobel'man II 1972 *Introduction to the theory of atomic spectra* (Pergamon Press, Oxford).

[Som19] Sommerfeld A 1919 Atombau und Spektrallinien, Vol. II, 4th edition (1956) (Vieweg, Braünschweig) p. 127.

[Spi82] Spieler H 1992 *IEEE Trans. Nucl. Sci.* **29** 1142.

[SSc95] Schwarzkopf O and Schmidt V 1995 *J. Phys. B: At. Mol. Opt. Phys.* **28** 2847.

[SSm91] Seah MP and Smith GC 1991 *Rev. Sci. Instrum.* **62** 62.

[SSt75] Samson JAR and Starace AF 1975 *J. Phys. B: At. Mol. Phys. B* 8 1806 and 1979 *J. Phys. B: At. Mol. Phys.* **12** 3993.

[Sta82a] Starace AF 1982 Theory of atomic photoionization, in *Corpuscles and radiation in matter* ed. W Mehlorn *Encyclopedia of Physics* XXXI ed in chief S. Flügge (Springer-Verlag, Berlin) p. 1.

[Sta82b] Stauffer AD 1982 *Phys. Lett.* **91A** 114.

[STa63] de Shalit A and Talmi I 1963 *Nuclear shell theory* (Academic Press, New York and London).

[Ste74] Stein R 1974 *Int. J. Mass Spectr.* **14** 205.

[Sto52] Stokes GG 1852 *Trans. Cambridge Phil. Soc.* **9** 399 (reprinted 1904 *Mathematical and physical papers*, Vol. 3 (Cambridge Univ. Press, Cambridge) p. 233.

[Sto83] Stolterfoht N 1983 Techniques of high resolution Auger electron and X-ray spectroscopy in energetic ion atom collisions, in: *Fundamental processes in energetic atomic collision*, NATO ASI Series Vol. 103, eds. HO Lutz, JS Briggs and H Kleinpoppen (Plenum Press, New York) p. 295.

[SVö89] Sandner W and Völkel M 1989 *Phys. Rev. Lett.* **62** 885.

[SWe63] Stewart AL and Webb TG 1963 *Proc. Phys. Soc. (London)* **82** 532.

[Swi] Sobottka SE and Williams MB 1988 *IEEE Transactions on Nucl. Science* **35** 348.

[SZi92] Sonntag B and Zimmermann P 1992 *Rep. Prog. Phys.* **55** 911.

[TAA87] Tulkki J, Armen GB, Åberg T, Crasemann B and Chen MH 1987 *Z. Phys. D – Atoms, Molecules and Clusters* **5** 241.

[TAM92] Tulkki J, Åberg T, Mäntykenttä and Aksela H 1992 *Phys. Rev. A* **46** 1357.

[TBH91] Theis W, Braun W and Horn K 1991 BESSY annual report, p. 462.

[TCR91] Tromp RM, Copel M, Reuter MC, Horn von Hoegen M, Speidell J and Koudijs R 1991 *Rev. Sci. Instrum.* **62** 2679.

[Tem74] Temkin A 1974 *J. Phys. B: At. Mol. Phys.* **7** L450.

[Tem82] Temkin A 1982 *Phys. Rev. Lett.* **49** 36h: *Comments At. Molec. Phys.* **11** 287.

[THa56] Tomboulian DH and Hartman PL 1956 *Phys. Rev.* **102** 1423.

[THa74] Temkin A and Hahn Y 1974 *Phys. Rev. A* **9** 708.

[The75] Theodosiou CE 1973 cited by U Fano and C D Lin, *IV. int. conf. on atomic physics (Heidelberg, FRG 1974)* eds G zu Putlitz, EW Weber and A Winnacker (Plenum Press, New York) p. 47.

[Tho26] Thomas LH 1926 *Nature* **117** 514.

[Tim81] Timothy JG 1981 *Rev. Sci. Instrum.* **52** 1131.

[TLR85]Toffoletto F, Leckey RCG and Riley JD 1985 *Nucl. Instr. Meth. B* **12** 282.

[UDM94] Ullrich J, Dörner R, Mergel V, Jagutzki O, Spielberger L and Schmidt-Böcking H 1994 *Comm. At. Mol. Phys.* **30** 285.

[Urc79] Urch DS 1979 S-ray Emission spectroscopy, in: *Electron spectroscopy: Theory, techniques and applications*, Vol. 3, eds CR Brundle and AD Baker (Academic Press, London) p. 1.

[VAC75] Vacuumschmelze Hanau, Trade Reort FS-M9 1975: Magnetische Abschirmungen.

[VBe92] Végh L and Becker RL 1992 *Phys. Rev. A* **46** 2445.

[VMa94] Végh L and Macek JH 1994 *Phys. Rev. A* **50** 4031.

[Vol68] Volz D 1968 PhD thesis University of Nebraska, USA cited by P Dahl, M Rodbro, B Fastrup and ME Rudd 1976 *J. Phys. B* **9** 1567.

[Völ88] Völkel M 1988 University of Freiburg, Germany, private communication.

[VSa83] Völkel M and Sandner W 1983 *J. Phys. E: Sci. Instrum.* **16** 456.

[VSS88] Völkel M, Schnetz M and Sandner W 1988 *J. Phys. B: At. Mol. Opt. Phys.* **21** 4249.

[Wac85] Wachter A 1985 Diplom-Thesis University of Freiburg, Germany (unpublished).

[WAD77] Wuilleumier F, Adam MY, Dhez P, Sandner N, Schmidt V and Mehlhorn W 1977 *Phys. Rev. A* **16** 646.

[Wan53] Wannier GH 1953 *Phys. Rev.* **90** 817.

[Wan55] Wannier GH 1955 *Phys. Rev.* **100** 1180.

[Wap66] Wapstra AH 1966 The coincidence method, in: *Alpha-, beta- and gamma-ray spectroscopy*, Vol. I, ed. K. Siegbahn (North-Holland, Amsterdam) p. 539.

[Was74] Waßmer E 1974 Zulassungsarbeit University of Freiburg, Germany (unpublished).

[WBS72] Werme LO, Bergmark T and Siegbahn K 1972 *Phys. Scripta* **6** 141.

[Wei69] Weiss AW 1969 *Phys. Rev.* **178** 82.

[Wen27] Wentzel G 1927 *Z. Physik.* **43** 524.

[Wen81] Wendin G 1981 Breakdown of the one-electron pictures in photoelectron spectra, *Structure and bonding* **45** (Springer, Berlin).

[Wen84] Wendin G 1984 Application of many-body problems to atomic physics, in: *New trends in atomic physics*, Vol. II, eds G Grynberg and R Stora (North-Holland, Amsterdam) p. 558.

[WHK94] Whitfield SB, Hergenhahn U, Kabachnik NM, Langer B, Tulkki J and Becker U 1994 *Phys. Rev. A* **50** R3569.

[Wig27] Wigner EP 1927 *Z. Physik* **43** 624.

[Wig48] Wigner EP 1948 *Phys. Rev.* **73** 1002.

[Wig51] Wigner EP 1951 On the matrices which reduce the Kronecker products of representations of simply-reducible groups (unpublished), cited in [App68].

[Wiz79] Wiza JL 1979 *Nucl. Instr. Meth.* **162** 587.

[WKe87] Wijesundera W and Kelly HP 1987 *Phys. Rev. A* **36** 4539.

[WKe89] Wijesundera W and Kelly HP 1989 *Phys. Rev. A* **39** 634.

[WKr74] Wuilleumier F and Krause MO 1974 *Phys. Rev. A* **10** 242.

[WLi] Wilson WS and Lindsay RB 1935 *Phys. Rev. A* **47** 681.

[WMc55] Wiley WC and McLaren IK 1955 *Rev. Sci. Instrum* **26** 1150.

[WOh76] Wendin G and Ohno M 1976 *Phys. Scripta* **14** 148.

[Wol90] de Wolf DA 1990 *Basics of electron optics* (John Wiley & Sons, New York).

[WRG79] White MG, Rosenberg RA, Gabor G, Poliakoff ED, Thornton G, Southworth SH and Shirley DA 1979 *Rev. Sci. Instrum.* **50** 1268.

[WTW] Woodruff PR, Troop L and West JB 1977 *J. Electron. Spectrosc. Relat. Phenom.* **12** 133.

[WWa73] Walker TEH and Waber JT 1973 *Phys. Rev. Lett.* **30** 307, and *J. Phys. B: At. Mol. Phys.* **6** 1165.

[Yan48] Yang CN 1948 *Phys. Rev.* **74** 764.

[YPN85] Yudin NP, Pavlitchenkov AV and Neudatchin VG 1985 *Z. Physik A – Atoms and Nuclei* **20** 565.

[Zug66] Zugenmaier P 1966 *Z. Angew. Phys.* **20** 184.

Index